BIOTECHNOLOGY IN AGRICULTURE SERIES

General Editor: Gabrielle J. Persley, Biotechnology Adviser, Environmentally Sustainable Development, The World Bank, Washington DC, USA.

For a number of years, biotechnology has held out the prospect for major advances in agricultural production, but only recently have the results of this new revolution started to reach application in the field. The potential for further rapid developments is, however, immense.

The aim of this book series is to review advances and current knowledge in key areas of biotechnology as applied to crop and animal production, forestry and food science. Some titles focus on individual crop species, others on specific goals such as plant protection or animal health, with yet others addressing particular methodologies such as tissue culture, transformation or immunoassay. In some cases, relevant molecular and cell biology and genetics are also covered. Issues of relevance to both industrialized and developing countries are addressed and social, economic and legal implications are also considered. Most titles are written for research workers in the biological sciences and agriculture, but some are also useful as textbooks for senior-level students in these disciplines.

BIOTECHNOLOGY IN AGRICULTURE SERIES

Titles Available:

1: Beyond Mendel's Garden: Biotechnology in the Service of World Agriculture *G.J. Persley*
2: Agricultural Biotechnology: Opportunities for International Development *Edited by G.J. Persley*
3: The Molecular and Cellular Biology of the Potato *Edited by M.E. Vayda and W.D. Park*
4: Advanced Methods in Plant Breeding and Biotechnology *Edited by D.R. Murray*
5: Barley: Genetics, Biochemistry, Molecular Biology and Biotechnology *Edited by P.R. Shewry*
6: Rice Biotechnology *Edited by G.S. Khush and G.H. Toenniessen*
7: Plant Genetic Manipulation for Crop Protection *Edited by A. Gatehouse, V. Hilder and D. Boulter*
8: Biotechnology of Perennial Fruit Crops *Edited by F.A. Hammerschlag and R.E. Litz*
9: Bioconversion of Forest and Agricultural Plant Residues *Edited by J.N. Saddler*
10: Peas: Genetics, Molecular Biology and Biotechnology *Edited by R. Casey and D.R. Davies*
11: Laboratory Production of Cattle Embryos *I. Gordon*
12: The Molecular and Cellular Biology of the Potato, 2nd edn *Edited by W.R. Belknap, M.E. Vayda and W.D. Park*
13: New Diagnostics in Crop Sciences *Edited by J.H. Skerritt and R. Appels*
14: Soybean: Genetics, Molecular Biology and Biotechnology *Edited by D.P.S. Verma and R.C. Shoemaker*
15: Biotechnology and Integrated Pest Management *Edited by G.J. Persley*
16: Biotechnology of Ornamental Plants *Edited by R.L. Geneve, J.E. Preece and S.A. Merkle*
17: Biotechnology and the Improvement of Forage Legumes *Edited by B.D. McKersie and D.C.W. Brown*

(a) and (b) Flowers from a control plant (a) or a CHS-antisense transgenic plant (b) of *Eustoma*. (c)-(f) Foliage and a flower from a control plant (c and e) and from a *Lc*-sense transgenic plant (d and f) of *Petunia*. (g) Florets from a *Dendrathema* flower fed a control (left) or dihydromyricetin-containing (right) solution.
Also reproduced in black and white as Fig. 15.3 (see p. 269).

Biotechnology of Ornamental Plants

Edited by

R.L. Geneve

Department of Horticulture and Landscape Architecture, University of Kentucky, USA

J.E. Preece

Department of Plant and Soil Science, Southern Illinois University, USA

and

S.A. Merkle

School of Forest Resources, University of Georgia, USA

CAB INTERNATIONAL

CAB INTERNATIONAL
Wallingford
Oxon OX10 8DE
UK

Tel: +44 (0)1491 832111
Fax: +44 (0)1491 833508
E-mail: cabi@cabi.org

CAB INTERNATIONAL
198 Madison Avenue
New York, NY 10016-4341
USA

Tel: +1 212 726 6490
Fax: +1 212 686 7993
E-mail: cabi-nao@cabi.org

A catalogue record for this book is available from the British Library, London, UK.
A catalogue record for this book is available from the Library of Congress, New York, USA.

ISBN 0 85199 110 6

Typeset in Baskerville by Wyvern Typesetting Ltd
Printed and bound in the UK at the University Press, Cambridge

Contents

Part III: Plant Growth and Development

Part IV: Abiotic and Biotic Stress

List of Contributors

R.G. Anthony, *Department of Biochemistry, Royal Holloway College, University of London, Egham, Surrey TW20 0EX, UK.*

R.E. Ballard, *Department of Biological Sciences, Clemson University, Clemson, South Carolina 29634, USA.*

M.R. Boase, *New Zealand Institute for Crop & Food Research Ltd, Levin Research Centre, Private Bag 4005, Levin, New Zealand.*

H. Bouman, *Centre for Plant Tissue Culture Research, PO Box 85, 2160 AB Lisse, The Netherlands.*

A.E. Clarke, *Plant Cell Biology Research Centre, School of Botany, University of Melbourne, Parkville, Victoria 3052, Australia.*

M.E. Daub, *Department of Plant Pathology, North Carolina State University, Raleigh, North Carolina 27695, USA.*

K.M. Davies, *New Zealand Institute for Crop & Food Research Ltd, Levin Research Centre, Private Bag 4005, Levin, New Zealand.*

G.-J. de Klerk, *Centre for Plant Tissue Culture Research, PO Box 85, 2160 AB Lisse, The Netherlands.*

S.C. Deroles, *New Zealand Institute for Crop & Food Research Ltd, Levin Research Centre, Private Bag 4005, Levin, New Zealand.*

K.A. Feldmann, *Department of Plant Sciences, University of Arizona, Tucson, Arizona 85721, USA.*

D.E.A. Florack, *DLO Centre for Plant Breeding and Reproduction Research (CPRO-DLO), PO Box 16, 6700 AA Wageningen, The Netherlands.*

R.L. Geneve, *Department of Horticulture and Landscape Architecture, University of Kentucky, N318 Agriculture Science North, Lexington, Kentucky 40546, USA.*

K.L. Giles, *Department of Horticulture Science, University of Saskatchewan, Saskatoon, Canada S7N 5A8.*

C.L. Guy, *Plant Molecular and Cellular Biology Program, Department of Environmental Horticulture, Institute of Food and Agricultural Sciences, University of Florida, Gainesville, Florida 32611, USA.*

W.P. Hackett, *Department of Horticultural Science, University of Minnesota, St Paul, Minnesota 55108, USA.*

E.R. Hart, *Departments of Entomology and Forestry, 403 Science II, Iowa State University, Ames, Iowa 50011, USA.*

M.A. Jenks, *Department of Plant Sciences, University of Arizona, Tucson, Arizona 85721, USA*; present address: *206 College of Agriculture Building, PO Box 1080, Arkansas State University, Arkansas 72467, USA.*

R.K. Jones, *Department of Plant Pathology, North Carolina State University, Raleigh, North Carolina 27695, USA.*

B.R. Jordan, *New Zealand Institute for Crop & Food Research Limited, Levin Research Centre, Private Bag 4005, Levin, New Zealand.*

N.B. Klopfenstein, *USDA Forest Service, Rocky Mountain Forest and Range Experiment Station, National Agroforestry Center, East Campus, University of Nebraska, Lincoln, Nebraska 68583, USA.*

I. Konczak, *New Zealand Institute for Crop & Food Research Ltd, Levin Research Centre, Private Bag 4005, Levin, New Zealand.*

H.J.M. Löffler, *DLO Centre for Plant Breeding and Reproduction Research (CPRO-DLO), PO Box 16, 6700 AA Wageningen, The Netherlands.*

B.H. McCown, *Department of Horticulture, University of Wisconsin-Madison, Madison, Wisconsin 53706, USA.*

M. McLaughlin, *Ag-West Biotech, Saskatoon, Saskatchewan, Canada.*

S.A. Merkle, *Daniel B. Warnell School of Forest Resources, University of Georgia, Athens, Georgia 30602, USA.*

J.W. Moyer, *Department of Plant Pathology, North Carolina State University, Raleigh, North Carolina 27695, USA.*

J.R. Murray, *Department of Horticultural Science, University of Minnesota, St Paul, Minnesota 55108, USA.*

N. Nass, *Plant Cell Biology Research Centre, School of Botany, University of Melbourne, Parkville, Victoria 3052, Australia.*

E. Newbigin, *Plant Cell Biology Research Centre, School of Botany, University of Melbourne, Parkville, Victoria 3052, Australia.*

J.E. Preece, *Department of Plant and Soil Science, Southern Illinois University, Carbondale, Illinois 62901-4415, USA.*

S. Rajapakse, *Department of Biological Sciences, Clemson University, Clemson, South Carolina 29634, USA.*

D.E. Riemenschneider, *Forestry Sciences Laboratory, 5985 Highway K, Rhinelander, Wisconsin 54501, USA.*

K.E. Schwinn, *New Zealand Institute for Crop & Food Research Ltd, Levin Research Centre, Private Bag 4005, Levin, New Zealand.*

R.R. Tripepi, *Plant Science Division, University of Idaho, Moscow, Idaho 83844, USA.*

A.R. van der Krol, *Department of Plant Physiology, Agricultural University Wageningen, Arboretumlaan 4, 6703 BD Wageningen, The Netherlands.*

O. Vorst, *Department of Plant Physiology, Agricultural University Wageningen, Arboretumlaan 4, 6703 BD Wageningen, The Netherlands.*

R. Whetten, *Department of Forestry, North Carolina State University, Raleigh, North Carolina 27695, USA.*

Preface

This volume attempts to provide a comprehensive review of strategies being employed either commercially or experimentally for the improvement of ornamental plants. For this book, we have defined biotechnology as the cellular and molecular techniques providing the foundation for direct transfer of genes into plants or molecular techniques that enhance genetic improvement through traditional plant breeding.

This book is unique in that it is a collection of chapters documenting the impact and future potentials of biotechnology on a diverse group of plants produced as commercial ornaments. Ornamental plant production is of significant importance worldwide. Its monetary value has significantly increased over the last two decades and there is potential for continued future growth in both domestic and international markets. These include herbaceous plants produced as bedding plants, greenhouse pot plants and cut flowers. Ornamentals also include bulbs, trees, shrubs and vines used for private and public landscapes.

Basic scientific research in the last several years has provided an excellent foundation to better understanding plant regeneration, genetics and aspects of growth and development. Taken together, this 'biotechnology' offers the potential for significant advances in the improvement of ornamental species. Many of these improvements are currently impacting commercial production of ornamental plants. A large number of landscape plants are being commercially propagated through tissue culture. Recently-identified genes controlling flower characteristics, pest resistance and hormones have been 'engineered' into ornamental species and in concert with classical breeding programs will significantly change landscape and cut flower crop improvement into the next century.

Rather than organizing this book based on major crop species, as is usual for works on edible food crops, this effort is arranged by five broad subject categories: (i) supporting technologies dealing with plant regeneration systems; (ii) genetics including genome mapping, cultivar identification, barriers to sexual reproduction, transformation protocols and techniques for induced variation; (iii) plant growth and development with special emphasis on flower characteristics as well as overall plant form; and finally (iv) strategies for coping with biotic and abiotic environmental stress.

Each chapter provides a comprehensive study of the current status for each topic area and the future prospectus for improving ornamental crops utilizing biotechnology. We believe the reader will be impressed with the large body of work dealing with these crops and we hope these reviews will inspire additional research toward the improvement of this important group of plants.

The editors would like to thank Patricia Battle for assistance with the index.

R.L. Geneve

The Commercialization of Biotechnology

1

Kenneth L. Giles[1] and Murray McLaughlin[2]
[1]*Department of Horticulture Science, University of Saskatchewan, Saskatoon, Canada S7N 5A8;* [2]*Ag-West Biotech, Saskatoon, Saskatchewan, Canada*

INTRODUCTION AND DEFINITION

The commercialization of any new product is as much a work of art as it is a science. This is particularly true of the products of biotechnology, since funding and investment was often sought to establish companies when there was only an idea of the final product to be developed. Most of the small companies that were started in the 1980s had access to a methodology that might be used in the development of a new product. Initially, however, these methodologies had not been moulded into a functional technology, which might reliably produce the final product that could be weighed, measured or counted. Venture capital groups, realizing the potential value of these projected products, invested heavily in small, start-up companies. Investors hoped that the final product would be available after eight to ten years and possibly some income seen after four or five years of operation. Some investors were satisfied; others were disappointed.

Before pursuing a detailed discussion of the commercialization activities within the industry, it is important to develop a working definition of biotechnology. It is without doubt the great 'Saranwrap' word of the 1980s and 1990s. Any technology wrapped in the cloak of biotechnology would last longer, keep fresher and raise more investment capital than might otherwise be the case. Biotechnology must be considered both in terms of the cutting-edge molecular biology and also the extension of the traditional breeding techniques that have been in place for decades. Perhaps one useful working definition of biotechnology is the selection, production or creation of new and improved strains of microorganisms, plants and animals, maximizing their economic usefulness for the public good. It is a very general definition but one that covers almost all of biotechnology's applications.

 Biotechnology of Ornamental Plants (eds R.L. Geneve, J.E. Preece and S.A. Merkle)

EXAMPLES

How, then, can biotechnology improve plant production? The first products of plant biotechnology were focused on improving crop characteristics by altering microorganisms associated with the plants and by controlling crop pests and viruses. More advanced products with even greater potential to change and improve plant production techniques are still on the horizon.

In one important environmental application of genetic engineering in plants, scientists have successfully transferred glyphosate resistance from microorganisms into a number of crop plant species (Cemai *et al.*, 1985). Glyphosate is a common herbicide that is both effective and environmentally safe, but because the compound indiscriminately kills crops as well as weeds it can only be used before the crop germinates. The development of glyphosate-tolerant crop cultivars provides growers with an additional weed control option, potentially reducing their reliance on more costly, less effective and more environmentally hazardous herbicides.

Microorganisms are associated with both millions of dollars of benefit and damage to plant crops every year. They are involved in the uptake of nitrogen in root nodules and their modification and improvement can cause faster vegetative growth in leguminous crops. Others help protect plants from the stress of acidity, salinity or high concentrations of toxic metals. Some microorganisms, such as bacteria and fungi, attack crops and can cause significant disease problems. A classic case is Dutch elm disease, which is an example of a fungus that has caused the loss of millions of trees in North America during this century.

Through molecular biology and genetic engineering techniques, scientists have improved the chances of high crop production by engineering certain microorganisms that compete with harmful bacteria or fungi, thus improving the plant's potential growth. Also, genetically engineered bacteria, 'the ice-minus' bacteria, have been developed with the potential to protect plants from frost damage by inhibiting the natural populations of microorganisms that help to promote frost formation on fruit trees. These bacteria have now been field tested (Baertlein *et al.*, 1992).

Insects that normally interfere with crop plants can also be inhibited using genetic engineering technology. A protein from both tomatoes and potatoes appears to repel horn worms, a predator of the tobacco plant. Work is under way to transfer these proteins to tobacco and to other fibre and field crops to test for its potential expression and effect in repelling insect predators. Large-scale trials of tobacco (Barton *et al.*, 1987) and tomatoes (Fischhoff *et al.*, 1987) containing the endotoxin gene of *Bacillus thuringiensis* (a naturally occurring insecticide) which confers resistance to lepidopteran insect species are under way.

IMPACT ON AGRI-BUSINESS

Because the influence of biotechnology has grown so rapidly to involve so many facets of agriculture, the impact of the growth of biotechnology on agri-business has been profound. Biotechnology products offer both new opportunities and new challenges for agri-business.

What are the implications of this vast array of biotechnological possibilities for the farming community? The US Office of Technology Assessment (1986) conducted a survey which indicated that 70% of all larger farms would adopt new biotechnologically derived plant materials, whereas only 10% of smaller farms thought it likely that they would adopt similar products. It is believed that the adoption of biotechnological products will bring down overall costs of production because of increased efficiency. Therefore, it will provide an advantage to larger farms which, having adopted the products, will have lower overhead costs and a greater economic advantage in producing higher qualities and/or larger amounts of product per hectare than smaller farms. This, it is felt, will probably lead to an increase in larger farms because smaller landholders will be priced out of the market. However, these conclusions are probably outdated. It seems likely now that many small farm holders will adopt this technology and that the anticipated drop in farm numbers with increasing land area per farm has been overestimated. The large number of niche areas into which biotechnology is now penetrating suggests that there will be equally numerous opportunities for small landholders and horticulturalists to make considerable use of the technology and quite probably, with astute marketing, remain commercially viable. The general feeling in the agricultural industry now is that biotechnology could possibly help slow the decline in farm numbers.

Over the past 15 years, small firms have been at a disadvantage in developing and marketing new biotechnology products because the research and development processes are long and costly. Because the industry is young, few products have reached the market to fund the development of further products. Frequently, new products have been met with resistance from the public due to lack of knowledge and often misinformation. The regulatory processes have been extremely rigorous and, to some extent, have inhibited the successful commercialization of products. In contrast, larger companies with existing viable products, not necessarily of biotechnological origin, have the financial backing and reserves to pursue commercialization of biotechnology products and to support the research, development and regulatory costs that have been involved with biotechnology.

Larger companies in agri-business, such as are represented in the seed industry, are likely to be first to benefit from biotechnology. The seed industry is large in terms of both volume and value, especially in North America and Europe. Plant breeding is research-intensive and the time-scale for product development is relatively long compared to many other related industries. Although the industry is international, with varieties being spread throughout the world in similar environments, the majority of the companies operating in the industry have a relatively low level of sustained profitability. There are notable exceptions, such as those companies involved with the sale of hybrid maize and sugar beet. In simple terms biotechnology allows the plant breeder to improve crops in two key ways: first, by increasing the efficiency of the plant selection process, and, second, by broadening the genetic base with which the plant breeder may work through direct or indirect genetic modification.

INVESTMENT

It is hardly surprising that with the huge potential earnings involved the venture capital community descended on embryonic, start-up biotechnology companies with great glee. A number of venture capitalists, while still enthusiastic today about both past and potential successes to be found in biotechnology start-up companies, have felt the sting and backlash of investments in small start-up companies. These companies promised great improvements but were often misled as to the market potential and the time-scale required for successful development of their products.

Before any venture capital company invests in a start-up company, it will determine the key factors for success and the new company's potential. These essential qualities may be in innovative proprietary technology or they might be in strong market potential for the product. They will almost certainly focus on the quality and potential of senior management. Most venture capital companies expect a time frame of four to five years for building value within the company, though some anticipate a longer term of perhaps seven or eight years.

It seems to be believed by many venture capitalists that agricultural biotechnology (or ag-biotech) companies have a better chance of fulfilling both product development and sales in a shorter time than pharmaceutical ventures. This is so because shorter testing times and fewer regulations are involved. Of the dozens of start-up ag-biotech companies established in the 1980s, only a few have gone public. Companies such as Agridyne Technologies, Inc. (Salt Lake City, Utah), Biosys Inc. (Palo Alto, California), Calgene Inc. (Davis, California), Crop Genetics International (Hannover, Michigan), D.N.A. Plant Technology (Cinnaminson, New Jersey), Ecogen Inc. (Langhorne, Pennsylvania), EcoScience Corp. (Worcester, Massachusetts), Embrex Inc. (Research Triangle Park, North Carolina), Mycogen (San Diego, California), EscaGenetics Corp. (San Carlos, California), IDEXX Laboratories (Westbrook, Maine), Neogen Corp. (Lansing, Michigan) and Ringer Corp. (Minneapolis, Minnesota) represent some of the successes that have gone public and are currently making their way in the commercial world. The first of these to achieve public status were Calgene, Crop Genetics International and EscaGenetics in 1986–1987. Most of the others went public in 1992 or later.

In general, venture capital companies wishing to invest in ag-biotech are looking for start-up companies with products appealing to specialty markets where product demand is strong and competition almost non-existent. They are also looking for relatively short product development times, such as those that might be found in the biopesticide and diagnostic product areas, in order that they may show a return on their investment in a three to five year time period. Because of the high-priced research ticket and the long term required for product development, many companies have moved away from venture capitalists to strategic partners for funding product development. The company Biosys Inc. produces a species of nematode used as a bioinsecticide. These nematodes prey on harmful insects but do not harm humans, livestock or beneficial insects. Biosys has decided to market its product with major players as partners. They have established marketing partnerships with CIBA-Geigy and Ortho. Biosys is entering a market currently dominated by synthetic chemical insecticides and their products require strong marketing inputs to compete with the established chemical companies.

INTELLECTUAL PROPERTY PROTECTION

The very significant research and development costs associated with biotechnology and the equally significant marketing costs mean that it is only with protection for intellectual property that companies can afford the risk of advancing into this field. Although humans have been manipulating the genetics of plants for thousands of years, it was not until Gregor Mendel's work was rediscovered at the end of the nineteenth century that the physical laws controlling inheritance were fully understood, and it is only since then that plant breeders have been able to base their breeding on scientific principles rather than on trial and error. It is as a result of these developments that the aggressive programme of selection of crop plant varieties continues to this day. Only within the last 15 years has the revolution of genetic engineering, molecular biology and associated biotechnology swept into the plant breeding industry. Selective crop improvement by genetic engineering represents a quantum leap for variety development. The protection of these technologies as proprietary property is the only safeguard small or even large companies can have for the large investments they have to make to enter this arena.

Plant patents were first granted by the US Congress in 1930, giving the 'right to exclude others from asexually reproducing a plant or selling or using the plants so reproduced.' To be patentable the plant variety had to be novel, distinct and non-obvious, and it must have been asexually reproduced. The purpose of this legislation was, 'to afford to agriculture the same opportunity to participate in the benefits of the patent system as has been given industry'. The scope given to plant patents was limited to plants alone and thus did not include bacteria or fungi. The patentee was limited to a single claim for the plant and could not expand the protection to several varieties. Protection was not extended to the products of the plant, such as cut flowers, fruit, etc., and because it was only applied to asexually reproduced plants, seed propagated plants were excluded. The creation of the Union for the Protection of New Varieties of Plants (UPOV) in 1961 provided the basis for breeders' rights legislation. Under the UPOV Agreement each nation member has its own national legislation and must conform to UPOV guidelines. To be given protection under plant breeders' rights, a new plant variety must be 'distinct, uniform and stable', it must not have been previously commercialized, and its development not previously described in a publication enabling a skilled worker to reproduce the variety. Under the legislation, the plant breeder cannot limit an individuals' propagation of the variety for his or her own use, nor may he or she bar the use of the variety as a source of variation for other breeders to use in the creation of new varieties.

In 1985 the US Board of Patent Appeals and Interferences held that claims for a genetically engineered maize plant, its seeds and tissue cultures were patentable under the utility patent statute. In so doing the board rejected the US Patent and Trademark Office case that the Plant Patent Act and the Plant Variety Protection Act were the exclusive forms of protection for human-altered plant life. The advantages of utility protection for plant biotechnology are clear. Patent claims can be drawn much more broadly, may include an entire genus or family of plants that is transformed through genetic engineering, and can protect the process as well as the product of the process. Utility patents also allow claims making it possible to protect valuable plants and derivatives of the plant. It also changes the plant breeders' rights rule for free pollen, in that patented plants cannot be used as additional sources of variation by other plant biotechnologists without the express permission of the patentee.

Probably the greatest threat to plant biotechnologists is that, even after having patented and protected his or her plant material, in many parts of the world these patents will not be recognized and no protection for plant varieties will be given. This is true in areas such as Brazil and among members of the African and Malagasy Unions. Some countries have even withdrawn protection that had previously been given for plant varieties.

Most developed Western nations are calling for an increase in the level of protection afforded for plant biotechnology through the patent system. Not unexpectedly, the major impetus for such a push comes from the commercial sector. The private sector cannot underwrite huge research and development costs without proper incentives, and a unified multinational protection system is required if a company is to enter global markets and anticipate global protection. There is no doubt that a patent is a powerful business tool that can spur innovation and new technology as well as the transfer of technology from company to company through license agreements. It seems, however, that generally speaking, underdeveloped nations distrust the motives and goals of developed nations in seeking these higher levels of protection for plant biotechnology. The distrust comes from the historic exploitation and the general failure of international patent systems to foster innovation and technology transfer to the Third World. Although underdeveloped nations are fully aware of the advantages presented by plant biotechnology, they are concerned by the cost of the new technology and simply cannot afford to subsidize the R & D costs required. Thus, they are concerned that they will not be able to access or develop their own superior grain crops that could be helpful in addressing the needs of food production for growing Third World populations. It is certain that underdeveloped countries will remain unwilling to extend monopoly protection to Western biotechnology companies unless they are certain that some transfer of technology will occur in the future.

REGULATIONS: A GLOBAL PERSPECTIVE

Agricultural biotechnology is coming to market at an interesting time in world history. As a biological science that will help us meet world food demands by providing healthier, safer and more nutritious food while helping protect the environment, it should be readily accepted. However, we are also moving into a time of a global economy with trade zones, developing countries becoming developed, increased consumer awareness/concerns, and a new technology that consumers are just starting to learn about. On top of this scenario, developed countries have been developing their regulations for biotechnology on a broad base, including agriculture. Countries such as the United States, the United Kingdom and Canada are leading the direction on this. However, because this is a global technology there are efforts being made to harmonize or develop consistent regulatory processes. International organizations such as OECD (Organization for Economic Co-operation and Development), UNIDO (United Nations Industrial Development Organization) and IBF (International Biotechnology Forum) are all working to create international understanding.

In the overall scheme, each country is developing their own regulatory process for biotechnology. The United States and Canada are possibly the most advanced in reaching a positive, well-balanced regulatory process. In the United States, there are products in the market-place now, and this helps with consumer understanding and acceptance. The Canadian system is probably one of the best regulatory processes in the world. It has evolved through six years of development and consultation, and will be complete in 1995. As it is now being finalized, it appears that it will use a regulatory act coupled with guidelines. By using the guideline process there will be more flexibility in the approval process as our knowledge base increases. The areas of agricultural biotechnology with regulations developed or being developed in Canada are seeds; transgenic plants; food; animal feed; microbials (natural products); animal health products; and biopesticides. By contacting the appropriate regulatory personnel, people can be linked to the appropriate regulatory agencies.

Canada and the United States do not differ greatly in the end result, only in the process to get there. Both use a tiered decision-making tree that can expedite the process once the base knowledge is understood. For developing countries that see the potential benefits of agricultural biotechnology for their economy it is probably beneficial to use something like the Canadian or American model, rather than developing a separate process. This would help ensure acceptance of their technologies on a global basis, as well as being cost efficient.

Biotechnology will be a global technology for the twenty-first century. It will create tremendous benefits for the agricultural industries, generating unforeseen products and global opportunities. It will help create and sustain viable agricultural industries. There will be opportunities in food and non-food industries – lubricants, biofuels, plastics, pharmaceuticals, cosmetics, sweeteners, animal feeds, forestry, etc. It will impact almost all industries.

Those countries that embrace the technology will benefit from it and see their agricultural industries and research community grow. We are just seeing the first products of biotechnology enter the food market: cheese produced with chymazin, milk from cows carrying the gene for recombinant bovine somatotropin, and FlavrSavr tomato. In the next two years we will see a number of new products enter the market in several countries.

In summary, biotechnology regulations are evolving and in some countries they are in place already. The general conclusion by those countries with regulations in place is that safety of food produced by biotechnology can be secured and evaluated by the principles used in existing regulations for agricultural food products.

BIOTECHNOLOGY AND SOCIETY

After investment and protection have been secured, marketing of plant biotechnology products is the next step. Market research in biotechnology is still a young, poorly developed sector of the market. Although there are companies that claim extensive experience in this area, the brief span of their operation raises questions as to the validity of their claim. Where the introduction of a biotechnology product is closely tied to food, there appears to be considerable public concern, largely based on a lack of clear understanding of the principles involved, about genetically engineered food products. If a company is anticipating the commercialization of a product for the food industry, considerable public relations efforts must be made. Consumer concerns tend to relate to safety factors and social and ethical considerations. The overall challenge for the biotechnology industry is to strike a balance between these concerns and their own interest in being progressive and innovative. Because the news media are the primary source from which the public gain expectations and ideas concerning technology and biotechnology, public knowledge tends to be fragmentary and incomplete. Studies have shown that even when public knowledge is more complete there can still be strong objections to genetically engineered foodstuffs. Perhaps this is not surprising because of the close association of food products and the human consumers, but it raises marketing problems for the commercialization of new food products developed utilizing genetically engineered plant material. There is, however, in countries exposed to products of ag-biotech, a higher level of acceptance, because there is a growing realization of the benefits and an acceptance of minimal risk.

COMMERCIAL OPPORTUNITIES

Investment, protection and marketing are the real problems that the commercial biotechnology industry should perhaps be reviewing more thoroughly. Is the industry really looking at the appropriate markets for this technology in anything more than the short term? The promise of a doubling in world population by 2020, and the fact that many of our planet's crop species evolved in what are now Third World tropical countries where this population growth will be most acute, strongly suggest that the real market for biotechnology products lies in the developing world. Certainly the challenge of introducing innovative biotechnology to solve specific problems in developing countries requires the talents of a well-trained work force who can define problems and devise strategies that work in their local environment. An infrastructure capable of maintaining biotechnological research and development is also necessary. The university structure and research facilities in most developing countries are not adequate for biotechnology development. The expensive equipment and supplies necessary for many forms of biotechnology are also lacking because of economic limitations. However, despite these difficulties the perceived benefits of biotechnology are such that several developing nations have put considerable effort into establishing biotechnology laboratories and programmes. China, India, Pakistan, the Philippines, Indonesia, Thailand and Brazil have all taken the initiative to build biotechnology capabilities. A trained cadre of manpower is being established in these countries. Through national programmes, students are selected and sent to Canada, the United States or European centres for training in modern biotechnology. Student numbers from each country may be as high as 1200 or 1500 per annum. There is considerable effort being expended to establish the manufacturing (in these countries) of the necessary restriction enzymes and materials used in molecular biology.

However, in many other developing countries resources are not sufficient for the investment required by a biotechnology programme, and manpower is not always adequately trained to establish such a programme. This raises the issue of whether it would not be more innovative and far-sighted if small biotechnology companies focused their attention on establishing technology exchange agreements with developing countries. Certainly much of the genetic material necessary for crop improvement has been identified in tropical and underdeveloped countries, since this is where a huge portion of the world's plant material is to be found. Up to now developing countries have not benefited from superior plant crop varieties developed from genetic material collected in their cropping areas. There is, however, a real opportunity for the commercialization of this underdeveloped crop material by the utilization of biotechnology techniques, if funded by and coordinated through developed countries. Up to now developing countries have effectively been exploited by companies from developed Western countries. With the ever present threat of massive population growth in Third World countries and the pressure that would put on the West, perhaps the time has come for young biotechnology companies to look at fostering commercial links with countries of the Third World rich in species and genetic material.

In African countries about 50% of the gross national product comes from agriculture, and that agriculture employs 70% of the continent's population, estimated at present to be approaching 500 million. Only 30% of the continent is arable, the rest being arid or semi-arid, and the population is increasing at an estimated rate of 3.3% annually. On the other hand, annual food production in the region is increasing at only 1.5%. There is a fear that replacement of existing tropical agricultural products with biotechnology products from the industrialized countries will occur, thus eroding the chances for African countries to earn foreign exchange.

Although most African countries have established centres for biotechnology, there are no international centres or institutes focused solely on biotechnology. Tissue culture techniques are probably the most commonly utilized biotechnological tool. In Kenya, work is under way with meristem culture and embryogenesis applied especially to coffee and horticultural crops to develop pest-resistant plants. Certified virus-free potato tubers utilizing tissue culture techniques have also been developed in Kenya. The elite clonal material is then passed on to farmers. Work on nitrogen fixation is underway, sponsored by UNESCO, and has already resulted in the commercial production of *Rhizobium* inoculants. In Ethiopia, an area with greater genetic diversity than any other country in sub-Saharan Africa, tissue culture is used for the conservation of germplasm at the gene bank in Addis Ababa. It is probably in Zimbabwe that the most advanced biotechnology on the African continent is found, with work on tomatoes, potatoes and coffee, as well as tobacco research.

FUTURE STRATEGIES

Although the challenge in Africa could be partially met by the transfer of technology and development of drought resistant crop lines or by reforestation technologies, it appears likely that African countries will have to build their own facilities with which to do this work. It seems that without re-structuring the current industrial framework, development of Western biotechnologies is likely to weaken African economies, basically by replacing some of the cash crops or tropical products grown in the region which would have earned foreign exchange. For instance, coffee exports from Uganda, Tanzania and Kenya may be affected by the application of research in Southeast Asia to increase coffee production. A tropical product targeted for substitution is vanilla, which represents over 25% of the export products from Madagascar and 34% of those from Comoros. Work by EscaGenetics Inc. in California produces phyto-vanilla *in vitro* at a very competitive price which threatens to replace this market.

Africa imports virtually all of the technologies it needs in the agricultural and industrial sectors and has, therefore, not built an indigenous technological capability. Most African countries also have limited purchasing power. However, in developing countries there is a tremendous germplasm base available for biotechnologists to use. Technology transfer from companies both large and small in the Western developed countries must be the answer to Africa's pressing technology vacuum. The opportunity for commercialization of biotechnology within the African continent, and indeed any other underdeveloped country, is a pressing economic, social and ethical requirement to which commercial biotechnology should respond.

This chapter has been presented with few specific references. This has been done because the field is fast moving and changing and many of the views expressed are those of the authors, developed from long exposure to the ag-biotech field. There is a list of additional reading listed below that should be referred to by those who wish for more information.

REFERENCES

Baertlein, D.A., Lindrow, S.E., Panopoulous, N.J., Lee, S.P., Mindrinos, M.N., Chen, T.H.H. (1992) Expression of a bacterial ice nucleation gene implants in tobacco and *Solanum commersonii* ice gene from *Pseudomonas syringae*. *Plant Physiology* 100, 1730–1736.

Barton, K.A., Whiteley, H.R. and Yang, N.S. (1987) *Bacillus thuringiensis* delta endotoxin expressed in *Nicotiana tabacum* provides resistance to lepidopteran insects. *Plant Physiology* 85, 1103–1109.

Cemai, L., Facciotti, D., Hiatt, W., Thompson, G., Rose, R. and Stalker, D. (1985) Expression in plants of a mutant *aroA* gene from *Salmonella typhimurium* confers tolerance to glyphosate. *Nature* 317, 741–744.

Fischhoff, D.A., Bowdish, K.S., Perlak, F.I., Marone, P.G., McCormick, S.M., Niedermeyer, J.G. *et al.* (1987) Insect tolerant transgenic tomato plants. *Bio/Technology* 5, 807–813.

FURTHER READING

BIOTOL (Biotechnology by Open Learning) (1991) *Biotechnological Innovations in Crop Improvement.*

Brown, B. (ed.) (1991) *Biotechnology in a Global Economy*. Congress of the US Office of Technology Assessment.

Burke, W.S. (ed.) (1993) *The Societal Issues of Food Biotechnology, Conference Proceedings.* Conference and Educational Facility, North Carolina Biotechnology Center.

Caswell, M. *et al.* (1994) *Agricultural Biotechnology: An Economic Perspective*. USDA, Agric. Econ. Report, Number 687.

Federal Coordinating Council for Science, Engineering, and Technology (1993) *Biotechnology or the 21st Century – Realizing the Promise.*

Habbelink, H. (1991) *Biotechnology and the Future of World Agriculture.*

MacDonald, J.F. (ed.) (1993) *Agricultural Biotechnology: A Public Conversation about Risk.* NABC Report 5.

OECD (1992) *Biotechnology, Agriculture and Food*. Organization for Economic Cooperation and Development.

Ono, R.D. (1991) *The Business of Biotechnology – From the Bench to the Street*. Butterworth-Heinemann, Stoneham, USA.

Somatic Embryogenesis in Ornamentals

2

Scott A. Merkle
Daniel B. Warnell School of Forest Resources, University of Georgia, Athens, Georgia 30602 USA

INTRODUCTION

Somatic embryogenesis is the formation of an embryo from a cell other than a gamete or the product of gametic fusion. Although the process has become most familiar as an *in vitro* phenomenon with potential application as a micropropagation system, it has long been known to occur in nature as adventitious embryony, a type of apomixis. A substantial amount of information on the process of somatic embryogenesis has been reported in the 37 years since the first reports of somatic embryogenesis in the laboratory (Reinert, 1958; Steward, 1958). In recent years, much has been written on the potential for application of embryogenic cultures for a number of purposes, including mass clonal propagation, genetic engineering, protoplast culture and production of potentially useful somaclonal variants. Just as many of these applications are beginning to affect breeding and propagation of agronomic, fruit-producing and forest tree species, so they are likely to have a major role in genetic improvement of a number of ornamental species. However, despite a constantly improving understanding of the factors controlling induction of somatic embryogenesis, development and maturation of somatic embryos, and conversion of embryos to free-living plantlets, these processes are not routine or efficient for the majority of ornamental species. Furthermore, the very nature of many ornamental species, especially woody perennials, makes them less than ideal model systems with which to investigate and optimize these fundamental developmental processes. Even with these limitations, however, somatic embryogenesis constitutes a powerful biotechnological tool for improvement of ornamental species. The manipulation of embryogenic cultures of some ornamental plants will undoubtedly make revolutionary new products available within a few years. Here, I will discuss the advantages and disadvantages of somatic embryogenesis relative to other vegetative propagation methods, including other micropropagation techniques. Then I will summarize the progress made to date with development of embryogenic regeneration systems for the major ornamental genera, focusing on the status of each with regard to clonal propagation. Finally, I will use specific examples to review potential applications of somatic embryogenesis, other than simple clonal propagation, to biotechnological improvement of ornamentals.

ADVANTAGES AND DISADVANTAGES OF SOMATIC EMBRYOGENESIS AS A PROPAGATION TOOL

In general, somatic embryogenesis as a tool for propagation of ornamental species has a number of advantages over both conventional vegetative propagation (macropropagation) and other micropropagation techniques (i.e. axillary shoot multiplication and organogenesis). It is also true, however, that somatic embryogenesis has some significant disadvantages in comparison to the alternative techniques.

A critical factor determining the relative usefulness of embryogenic cultures for propagation or other applications is whether the character of the embryogenesis is 'direct' or 'indirect.' Although there are currently varying definitions as to what constitutes direct versus indirect embryogenesis, for the purposes of this review, I will simply define direct embryogenesis as somatic embryo production from cells of the explant tissue, with no intervening callus. Indirect embryogenesis, on the other hand, is characterized by the production of callus from the explant tissue, from which somatic embryos form. Some have expressed the idea that indirect embryogenic cultures have a higher potential for such applications as scaled-up propagation and gene transfer, since they may be easily converted to fast-growing cell suspensions (e.g. Caligari and Shohet, 1993). By contrast, direct embryogenic cultures are usually characterized by clumps of embryos arising directly from explant tissue, making them more difficult to manipulate, for example, as suspensions. However, there are numerous cases of direct embryogenic cultures that proliferate as proembryogenic masses (PEMs; Halperin, 1966), which can be distinguished from embryogenic callus by their structure (Halperin and Jensen, 1967), and are as amenable to scale-up and gene transfer as indirect embryogenic cultures (e.g. Merkle *et al.*, 1990; Wilde *et al.*, 1992).

In reviewing the relative advantages and disadvantages of embryogenic cultures versus other micropropagation techniques and direct versus indirect embryogenic cultures, it should be kept in mind that cultures of each species have their own sets of characteristics that may make them more or less suitable for different applications. Indeed, even different cultivars or genotypes of a given species can vary widely in their responses to a given cultural treatment (e.g. Kielly and Bowley 1992; Bailey *et al.*, 1993).

In a general comparison of *in vitro* propagation techniques, one important advantage of some embryogenic systems over axillary and adventitious shoot-based systems is the very high multiplication rates possible. Depending on species, virtually unlimited numbers of embryos can be generated from a single explant. In addition, embryogenic cultures of many species can be grown in liquid culture, making possible the production and handling of thousands of embryos at a time. Thus, in comparison to other micropropagation techniques, somatic embryos offer the potential for high-volume, large-scale propagation that can be translated into significant labour savings. Even greater economies of scale may be possible if bioreactor and continuous culture technologies can be applied to embryogenic systems. The feature of embryogenic cultures that may ultimately have the most impact for mass propagation of ornamental species, however, is the fact that the product is an embryo. The morphological and physiological similarity of somatic embryos to zygotic embryos means that they are complete propagules in themselves, with embryonic roots, shoots and leaves, and, most importantly, the 'program' to make a complete plant. Thus, unlike other *in vitro* propagation systems, no separate shoot elongation or rooting steps are required for plantlet production. Furthermore, many embryogenic systems can be manipulated to produce singularized embryos, requiring no excision from source tissues or other embryos to be handled. Not only do these characteristics further lower labour inputs, but they also give somatic embryos the potential for direct delivery to greenhouse or field as 'synthetic seeds', thereby eliminating the need for transplanting (see section on 'Synthetic seeds').

One major disadvantage of embryogenic cultures, especially with regard to propagation of woody perennial ornamentals, is that many of the embryogenic systems reported for these species to date have relied on immature tissues (i.e. seeds or seedlings) for explanting material. Most reports of somatic embryogenesis in woody perennials describe direct embryogenesis followed by repetitive embryogenesis or 'embryo cloning', in which the zygotic embryo is induced to replicate itself indefinitely. Thus the material being propagated is of unproven genetic value. Another problem with using somatic embryogenesis for mass propagation, as illustrated by the generic summaries below, is that while reports of somatic embryogenesis in these species are numerous, for many the ability to generate large numbers of normal, free-living plantlets remains a major bottleneck. Barriers to high-frequency plantlet production may occur at any of a number of points between induction and the production of a plantlet capable of surviving transfer to *ex vitro* conditions, and may include such problems as low-frequency embryo production, production of malformed embryos, incomplete embryo maturation, unbreakable embryo quiescence or dormancy, or low plantlet vigour. Finally, some embryogenic regeneration systems, especially those characterized by indirect embryogenesis with an intervening callus phase, may be associated with unwanted new variation in the regenerated plants. While levels of variation may be less than organogenic systems, they still may be unacceptable (see section on 'Somaclonal variation').

Below are summaries of progress with somatic embryogenesis research for the major ornamental species for which this phenomenon has been reported. It is evident that for some of these species, application of embryogenic cultures for commercial propagation is close to being economically feasible, while others await additional progress before they can be applied to operational mass propagation. For convenience, I have divided ornamental species into floricultural crops (dicots and monocots) and woody perennials, and within each category, I have alphabetized the plants by genus.

PROGRESS WITH DEVELOPMENT OF EMBRYOGENIC REGENERATION SYSTEMS

FLORICULTURAL CROPS

Dicots

Among dicotyledonous floricultural crops, some species have been extensively investigated over several years with regard to their potential for propagation via somatic embrygogenesis, while research with other species may only be represented by a single report.

Cyclamen (*Cyclamen persicum*) is an example of a relatively new addition to the list of floricultural crops regenerated via somatic embrygogenesis. Somatic embryos of this species, produced from callus derived from anthers, ovaries and zygotic embryos cultured on medium with 2,4-dichlorophenoxcyacetic acid (2,4-D) and coconut water, germinated into plantlets that flowered normally (Kiviharju *et al.*, 1992).

Progress in initiating embryogenic cultures and generating somatic embryo-derived plants in both chrysanthemum (*Dendranthema*) and carnation (*Dianthus*) has also been relatively recent. Leaf midrib explants of 12 of 23 chrysanthemum cultivars produced somatic embryos when cultured on MS (Murashige and Skoog, 1962) medium supplemented with 2,4-D and benzyladenine (BA), but plants were regenerated from only five of the cultivars (May and Trigiano, 1991). Light regime, sucrose concentration and genotype all apparently influenced the ability of a culture to produce morphologically mature somatic embryos. Somatic embryogenesis was induced by culturing young internodes of three cultivars of carnation in liquid MS medium supplemented with 2,4-D for callus production, followed by transfer of the callus to basal medium. A few plantlets were produced from the somatic embryos (Frey *et al.*, 1992). An alternative protocol for somatic embryogenesis in carnation employed antibiotics instead of plant growth regulators to induce somatic embryo production. Cephalosporin, penicillin and carbenicillin all induced somatic embryogenesis from leaf explants of different *Dianthus* cultivars, while naphthaleneacetic acid (NAA), 2,4-D and picloram failed to do so. Antibiotic-induced somatic embryos transferred to basal medium developed into plantlets (Nakano and Mii, 1993).

Cell suspension cultures of poinsettia (*Euphorbia pulcherrima*) were induced to produce somatic embryos by culturing the cells in medium containing 4-chlorophenoxyacetic acid (CPA) and BA (Kleffel and Preil, 1986). These suspensions later were employed as test material for bioreactor experiments (see section on 'Bioreactor culture').

Probably the most extensive progress in initiating and manipulating embryogenic cultures of any floricultural crop has been made with geranium (*Pelargonium*). Somatic embryos were produced from petioles and hypocotyls of zonal geranium (*P.* × *hortorum*) and from petioles of regal geranium (*P.* × *domesticum*) cultured on different media with different auxin types and concentrations (Marsolais *et al.*, 1991). All 30 cultivars of zonal geranium tested were capable of producing somatic embryos from hypocotyl explants. Embryos from these experiments were tested as desiccated synthetic seeds (see section on 'synthetic seeds'). Although combinations of indoleacetic acid (IAA) or phenylacetic acid (PAA) and benzyladenine (BA) used in another study were effective in inducing embryogenesis from zonal geranium hypocotyls (Slimmon *et al.*, 1991), later studies showed that thidiazuron (TDZ) at levels as low as 0.2 μM was much more effective. Exposures to TDZ for as little as two days induced embryogenesis from hypocotyl explants (Qureshi and Saxena, 1992; Visser *et al.*, 1992). Further improvements in geranium somatic embryo production were made by employing thin-layer explants from hypocotyls, which yielded up to eight times more embryos than were obtained from whole hypocotyl explants, and by addition of glutamine to the medium (Gill *et al.*, 1993). A chance contamination of an explant of zonal geranium 'Ringo Rose' led to the novel finding that the contaminating bacterium actually improved somatic embryo production by this normally morphogenetically unresponsive cultivar (Visser-Tenyenhuis *et al.*, 1994).

Another dicotyledonous floricultural crop (that also could be classified as a woody perennial) for which considerable progress with somatic embryogenesis has been reported is rose (*Rosa hybrida* and other *Rosa* species). Although hybrid tea roses have been bred and selected for over 120 years, their genetic improvement is still challenging. This is in large part because roses are perennials characterized by a high degree of sterility due to varying chromosome numbers among species (Noriega and Sondahl, 1991). Thus, the primary goal of somatic embryogenesis research with this genus has been to develop a system that is amenable to improvement via cellular and molecular techniques.

The first embryogenic rose cultures were obtained by culturing leaf explants from micropropagated plantlets of two cultivars on medium supplemented with kinetin and either NAA or naphthloxyacetic acid (NOA) (de Wit *et al.*, 1990). Embryos, which developed from callus derived from the leaf base or the midrib, could be grown into phenotypically normal plants that produced flowers. Embryogenic cultures of another rose cultivar were initiated from immature leaf and internode segments. Explants were initially cultured on medium with BA, NAA and 2,4-D to produce callus, which was subcultured to medium with BA, NAA, gibberellic acid (GAR_3) and L-proline to generate a friable embryogenic callus and somatic embryos (Rout *et al.*, 1991). Repetitive embryogenesis was obtained by transferring primary embryos to medium lacking L-proline. However, even cold treatment was ineffective in inducing root elongation by the somatic embryos. The embryogenic rose system that has proved useful for genetic engineering applications (see section on 'Gene transfer') was that reported by Noriega and Sondahl (1991), in which embryogenic cultures were produced from callus from filament explants cultured on medium supplemented with 2,4-D and zeatin. Somatic embryos from this callus were in turn used to initiate friable embryogenic tissue on medium with a low NAA/zeatin ratio, followed by transfer to medium with a high 2,4-D/zeatin ratio. Normal plantlets were obtained from these somatic embryos.

In another study, embryogenic callus cultures initiated from rose leaf explants were still capable of regenerating embryos and plants after more than two years in culture, while suspensions had apparently lost that ability (Moyne *et al.*, 1993). Cytogenetic examination of the two culture types revealed that the suspensions were characterized by a high frequency of aneuploid and polyploid cells, while the majority of callus cells had the tetraploid karyotype of the explants from which they were initiated.

Monocots

Somatic embryogenesis has been reported for a number of monocotyledonous floricultural genera, each of which has shown its own peculiar character with regard to the factors involved in initiation of embryogenic cultures and somatic embryo production. As with the dicots, embryogenic cultures of some species have only recently been reported, while others have been the basis of extensive research programmes.

Leaf explants from *in vitro*-grown plants of four *Anthurium andreanum* cultivars formed embryogenic calluses at their bases on medium supplemented with 2,4-D and kinetin. Somatic embryos converted to plantlets following transfer to medium with BA (Kuehnle *et al.*, 1992).

Inflorescence stalks, cormel slices and tissue-cultured plantlet bases of some *Gladiolus* cultivars cultured on medium supplemented with NAA or 2,4-D produced a friable callus with embryo-like structures from which plantlets were regenerated (Kamo *et al.*, 1990). Curiously, some later studies by the same laboratory, including one in which suspension cells and callus were transformed using particle bombardment, did not mention that the regenerated plantlets were of somatic embryo origin (Kamo, 1994; Kamo *et al.*, 1995). However, a more recent study comparing numbers of plants regenerated from suspension cultured cells, callus and cormel slices of six *Gladiolus* cultivars, once again described regenerable callus as being covered with globular structures or somatic embryos (Kamo, 1995). A study by another laboratory showed that friable embryogenic callus initiated from corm slices, young leaf bases and intact plantlets of four *Gladiolus* cultivars could still regenerate somatic embryos and plantlets after more than two years in culture (Stefaniak, 1994).

Daylily (*Hemerocallis*) cell suspension cultures were generated from larger tissue masses derived from excised shoot tips by replacing synthetic cytokinins with coconut water. Lowering the 2,4-D concentration in the medium gave rise to clusters that generated embryonic structures capable of forming plantlets

(Krikorian and Kann, 1981). In a subsequent study, daylily cultures initiated in this manner were used to demonstrate that embryogenic cultures could be maintained as preglobular stage proembryos in the absence of plant growth regulators at a low pH (between 4 and 4.5). These proliferating preglobular embryos were released to form later developmental stages and plantlets by raising the pH to 5.8 and culturing them on filter paper impregnated with activated charcoal (Smith and Krikorian, 1991).

Embryogenic cultures have been initiated from at least five different species of *Iris*. Explants have included flower parts (Laublin and Cappadocia, 1990; Jehan *et al.*, 1994), roots (Laublin and Cappadocia, 1990; Vieth *et al.*, 1992) and mature seeds (Reuther, 1977; Radojevic *et al.*, 1987; Radojevic and Subotic, 1992).

Lily (*Lilium*) embryogenic cultures were established from seeds on medium supplemented with picloram. The embryogenic calluses were used to establish suspension cultures of meristematic nodules, which in turn were used as a source of protoplasts (Mii *et al.*, 1994; see section on 'Protoplast culture'). *Nerine* somatic embryos developed from meristematic clusters in liquid cultures (Lilien-Kipnis *et al.*, 1992) and later in bioreactors (Lilien-Kipnis *et al.*, 1994; see section on 'Bioreactor culture').

WOODY PERENNIALS

As stated earlier, one major disadvantage of using embryogenic cultures for propagation of woody perennial ornamentals is that most of the embryogenic systems reported for these species have been initiated from immature (i.e. genetically unproven) explants. Horsechestnut (*Aesculus hippocastanum*), however, presents a special case with regard to somatic embryogenesis in woody ornamentals. Not only have embryogenic cultures been initiated from excised immature embryos (Radojevic, 1988) and seedling leaves (Dameri *et al.*, 1986), but somatic embryos have been produced from callus derived from filaments of trees up to 100 years old (Jorgensen, 1989). In addition, horsechestnut has also been regenerated via androgenesis (Radojevic, 1978). Further investigations on repetitive embryogenesis in cultures established from each of these explant sources was carried out by Kiss *et al.* (1992). Thus, horsechestnut is currently the only major arborescent ornamental in which mature individuals can be propagated via somatic embryogenesis. Radojevic (1988) achieved regeneration of horsechestnut somatic embryos by culturing them in medium supplemented with glutamine, IAA or indolebutyric acod (IBA) and GA_3. Profumo *et al.* (1991) found that cold treatment of somatic embryos for six months in the dark was needed for vigorous plantlet regeneration.

Among woody perennial genera, the largest number of reports on somatic embryogenesis has been for *Camellia*. *Camellia* embryogenic cultures have been induced from a variety of explant sources, including zygotic embryo cotyledons and axes (Kato, 1986; Plata and Vieitez, 1990; Vieitez and Barciela, 1990; Barciela and Vieitez, 1993; Pedroso and Pais, 1995a), roots (Vieitez *et al.*,

1991), stems from *in vitro* shoot cultures (Pedroso and Pais, 1995a) and leaves from *in vitro* shoot cultures or leaf fragments (San-José and Vieitez, 1993; Pedroso and Pais, 1995a). Embryo production has also been reported from anthers and petaloids of two varieties of *C. japonica*, although the anther tissue from which they originated could not be identified (Pedroso and Pais, 1993). Early reports indicated that somatic embryos from hypocotyls and cotyledons were of epidermal or subepidermal origin (Plata and Vieitez, 1990; Vieitez *et al.*, 1991; Barciela and Vieitez, 1993), while a more recent study reported that embryos formed directly from leaf, stem and cotyledon explants were all derived from subepidermal parenchyma cells, as was embryogenic callus derived from leaves (Pedroso and Pais, 1995a). Studies applying scanning electron microscopy (SEM) and electron probe microanalysis (EPMA) to study induction of embryogenesis from *Camellia* leaves indicated that embryo formation was preceded by the deposition of callose on surfaces of induced cells, and that embryogenic and non-embryogenic leaf regions could be distinguished by their X-ray spectra (Pedroso and Pais, 1992, 1994).

Camellia reticulata and *C. japonica* zygotic embryos and roots of *in vitro*-grown plantlets could be induced to produce somatic embryos directly by explanting on medium with no growth regulators, although addition of IBA to the medium improved embryo production (Plata and Vieitez, 1990; Vieitez and Barciela, 1990; Vieitez *et al.*, 1991). In another study, juvenile leaf fragments of *C. japonica* and *C. reticulata* required BA and IBA or NAA supplements to produce an embryogenic callus (San-José and Vieitez, 1993), while other researchers used 2,4-D and BA or kinetin to induce direct and indirect embryogenesis from *C. japonica* leaves (Pedroso and Pais, 1995a). Similarly, indirect somatic embryogenesis from *C. reticulata* cotyledon explants required medium supplemented with IBA (Plata and Vieitez, 1990).

Once induced, embryogenic *Camellia* cultures were amenable to long-term maintenance of repetitive embryogenesis (Vieitez *et al.*, 1991; Pedroso and Pais, 1995a). Leaf-derived callus or embryogenic cells from leaves cultured in liquid medium could be used to initiate embryogenic suspension cultures of *C. japonica* , from which plantlets were regenerated (Pedroso and Pais, 1995b). Gibberellic acid (GA_3) has been routinely employed to promote germination of *Camellia* somatic embryos (Plata and Vieitez, 1990; Vieitez *et al.*, 1991; San-José and Vieitez, 1993).

In addition to the problem of working with genetically unproven material, embryogenic cultures of some woody perennials have been characterized by low conversion rates. For example, eastern redbud (*Cercis canadensis*) somatic embryos were produced directly from immature zygotic embryos cultured on medium supplemented with 2,4-D (Trigiano *et al.*, 1988; Geneve and Kester, 1990). However, only a low percentage of embryos transferred to basal medium germinated and were successfully acclimatized to greenhouse conditions (Geneve and Kester, 1990). Cotyledons excised from immature zygotic embryos of yellowwood (*Cladrastis lutea*) and cultured on 2,4-D-supplemented medium produced somatic embryos directly, but very few plantlets were produced (Weaver and Trigiano, 1991). Flowering dogwood (*Cornus florida*) somatic embryos formed either directly on detached cotyledons of immature zygotic embryos or indirectly from zygotic embryo-derived callus following culture on medium supplemented with 2,4-D (Trigiano *et al.*, 1989). Although a few plantlets were produced, none continued to develop.

Probably the earliest report of somatic embryogenesis in a woody ornamental was for hollies (*Ilex* spp.). Clusters of somatic embryos formed on unexpanded cotyledons of excised heart-stage seed embryos of English holly (*Ilex aquifolium*) cultured on plant growth regulator-free medium (Hu and Sussex, 1971; Hu *et al.*, 1978). Embryos removed from the clusters and cultured individually germinated and developed into plantlets. Somatic embryos of *Ilex cornuta* and *Ilex opaca* were also obtained.

Somatic embryogenesis has been achieved in six North American *Magnolia* species to date. With all the magnolias, immature seeds were initially cultured on medium supplemented with 2,4-D to induce repetitive embryo production. However, while some species (*M. cordata, M. fraseri, M. pyramidata)* could continually proliferate on 2,4-D-supplemented medium, others (*M. virginiana, M. macrophylla)* declined with continued exposure to auxin, but maintained repetitive embryogenesis for years on plant growth regulator-free medium (Merkle and Wiecko, 1990; Merkle and Watson-Pauley, 1993, 1994). The same technique has been used to generate embryogenic cultures of two *M. grandiflora* cultivars (Merkle, unpublished; Fig. 2.1). Plantlets were regenerated from embryos of all the tested magnolias following germination on basal medium lacking casein hydrolysate.

Finally, embryogenic callus of empress tree (*Paulownia tomentosa*) was obtained either by culturing fertilized ovules on medium supplemented with 2,4-D or by culturing zygotic embryos on medium with IAA and kinetin (Radojevic, 1979). Removal of 2,4-D from the medium allowed plantlet formation to proceed.

PROPAGATION OF RARE OR UNUSUAL MATERIAL

The power of somatic embryogenesis as a propagation tool becomes especially apparent when the goal is to generate large numbers of propagules of an ornamental that is either rare or has some unusual feature that makes clonal multiplication highly desirable. As outlined earlier, the advantages of high multiplication rates and fewer steps to obtain plantlets may make somatic embryogenesis the propagation method of choice for these special cases.

One example of the use of somatic embryogenesis to propagate unusual material is its application to generate polyploid plants. Polyploid plantlets of *Camellia japonica* were generated by treating somatic embryos with colchicine for one week. Secondary embryos were then induced to form on the hypocotyls of the primary embryos, which in turn produced tertiary embryos on their hypocotyls. Tertiary embryos were diploid, tetraploid, triploid, hexaploid or mixoploid. Plants regenerated from the tetraploid somatic embryos were shorter and had fewer leaves than diploid plants (Kato, 1989).

Fig. 2.1. Somatic embryos of *Magnolia grandiflora*. Bar = 1 mm.

Another case where somatic embryogenesis has been used with special material is for propagation of hybrids. The potential value of embryogenic cultures to produce large numbers of certain hybrids was first recognized by Hu and Sussex (1971), who noted that interspecific hybrids of *Ilex* were desirable as ornamental plants, but that hybrid embryos were only produced in low numbers in many crosses, germinated slowly and were slow to propagate as cuttings. Although these researchers apparently did not extend their work to propagate *Ilex* hybrids, other hybrid ornamentals have been propagated using somatic embryogenesis. A tetraploid cultivar of *Alstroemeria*, a complex hybrid between *A. pelegrina* and *A. psittacina*, was propagated via somatic embryogenesis by culturing mature zygotic embryos on medium supplemented with kinetin and NAA (Hutchinson *et al*., 1994). Plantlets regenerated from the somatic embryos were successfully acclimatized and flowered normally in the greenhouse. Propagation of this hybrid via somatic embryogenesis is expected to overcome the problems of slow multiplcation rate and disease associated with the conventional propagation system of rhizome division. Another example of propagation of a rare hybrid with potential ornamental use is the embryogenic lines of hybrid *Liriodendron* generated by crossing yellow poplar (*L. tulipifera)* and Chinese tuliptree (*L. chinense*) (Merkle *et al*., 1993; Fig. 2.2).

Finally, somatic embryogenesis may be a relatively easy method to propa-

Fig. 2.2. Hybrid *Liriodendron* trees derived from somatic embryos. Reprinted by permission of Kluwer Academic Publishers from Merkle *et al.* (1993).

gate rare species with potential ornamental value. The yellow cucumber magnolia (*Magnolia cordata*; Fig. 2.3) and the pyramid magnolia (*Magnolia pyramidata*; Fig. 2.4) are both highly desirable ornamentals only found in a few, scattered populations. Both were easily propagated via somatic embryogenesis, although immature (seed) tissues were used as explants (Merkle and Wiecko, 1990; Merkle and Watson-Pauley, 1994).

OTHER APPLICATIONS OF EMBRYOGENIC CULTURES OF ORNAMENTAL SPECIES

GENE TRANSFER

Embryogenic cultures of ornamentals are increasingly finding application as a source of material amenable to production of genetically transformed plants. Indeed, for some ornamentals, one of the primary objectives in producing embryogenic cultures has been for use in gene transfer experiments, with the long-term goal of producing plants with characteristics unattainable via traditional breeding and selection. The prime example of such a goal is the introduction of genes controlling flower colour into rose. As excellent reviews of progress with genetic transformation of ornamentals are available (see

Fig. 2.3. *Magnolia cordata* plantlets derived from somatic embryos.

Fig. 2.4. *Magnolia pyramidata* plantlets derived from somatic embryos. From Merkle and Watson-Pauley (1994).

Deroles *et al.*, Chapter 6, this volume; Hutchinson *et al.*, 1992; Robinson and Firoozabady, 1993), I will not summarize the different strategies currently in use for transformation of ornamentals, but I will emphasize the advantages of using embryogenic cultures for this goal and discuss the progress that has been made in use of embryogenic cultures to date.

As with clonal propagation, embryogenic cultures have some strong advantages as target material for transformation by *Agrobacterium* Ti plasmid mediated gene transfer or by biolistic gene transfer, in comparison to other culture systems. Of course, as with propagation, one of these advantages is the high multiplication rates once transformed cells are obtained. This is apparent with indirect embryogenic systems, where transformed cells are selectively multiplied to form a callus, from which theoretically unlimited numbers of embryos can be produced. Since many ornamentals are characterized by direct-repetitive embryogenic systems, however, another critical feature may be the fact that embryos formed during direct repetitive embryogenesis can originate from single cells. Thus, the process of repetitive embryogenesis allows recovery of non-chimeric embryos from transformed sectors within a primary embryo. Repetitive cycles of direct embryogenesis effectively substitute for the callus phase found in indirect embryogenic systems, and make repetitive embryogenesis a potentially powerful method for generating transformed plants (e.g. McGranahan *et al.*, 1989).

Recently, embryogenic cultures have been used in the generation of transformed plants of two major ornamental species. Filament-derived embryogenic cultures of rose were transformed with genes encoding the selectable marker neomycin phosphotransferase II (NPT II) and the reporters β-glucuronidase (GUS) or firefly luciferase (LUC), using *Agrobacterium tumefaciens* or *A. rhizogenes*. Transformed embryogenic calluses formed somatic embryos, which produced plants that were established in the greenhouse, where they flowered (Firoozabady *et al.*, 1994). This research established a system whereby genes controlling such characteristics as flower colour can be introduced into roses. Similarly, embryogenic cultures established from leaf segments of chrysanthemum cultivar 'White Snowdon' were transformed using *A. tumefaciens* carrying genes encoding NPT II and GUS. These cultures produced transformed plantlets expressing the GUS gene which were established in the greenhouse (Pavingerova *et al.*, 1994).

PROTOPLAST CULTURE

Embryogenic callus and suspension cultures, as well as somatic embryos themselves, have been employed as a source of protoplasts for a range of species. The logic of this approach is that isolation of protoplasts from cells or tissues that are themselves regenerable will likely yield protoplast cultures capable of forming whole plants. Protoplasts, in turn, can be used with such gene transfer techniques as electroporation and microinjection, although these techniques have been largely obviated by the development of *Agrobacterium* co-cultivation techniques with intact tissues or cells and by microprojectile-mediated gene transfer.

Another potential use of embryogenic cell-derived protoplasts of particular interest for ornamentals is for somatic hybridization or cybridization. While such fusion products have yet to be described for an ornamental, embryogenic cultures of a few ornamentals have been employed as sources of regenerable protoplasts, thus laying the foundation for protoplast fusion research. Totipotent cell suspensions of daylily were digested to produce protoplasts that were induced to form cell clusters from which somatic embryos and plantlets were regenerated (Fitter and Krikorian, 1981). Protoplasts isolated from embryogenic suspension cultures of *Rosa persica* × *xanthina* and *Rosa wichuraiana* developed into colonies in agarose beads. The *R. persica* × *xanthina* colonies were induced to produce somatic embryos, which formed plantlets (Matthews *et al.*, 1991). Picloram-induced embryogenic calluses derived from seeds of *Lilium* × *formolongi* formed meristematic nodules in suspension, from which protoplasts were isolated (Mii *et al.*, 1994). The protoplasts formed colonies in gellan gum-solidified medium and plantlets were regenerated from the colonies, although it was unclear whether they were of somatic embryo origin.

BIOREACTOR CULTURE

The idea of scaling-up of embryogenic cultures via automated, continuous culture systems has been discussed for some time, and a number of bioreactor configurations have been tested. However, since the emphasis of this research has been on optimizing bioreactor design, the majority of this research has used model systems such as carrot (e.g. Styer, 1985) or agronomic crops such as alfalfa (e.g. Stuart *et al.*, 1987), rather than ornamentals. Only a few embryogenic cultures of ornamental species have been employed in testing bioreactor designs. Preil *et al.* (1988) reported that a vibro-mixer design equipped with silicone tubes for bubble-free aeration was capable of producing 100,000 poinsettia somatic embryos per litre of suspension, if the suspension was plated on filter paper and allowed to develop for four weeks. Raising the dissolved oxygen level in the bioreactor from 10 to 60% dramatically increased cell proliferation rates, although dissolved oxygen levels for somatic embryo production were not optimized. In a later report, Luttman *et al.* (1994) employed poinsettia and *Clematis tangutica* embryogenic cultures to optimize silicone tubing systems for oxygenation of different bioreactor types. Proliferating meristematic clusters of *Nerine* maintained in bubble bioreactors produced proembryogenic structures upon removal of paclobutrazol (PAC) from the medium (Ziv *et al.*, 1994). Mature embryos capable of forming plantlets were produced following transfer of the proembryogenic clusters to medium with N^6-(isopentenyl) adenine (2iP; Lilien-Kipnis *et al.*, 1994). As bioreactor and other scale-up and automation technologies are optimized for use with embryogenic cultures, no doubt these technologies will be adapted for other ornamentals characterized by highly proliferative embryogenic cultures.

SYNTHETIC SEEDS

As noted earlier in the discussion of the relative advantages of somatic embryos for clonal propagation, one of the relative strengths of somatic embryos as propagules is their resemblance to nature's own propagule-delivery system, the seed. Thus, one much publicized spin-off of somatic embryogenesis research has been development of techniques for direct delivery of somatic embryos to greenhouse, nursery or field as synthetic seeds. If successful, the benefits would be savings in time and labour, as the steps of *in vitro* germination and plantlet development and acclimatization of plantlets would all be eliminated. At least one extensive review (Gray and Purohit, 1991) and an entire book (Redenbaugh, 1993) devoted to the subject of synthetic seeds have been published. As was the case with application of bioreactor technology, however, ornamental species are not well represented in this type of research. Two of the four types of synthetic seeds described by Redenbaugh *et al.* (1991) (desiccated, uncoated synthetic seeds and hydrated, coated synthetic seeds) have been tested with somatic embryos of an ornamental plant, geranium. Although desiccated geranium somatic embryos turned green and began to germinate upon rehydration, few free-living plants were recovered from most cultivars tested (Marsolais *et al.*, 1991). Up to 90% of calcium alginate hydrogel-encapsulated geranium somatic embryos germinated normally on basal medium and produced flowering plants within 14 weeks (Gill *et al.*, 1994). Plantlet recovery rates from encapsulated somatic embryos stored at 4°C for 45 days were higher for embryos derived from cultures induced using TDZ than for those from IAA-induced cultures.

SOMACLONAL VARIATION

Somaclonal variation refers to heritable variation either pre-existing in explant material or generated in culture, which frequently results in regenerants that are phenotypically different from the source plant. It has been attributed to a range of causes including karyotype changes, cryptic chromosome rearrangements, the action of transposable elements, somatic gene rearrangements and gene amplification (Larkin and Scowcroft, 1981). As with other types of crops, in ornamentals the phenomenon can be regarded either as a problem, if the goal of *in vitro* culture is clonal fidelity, or as providing an opportunity to make selections of useful new variants. The opportunistic view is especially relevant with ornamental species, where rare new variants in flower colour, leaf variegation, dwarfism, etc. may be highly desirable. While somaclonal variation affecting significant traits has been reported in a number of important ornamental species (see Bouman and de Klerk, Chapter 10, this volume; and review by Hutchinson *et al.*, 1992), almost none of these reports has dealt with somaclonal variation from embryogenic cultures. The lack of reports on somaclonal variants arising from embryogenic cultures of ornamentals may not be surprising given the common conception that plants regenerated via somatic embryogenesis are less likely to display within-clone variation than plants derived from organogenesis (Vasil, 1987). However, this idea has not been supported by data (see discussion in Merkle *et al.*, 1995). A more likely explanation for the absence of such reports in ornamentals is the fact that the majority of embryogenic cultures of ornamental species are of the direct type, without the intervening callus phase that has been associated with somaclonal variation (Maheswaran and Williams, 1987; De Klerk, 1990). A report comparing the level of somaclonal variation in rose plants regenerated from callus versus directly from different tissues found that plants regenerated from direct adventitious shoots were as stable as reference plants propagated by rooted cuttings, while plants regenerated from callus were unstable, when evaluated for petal number, form and colour, and plant growth habit (Arene *et al.*, 1993). Unfortunately, plantlets derived from callus via organogenesis were not distinguished from those derived from callus via embryogenesis in the analysis in this study, so the role of embryogenesis in screening out somaclonal variants could not be evaluated. However, it could be inferred from the discussion that at least some proportion of the variant rose plants were derived from somatic embryos.

CONCLUSIONS

Somatic embryogenesis constitutes a powerful tool for improvement of ornamentals, not only with regard to clonal propagation, but for other biotechnological applications as well. While for most ornamental species, use of embryogenic cultures for such applications as operational mass propagation or gene transfer awaits improvements in embryo and plantlet production efficiency, it may not be long before somatic-embryo-derived floricultural crops or perennials representing rare, hybrid, variant or engineered phenotypes appear on the market. Undoubtedly, the ornamentals industry will benefit from the advances in propagation and genetic manipulation technology made possible by embryogenic regeneration systems.

REFERENCES

Arene, L., Pellegrino, C. and Gudin, S. (1993) A comparison of the somaclonal variation level of *Rosa hybrida* L. cv Meirutral plants regenerated from callus or direct induction from different vegetative and embryonic tissues. *Euphytica* 71, 83–90.

Bailey, M.A., Boerma, H.R. and Parrott, W.A. (1993) Genotype effects on proliferative embryogenesis and plant regeneration of soybean. *In Vitro Cellular and Developmental Biology* 29P, 102–108.

Barciela, J. and Vieitez, A.M. (1993) Anatomical sequence and morphometric analysis during somatic embryogenesis on cultured cotyledon explants of *Camellia japonica* L. *Annals of Botany* 71, 395–404.

Caligari, P.D.S. and Shohet, S. (1993) Variability in somatic embryos. In: Redenbaugh,

K. (ed.), *Synseeds: Applications of Synthetic Seeds to Crop Improvement*. CRC Press, Boca Raton, FL, pp. 163–174.

Dameri, R.M., Caffaro, L., Castaldo, P. and Profumo, P. (1986) Callus formation and embryogenesis with leaf explants of *Aesculus hippocastanum* L. *Journal of Plant Physiology* 126, 93–96.

De Klerk, G.J. (1990) How to measure somaclonal variation. *Acta Botanica Neerlandica* 39, 129–144.

de Wit, J.C., Esendam, H.F., Honkanen, J.J. and Tuominen, U. (1990) Somatic embryogenesis and regeneration of flowering plants in rose. *Plant Cell Reports* 9, 456–458.

Firoozabady, E., Moy, Y., Courtney-Gutterson, N. and Robinson, K. (1994) Regeneration of transgenic rose (*Rosa hybrida*) plants from embryogenic tissue. *Bio/Technology* 12, 609–613.

Fitter, M.S. and Krikorian, A.D. (1981) Recovery of totipotent cells and plantlet production from daylily protoplasts. *Annals of Botany* 48, 591–597.

Frey, L., Saranga, Y. and Janick, J. (1992) Somatic embryogenesis in carnation. *HortScience* 27, 63–65.

Geneve, R.L. and Kester, S.T. (1990) The initiation of somatic embryos and adventitious roots from developing zygotic embryo explants of *Cercis canadensis* L. cultured *in vitro*. *Plant Cell, Tissue and Organ Culture* 22, 71–76.

Gill, R., Gerrath, J.M. and Saxena, P.K. (1993) High-frequency direct somatic embryogenesis in thin-layer cultures of hybrid seed geranium (*Pelargonium* × *hortorum*). *Canadian Journal of Botany* 71, 408–413.

Gill, R., Senaratna, T. and Saxena, P.K. (1994) Thidiazuron-induced somatic embryogenesis enhances viability of hydrogel-encapsulated somatic embryos of geranium. *Journal of Plant Physiology* 143, 726–729.

Gray, D.J. and Purohit, A. (1991) Somatic embryogenesis and development of synthetic seed technology. *Critical Reviews in Plant Sciences* 10(1), 33–61.

Halperin, W. (1966) Alternative morphogenetic events in cell suspensions. *American Journal of Botany* 53, 443–453.

Halperin, W. and Jensen, W.A. (1967) Ultrastructural changes during growth and embryogenesis in carrot cell cultures. *Journal of Ultrastructure Research* 18, 428–443.

Hu, C.Y. and Sussex, I.M. (1971) *In vitro* development of embryoids on cotyledons of *Ilex aquifolium*. *Phytomorphology* 21, 103–107.

Hu, C.Y., Ochs, J.D. and Mancini, F.M. (1978) Further observations on *Ilex* embryoid production. *Zeitschrift für Pflanzenphysiologie* 89, 41–49.

Hutchinson, J.F., Kaul, V., Maheswaran, G., Moran, J., Graham, M.G. and Richards, D. (1992) Genetic improvement of floricultural crops using biotechnology. *Australian Journal of Botany* 40, 765–787.

Hutchinson, M.J., Tsujita, J.M. and Saxena, P.K. (1994) Callus induction and plant regeneration from mature zygotic embryos of a tetraploid *Alstroemeria* (*A. pelegrina* × *A. psittacina*). *Plant Cell Reports* 14, 184–187.

Jehan, H., Courtois, D., Ehret, C., Lerch, K. and Petiard, V. (1994) Plant regeneration of *Iris pallida* Lam. and *Iris germanica* L. via somatic embryogenesis from leaves, apices and young flowers. *Plant Cell Reports* 13, 671–675.

Jorgensen, J. (1989) Somatic embryogenesis in *Aesculus hippocastanum* L. by culture of filament callus. *Journal of Plant Physiology* 135, 240–241.

Kamo, K. (1994) Effect of phytohormones on plant regeneration from callus of *Gladiolus* cultivar Jenny-Lee. *In Vitro Cellular and Developmental Biology* 30, 26–31.

Kamo, K. (1995) A cultivar comparison of plant regeneration from suspension cells, callus and cormel slices of *Gladiolus*. *In Vitro Cellular and Developmental Biology* 31, 113–115.

Kamo, K., Chen, J. and Lawson, R. (1990) The establishment of cell suspension cultures of *Gladiolus* that regenerate plants. *In Vitro Cellular and Developmental Biology* 26, 425–430.

Kamo, K., Blowers, A., Smith, F., Vaneck, J. and Lawson, R. (1995) Stable transformation of *Gladiolus* using suspension cells and callus. *Journal of the American Society for Horticultural Science* 120, 347–352.

Kato, M. (1986) Micropropagation through cotyledon culture in *Camellia japonica* L. and *Camellia sinensis* L. *Japanese Journal of Breeding* 36, 82–83.

Kato, M. (1989) Polyploids of *Camellia* through culture of somatic embryos. *HortScience* 24, 1023–1025.

Kielly, A. and Bowley, S.R. (1992) Genetic control of somatic embryogenesis in alfalfa. *Genome* 35, 474–477.

Kiss, J., Heszky, L.E., Kiss, E. and Gyulai, G. (1992) High efficiency adventive embryogenesis in somatic embryos of anther, filament and immature proembryo origin in horse-chestnut (*Aesculus hippocastanum* L.) tissue culture. *Plant Cell, Tissue and Organ Culture* 30, 59–64.

Kiviharju, E., Tuominen, U. and Tormala, T. (1992) The effect of explant material on somatic embryogenesis of *Cyclamen persicum* Mill. *Plant Cell, Tissue and Organ Culture* 28, 187–194.

Kleffel, B. and Preil, W. (1986) Studies on embryogenesis in suspension culture of poinsettia (*Euphorbia pulcherrima*). In: *Abstracts VI International Congress of Plant Tissue and Cell Culture*, 3–8 August 1986, Minneapolis, Minnesota, p. 294.

Krikorian, A.D. and Kann, R.P. (1981) Plantlet production from morphogenetically competent cell suspensions of daylily. *Annals of Botany* 47, 679–686.

Kuehnle, A., Chen, F.C. and Sugii, N. (1992) Somatic embryogenesis and plant regeneration in *Anthurium andraeanum*. *Plant Cell Reports* 11, 438–442.

Kunitake, H., Imamizo, H. and Mii, M. (1993) Somatic embryogenesis and plant regeneration from immature seed-derived calli of rugosa rose (*Rosa rugosa* Thunb.). *Plant Science* 90, 87–194.

Larkin, P.J. and Scowcroft, W.R. (1981) Somaclonal variation – a novel source of variability from cell cultures for plant improvement. *Theoretical and Applied Genetics* 60, 197–214.

Laublin, G. and Cappadocia, M. (1990) Plant regeneration via somatic embryogenesis in several *Iris* species. *In: Abstracts VIIth International Congress on Plant Tissue and Cell Culture*, June 24–29, 1990, Amsterdam, p. 93.

Lilien-Kipnis, H., Ziv, M., Kahany, S. and Azizbekova, N. (1992) Proliferation and regeneration of *Nerine* in liquid culture. *Acta Horticulturae* 325, 467–473.

Lilien-Kipnis, H., Azizbekova, N. and Ziv, M. (1994) Scaled-up proliferation and regeneration of *Nerine* in liquid cultures. Part II. Ontogeny of somatic embryos and bulblet regeneration. *Plant Cell, Tissue and Organ Culture* 39, 117–123.

Luttman, R., Florek, P. and Preil, W. (1994) Silicone-tubing aerated bioreactors for somatic embryo production. *Plant Cell, Tissue and Organ Culture* 39, 157–170.

Maheswaran, G. and Williams, E.G. (1987) Uniformity of plants regenerated by direct embryogenesis from zygotic embryos of *Trifolium repens*. *Annals of Botany* 59, 93–97.

Marsolais, A.A., Wilson, D.P.M., Tsujita, M.J. and Senaratna, T. (1991) Somatic embryogenesis and artificial seed production in Zonal (*Pelargonium* × *hortorum*) and Regal (*Pelargonium* × *domesticum*) geranium. *Canadian Journal of Botany* 69, 1188–1193.

Matthews, D., Mottley, J., Horan, I. and Roberts, A.V. (1991) A protoplast to plant system in roses. *Plant Cell, Tissue and Organ Culture* 24, 173–180.

May, R.A. and Trigiano, R.N. (1991) Somatic embryogenesis and plant regeneration from leaves of *Dendranthema grandiflora*. *Journal of the American Society for Horticultural Science* 116, 366–371.

McGranahan, G.H., Leslie, C.A., Uratsu, S.L., Martin, L.A. and Dandekar, A.M. (1989) *Agrobacterium*-mediated transformation of walnut somatic embryos and regeneration of transgenic plants. *Bio/Technology* 6, 800–804.

Merkle, S.A. and Watson-Pauley, B.A. (1993) Regeneration of bigleaf magnolia by somatic embryogenesis. *HortScience* 28, 672–673.

Merkle, S.A. and Watson-Pauley, B.A. (1994) Ex vitro conversion of pyramid magnolia somatic embryos. *HortScience* 29, 1186–1188.

Merkle, S.A. and Wiecko, A.T. (1990) Somatic embryogenesis in three magnolia species. *Journal of the American Society for Horticultural Science* 115, 858–860.

Merkle, S.A., Wiecko, A.T., Sotak, R.J. and Sommer, H.E. (1990) Maturation and conversion of *Liriodendron tulipifera* somatic embryos. *In Vitro Cellular and Developmental Biology* 26, 1086–1093.

Merkle, S.A., Hoey, M.T., Watson-Pauley, B.A. and Schlarbaum, S.E. (1993) Propagation of *Liriodendron* hybrids via somatic embryogenesis. *Plant Cell, Tissue and Organ Culture* 34, 191–198.

Merkle, S.A., Parrott, W.A. and Flinn, B.S. (1995) Morphogenic aspects of somatic embryogenesis. In: Thorpe, T.A. (ed.), *In Vitro Embryogenesis in Plants*. Kluwer Academic Publishers, Dordrecht, pp. 155–203.

Mii, M., Yuzawa, Y., Suetomi, H., Motegi, T. and Godo, T. (1994) Fertile plant regeneration from protoplasts of a seed-propagated cultivar of *Lilium* × *formolongi* by utilizing meristematic nodular cell clumps. *Plant Science* 100, 221–226.

Moyne, A.L., Souq, F., Yean, L.H., Brown, S.C., Boulay, M. and Sangwannorreel, B.S. (1993) Relationship between cell ploidy and regeneration capacity of long-term *Rosa hybrida* cultures. *Plant Science* 93, 159–168.

Murashige, T. and Skoog, T. (1962) A revised medium for rapid growth and bioassays with tobacco tissue culture. *Physiologia Plantarum* 15, 473–497.

Nakano, M. and Mii, M. (1993) Antibiotics stimulate somatic embryogenesis without plant growth regulators in several *Dianthus* cultivars. *Journal of Plant Physiology* 141, 721–725.

Noriega, C. and Sondahl, M. (1991) Somatic embryogenesis in hybrid tea roses. *Bio/Technology* 9, 991–993.

Pavingerova, D., Dostal, J., Biskova, R. and Benetka, V. (1994) Somatic embryogenesis and *Agrobacterium*-mediated transformation of chrysanthemum. *Plant Science* 97, 95–101.

Pedroso, M.C. and Pais, M.S. (1992) A scanning electron microscropy and X-ray microanalysis study during induction of morphogenesis in *Camellia japonica* L. *Plant Science* 87, 99–108.

Pedroso, M.C. and Pais, S. (1993) Regeneration from anthers of adult *Camellia japonica* L. *In Vitro Cellular and Developmental Biology* 29P, 155–159.

Pedroso, M.C. and Pais, M.S. (1994) Early detection of embryogenic competence and of polarity in *Camellia japonica* by electron probe X-ray microanalysis. *Plant Science* 96, 189–201.

Pedroso, M.C. and Pais, M.S. (1995a) Explant region-specific embryogenic competence and plant recovery in *Camellia japonica*. *In Vitro Cellular and Developmental Biology* 31, 8–14.

Pedroso, M.C. and Pais, M.S. (1995b) Plant regeneration from embryogenic suspension cultures of *Camellia japonica*. *In Vitro Cellular and Developmental Biology* 31, 31–35.

Plata, E. and Vieitez, A.M. (1990) *In vitro* regeneration of *Camellia reticulata* by somatic embryogenesis. *Journal of Horticultural Science* 65, 707–714.

Preil, W., Florek, P., Wix, U. and Beck, A. (1988) Towards mass propagation by use of bioreactors. *Acta Horticulturae* 226, 99–106.

Profumo, P., Gastaldo, P., Bevilacqua, L. and Carli, S. (1991) Plant regeneration from cotyledonary explants of *Aesculus hippocastanum* L. *Plant Science* 76, 139–142.

Qureshi, J.A. and Saxena, P.K. (1992) Adventitious shoot induction and somatic embryogenesis with intact seedlings of several hybrid seed geranium (*Pelargonium* × *hortorum* Bailey) varieties. *Plant Cell Reports* 11, 443–448.

Radojevic, L. (1978) *In vitro* induction of androgenic plantlets in *Aesculus hippocastanum* L. *Protoplasma* 96, 369–374.

Radojevic, L. (1988) Plant regeneration of *Aesculus hippocastanum* (horsechestnut) through somatic embryogenesis. *Journal of Plant Physiology* 132, 322–326.

Radojevic, L. (1979) Somatic embryos and plantlets from callus cultures of *Paulownia tomentosa* Steud. *Zeitschrift für Pflanzenphysiologie* 91, 57–62.

Radojevic, L. and Subotic, A. (1992) Plant regeneration of *Iris setosa* Pall. through somatic embryogenesis and organogenesis. *Journal of Plant Physiology* 139, 690–696.

Radojevic, L., Sokic, O. and Tucic, B. (1987) Somatic embryogenesis in tissue culture of iris (*Iris pumila* L.). *Acta Horticulturae* 212, 719–723.

Redenbaugh, K. (1993) *Synseeds: Applications of Synthetic Seeds to Crop Improvement*. CRC Press, Boca Raton.

Redenbaugh, K., Fujii, J.A. and Slade, D. (1991) Synthetic seed technology. In: Vasil, I.K. (ed.), *Scale-up and Automation in Plant Propagation, Cell Culture and Somatic Cell Genetics of Plants, Vol. 8*. Academic Press, San Diego. pp. 35–74.

Reinert, J. (1958) Morphogenese und ihre Kontrolle an Gewenbekulturen aus Carotten. *Naturwissenschaften* 45, 344–345.

Reuther, G. (1977) Adventitious organ formation and somatic embryogenesis in callus of asparagus and iris and its possible applications. *Acta Horticulturae* 78, 217–224.

Robinson, K.E.P. and Firoozabady, E. (1993) Transformation of floricultural crops. *Scientia Horticulturae* 55, 83–99.

Rout, G.R., Debata, B.K. and Das, P. (1991) Somatic embryogenesis in callus cultures of *Rosa hybrida* L. cv. Landora. *Plant Cell, Tissue and Organ Culture* 27, 65–69.

San-José, M.C. and Vieitez, A.M. (1993) Regeneration of *Camellia* plantlets from leaf explant cultures by embryogenesis and caulogenesis. *Scientia Horticulturae* 54, 303–315.

Sharp, W.R., Sondahl, M.R., Caldas, L.S. and Maraffa, S.B. (1980) The physiology of *in vitro* asexual embryogenesis. *Horticultural Reviews* 2, 268–310.

Slimmon, T., Qureshi, J.A. and Saxena, P.K. (1991) Phenylacetic acid-induced somatic embryogenesis in cultured hypocotyl explants of geranium (*Pelargonium* × *hortorum* Bailey). *Plant Cell Reports* 10, 587–589.

Smith, D.L. and Krikorian, A.D. (1991) Growth and maintenance of an embryogenic cell culture of daylily (*Hemerocallis*) on hormone-free medium. *Annals of Botany* 67, 443–449.

Stefaniak, B. (1994) Somatic embryogenesis and plant regeneration of gladiolus. *Plant Cell Reports* 13, 386–389.

Steward, F.C. (1958) Growth and development of cultivated cells. III. Interpretations of the growth from free cell to carrot plant. *American Journal of Botany* 45, 709–713.

Stuart, D.A., Strickland, S.G. and Walker, K.A. (1987) Bioreactor production of alfalfa somatic embryos. *HortScience* 22(5), 800–803.

Styer, D.J. (1985) Bioreactor technology for plant propagation. In: Henke, R.R., Hughes, K.W., Constantin, M.J. and Hollaender, A. (eds), *Tissue Culture in Forestry and Agriculture*. Plenum Press, New York. pp. 117–130.

Trigiano, R.N., Beaty, R.M. and Graham, E.T. (1988) Somatic embryogenesis from immature embryos of redbud. *Plant Cell Reports* 7, 148–150.

Trigiano, R.N., Beaty, R.M. and Dietrich, J.T. (1989) Somatic embryogenesis and plantlet regeneration in *Cornus florida*. *Plant Cell Reports* 8, 270–273.

Vasil, I.K. (1987) Developing cell and tissue culture systems for the improvement of cereal and grass crops. *Journal of Plant Physiology* 128, 193–218.

Vieitez, A.M. and Barciela, J. (1990) Somatic embryogenesis and plant regeneration from embryonic tissues of *Camellia japonica*. L. *Plant Cell, Tissue and Organ Culture* 21, 267–274.

Vieitez. A.M., San Jose, C., Vieitez, F.J. and Ballester, A. (1991) Somatic embryogenesis from roots of *Camellia japonica* plantlets cultured *in vitro*. *Journal of the American Society for Horticultural Science* 116, 753–757.

Vieth, J., Laublin, G., Morisset, C. and Cappadocia, M. (1992) Multiplication *in vitro* of some iris plants from roots – histological aspects of somatic embryogenesis. *Canadian Journal of Botany* 70, 1809–1814.

Visser, C., Qureshi, J.A., Gill, R. and Saxena, P.K. (1992) Morphoregulatory role of thidiazuron. *Plant Physiology* 99, 1704–1707.

Visser-Tenyenhuis, C., Murthy, B.N.S., Odumeru, J. and Saxena, P.K. (1994) Modulation of somatic embryogenesis in hypocotyl-derived cultures of geranium (*Pelargonium* × *hortorum* Bailey) cv Ringo-Rose by a bacterium. *In Vitro Cellular and Developmental Biology* 30P, 140–143.

Weaver, L.A. and Trigiano, R.N. (1991) Regeneration of *Cladrastis lutea* (Fabaceae) via somatic embryogenesis. *Plant Cell Reports* 10, 183–186.

Wicart, G.A., Mouras, A. and Lutz, A. (1984) Histological study of organogenesis and embryogenesis in *Cyclamen persicum* Mill. tissue cultures, evidence for a single organogenetic pattern. *Protoplasma* 119, 159–157.

Wilde, H.D., Meagher, R.B. and Merkle, S.A. (1992) Expression of foreign genes in transgenic yellow-poplar plants. *Plant Physiology* 98, 114–120.

Ziv, M., Kahany, S. and Lilien-Kipnis, H. (1994) Scaled-up proliferation and regeneration of *Nerine in* liquid cultures. Part I. The introduction and maintenance of proliferating meristematic clusters by paclobutrazol in bioreactors. *Plant Cell, Tissue and Organ Culture* 39, 109–115.

Axillary Shoot Proliferation 3

John E. Preece
Department of Plant and Soil Science, Southern Illinois University, Carbondale, Illinois 62901-4415, USA

INTRODUCTION

Several years ago, McCown (1985) published a paper in which he described uses of shoot cultures for clonal propagation and genetic improvement. At that time the use of axillary shoot proliferation for clonal micropropagation was commercially viable. However, use of shoot cultures for biotechnology was a relatively new idea. He wrote of the use of shoot cultures as a source of explant material to yield protoplasts for genetic transformation.

In this chapter, methodology for establishing and maintaining axillary shoot cultures is described and the use of this technology to support various aspects of biotechnological manipulation of ornamental plants is discussed. It is clear that uses of axillary shoot cultures that were predicted by McCown (1985) are now quite important and have expanded beyond what was described in his paper.

AXILLARY SHOOT CULTURES

Axillary shoots are those that originate in the axils where leaves are attached to stems (at the nodes). This is the 'normal' location where shoots are expected to form on a plant, except for the shoot tip that grows from a seed. When a shoot forms on a plant at a location that is other than a node, for example from internodes, leaves, roots or callus, this is not normally expected and these shoots are considered to be adventitious.

An advantage of axillary shoot proliferation over production of adventitious shoots is the reduction in somaclonal variation with axillary shoots. Axillary shoots form from preformed meristems at nodes. The chances of the organized shoot meristem undergoing a mutation is relatively low. Therefore, axillary shoot proliferation is used to propagate plants clonally.

Adventitious shoots can form from the explant directly (direct regeneration) or from callus tissue that forms from the explant (indirect regeneration). While more common with indirect regeneration, somaclonal variation is more likely to occur when shoots form adventitiously than when they are the result of axillary shoot outgrowth.

ESTABLISHMENT

Initial explants used for axillary shoot proliferation must contain at least one node. This includes shoot tips, apical meristems, single nodes, shoot segments that contain nodes, embryos and seeds. Although more difficult to establish *in vitro*, it is preferable if explants are from adult plants if the genotype does not normally 'come true' from seed. By beginning with adult plants, the phenotype of the plant is known. With many perennial ornamental plants, seedlings segregate for many traits, making use of juvenile stock plants undesirable because the ultimate plant phenotype is unknown.

Care and maintenance of the stock plant is important for establishing shoot cultures. Often, field-grown plants yield explants that are difficult or impossible to surface disinfest. With perennial plants, especially woody plants, it is usually preferable that the explant material be from very young, vigorous softwood shoots. After they set buds, the bud scales can trap bacteria and fungal spores making establishment of aseptic cultures impossible.

If possible, stock plants should be grown in a greenhouse and the irrigation water should be applied in such a manner that it does not contact the new shoots. If stock plants cannot be grown in a greenhouse because of lack of facilities or the plant is too large and cannot be otherwise propagated vegetatively, an alternative method of producing explants must be used. A method that can result in production of aseptic cultures is the forcing of shoots from dormant or non-dormant stems within the laboratory (Yang and Read, 1992, 1993). Stems can be cut from the stock plant and their bases placed in appropriate solutions under electric lamps in a laboratory. New growth from the buds is soft, succulent and can be relatively easy to surface disinfest.

In addition to problems with surface disinfestation, explants from many ornamental (especially woody) plants produce an exudate that can discolour the medium and can lead to explant browning or blackening (Compton and Preece, 1986; Preece and Compton, 1991). These exudates can be inhibiory or lethal to explant tissues. Various techniques are used to overcome inhibitory exudates, including frequent transfers to fresh medium, use of antioxidants, changing plant growth regulators, and changing the conditions of incubation.

Establishment of shoot cultures can occur rapidly, or may take months or even years with some adult woody species. During establishment there should be elongation of axillary shoots and this new growth will be acclimatized to the *in vitro* conditions.

If the advantages of axillary shoot proliferation (lack of somaclonal variation) are to be realized, there should be caution that adventitious shoots are not produced. Adventitious shoots can be avoided by not raising the levels of plant growth regulators (particularly cytokinins) to too high a level in the medium. This may mean that shoot numbers are not at their maximum level, but will ensure clonal uniformity.

MAINTENANCE

Over time, shoot cultures will become stabilized (McCown and McCown, 1986). This means that if they are maintained properly, they will have a steady and high rate of shoot production. Branching will occur, and axillary shoots will elongate at a regular rate.

Stabilized shoot cultures are predictable and are theoretically sterile. They will produce a steady supply of shoots and leaves that can be used in other studies that support biotechnology (see Table 3.1) or for producing clonal plants.

The conditions of maintenance are critical for successfully supporting the growth of stabilized shoot cultures. These conditions will vary with the species, but there are several factors that are usually important. There normally is a defined amount of cytokinin in the medium. However, this may be adjusted if symptoms of toxicity occur. Signs of toxicity include excessive callus formation and short stunted shoots that either fail to elongate normally or elongate very slowly.

Cultures must be transferred to fresh medium on a regular basis. This is normally monthly. Sometimes, rather than transferring the cultures, a liquid overlay may be added (Maene and Debergh, 1985). This consists of adding a quantity of a liquid formulation of a nutrient medium over the top of the solidified medium. The depth of the liquid medium should not be too great because the cultures could become anaerobic. With proliferating shoot cultures, with good elongation, the liquid often can be added to a depth of 1 cm.

If the uniformity provided by axillary shoot proliferation is the goal of this supporting technology, cultures must be monitored to ensure that adventitious shoots are not being produced. If some shoots appear to be adventitious, they should be discarded, and axillary shoots should be excised and subcultured on medium with adjusted plant growth regulators to avoid adventitious shoot organogenesis.

AXILLARY SHOOT PROLIFERATION AS SUPPORTING TECHNOLOGY

For biotechnology, axillary shoot proliferation is used as supporting technology at both the beginning and the end of the process. At the beginning the *in vitro*

Table 3.1. Examples where explants from shoot cultures were used in biotechnology-related studies with ornamental plants.

Species	Explant	Objective	Final result	Reference
Anthurium andraeanum	Leaf	Somatic embryogenesis and plant regeneration	Indirect embryogenesis and plants regenerated	Kuehnle *et al.* (1992)
Betula pendula	Leaf	Adventitious shoots	Regenerated shoots	Valobra and James (1990)
B. pendula	Leaf	Adventitious shoots	Regenerated plants	Leege and Tripepi (1993)
Dendranthema grandiflora	Leaf blade, petiole and internode	Adventitious shoots	Regenerated plants	Kaul *et al.* (1990)
D. grandiflora	Leaf	Adventitious shoots and transformation	Regenerated shoots. GUS-positive spots on explants	De Jong *et al.* (1993)
D. grandiflora	Internode	Produce transformed plants	Transformed callus	Lowe *et al.* (1993)
D. zawadskii × *D. grandiflora*	Leaf, shoot tip	Isolate protoplasts and regenerate plants	Regenerated plants from protoplasts	Lindsay and Ledger (1993)
Dianthus caryophyllus	Leaf	Establish selection system for resistance to *Fusarium* toxins	Susceptible cultivar was sensitive and resistant cv. was more tolerant	Arai and Takeuchi (1993)
D. caryophyllus	Leaf, internode	Shoot regeneration	Regenerated shoots poorly	Nakano *et al.* (1994)
Dianthus spp.	Leaf	Isolate protoplasts, produce callus, regenerate shoots	Obtained protoplasts and callus, species-specific shoot regneration	Nakano and Mii (1992)

Eucalyptus grandis	Leaf	Callus and adventitious shoots	Regeneated plants	Laine and David (1994)
Gerbera hybrida	Leaf	Adventitious shoots	Regenerated plants	Reynoird *et al.* (1993)
Gladiolus hortulanus	Whole plantlet, sliced corm, leaf	Somatic embryogenesis and plant regeneration	Embryos from callus and conversion to plants	Stefaniak (1994)
Gladiolus 'Jenny Lee'	Leaf, cormel	Callus and adventitious shoots	Regenerated plants	Kamo (1994)
Liquidambar styraciflua	Leaf	Transformation and regeneration	Regenerated transgenic plants	Sullivan and Lagrimini (1993)
Lonicera japonica	Leaf, stem, root	Callus that would regenerate shoots	Regenerated plants from callus	Georges *et al.* (1993)
Morus bombycis	Internode	Callus and adventitious shoots	Regenerated plants	Jain and Datta (1992)
Populus nigra	Leaf	Transformation	Transgenic plants	Confalonieri *et al.* (1994)
P. tremuloides	Leaf	Somaclonal variation	Pigment and isozyme variants	Noh and Minocha (1990)
P. alba × *P. grandidentata*	Leaf	Cell cultures for transformation	Transgenic plants	Howe *et al.* (1994)
P. nigra × *P. maximowiczii*	Leaf	Protoplasts for regeneration	Regenerated plants	Park and Son (1992)
P. tremula × *P. alba*	Leaf, internode	Transformation	Transgenic plants	Leple *et al.* (1992)
Rauvolfia serpentina	Shoot	Transformation	Transgenic plants	Benjamin *et al.* (1993)
Rhododendron 'P.J.M. Hybrids'	Leaf	Plant regeneration	Regenerated plants	Preece and Imel (1991)
Weigela × *florida*	Root, leaf, internode from *in vitro* rooted shoots	Protoplasts for regeneration	Regenerated plants	Ochatt (1993)
Zingiber officinale	Buds	Produce encapsulated disease-free buds	Encapsulated buds in alginate-produced plants	Sharma *et al.* (1994)

stems and leaves serve as a source of explants and, at the end, axillary cultures are used to propagate clonally the special plants that are the result of the biotechnological manipulations.

AXILLARY SHOOT CULTURES AS A SOURCE OF EXPLANTS

McCown (1985, p.18) stated that shoot cultures had uses in biotechnology as 'a method of maintaining genetic stocks in a disease-free, reproducible, easily accessible form'. There are additional reasons for using shoot cultures as a source of explants, including the year-round uniformity and predictability of the explant material, the increased responsiveness of explants from shoot cultures compared to explants from field- or greenhouse-grown plants, any rejuvenation that occurs in the shoot cultures, the fact that shoot cultures are exposed to plant growth regulators that may programme their explants for success, the clonal uniformity that axillary shoot cultures maintain, and the minimal space that shoot cultures require compared to whole stock plants.

Perhaps one of the main reasons that people use shoot cultures as a source of explants is because they are free from microbial contamination. Indeed, Jain and Datta (1992) used shoot cultures specifically to avoid contamination. Methods of surface disinfestation on explants are harsh. Explants are usually soaked in a hypochlorite solution until all fungal spores and bacteria are killed, but not quite to the point that the plant part is lethally injured. Although not dead, damage is done to plant materials exposed to the disinfestant solution. This may have a negative impact on *in vitro* performance. Use of shoot cultures that are growing aspectically means that explants do not require surface disinfestation. This facilitates the process of preparing explants for manipulation using biotechnology and may enhance their response *in vitro*.

The genotypic uniformity and high explant quality is of great benefit to people who are using biotechnology to manipulate ornamental plants. Many ornamental selections do not produce genetically uniform seedlings; therefore, shoot cultures that produce clonal replicas of a selected phenotype offer the advantage of producing only slightly altered phenotypes after they are manipulated using biotechnology.

Table 3.1 demonstrates that use of explants from shoot cultures is a common practice by people who wish to regenerate plants adventitiously and by those who are interested in gene transfer then regeneration. Callus is even stimulated to form from explants from shoot cultures (Jain and Datta, 1992; Nakano and Mii, 1992; Georges *et al.* 1993; Lowe *et al.*, 1993; Kamo, 1994; Stefaniak, 1994). This callus is in turn used for various purposes including adventitious regeneration and transformation.

AXILLARY SHOOT CULTURES TO CLONE PRODUCTS OF BIOTECHNOLOGY

When annuals are manipulated using biotechnology, inheritance can be studied in subsequent generations and the genetic trait(s) can be passed on to other generations via seeds. Many ornamental plants segregate when propagated sexually and woody plants have long juvenility periods, making seed propagation impractical. Axillary shoot proliferation can be used to maintain new genotypes and the desirable phenotypic traits for which the plant was originally selected.

When somaclonal variation or genetic transformation is obtained with ornamental plants it is desirable to have several plants of each new clone for destructive testing, field and greenhouse evaluation, or release. Clonal micropropagation via axillary shoot proliferation is an ideal technique for rapidly increasing plant numbers and maintaining the new genotype.

Han *et al.* (1993) regenerated adult *Robinia pseudoacacia* L. (black locust) from callus. They placed the regenerated shoots onto a shoot multiplication medium and achieved axillary shoot proliferation from these adventitious shoots. They did not report any somaclonal variation, but this method would have provided sufficient replicates of each adventitious shoot for a statistically designed field study.

McGranahan *et al.* (1988) germinated somatic embryos that grew and proliferated from other somatic embryos on medium containing kanamycin. These were the result of a previous inoculation with *Agrobacterium tumefaciens*. Nodes from the germinated somatic embryos were micropropagated via axillary shoot proliferation. This was done to increase numbers before rooting and acclimatization of transgenic clones.

Perennial ornamental plants that have been genetically transformed should be handled in the same manner as by McGranahan *et al.* (1988). That will provide sufficient replicates for further studies or for release.

CONCLUSIONS

Although axillary shoot proliferation is used primarily to clone selected individuals, it has much potential as supporting technology for biotechnology of ornamental plants. The predictability of the shoot cultures throughout the year and the circumvention of surface disinfestation procedures with greater ensurance of non-contaminated cultures makes them ideal as sources of explants.

When new ornamental plants are developed via biotechnology, axillary shoot proliferation will be a major method of clonally multiplying them. The resulting plants can be tested or sold with assurance that they will all be genotypically uniform.

Frequently people are interested in the more ‘glamorous’ techniques of transformation, regeneration and somaclonal variation. Knowledge of axillary shoot proliferation is also important to support the biotechnology of ornamental plants.

REFERENCES

Arai, M., and Takeuchi, M. (1993) Influence of Fusarium wilt toxin(s) on carnation cells. *Plant Cell, Tissue and Organ Culture* 34, 287–293.

Baker, B.S. and Bhatia, S.K. (1993) Factors effecting adventitious shoot regeneration from leaf explants of quince (*Cydonia oblonga*). *Plant Cell, Tissue and Organ Culture* 35, 273–277.

Benjamin, B.D., Roja, G., and Heble, M.R. (1993) *Agrobacterium rhizogenes* mediated transformation of *Rauvolfia serpentina*: regeneration and alkaloid synthesis. *Plant Cell, Tissue and Organ Culture* 35, 253–257.

Compton, M.E. and Preece, J.E. (1986) Exudation and explant establishment. *Newsletter International Association for Plant Tissue Culture* 50, 9–18.

Confalonieri, M., Balestrazzi, A. and Bisoffi, S. (1994) Genetic transformation of *Populus nigra* by *Agrobacterium tumefaciens*. *Plant Cell Reports* 13, 256–261.

De Jong, J., Rademaker, W. and Van Wordragen, M.F. (1993) Restoring adventitious shoot formation on chrysanthemum leaf explants following cocultivation with *Agrobacterium tumefaciens*. *Plant Cell, Tissue and Organ Culture* 32, 263–270.

Georges, D., Chenieux, J.-C. and Ochatt, S.J. (1993) Plant regeneration from aged-callus of the woody ornamental species *Lonicera japonica* cv. Hall's Prolific. *Plant Cell Reports* 13, 91–94.

Han, K.H., Keathley, D.E. and Gordon, M.P. (1993) Cambial tissue culture and subsequent shoot regeneration from mature black locust (*Robinia pseudoacacia* L.). *Plant Cell Reports* 12, 185–188.

Howe, G.T., Goldfarb, B. and Strauss, S.H. (1994) *Agrobacterium*-mediated transformation of hybrid poplar suspension cultures and regeneration of transformed plants. *Plant Cell, Tissue and Organ Culture* 36, 59–71.

Jain, A.K. and Datta, R.K. (1992) Shoot organogenesis and plant regeneration in mulberry (*Morus bombycis* Koidz): factors influencing morphogenetic potential in callus cultures. *Plant Cell, Tissue and Organ Culture* 29, 43–50.

Kamo, K. (1994) Effect of phytohormones on plant regeneration from callus of *Gladiolus* cultivar Jenny Lee. *In Vitro Cellular and Developmental Biology* 30P, 26–31.

Kaul, V., Miller, R.M., Hutchinson, J.F. and Richards, D. (1990) Shoot regeneration from stem and leaf explants of *Dendranthema grandiflora* Tzvelev (syn. *Chrysanthemum morifolium* Ramat.). *Plant Cell, Tissue and Organ Culture* 21, 21–30.

Kuehnle, A.R., Chen, F.C. and Sugii, N. (1992) Somatic embryogenesis and plant regeneration in *Anthurium andraeanum* hybrids. *Plant Cell Reports* 11, 438–442.

Laine, E. and David, A. (1994) Regeneration of plants from leaf explants of micropropagated clonal *Eucalyptus grandis*. *Plant Cell Reports* 13, 473–476.

Leege, A.D. and Tripepi, R.R. (1993) Rapid adventitious shoot regeneration from leaf explants of European birch. *Plant Cell, Tissue and Organ Culture* 32, 123–129.

Leple, J.C., Brasileiro, A.C.M., Michel, M.F., Delmotte, F. and Jouanin, L. (1992) Transgenic poplars: expression of chimeric genes using four different constructs. *Plant Cell Reports* 11, 137–141.

Lindsay, G.C. and Ledger, S.E. (1993) A protoplast to plant system for the chrysanthemum *Dendranthema zawadskii* × *D. grandiflora*. *Plant Cell Reports* 12, 278–280.

Lowe, J.M., Davey, M.R., Power, J.B. and Blundy, K.S. (1993) A study of some factors affecting *Agrobacterium* transformation and plant regeneration of *Dendranthema grandiflora* Tzvelev (syn. *Chrysanthemum morifolium* Ramat.). *Plant Cell, Tissue and Organ Culture* 33, 171–180.

Maene, L. and Debergh, P. (1985) Liquid medium additions to established tissue cultures to improve elongation and rooting *in vivo. Plant Cell, Tissue and Organ Culture* 5, 23–33.

McCown, B.H. 1985. From gene manipulation to forest establishment: shoot cultures of woody plants can be a central tool. *Tappi Journal* 68, 116–119.

McCown, D.D. and McCown, B.H. (1986) North American hardwoods. In: Bonga, J.M. and Durzan, D.J. (eds), *Cell and Tissue Culture in Forestry, Vol. 3, Case Histories: Gymnosperms, Angiosperms and Palms* 2nd edn. Martinus Nijhoff, The Netherlands, pp. 247–260.

McGranahan, G.H., Leslie, C.A., Uratsu, S.L., Martin, L.A. and Dandekar, A.M. (1988) *Agrobacterium*-mediated transformation of walnut somatic embryos and regeneration of transgenic plants. *Bio/Technology* 6, 800–804.

Nakano, M. and Mii, M. (1992) Protoplast culture and plant regeneration of several species in the genus *Dianthus*. *Plant Cell Reports* 11, 225–228.

Nakano, M., Hoshino, Y. and Mii, M. (1994) Adventitious shoot regeneration from cultured petal explants of carnation. *Plant Cell, Tissue and Organ Culture* 36, 15–19.

Noh, E.W. and Minocha, S.C. (1990) Pigment and isozyme variation in aspen shoots regenerated from callus culture. *Plant Cell, Tissue and Organ Culture* 23, 39–44.

Ochatt, S.J. (1993) An efficient protoplast-to-plant system for hybrid ornamental shrub, *Weigela* ×*florida* cv. Bristol Ruby (*Caprifoliaceae*). *Plant Cell, Tissue and Organ Culture* 33, 315–320.

Park, Y.G. and Son, S.H. (1992) *In vitro* shoot regeneration from leaf mesophyll protoplasts of hybrid poplar (*Populus nigra* × *P. maximowiczii*). *Plant Cell Reports* 11, 2–6.

Preece, J.E. and Compton M.E. (1991) Problems with explant exudation. In: Bajaj, Y.P.S. (ed.). *Biotechnology in Agriculture and Forestry. Vol 17. High-Tech and Micropropagation I.* Springer-Verlag, Berlin, pp. 168–189.

Preece, J.E. and Imel, M.R. (1991) Plant regeneration from leaf explants of *Rhododendron* 'P.J.M. Hybrids'. *Scientia Horticulturae.* 48, 159–170.

Preece, J.E., Navarrete, N., Van Sambeek, J.W. and Gaffney, G.R. (1991) An *in vitro* microplant bioassay using clonal white ash to test for tall fescue allelopathy. *Plant Cell, Tissue, and Organ Culture* 27, 203–210.

Reynoird, J.P., Chriqui, D., Noin, M., Brown, S. and Marie, D. (1993) Plant regeneration from *in vitro* leaf culture of several *Gerbera* species. *Plant Cell, Tissue and Organ Culture* 33, 203–210.

Sharma, T.R., Singh, B.M. and Chauhan, R.S. (1994) Production of disease-free encapsulated buds of *Zingiber officinale* Rosc. *Plant Cell Reports* 13, 300–302.

Stefaniak, B. (1994) Somatic embryogenesis and plant regeneration of gladiolus (*Gladiolus hort.*). *Plant Cell Reports* 13, 386–389.

Sullivan, J. and Lagrimini, L.M. (1993) Transformation of *Liquidambar styraciflua* using *Agrobacterium tumefaciens*. *Plant Cell Reports* 12, 303–306.

Valobra, C.P. and James, D.J. (1990) *In vitro* shoot regeneration from leaf discs of *Betula pendula* 'Dalecarlica' EM 85. *Plant Cell, Tissue and Organ Culture* 21, 51–54.

Yang, G. and Read, P.E. (1992) Pre-forcing treatments influence bud break and shoot elongation in forced woody species. *Journal of Environmental Horticulture* 10, 101–103.

Yang, G. and Read, P.E. (1993) *In vitro* culture of Vanhoutte's spirea explants from 'secondary cultures' and dormant stems forced in solutions containing plant growth regulators. *Plant Cell, Tissue and Organ Culture* 33, 25–30

Adventitious Shoot Regeneration

4

Robert R. Tripepi
Plant Science Division, University of Idaho, Moscow, Idaho 83844, USA

INTRODUCTION

Adventitious shoot organogenesis is a type of *in vitro* propagation that can be used for clonal reproduction or mass propagation of plants. Organogenesis involves *de novo* production of adventitious shoots on explants from many different sources, and this tissue culture technique is widely used in horticulture and forestry for production of ornamental plants and timber species. Many ornamental species can be propagated by shoot organogenesis (see Table 4.1 for a selected list), and additional species, those untried or thought to be recalcitrant, are being successfully multiplied by this technique as time passes. Plants are multiplied by organogenesis for various reasons including improved multiplication rates, genetic engineering, and, perhaps most importantly to maintain clonal characteristics. Although organogenesis does have a few inherent liabilities as a tissue culture technique, its widespread use will remain important because of its application to future research needs. This chapter describes assets and liabilities of adventitious shoot organogenesis with ornamental plants, stages of shoot formation, factors affecting organogenesis, and future applications of this technology.

Adventitious shoot organogenesis is used as a propagation technique for ornamental plants for several reasons. First and foremost, organogenesis can be used for reliable clonal propagation of plants. Since many ornamental species are cultivars that are propagated for one or more unique characteristics, plants produced via organogenesis usually retain the traits of the parent plant. Second, propagation rates via organogenesis can be much higher than by axillary shoot proliferation (Chun, 1993). Therefore, organogenesis can be used for mass propagation if the shoot initiation protocol is highly efficient. Other reasons to use organogenesis include

Table 4.1. Selected herbaceous and woody ornamental species that can be propagated via adventitious shoot organogenesis.

Species	Explant	Medium	Plant growth regulators (μM)	Reference
Herbaceous				
Alstroemeria spp.	Peduncle	½ MS	BA 4.44–71.04 Kin 4.65–74.4	King and Bridgen (1987)
Anthurium scherzerianum	Leaf	Modified MS	2,4-D 0.45 BA 4.44	Geier (1986)
Antirrhinum majus	Hypocotyl, stem segments	MS	NOA 1.24	Atkinson *et al.* (1989)
Begonia semperflorens	Leaf, petiole	White's	Kin various conc.	Sehgal (1975)
Begonia spp.	Petiole	MS	NAA 0.54 BA 2.22	Simmonds (1984)
Begonia venosa	Inflorescence	MS	NAA 2.69 BA 2.22 & 0.44	Pierik and Tetteroo (1987)
Dendranthema grandiflora	Stem, leaf	MS	BA 5 NAA 5	Kaul *et al.* (1990)
D. indicum	Leaf	MS	IAA 1.14 BA 13.3–22.19	Ledger *et al.* (1991)
Episcia cupreata	Leaf	MS	NAA and IAA various conc.	Johnson (1978b)
Euphorbia flanaganii	Stem segments	MS	IAA, NAA and 2,4-D various conc. Kin 0.46	Silberstein *et al.* (1983)
E. hirta	Leaf	MS	IAA 11.42 NAA 2.69–5.37 BA 2.22–13.30 GA 5.77	Baburaj *et al.* (1987)
E. pulcherrima	Decoated seed	White's	IAA 5.71 2,4-D 4.5 Kin 4.65 GA 2.89	Nataraja *et al.* (1974)
E. pulcherrima	Petiole, stem internode	MS and ½ MS	2,4-D, NAA, IBA, BA, Kin, Zeatin, and 2iP various conc.	De Lange *et al.* (1974)
Gladiolus spp.	Cormel stem tips	Modified MS	NAA 0.54	Simonsen and Hildebrandt (1971)

Table 4.1. (*cont.*)

Species	Explant	Medium	PGR (μM)	Reference
Gladiolus spp.	Inflorescence stalks	Modified MS	NAA and Kin various conc.	Ziv *et al.* (1970)
Gladiolus spp.	Various parts	MS	IAA, NAA and 2,4-D various conc.	Hussey (1975)
Gloxinia spp.	Leaf	MS	Kin and IAA various conc.	Johnson (1978a)
Hyacinthus orientalis	Bulb scales	½ Knops	IAA, NAA, IBA various conc.	Pierik and Steegmans (1975b)
Hyacinthus spp.	Bulb scales	MS	NAA, IAA and 2,4-D various conc.	Hussey (1975)
Iris hollandica 'Prof. Blaauw'	Bulb scales	½ MS	None	Van der Linde *et al.* (1988)
Laeliocattleya 'Mayflower'	Leaf tips	Heller's	BA 2.22 2,4-D 4.52	Churchill *et al.* (1973)
Lycoris squamigera	Bulb scales	MS	NAA, 2,4-D and BA various conc.	Huang *et al.* (1992)
Pelargonium spp.	Callus	MS	Zeatin 9.12 IAA 10.84 BA 8.87 NAA 10.74	Dunbar and Stephens (1989)
Pelargonium × *hortorum*	Seedling	MS	TDZ 1–10	Qureshi and Saxena (1992)
Petunia hybrida	Leaf sections	MS	BA 4.4 or 44	Auer *et al.* (1992)
Saintpaulia ionantha	Floral parts, leaf discs	MS	NAA 5.4 BA 4.4	Vazquez *et al.* (1977)
Saintpaulia spp.	Petiole	MS	NAA 2.69 Kin 2.32	Harney and Knap (1979)
Senecio × *hybridus*	Petiole	MS	BA, NAA, IAA, Zeatin, Kin, and 2iP various conc.	Gertsson (1986)
Tulipa gesneriana	Floral stem segments	MS	NAA 5.37 2iP 7.38	Wilmink *et al.* (1995)

Table 4.1. (*cont.*)

Species	Explant	Medium	PGR (μM)	Reference
Woody				
Betula pendula	Leaf sections	DKW	BA 7.5, 15 or 22.5	Leege and Tripepi (1993)
B. pendula	Roots, leaves	Modified MS	Leaves: IAA 11.42 Zeatin 22.81 Roots: NAA 2.69 2iP 24.61	Srivastava *et al.* (1985)
B. pendula	Catkins	Modified White, DBM	IAA 11.42 Kin 23.23	Srivastava and Steinhauer (1981)
B. platyphylla var. *japonica*	Petioles	Modified IS	BA 3.55	Saito and Ide (1985)
Chamaecyparis obtusa	Leaf	Campbell and Durzan	NAA 0.03 BA 9.8	Ishii (1991)
Cryptomeria japonica	Hypocotyl segments	WS	BA 8.9	Isikawa (1974)
Ficus lyrata	Leaves	NN	IBA 12.5 BA and 2iP 24.0 each	Jona and Gribaudo (1987)
F. religiosa	Leaves	MS	2,4-D 2.3 BA 2.2	Narayan and Jaiswal (1986)
Fraxinus americana	Callus, cotyledons, hypocotyls	Modified MS	TDZ 10	Bates *et al.* (1992)
F. excelsior	Hypocotyl	Modified MS	TDZ 0.0045–22.71	Tabrett and Hammatt (1992)
Larix decidua	Immature female cones	Litvay	BA 10	Bonga (1984)
L. decidua	Needles, vegetative buds	WPM, Quoirin and LePoivre	BA 13.30 IBA 0.49	Chalupa (1989)
Liquidambar styraciflua	Hypocotyl sections	Modified RW and BL	NAA 0.054 BA 2.22	Wetzstein and Sommer (1982)
L. styraciflua	Leaf	WPM	NAA 0.5 BA 11.1	Brand and Lineberger (1988)

Table 4.1. (*cont.*)

Species	Explant	Medium	PGR (μM)	Reference
Malus hupehensis	Leaf	MS	2,4-D 50 Kin 2.3	Bates (1986)
Paulownia tomentosa	Shoot tip	MS	IBA 0.5 BA 4.4	Marcotrigiano and Stimart (1983)
P. tomentosa	Seedling stem	½ MS	IAA 30 Kin 12	Fu (1978)
Picea abies	Hypocotyl, needle segments	WS	NAA 0.5 BA 20	Chalupa (1977)
P. engelmannii	Hypocotyls, epicotyls	AE	BA 5–10	Patel and Thorpe (1986)
P. engelmannii	Mature embryos	AE	NAA 10 BA 25	Lu and Thorpe (1988)
P. glauca	Embryo, epicotyl	WPM, ½ SH	TDZ 0.01–0.1	Ellis *et al.* (1991)
P. glauca	Cotyledons	GMD	BA 2	Toivonen and Kartha (1988)
Pinus brutia	Embryos	Modified SH	BA 13.3	Abdullah *et al.* (1985)
P. eldarica	Seedling cotyledons	Modified MS	BA 4.44–8.87 IBA 0.246–2.46	Gladfelter and Phillips (1987)
P. elliottii	Embryos	Modified RW	BA 22	Perez-Bermudez and Sommer (1987)
P. oocarpa	Cotyledons	Modified MS	NAA 0 – 0.025 BA 25, 50	Franco and Schwarz (1985)
P. oocarpa	Embryos	S	NAA 10.74 BA 22.19	Floh and Handro (1986)
P. radiata	Embryo cotyledons	SH	BA 22.2	Reilly and Washer (1977)
P. strobus	Cotyledon, hypocotyl	Modified MS and Litvey	NAA 1.07 BA 8.87	Kaul (1987)
P. strobus	Mature embryo	Modified MS	BA 10	Schwartz *et al.* (1988)
P. sylvestris	Seedling plumule	S	BA and 2iP each at 5–10	Bornman and Jansson (1980)
P. sylvestris	Embryos	Lin and Staba	BA various conc.	Tranvan (1979)

Table 4.1. (*cont.*)

Species	Explant	Medium	PGR (μM)	Reference
P. taeda	Cotyledons	GD	Zeatin 4.56 NAA 0.054	Mehra-Palta *et al.* (1978)
Populus alba	Stem internode	Modified MS	None	Douglas (1984)
P. deltoides	Stem internode	Modified MS	None	Douglas (1984)
Populus hybrid TT32	Stem internode	Modified MS	None	Douglas (1984)
P. nigra	Stem internode	Modified MS	None	Douglas (1984)
P. tremula	Callus	Modified LS	BA 0.66–3.11	Chalupa (1974)
P. tremuloides	Callus	Modified Wolter and Skoog	BA 0.66	Winton (1970), Wolter (1968)
Pseudotsuga menziesii	Cotyledon	SH	TDZ 100–800	Goldfarb *et al.* (1991)
P. menziesii	Callus	$\frac{1}{4}$ MS	BA 500–1000	Cheng (1975)
Quercus robur	Stem node	WPM	TDZ 0.045 and 0.091	Chalupa (1988)
Rhododendron 'PJM Hybrids'	Leaf	Anderson	TDZ 0.01–10 IBA 0.98	Preece and Imel (1991)
Rhododendron spp. 6 cultivars	Leaf	Anderson	IBA 4.9 2iP 73.8	Iapichino *et al.* (1992)
Robinia pseudoacacia	Cotyledon	WPM	BA 22.2 2,4-D 0.4	Arrilaga and Merkle (1993)
R. pseudoacacia	Leaf sections	MS	BA 4.4 IBA 24.6	Arrilaga and Merkle (1993)
Rosa hybrida	Stem internodes, leaves	$\frac{1}{2}$ MS	BA 8.87 NAA 0.054 GA 28.89	Rout *et al.* (1992)
R. hybrida	Embryos	$\frac{1}{2}$ MS	BA 1.0 NAA 0.05	Burger *et al.* (1990)
Rosa spp.	Callus, leaves, roots	MS	BA and NAA various conc.	Lloyd *et al.* (1988)
Sapium sebiferum	Stem explants	MS	Kin 2.32 BA 4.44	Kotwal *et al.* (1983)

Table 4.1. (*cont.*)

Species	Explant	Medium	PGR (μM)	Reference
Thuja occidentalis	Embryo	½ Quorin and LePoivre	BA and 2iP 10	Harry *et al.* (1987)
T. orientalis	Hypocotyl, cotyledon, epicotyl	Lin and Staba	BA 2.2	Thomas *et al.* (1977)
Ulmus americana	Leaf	DKW	BA or TDZ 7.5, 15 or 22.5	George and Tripepi (1994)
U. americana	Leaf	Modified MS	TDZ 0.1	Bolyard *et al.* (1991)

Abbreviations: AE, Arnold and Eriksson; BA, 6-benzylaminopurine; BL, Blaydes improved salts; DBM, Defined basal medium; DKW, Driver Kuniyuki Walnut; GA, gibberellic acid; GD, Gresshof and Doy; GMD, Mohammed and Dunstan; IAA, indole-3-acetic acid; IBA, indole-3-butyric acid; 2iP, 6-(γ,γ-dimethyl-allylamino) purine; IS, Saito and Ide; Kin, kinetin; LS, Linsmaier and Skoog; MS, Murashige and Skoog; NAA, α-naphthaleneacetic acid; NOA, β-naphthoxyacetic acid; NN, Nitsch and Nitsch; RW, Risser and White; S, Sommer; SH, Schenk and Hilderbrandt; TDZ, thidiazuron (1-phenyl-3-[1,2,3-thiadiazol-5-yl] urea); WPM, Woody Plant Medium; WS, Wolter and Skoog; 2,4-D, 2,4-dichlorophenoxyacetic acid.

overcoming reproduction barriers caused by sterile male or female plants, difficulty propagating plants by seeds, conventional asexual techniques (cutting, grafting, etc.) or micropropagation via axillary shoot proliferation, and decreasing time to flowering for some plants. Selected cultivars of *Senecio* × *hybridus* (cineraria) that are reproduced by organogenesis develop flowers one month earlier than seedling populations (Gertsson, 1988a). Organogenesis can also be used to induce variability and obtain new plant phenotypes or genotypes. For instance, Brand and Lineberger (1988) obtained three new variegated forms of *Liquidambar stryaciflua* via shoot organogenesis from leaves of the cultivar Variegata. Therefore, organogenesis may be useful for improving plant phenotype or genotype.

Organogenesis is an indispensable tool in many types of scientific studies, including plant morphology, or attempting genetic engineering of ornamental species. Jungnickel and Zaid (1992) summarized a number of morphogenic experiments involving adventitious shoot initiation with *Saintpaulia ionantha* (African violet). Based on their experience, they claim that '. . . most morphogenetic reactions can be demonstrated more easily or even better with saintpaulias in comparison to tobacco cultures'. Organogenesis is also useful for genetic transformation of plants. Transformation requires plant regeneration techniques that result in production of large numbers of shoots (James *et al.*, 1989) that arise from one or a few cells. Although the number of cells involved in genesis of individual shoots is species specific and still debatable (Jungnickel and Zaid, 1992; Thorpe, 1994), organogenesis is an important regeneration technique for genetic transformation.

Even though shoot organogenesis has proven quite valuable for clonal propagation and as an investigative tool, it does have intrinsic liabilities that affect its usefulness. Thorpe (1994) stated that, in general, plant production via organogenesis is considered less efficient than other types of *in vitro* production, particularly somatic embryogenesis. His reasons include fewer steps to produce plantlets from explants, potential to produce higher numbers of plants, and better uniformity of plants produced, when comparing somatic embryogenesis to organogenesis. Thorpe's assertion must be considered on a case by case basis, since requirements for plant multiplication via somatic embryogenesis and organogenesis vary by species. A case in point against this assertion is plant regeneration via somatic embryogenesis (May and Trigiano, 1991) or organogenesis (Trigiano and May, 1994) from *Dendranthema grandiflora* 'Iridon' (chrysanthemum). This chrysanthemum cultivar is much easier to regenerate by organogenesis than somatic embryogenesis. Another potential drawback of organogenesis is that growth regulators can increase developmental time for flower formation. Although organogenesis can reduce time to flowering for regenerated *Senecio* × *hybridus* plants compared to seedlings, high benzyladenine (BA) concentrations in the tissue culture medium can delay flowering (Gertsson, 1988a). Perhaps the major shortcoming of organogenesis, however, is the potential for induced variation (somaclonal variation) among the new plants.

Many factors can affect mutation rate of ornamental plants produced by organogenesis, including genotype, plant growth regulators, and explant source. Bulbous species (e.g. lily and hyacinth) and some *Saintpaulia* cultivars tend to be stable when reproduced by adventitious shoot initiation, but the mutation rates of some *Freesia* hybrids and selected begonia cultivars can be high (Pierik, 1987). Likewise, *Populus nigra* plants produced via organogenesis varied significantly in height after one year in the field (Lester and Berbee, 1977), but *Betula pendula* plants produced from adventitious shoots were quite stable (Welander, 1993). Plant growth regulators, particularly cytokinins, appear to increase the rate of plant mutation (Ando *et al.*, 1986). The chemical structure of the growth regulator is apparently important since synthetic auxins and cytokinins can increase the chances of mutation when plants are produced via organogenesis (Pierik, 1987). The type of tissue or explant used can also influence the chances of genetic variation. Because of the non-uniform nature of callus tissue (Thorpe, 1994), genetic mutations can appear more frequently in shoots regenerated from callus, particularly with prolonged subculturing, than from other types of tissues (D'Amato, 1977; Ahuja, 1993a). If the explant consists of undifferentiated tissue (pericycle or cambium), the chance of mutation is reduced when compared with differentiated tissue such as phloem or pith since the cell differentiation process *in vitro* can lead to production of polyploid cells (D'Amato, 1977). Although the chance of genetic variation and mutation increases if multiplying plants via organogenesis compared to axillary shoot proliferation, the benefits of the former tissue culture system merit its use for multiplying numerous species under specific situations.

STAGES OF ORGANOGENESIS

Organogenesis is a multi-step process and consists of three stages including shoot bud formation, shoot development and finally rooting of shoots (Thorpe, 1994). The first step, shoot bud formation, is perhaps the most critical part of the process since the other two steps are usually of little consequence if cells capable of eventually forming a meristem fail to respond. Many factors can affect shoot initiation (see below), and these factors, either individually or in concert with other factors, induce a sequence of changes in morphogenically competent cells.

Thorpe (1994) reviewed the early stages of organogenesis. An adventitious shoot can develop from a single cell or several cells, depending on the species and explant. Some questions, however, arise about the unicellular or multicellular origin of adventitious shoots. In particular, does the new shoot arise from a single cell or do surrounding cells divide and become incorporated in the new structure? None the less, the first step begins with changes in morphogenically competent cells responding to an inductive signal. The signal can cause quiescent or actively proliferating cells to begin an altered path of differentiation. Differentiated cells must be capable of dedifferentiation so that the cells can be redetermined (Pierik, 1987). The competent cells must undergo precise patterns of division leading to formation of a shoot meristem or shoot bud. If the adventitious shoot has a multicellular origin, the cells must have some sort of interaction or coordination as they divide and subsequently form a competent meristem (Thorpe, 1994). Once the meristem and, later, shoot bud are formed, adventitious shoots of most species readily develop and can be multiplied before rooting.

The first step of organogenesis, inducing changes in morphogenically competent cells, is the key to adventitious shoot regeneration but seems blocked in some species for unknown reasons. Halperin (1986) considered three possibilities for explant failure to regenerate adventitious shoots. First, cells in the explant may lack totipotency (competent cells), or their chromosomes may be incapable of changes as the cells mature. Second, differentiated cells may lack the ability to undergo epigenetic changes (altered gene expression) and are unable to reverse the effects of genes involved in cellular growth and development. Finally, the explant may have a physiological block to differentiation or dedifferentiation because of a lack of essential induction signals such as wrong type or concentration of plant growth regulators. Determining the proper or requisite stimuli for inducing morphogenically competent cells to begin organogenesis is the one block to regeneration that can be empirically overcome. Since induction signals, whether environmental or biochemical in nature, can be manipulated to start differentiation or dedifferentiation, a discussion of various factors affecting organogenesis is important.

FACTORS AFFECTING ORGANOGENESIS

Many factors can affect shoot organogenesis, and most of these factors can be placed into roughly three categories. The categories include environmental conditions, such as light and temperature, culture medium including components from salts to plant growth regulators, and finally explant material including explant type and physiological development of the tissue. Perhaps the most important idea to keep in mind as each factor is considered is that it can affect species, even cultivars within a species, differently.

ENVIRONMENT

The physical environment, particularly light and temperature, can affect shoot organogenesis. Light quality and photoperiod appear to influence shoot formation. Low levels of light (around 40 μmol $m^{-2}s^{-1}$) and blue light appear to promote shoot regeneration from *Pinus radiata* (Thorpe, 1994) and *Liquidambar stryaciflua* (Brand and Lineberger, 1992). In contrast, red and far-red light helped promote organogenesis from zygotic embryos of *Pinus banksiana*, since combining incandescent bulbs with fluorescent lamps stimulated shoot formation (Chesick and Bergmann, 1991). Red light promoted shoot regeneration from *Petunia hybrida* leaf explants grown on a cytokinin-free medium (Economou and Read, 1986).

Explants from some species need a dark treatment, from several days to several weeks, before they can form shoots. At one extreme, *Freesia* flower buds required darkness for adventitious shoot formation (Pierik and Steegmans, 1975a). At the other end of the spectrum, 'Iridon' chrysanthemum (Trigiano and May, 1994) and *Rhododendron* cultivars 'Cunningham's White' (Iapichino *et al.*, 1992) and 'PJM' (Preece and Imel, 1991) can form shoots on leaves placed immediately into light after placing the explant on an induction medium. Likewise in our laboratory, *Betula pendula* and *Ulmus americana* produced numerous shoots from leaves if a dark period was omitted after explants are placed on induction medium (Tripepi, unpublished data). Many species, however, need some type of dark treatment. Placing *Iris* × *hollandica* scale explants in light immediately after placing explants on induction medium reduced regeneration capacity (Van der Linde *et al.*, 1988). If a two-week dark treatment precedes the light treatment, however, regeneration capacity was not decreased, and shoots developed more uniformly than if a dark period is omitted (Van der Linde *et al.*, 1988). Regeneration percentage of petiole explants from *Senecio* × *hybridus* increased from 40 to 62% as the duration of the dark treatment increased from 0 to 21 days (Gertsson, 1988b). Photoperiod may also influence shoot organogenesis indirectly. For *Betula*, short days were more effective than long days for preparing shoot explants for *in vitro* growth (Welander, 1988).

Temperature can affect shoot organogenesis. The optimum range for shoot formation for most species is between 22 and 28°C (Pierik, 1987). For bulbous species, however, lower temperatures often promote organogenesis. *Iris* × *hollandica* scale explants cultured between 15 and 20°C formed adventitious shoots, but those grown at 25°C failed to form shoots (Van der Linde *et al.*, 1988). The optimum temperature for bulb differentiation of *Hyacinthus orientalis* was 13°C, but further growth of the bulblets required temperatures between 21 and 25°C (Pierik and Ruibing, 1973). Temperature can also cause donor plants to produce larger shoots or bulbs in shorter periods of time. Storing parent bulbs at 2°C promoted organogenesis from *Hyacinthus orientalis* (Bach, 1992).

MEDIUM

Shoot organogenesis can be affected by tissue culture medium, particularly formulation, carbohydrate sources and plant growth regulators. Concentrations of some or all salts in the medium can influence shoot formation. Increasing the concentration of Murashige and Skoog (MS) salts in the medium decreased the percentage of regeneration from *Iris* × *hollandica* bulb scales (Van der Linde *et al.*, 1988) but significantly increased the number of shoots per regenerating explant (Van der Linde and Schipper, 1992). Reduced concentrations of N and K in MS medium tripled the number of *Senecio* × *hybridus* shoots formed from petiole explants (Gertsson, 1988b). Reducing medium concentration to half-strength can promote adventitious shoot formation for some species such as *Pinus banksiana* (Chesick and Bergmann, 1991). Medium formulation can also affect regenerative capacity. Zygotic embryo explants from *Pinus banksiana* had better regeneration capacity on Gresshoff and Doy (GD) or LePoivre (LP) media than on MS or Schenk and Hildebrandt (SH) media (Chesick and Bergmann, 1991). In contrast, explants from mature tissue produced more adventitious shoots if placed on GD or Gupta and Durzan media than on MS, SH, or LP media.

Organic constituents such as carbohydrates and vitamins can affect shoot regenerative capacity of explants. With most species, sucrose is the carbohydrate of choice and is used at a concentration around 3% (w/v). *Pinus banksiana* explants produced more bud primordia when placed on 2 or 3% sucrose compared to 1 or 4% sucrose (Chesick and Bergmann, 1991). In addition to being an energy source, carbohydrates can affect organogenesis by acting as an osmoticum (Thorpe, 1994). Vitamins can promote shoot initiation on explants from herbaceous or woody species (Seeliger, 1959; Stichel, 1959).

Of all the media components, plant growth regulators appear to have the most influential role on adventitious shoot initiation. Some species in a few genera, such as *Populus* (Douglas, 1984), *Streptocarpus* (Appelgren and Heide, 1972) and *Saintpaulia* (Jungnickel and Zaid, 1992), can form adventitious shoots without plant growth regulators added to the medium. Most species, however, require plant growth regulators in the medium to induce shoot organogenesis. Moreover, growth regulator requirements can vary among species and even cultivars, but most species require a cytokinin, auxin, or both.

The type of plant growth regulator needed for shoot organogenesis is species specific, but BA and naphthaleneacetic acid (NAA) are among the growth regulators used most often. Thorpe (1994) stated that 75% of the species that are capable of forming shoots are induced by use of 0.05 to 46 μM BA or kinetin. For some gymnosperms, BA is more effective than the other cytokinins (Bonga and Von Aderkas, 1992). Zeatin, however, was more effective than BA for promoting shoot initiation from leaf tips on adult *Chamaecyparis obtusa* (Ishii, 1991). Zeatin has also been the most effective cytokinin for inducing shoot formation from *Populus nigra* 'Italica' (Venverloo, 1973). The amount of cytokinin needed for shoot initiation varies depending on the species or even cultivar. For instance, 'Fluffy Ruffles' and 'Todeoides' cultivars of *Nephrolepis exaltata* required at least 4.4 μM BA to induce proliferation of adventitious shoots, but the cultivars 'Scottie' and 'Smithii' require less than 1.3 μM (Amaki and Higuchi, 1992).

A high cytokinin to auxin ratio promotes shoot formation, however, exogenous auxin may be unnecessary for shoot organogenesis. *Pinus strobus* (Flinn *et al.*, 1986) and other conifers (Thorpe *et al.*, 1991) required only exogenous cytokinin for shoot formation. The presence of exogenous auxin, however, may promote organogenesis (Kaul, 1987). In contrast, other species, such as *Betula pendula* (Leege and Tripepi, 1993), required only exogenous cytokinin. With these species exogenous auxin must be excluded, since its addition changes the cytokinin:auxin ratio, preventing shoot formation. If exogenous auxin is required, 0.06 to 27 μM indole-3-acetic acid or NAA is used for most species (Thorpe,1994). The auxin 2,4-dichlorophenoxyacetic acid (2,4-D) is usually associated with promoting embryogenesis on explants from many different species (Pierik, 1987). In our laboratory (Tripepi, unpublished data), 2,4-D stimulated shoot organogenesis without root formation on leaf explants from *Rosa hybrida* 'Golden Emblem', whereas NAA stimulated root organogenesis before adventitious buds formed on the roots. Several bulbous species, including *Hyacinthus orientalis*, formed adventitious shoots in the presence of exogenous auxin only (Pierik and Steegmans, 1975b).

Other cytokinin-like plant growth regulators, most notably thidiazuron, 1-phenyl-3-(1,2,3-thiadiazol-5-yl) urea (TDZ), have also been used to induce shoot organogenesis from various kinds of explants. Thidiazuron, a urea derivative, has dramatically increased micropropagation efficiency of many recalcitrant woody species (Huetteman and Preece, 1993), and this growth regulator also promotes shoot regeneration from many species of plants (Huetteman and Preece, 1993; Lu, 1993). For example, leaf explants from American elm on medium supplemented with TDZ had higher percentages of shoot regeneration (Table 4.2) and more shoots per explant (Fig. 4.1) than those on media supplemented with BA (George and Tripepi, 1994). Reviews by Huetteman and Preece (1993) and Lu (1993) listed many species, including numerous ornamental plants, that form shoots when TDZ is included in the medium. In fact, TDZ often induces shoot organogenesis when other cytokinins fail. For instance, rose receptacles failed to form shoots on BA-supplemented medium but formed shoots when TDZ was used (Lu, 1993). Potential liabilities with using TDZ for organogenesis include problems with shoot fasciation

Table 4.2. Effect of donor plants and growth regulators on the percentage shoot regeneration from *Ulmus americana* leaf explants after six weeks on regeneration medium. All percentages based on 15 leaf explants per treatment except where noted in parentheses. Data taken from George and Tripepi (1994).
(a) Percentage regeneration

Plant	Growth regulator	Concentration (μM) 0	7.5	15.0	22.5
DEL2-1	BA	0	27	50 (10)	47
	TDZ	0	80 (10)	78	80 (10)
DEL2-3	BA	0	80	80	53
	TDZ	0	93	87	87
DEL2-5	BA	0	27	93	67
	TDZ	0	67	90 (10)	100 (10)
DEL2-6	BA	0	20	20	27
	TDZ	0	50 (10)	80 (10)	60
DEL2-7	BA	0	33	60 (10)	73
	TDZ	0	93	100 (5)	100 (10)
DEL2-12	BA	0 (7)	20 (10)	60 (10)	30 (10)
	TDZ	0	60	87	90 (10)

(b) Statistical analysis[1]

Source	DF	Wald's χ^2
Plant	5	21.42***
Growth regulator	1	15.66***
Plant × growth regulator	5	6.26
Concentration	2	2.51
Plant × concentration	10	9.32
Growth regulator × concentration	2	2.03
Plant × growth regulator × concentration	10	7.06

[1] The 0 μM level of growth regulator was excluded from the statistical analysis since leaf explants receiving this treatment failed to produce shoots.*** significant at $P < 0.001$

(Huetteman and Preece, 1993), very short internodes (Preece and Imel, 1991; Bates *et al.*, 1992; George and Tripepi, 1994), and lower survival rates upon shoot removal from the explant (George and Tripepi, 1994). Other cytokinin-like plant growth regulators include *N*-(4-pyridyl)-*N*′-phenylurea (4PU), *N*-*N*′-diphenylurea (DPU), 6-(2-methylbenzylamino) purine (MBP), *N*-(2-chloro-4-pyridyl)-*N*′-phenylurea (CPPU), *N*-(2-chloro-4-pyridyl)-*N*′-(3,5-difluorophenyl) urea (CPFPU), *N*-furfuryl-9(2-tetrahydropyranyl) 9H purine (FPA), and *O*-acetyl zeatin. Of these compounds, only a few have been tested for their effects on shoot induction on ornamental species (Fellman *et al.*, 1987).

Other types of plant growth regulators can promote or inhibit shoot organ-

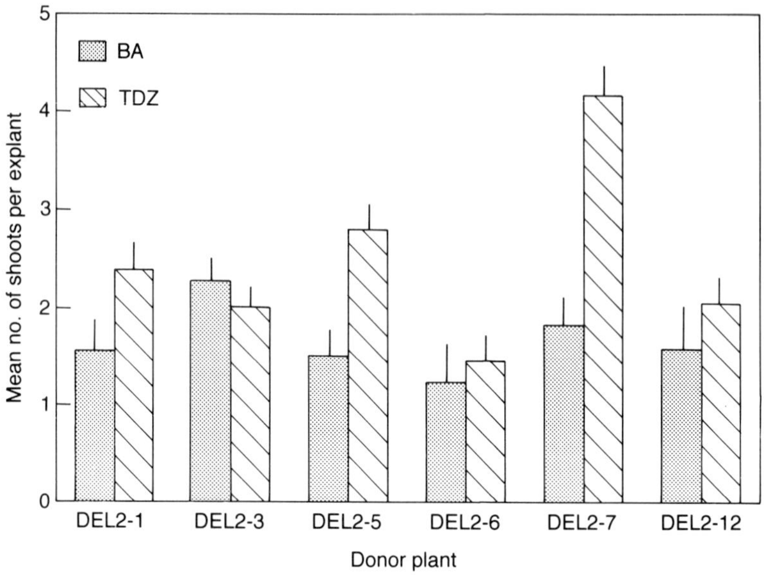

Fig. 4.1. Mean number of shoots per regenerating leaf explant as affected by the plant growth regulators, BA or TDZ, and *Ulmus americana* donor plants. Means were averaged across 7.5, 15 and 22.5 μM TDZ or BA. Vertical lines indicate ± SE. Data taken from George and Tripepi (1994).

ogenesis from explants of ornamental plants. Although abscisic acid (ABA) is usually thought to inhibit shoot regeneration (Pierik, 1987), it promoted organogenesis from cotyledons of *Pinus taeda* (Sen *et al*., 1989). Gibberellic acid (GA_3) can inhibit shoot formation from *Hyacinthus* (Pierik and Steegmans, 1975b), *Streptocarpus* (Appelgren and Heide, 1972), and *Saintpaulia* (Margara and Piollat, 1985), but this compound was essential for promoting shoot regeneration from callus explants of *Rosa hybrida* 'Landora' (Rout *et al*., 1992). Triiodobenzoic acid (TIBA), an inhibitor of auxin transport, reduced shoot formation from *Saintpaulia* (Margara and Piollat, 1985) but stimulated shoot organogenesis of lily (Van Aartrijk *et al*., 1986). Ethylene also increased adventitious shoot regeneration of lily (Van Aartrijk *et al*., 1986). The differing results obtained when using plant growth regulators to induce organogenesis demonstrate that their role in shoot regeneration is somewhat ambiguous and complicated.

GENOTYPE

Genotype, source, stage of development and even orientation of an explant can affect a species' ability to form shoots. These factors are distinctly different yet interrelated in that genotype can influence the type (leaf, inflorescence, or root) of explant tissue used, optimum stage of development for regeneration, and orientation of the explant on the medium.

Shoot organogenesis is strongly influenced by the genetic component of the explant and its resulting characteristics. Explants taken from embryogenic tissues (seeds) or very young (recently germinated) conifers are easily manipulated to undergo organogenesis or embryogenesis (Thorpe, 1994). Species plays an important role in organogenesis. Explants taken from mature *Larix decidua* formed adventitious shoots, but those from *Larix leptolepis* failed to form shoots when components in the medium were identical (Bonga and Pond, 1991). Hybrids from conventional breeding of these two species yielded offspring that had the ability to undergo shoot organogenesis, indicating that the capacity for shoot regenerability may be heritable. Even though the sex of the plant mildly affects shoot regeneration potential of *Populus tremula*, genotype is the predominant factor (Ahuja, 1993b).

The ability of explants to undergo shoot organogenesis can also vary from plant to plant or by cultivar. For example, regenerability of callus explants from *Robinia pseuodacacia* varied from plant to plant (Han *et al.*, 1990). In our work with American elm (George and Tripepi, 1994), individual seedlings differed in their ability to regenerate shoots from leaves (Table 4.1). These differences in regenerability were still evident even if different plant growth regulators (BA versus TDZ) were used.

EXPLANT EFFECTS

Many types of organs can be used as explants for shoot organogenesis, and the organ of choice depends on the plant species. Floral parts, leaves, petioles, roots, seeds or specific parts of seeds as well as stems can serve as explants for adventitious shoot formation from ornamental plants. Leaves and stem segments, either nodes or internodes, are often used as explants since they are readily available and easily manipulated if necessary. The source or donor plant that the explant is taken from can influence shoot regeneration capacity. For instance, Brand and Lineberger (1992) reported that leaf explants from intact (presumably stock) plants of *Liquidambar stryaciflua* formed more shoots than those from proliferating shoot cultures. In our work with *Betula pendula* (Leege and Tripepi, 1993) and *Ulmus americana* (George and Tripepi, 1994), however, we found that leaf explants from microshoots subcultured for about one year produced more shoots per explant than leaves from stock plants. If seeds or specific parts (e.g. cotyledons) are used as the explant source, they should be hydrated or stratified before placing in tissue culture since they may be more responsive for organogenesis than are dry seeds placed *in vitro* (Bonga and Von Aderkas, 1992).

Callus from various parts of ornamental plants is often used as an explant for shoot organogenesis. Shoot formation on callus is considered indirect regeneration, and callus used for organogenesis may be directly on the explant (George and Tripepi, 1994), removed from the explant and used immediately for regeneration (Rout *et al.*, 1992), or subcultured several times after removal from the explant before using for organogenesis (Isikawa, 1974). Usually direct regeneration, a shoot forming directly on an explant without an intervening callus stage, is preferred since extensive callus formation and long-term callus culture can lead to somaclonal variation (Ahuja, 1993a). Many plant species regenerated from calluses, however, are stable (Pierik, 1987).

The ability of callus to form shoots depends, in part, on its source (Chalupa, 1991) and cellular composition (Thorpe, 1994). In general, callus is composed of unorganized but somewhat differentiated cells (Torrey, 1966). Callus cells differ in shape, size, amount of vacuolization and cytoplasmic contents, as well as having biochemical and physiological differences (Thorpe, 1994). Callus morphology often varies by species, and callus growth can be affected by original position of the explant on the donor plant and growth conditions (Pierik, 1987). For conifers, Thorpe (1994) stated that '. . . a true callus is not induced, rather embryogenic tissue, which consists of highly vacuolated parenchymatous cells, interspersed with single or small groups of densely cytoplasmic cells and filamentous immature embryos'. For this reason, callus from some conifer species is incapable of shoot organogenesis (Chalupa, 1991) and more amenable to somatic embryogenesis.

Callus used for shoot organogenesis should retain its ability to form shoots after repeated subculturing and remain genetically stable (Pierik, 1987). Although calluses from some species maintain their potential to form shoots, organogenic capacity usually declines with successive subculturing (Chalupa, 1991; Thorpe, 1994). In addition, continued subculturing of callus can lead to habituation (loss of explant requirement for exogenously supplied growth regulators) and spontaneous appearance of altered callus morphology (i.e. callus becomes friable) (Thorpe, 1994). Even with its potential liabilities, callus is important for organogenesis of ornamental plants and in many cases may be the only source for regenerating adventitious shoots.

Other factors associated with explant effects can influence organogenic potential. Size of the explant and quality or condition of the donor plant are among the factors that can affect shoot organogenesis. These subjects are reviewed elsewhere (Pierik, 1987; Bonga and Von Aderkas, 1992; Thorpe, 1994).

The physiological state of development of the explant or donor plant that the explant was taken from can influence shoot organogenesis (Ahuja, 1993a). In general, explants from herbaceous species or juvenile plants have a better regenerative capacity than those from woody species or adult plants (Pierik, 1987; Bonga and Von Aderkas, 1992). Although mature plants may lack morphogenic potential to form shoots, they may be invaluable sources of superior species characteristics and warrant asexual cloning. To improve shoot regeneration potential of explants from mature plants, donor tissue can be rejuvenated. The most common method to rejuvenate tissue is by serial subculturing of axillary shoots on a cytokinin-rich medium (Franclet *et al.*, 1987; Bonga and Von Aderkas, 1992). Although regenerative capacity may improve after only a few subcultures (Bonga and Von Aderkas, 1992), we found that at least seven subcultures for *Betula pendula* (Leege and Tripepi, 1993) and ten subcultures for *Ulmus americana* (George and Tripepi, 1994) were necessary before shoot organogenesis from leaf explants increased considerably.

Habituation of donor plants or tissues used for explants can affect shoot organogenesis from ornamental plants. Plant tissue cultures that require lower concentrations or even no exogenous growth regulators, particularly auxin and cytokinin, as time *in vitro* increases are thought to be habituated (Pierik, 1987). Although habituation is often associated with callus cultures, shoot cultures can also be affected (Pierik, 1987). In a way, habituation and rejuvenation are probably interrelated since both are brought about by serial subculturing.

Habituation can promote or hinder organogenesis from explants, depending on the plant species. For instance, *Saintpaulia* leaf explants isolated from *ex vitro* stock plants formed shoots only if exogenous auxin and cytokinin were added to the regeneration medium, but leaves from *in vitro* shoot cultures formed shoots when placed on medium without exogenous growth regulators (Margara and Piollat, 1985). In contrast, habituation can have a negative impact on shoot regeneration. Leaf explants from microshoots of *Betula pendula* that had been in culture for almost two years produced fewer adventitious shoots than those taken from shoot cultures that were axenically grown for one year (Tripepi, unpublished results). Furthermore, callus and roots formed on leaf explants taken from *B. pendula* microshoots that were grown *in vitro* for almost four years even though the explants were placed on a regeneration medium without auxin (Tripepi, unpublished results). Leege and Tripepi (1993), however, found that leaf explants from microshoots that had been in culture up to one year lacked callus and roots when placed on auxin-free medium. These results indicate that *B. pendula* microshoots may have contained increased levels of endogenous auxin as time in culture increased, and these levels may affect shoot regenerability. Similarly, shoot regeneration of *Betula platyphylla* from leaf explants decreased if explants were taken from shoot cultures that had been grown *in vitro* for more than one and a half years (J.P. Schnurr, North Dakota, 1995, personal communication). Habituation, therefore, appears to change concentrations of endogenous growth regulators, which in turn influence shoot organogenesis.

Inducing shoot formation on explants taken from mature plants can be difficult for many conifers, but embryogenic tissues of these plants are usually easy to manipulate to form shoots (Thorpe, 1994). For this reason, the embryo, hypocotyl, or cotyledons from conifer seeds or young seedlings (up to several weeks after germination) are often used for shoot organogenesis (Zel, 1993). A number of gymnosperms, however, can be multiplied via organogenesis from explants taken from mature plants. For instance, buds from 12-year-old plants of *Larix × eurolepis* (Laliberte and Lalonde, 1988) and leaf tips from 13-year-old plants of *Chamaecyparis obtusa* (Ishii, 1991) formed adventitious shoots.

Tissues from mature plants may lack the capacity to form adventitious shoots because of one or more factors. As stated above, cells in mature plants may be unable to undergo epigenetic changes (Thorpe, 1994), thereby preventing redetermination of these tissues (Pierik, 1987). Phenolic compounds, particularly in woody species, can present serious problems for *in vitro* growth (Welander, 1993). Initial phenolic exudation by *Betula pendula* explants taken from adult plants varies dependent upon the type of explant taken and the physiological state of the tissue (Welander, 1988). Therefore, the amount and type of phenolic compounds present may interfere with or prevent organogenesis. Even though explants taken from mature plants can be recalcitrant to shoot organogenesis, Bonga and Von Aderkas (1992) have described several techniques useful for improving shoot formation on these types of explants. These techniques include obtaining adventitious shoots from roots, sphaeroblasts (wood tumours), or latent buds on stumps (epicormic origin) and miniaturizing explants to eliminate effects of adjacent tissues.

The physiological state of the explant can be affected by the time of year the tissue is taken, which in turn may be related to dormancy. Bud explants collected from *Picea sitchensis* shortly before bud break (March) readily formed adventitious shoots (Selby and Harvey, 1985). In contrast, stem explants taken from *Larix occidentalis* during the growing season formed more shoots than those taken during the dormant season (Chesick *et al.*, 1990). The relationship between organogenesis and date of explant collection, then, is probably influenced by levels of endogenous growth regulators in the donor plant since onset and release from dormancy are affected by these substances (Pierik, 1987).

If leaves are used as explants, those that are almost or just fully expanded are generally considered to be at the best stage of development for shoot organogenesis. Differentiated cells, such as parenchyma, in leaves that have recently expanded fully may be temporarily capable of dedifferentiating and becoming redetermined (Pierik, 1987). The dedifferentiated cells subsequently divide and eventually form a meristem (Pierik, 1987; Thorpe, 1994). Cells with organogenic potential in leaves that are too old or physiologically mature presumably are unable to dedifferentiate or respond to induction signals. In our work with *Betula pendula* (Leege and Tripepi, 1993) and *Ulmus americana* (George and Tripepi, 1994), leaves that have recently expanded fully, readily formed adventitious shoots. The optimum stage of leaf development for organogenesis is probably species specific, however, since ‘Iridon’ chrysanthemum leaves fully expanded for up to four weeks formed shoots as readily as those that more recently reached full size (Tripepi, unpublished results).

Orientation of the explant on the medium can affect shoot organogenesis. The best orientation for maximum shoot regeneration, however, depends on the species and type of explant used. Pierik (1987) stated that leaf explants of *Kalanchoe prinacea* had better shoot formation (presumably meaning more shoots) if the abaxial side of the leaf contacted the medium compared to the adaxial side. He further stated that flat rather than vertical orientation resulted in more shoots being formed. For *Betula pendula*, however, shoot regeneration percentage of leaf explants almost doubled when the adaxial side of the explant contacted the medium compared to the abaxial surface (Leege and Tripepi, 1993); the number of shoots formed per leaf was unaffected by orientation.

Vertical orientation of explants can have significant effects on adventitious shoot formation. Placing excised embryo or cotyledon explants from *Pinus strobus* (Flinn *et al.*, 1986) or *Larix decidua* (Karnosky *et al.*, 1988) upside down on the medium improved organogenesis. Likewise, embryo explants from *Pinus sylvestris* formed twice as many adventitious buds if placed vertically rather than horizontally on the medium (Zel, 1993). In contrast, placing stem explants of *Rhododendron* × 'Rose Elf' horizontally on the medium increased the number of adventitious shoots formed by 56% compared to those oriented vertically (Anderson, 1984). Explant orientation appears to affect organogenesis by changing distribution of nutrients and plant growth regulators within the explant (Bonga and Von Adrerkas, 1992). Even though changes in explant orientation can improve regeneration, orientation appears to have little effect on inducing recalcitrant species to form shoots.

FUTURE APPLICATIONS OF ORGANOGENESIS

The use of adventitious shoot organogenesis with ornamental species will most likely increase in the future. This increase will be related to plant improvements that are scientifically and commercially useful. Mass clonal propagation of superior ornamental plants by organogenesis may be used in the future if the regeneration system can be automated to keep costs low for each finished plantlet. An automated organogenic regeneration system, however, would require highly efficient regeneration methods and resulting plantlets that are phenotypically and genetically uniform and stable. Organogenesis will be used in the future to induce somaclonal variation in ornamental species. This variation may improve plant resistance to pests or cause commercially valuable changes in phenotype, such as new variegated leaf patterns (Brand and Lineberger, 1988; Michler, 1993). Future research in plant morphology and physiology will use organogenesis in the study of phenomena related to rejuvenation, maturation and stress tolerance (Meier-Dinkel, 1992; Libby and Ahuja, 1993).

Perhaps the most important future application of organogenesis will be its use in conjunction with genetic engineering of ornamental species. New molecular techniques and isolated genes responsible for desired plant traits have raised hopes and expectations of improved ornamental species via genetic transformation. Improved plants that require less pesticides, fertilizers or water to thrive in their environment are commercially desirable. The lack of a regeneration system is a major obstacle for successful application of biotechnology to plants, particularly woody species (Ostry and Michler, 1993).

SUMMARY

Organogenesis is used with ornamental species as a means of asexual propagation that involves *de novo* production of adventitious shoots on explants. Many kinds of tissues or organs can be used as explants. The ability of an explant to form adventitious shoots, however, is highly dependent on plant genotype, as demonstrated by organogenic potential differing from species to species and even among cultivars within a species. If the explant contains cells amenable to differentiation or redetermination, environmental conditions for culturing, tissue culture media, or even the explant itself can be manipulated to induce or improve shoot regeneration. Manipulations are usually determined experimentally depending on the type of explant used and plant genotype. Plant growth regulators, whether endogenous or exogenous, have a profound impact on shoot regeneration. The physiological state of the explant or donor plant also strongly influence organogenesis. Explants from recalcitrant species may be unable to form shoots because of an inability to respond to induction signals or an inability to undergo requisite genetic or epigenetic changes for forming a shoot meristem. Despite some inherent liabilities with the process, numerous benefits and potential improvements, both commercial and scientific, brought about by adventitious shoot organogenesis will ensure its future use with ornamental species.

ACKNOWLEDGEMENTS

Many thanks to Manoj Singlachar for assembling and organizing the information used in Table 4.1.

REFERENCES

Abdullah, A.A., Yeoman, M.M. and Grace, J. (1985) *In vitro* adventitious shoot formation from embryonic and cotyledonary tissues of *Pinus brutia* Ten. *Plant Cell, Tissue and Organ Culture* 5, 35–44.

Ahuja, M.R. (1993a) Micropropagation à la carte. In: Ahuja. M.R. (ed.), *Micropropagation of Woody Plants*. Kluwer Academic Publishers, Dordrecht, pp. 3–9.

Ahuja, M.R. (1993b) Regeneration and germplasm preservation in aspen – *Populus*. In: Ahuja, M.R. (ed.), *Micropropagation of Woody Plants*. Kluwer Academic Publishers, Dordrecht, pp. 187–194.

Amaki, W. and Higuchi, H. (1992) Micropropagation of Boston ferns (*Nephrolepis* spp.). In: Bajaj, Y.P.S. (ed.), *Biotechnology in Agriculture and Forestry. Vol. 20. High-tech and Micropropagation IV*. Springer-Verlag, Berlin, pp. 484–494.

Anderson, W.C. (1984) A revised tissue culture medium for shoot multiplication of rhododendron. *Journal of the American Society for Horticultural Science* 109, 343–347.

Ando, T., Akiyama, Y. and Yokoi, M. (1986) Flower colour sports in *Saintpaulia* cultivars. *Scientia Horticulturae* 29, 191–197.

Appelgren, M. and Heidi, O.M. (1972) Regeneration in *Streptocarpus* leaf discs and its regulation by temperature and growth substances. *Physiologia Plantarum* 27, 417–423.

Arrillaga, I. and Merkle, S.A. (1993) Regenerating plants from *in vitro* culture of black locust cotyledon and leaf explants. *HortScience* 28, 942–945.

Atkinson, N.J., Ford-Lloyd, B.V. and Newbury, H.J. (1989) Regeneration of plants from *Antirrhinum majus*. L. callus. *Plant Cell, Tissue and Organ Culture* 17, 59–70.

Auer, C.A., Cohen, J.D., Laloue, M. and Cooke, T.J. (1992) Comparison of benzyl adenine metabolism in two *Petunia hybrida* lines differing in shoot organogenesis. *Plant Physiology* 98, 1035–1041.

Baburaj, S., Damotharan, R. and Santhaguru, K. (1987) Regeneration in leaf callus cultures of *Euphorbia hirta* Linn. *Current Science* 56, 194.

Bach, A. (1992) Micropropagation of hyacinths (*Hyacinthus orientalis* L.). In: Bajaj, Y.P.S. (ed.), *Biotechnology in Agriculture and Forestry. Vol. 20. High-tech and Micropropagation IV*. Springer-Verlag, Berlin, pp. 144–159.

Bates, R.M. (1986) Callus organogenesis and protoplast isolation in *Malus hupehensis*. Unpublished MS thesis, West Virginia University, Morgantown.

Bates, S., Preece, J.E., Navarrete, N.E., Van Sambeek, J.W. and Gaffney, G.R. (1992) Thidiazuron stimulates shoot organogenesis and somatic embryogenesis in white ash (*Fraxinus americana* L.). *Plant Cell, Tissue and Organ Culture* 31, 21–29.

Bolyard, M.G., Srinivasan, C., Cheng, J. and Sticklen, M. (1991) Shoot regeneration from leaf explants of American and Chinese elm. *HortScience* 26, 1554–1555.

Bonga, J.M. (1984) Adventitious shoot formation in cultures of immature female strobili of *Larix decidua*. *Physiologia Plantarum* 62, 416–421.

Bonga, J.M. and Pond, S.E. (1991) Adventitious shoot formation in cultures of 30-year-old *Larix decidua*, *L. leptolepis*, *L. eurolepis*, and *L. laricina* trees. *Plant Cell, Tissue and Organ Culture* 26, 45–51.

Bonga, J.M. and Von Aderkas, P. (1992) *In Vitro Culture of Trees*. Kluwer Academic Publishers, Dordrecht.

Bornman, C.H. and Jansson, E. (1980) Organogenesis in cultured *Pinus sylvestris* tissue. *Zeitschrift für Pflanzenphysiologie* 96, 1–6.

Brand, M.H. and Lineberger, R.D. (1988) *In vitro* adventitious shoot formation on mature-phase leaves and petioles of *Liquidambar stryraciflua* L. *Plant Science* 57, 173–179.

Brand, M.H. and Lineberger, R.D. (1992) Micropropagation of American sweetgum (*Liquidambar styraciflua* L.). In: Bajaj, Y.P.S. (ed.), *Biotechnology in Agriculture and Forestry, Vol. 18. High-tech and Micropropagation II*. Springer-Verlag, Berlin, pp. 3–24.

Burger, D.W., Liu, L., Zary, K.W. and Lee, C.I. (1990) Organogenesis and plant regeneration from immature embryos of *Rosa hybrida* L. *Plant Cell, Tissue and Organ Culture* 21, 147–152.

Chalupa, V. (1974) Control of root and shoot formation and production of trees from poplar callus. *Biologia Plantarum* 16, 316–320.

Chalupa, V. (1977) Organogenesis in Norway spruce and Douglas fir tissue cultures. *Communicationes Instituti Forestalis Cechosloveniae* 10, 79–87.

Chalupa, V. (1988) Large scale micropropagation of *Quercus robur* L. using adenine-type cytokinins and thidiazuron to stimulate shoot proliferation. *Biologia Plantarum* 30, 414–421.

Chalupa, V. (1991) Larch (*Larix decidua* Mill.). In: Bajaj, Y.P.S. (ed.), *Biotechnology in Agriculture and Forestry, Vol. 16. Trees III*. Springer-Verlag, Berlin, pp. 446–470.

Cheng, T.-Y. (1975) Adventitious bud formation in culture of Douglas fir (*Pseudotsuga menziesii* (Mirb) Franco). *Plant Science Letters* 5, 97–102.

Chesick, E.E. and Bergmann, B.A. (1991) Jack Pine (*Pinus banksiana* Lamb.). In: Bajaj, Y.P.S. (ed.), *Biotechnology in Agriculture and Forestry, Vol. 16. Trees III*. Springer-Verlag, Berlin, pp. 241–253.

Chesick, E.E., Bilderback, D.E. and Blake, G.M. (1990) *In vitro* multiple bud formation by 20-year-old western larch buds and stems. *HortScience* 25, 114–116.

Chun, Y.W. (1993) Clonal propagation in non-aspen poplar hybrids. In: Ahuja. M.R. (ed.), *Micropropagation of Woody Plants*. Kluwer Academic Publishers, Dordrecht, pp. 209–222.

Churchill, M.E., Ball, E.A. and Arditti, J. (1973) Tissue culture of orchids, method for leaf tips. *New Phytologist* 72, 161–166.

D'Amato, F. (1977) Cytogenetics of differentiation in tissue and cell cultures. In: Reinert, J. and Bajaj, Y.P.S. (eds), *Applied and Fundamental Aspects of Plant Cell, Tissue and Organ Culture*. Springer-Verlag, Berlin, pp. 343–357.

De Lange, E., Debergh, P. and Van Rijk, R. (1974) *In vitro* culture as a method for vegetative propagation of *Euphorbia pulcherrima*. *Zeitschrift für Pflanzenphysiologie* 71, 271–274.

Douglas, G.C. (1984) Formation of adventitious buds in stem internodes of *Populus* spp. cultured *in vitro* on basal medium: influence of endogenous properties of explants. *Journal of Plant Physiology* 116, 313–321.

Dunbar, K.B. and Stephens, C.T. (1989) Shoot regeneration of hybrid seed geranium (*Pelargonium* × *hortorum*) and regal geranium (*Pelargonium* × *domesticum*) from primary callus culture-plant regeneration from cotyledon culture and leaf culture. *Plant Cell, Tissue and Organ Culture* 19, 13–21.

Economou, A.S. and Read, P.E. (1986) Light quality control of shoot and root formation *in vitro* in petunia cultures. *HortScience* 21, 761.

Ellis, D.D., Barczynska, H., McCown, B.H. and Nelson, N. (1991) A comparison of BA, zeatin and thidiazuron for adventitious bud formation from *Picea glauca* embryos and epicotyl explants. *Plant Cell, Tissue and Organ Culture* 27, 281–287.

Fellman, C.D., Read, P.E. and Hosier M.A. (1987) Effects of thidiazuron and CPPU on meristem formation and shoot proliferation. *HortScience* 22, 1197–1200.

Flinn, B.S., Webb, D.T. and Georgis, W. (1986) *In vitro* control of caulogenesis by growth regulators and media components in embryonic explants of Eastern white pine (*Pinus strobus*). *Canadian Journal of Botany* 64, 1948–1956.

Floh, E.I.S. and Handro, W. (1986) Tissue and cell culture of *Pinus* species. In: Somer, D.A., Gengenbach, B.G., Biesboer, D.D., Hackett, W.P. and Green, C.E. (eds), *6th International Congress of Plant Tissue and Cell Culture*, IAPTC. University of Minnesota, Minneapolis, p. 393.

Franclet, A., Boulay, M., Bekkaoui, F., Fouret, Y., Verschoore-Martouzet, B. and Walker, N. (1987) Rejuvenation. In: Bonja, J.M. and Durzan, D.J. (eds), *Cell and Tissue Culture in Forestry, General Principles and Biotechnology*, Vol. 1. Martinus Nijhoff Publishers, Dordrecht, pp. 232–248.

Franco, E.O. and Schwarz, O.J. (1985) Micropropagation of two tropical conifers: *Pinus oocarpa* Schiede and *Cupressus lusitanica* Miller. In: Henke, R.R., Hughes, K.W., Constantin, M.J. and Hollaender, A. (eds), *Tissue Culture in Forestry and Agriculture*. Plenum, New York, pp. 195–213.

Fu, M.L. (1978) Plantlets from *Paulownia* tissue culture. *Gardens Bulletin* (Singapore) 31, 61–66.

Geier, T. (1986) Factors affecting plant regeneration from leaf segments of *Anthurium scherzerianum* Schott (Araceae) cultured *in vitro*. *Plant Cell, Tissue and Organ Culture* 6, 115–125.

George, M.W. and Tripepi, R.R. (1994) Cytokinins, donor plants and time in culture affect shoot regenerative capacity of American elm leaves. *Plant Cell, Tissue and Organ Culture* 39, 27–36.

Gertsson, U.E. (1986) Influence of explant orientation and media composition on organ formation, growth and nitrogen content in petiole explants of *Senecio* × *hybridus* Hyl. *Journal of Horticultural Science* 61, 121–128.

Gertsson, U.E. (1988a) Development of micropropagated plants from different clones of *Senecio* × *hybridus* in relation to BAP concentration and temperature *in vitro*. *Journal of Horticultural Science* 63, 489–496.

Gertsson, U.E. (1988b) Influence of macronutrient composition, TIBA and dark treatment on shoot formation and nitrogen content in petiole explants of *Senecio* × *hybridus*. *Journal of Horticultural Science* 63, 497–502.

Gladfelter, H.J. and Phillips, G.C. (1987) *De novo* shoot organogenesis of *Pinus eldarica* Medw. *in vitro*. I. Reproducible regeneration of long-term callus cultures. *Plant Cell Reports* 6, 163–166.

Goldfarb, B., Howe, G.T., Bailey, L.M., Strauss, S.H. and Zaerr, J.B. (1991) A liquid cytokinin pulse induces adventitious shoot formation from Douglas-fir cotyledons. *Plant Cell Reports* 10, 156–160.

Halperin, W. (1986) Attainment and retention of morphogenetic capacity *in vitro*. In: Vasil, I.K. (ed.), *Cell Culture and Somatic Cell Genetics of Plants, Vol. 3.* Academic Press, New York, pp. 3–47.

Han, K.H., Davis, J.M. and Keathley, D.E. (1990) Differential responses persist in shoot explants regenerated from callus of two mature black locust trees. *Tree Physiology* 6, 235–240.

Harney, P.M. and Knap, A. (1979) A technique for the *in vitro* propagation of African violets using petioles. *Canadian Journal of Plant Science* 59, 263–266.

Harry, I.S., Thompson, M.R., Lu, C.Y. and Thorpe, T.A. (1987) *In vitro* plantlet formation from embryonic explants of eastern white cedar (*Thuja occidentalis* L.). *Tree Physiology* 3, 273–283.

Huang, F.H., Klingaman, G.L. and Chen, H.H. (1992) Micropropagation of surprise lily. In: Bajaj, Y.P.S. (ed.), *Biotechnology in Agriculture and Forestry. Vol. 20. High-Tech and Micropropagation IV.* Springer-Verlag, Berlin, pp. 198–212.

Huetteman, C.A. and Preece, J.E. (1993) Thidiazuron: A potent cytokinin for woody plant tissue culture. *Plant Cell, Tissue and Organ Culture* 33, 105–119.

Hussey, G. (1975) Totipotency in tissue explants and callus of some members of Liliaceae, Iridaceae and Amaryllidaceae. *Journal of Experimental Botany* 26, 253–262.

Iapichino, G., McCulloch, S. and Chen, T.H.H. (1992) Adventitious shoot formation from leaf explants of *Rhododendron*. *Plant Cell, Tissue and Organ Culture* 30, 237–241.

Ishii, K. (1991) Hinoki cypress (*Chamaecyparis obtusa*). In: Bajaj, Y.P.S. (ed.), *Biotechnology in Agriculture and Forestry. Vol 16. Trees III.* Springer-Verlag, Berlin, pp. 471–478.

Isikawa, H. (1974) *In vitro* formation of adventitious buds and roots on the hypocotyl of *Cryptomeria japonica*. *Botanical Magazine* (Tokyo) 87, 73–77.

James, D.J., Passey, A.J., Barbara, D.J. and Bevan, M. (1989) Genetic transformation of apple (*Malus pumila* Mill.) using a disarmed Ti-binary vector. *Plant Cell Reports* 7, 658–661.

Johnson, B.B. (1978a) *In vitro* propagation of *Gloxinia* from leaf explants. *HortScience* 13, 149–150.

Johnson, B.B. (1978b) *In vitro* propagation of *Episcia cupreata*. *HortScience* 13, 596.

Jona, R. and Gribaudo, I. (1987) Adventitious bud formation from leaf explants of *Ficus lyrata*. *HortScience* 22, 651–653.

Jungnickel, F. and Zaid, S. (1992) Micropropagation of African violets (*Saintpaulia* spp. and cvs.). In: Bajaj, Y.P.S. (ed.), *Biotechnology in Agriculture and Forestry. Vol 20. High-tech and Micropropagation IV.* Springer-Verlag, Berlin, pp. 366–395.

Karnosky, D.F., Diner, A.M. and Barnes, W.M. (1988) A model system for gene transfer in conifers: European larch and *Agrobacterium*. In: Ahuja. M.R. (ed.), *Somatic Cell Genetics of Woody Plants*. Kluwer Academic Publishers, Dordrecht, pp. 55–63.

Kaul, K. (1987) Plant regeneration from cotyledon-hypocotyl explants of *Pinus strobus* L. *Plant Cell Reports* 6, 5–7.

Kaul, V., Miller, R.M., Hutchinson, J.F. and Richards, D. (1990) Shoot regeneration from stem and leaf explants of *Dendranthema grandiflora* Tzvelev (syn. *Chrysanthemum morifolium* Ramat.). *Plant Cell, Tissue and Organ Culture* 21, 21–30.

King, J.J. and Bridgen, M.P. (1987) *In vitro* induction of axillary and adventitious shoots in *Alstroemeria*. *HortScience* 22, 1149.

Kotwal, M.M., Gupta, P.K. and Mascarenhas, A.F. (1983) Rapid multiplication of *Sapium sebiferum* Roxb. *Plant Cell, Tissue and Organ Culture* 2, 133–139.

Laliberté, S. and Lalonde, M. (1988) Sustained organogenesis in callus cultures of *Larix* × *eurolepis* initiated from short shoot buds of a 12-year-old tree. *American Journal of Botany* 75, 767–777.

Ledger, E., Deroles, S.C. and Given, N.K. (1991) Regeneration and *Agrobacterium*-mediated transformation of chrysanthemum. *Plant Cell Reports* 10, 195–199.

Leege, A.D. and Tripepi, R.R. (1993) Rapid adventitious shoot regeneration from leaf explants of European birch. *Plant Cell, Tissue and Organ Culture* 32, 123–129.

Lester, D.T. and Berbee, J.G. (1977) Within-clone variation among black poplar trees derived from callus culture. *Forestry Science* 23, 122–131.

Libby, W.J. and Ahuja, M.R. (1993) Micropropagation and clonal options in forestry. In: Ahuja. M.R. (ed.), *Micropropagation of Woody Plants*. Kluwer Academic Publishers, Dordrecht, pp. 425–442.

Lloyd, D., Roberts, A.V. and Short, K.C. (1988) The induction *in vitro* of adventitious shoots in *Rosa*. *Euphytica* 37, 31–36.

Lu, C.Y. (1993) The use of thidiazuron in tissue culture. *In Vitro Cellular and Developmental Biology* 29, 92–96.

Lu, C.Y. and Thorpe, T.A. (1988) Shoot-bud regeneration in subcultured callus of Engelmann spruce. *In Vitro Cellular and Developmental Biology* 24, 239–242.

Marcotrigiano, M. and Stimart, D.P. (1983) *In vitro* organogenesis and shoot proliferation of *Paulownia tomemtosa* Steud. (Empress tree). *Plant Science Letters* 31, 303–310.

Margara, J. and Piollat, M.T. (1985) Evolution de l'apitude à l'organogenèse *in vitro* à partir des feuilles de *Saintpaulia ionantha* au cours des cultures successives. *Comptes Rendus des Séances de l'Academie des Sciences, Serie III: Sciences de la Vie* 301, 265–268.

May, R.A. and Trigiano, R.N. (1991) Somatic embryogenesis and plant regeneration from leaves of *Dendranthema grandiflora*. *Journal of the American Society for Horticultural Science* 116, 366–371.

Mehra-Palta, A., Smeltzer, R.H. and Mott, R.L. (1978) Hormonal control of induced organogenesis experiments with excised plant parts of loblolly pine. *TAPPI Journal* 61, 37–40.

Meier-Dinkel, A. (1992) Micropropagation of birches (*Betula* spp.). In: Bajaj, Y.P.S. (ed.), *Biotechnology in Agriculture and Forestry. Vol. 18. High-tech and Micropropagation II*. Springer-Verlag, Berlin, pp. 40–81.

Michler, C.H. (1993) *In vitro* genetic selection for woody plant improvement. In: Ahuja. M.R. (ed.), *Micropropagation of Woody Plants*. Kluwer Academic Publishers, Dordrecht, pp. 443–455.

Narayan, P. and Jaiswal, V.S. (1986) Differentiation of plantlets from leaf callus of *Ficus religiosa* L. *Indian Journal of Experimental Biology* 24, 193–194.

Nataraja, K., Chennaveeraiah, M.S. and Girigowda, P. (1973) *In vitro* production of shoot buds in *Euphorbia pulcherrima*. *Current Science* 42, 577–578.

Ostry, M.E. and Michler, C.H. (1993) Use of biotechnology for tree improvement in *Populus* model systems. In: Ahuja. M.R. (ed.), *Micropropagation of Woody Plants*. Kluwer Academic Publishers, Dordrecht, pp. 471–483.

Patel, K.R. and Thorpe, T.A. (1986) *In vitro* regeneration of plantlets from embryonic and seedling explants of Engelmann spruce (*Picea engelmannii*) Parry. *Tree Physiology* 1, 289–301.

Perez-Bermudez, P. and Sommer, H.E. (1987) Factors affecting adventitious bud induction in *Pinus elliottii* (Engelm.) embryos cultured in *vitro*. *Plant Cell, Tissue and Organ Culture* 11, 25–35.

Pierik, R.L.M. (1987) *In Vitro Culture of Higher Plants*. Martinus Nijhoff Publishers, Dordrecht.

Pierik, R.L.M. and Ruibing, M.A. (1973) Regeneration of bulblets on bulb scale segments of hyacinth *in vitro*. *Netherlands Journal of Agricultural Science* 21, 129–138.

Pierik, R.L.M. and Steegmans, H.H.M. (1975a) *Freesia* plantlets from flower buds cultivated in *vitro*. *Netherlands Journal of Agricultural Science* 23, 334–337.

Pierik, R.L.M. and Steegmans, H.H.M. (1975b) Effects of auxins, cytokinins, gibberellins, abscisic acid and ethephon on regeneration and growth of bulblets on excised bulb scale segments of hyacinth (*Hyacinthus orientalis*). *Physiologia Plantarum* 34, 14–17.

Pierik, R.L.M. and Tetteroo, F.A.A. (1987) Vegetative propagation of *Begonia venosa* Skan *in vitro* from inflorescence explants. *Plant Cell, Tissue and Organ Culture* 10, 135–142.

Preece, J.E. and Imel, M.R. (1991) Plant regeneration from leaf explants of *Rhododendron* 'PJM Hybrids'. *Scientia Horticulturae* 48, 159–170.

Qureshi, J.A. and Saxena, P.K. (1992) Adventitious shoot induction and somatic embryogenesis with intact seedlings of several hybrid seed geranium (*Pelargonium × hortorum* Bailey) varieties. *Plant Cell Reports* 11, 443–448.

Reilly, K. and Washer, J. (1977) Vegetative propagation of radiata pine by tissue culture: plantlet formation from embryonic tissue. *New Zealand Journal of Forestry Science* 7, 199–206.

Rout, G.R., Debata, B.K. and Das, P. (1992) *In vitro* regeneration of shoots from callus cultures of *Rosa hybrida* L. cv. Landora. *Indian Journal of Experimental Biology* 30, 15–18.

Saito, A. and Ide, Y. (1985) *In vitro* plantlet regeneration from adventitious buds induced on cuttings of peeled twigs of Japanese white birch. *Journal of Japanese Forestry Society* 67, 282–284.

Schwartz, O.J., Schlarbaum, S.E. and Beaty, R.M. (1988) Plantlet regeneration from mature zygotic embryos of eastern white pine (*Pinus strobus* L.). *Plant Cell Reports* 7, 174–177.

Seeliger, I. (1959) Zur entwicklungsgeschichte und anatomie wurzelbuertiger sprosse an isolierten Robinienwurzeln. *Flora* 148, 119–124.

Sehgal, C.B. (1975) *In vitro* differentiation of foliar embryo and adventitious buds from leaves of *Begonia semperflorens* Link & Otto. *Indian Journal of Experimental Biology* 13, 486–488.

Selby, C. and Harvey, B.M.R. (1985) The influence of natural and *in vitro* bud flushing on adventitious bud production in Sitka spruce (*Picea sitchensis* (Bong.) Carr.) bud and needle cultures. *New Phytologist* 100, 549–562.

Sen, S., Newton, R.J., Fong, F. and Neuman, P. (1989) Abscisic acid: a role in shoot enhancement from loblolly pine (*Pinus taeda* L.) cotyledon explants. *Plant Cell Reports* 8, 191–194.

Silberstein, S.P., Huang, F.H. and Klingaman, G.L. (1983) *In vitro* propagation of succulent *Euphorbia flanaganii* N.E.Br. *Cactus and Succulent Journal* 55, 80–83.

Simmonds, J. (1984) Induction of growth and direct rooting of adventitious shoots of *Begonia* × *hiemalis*. *Plant Cell, Tissue and Organ Culture* 3, 283–289.

Simonsen, J. and Hildebrandt, A.C. (1971) *In vitro* growth and differentiation of gladiolus plants from callus cultures. *Canadian Journal of Botany* 49, 1817–1819.

Srivastava, P.S. and Steinhauer, A. (1981) Regeneration of birch plants from catkin tissue cultures. *Plant Science Letters* 22, 379–386.

Srivastava, P.S., Steinhauer, A. and Glock, H. (1985) Plantlet differentiation in leaf and root cultures of birch (*Betula pendula* Roth.). *Plant Science* 42, 209–214.

Stichel, E. (1959) Simultaneous induction of roots and shoots in excised tissues of *Cyclamen persicum*. *Planta* 53, 293–317.

Tabrett, A.M. and Hammatt, N. (1992) Regeneration of shoots from embryo hypocotyls of common ash (*Fraxinus excelsior*). *Plant Cell Reports* 11, 514–518.

Thomas, M.J., Duhoux, E. and Vazart, J. (1977) *In vitro* organ initiation in tissue cultures of *Biota orientalis* and other species of the Cupressaceae. *Plant Science Letters* 8, 395–400.

Thorpe, T.A. (1994) Morphogenesis and regeneration. In: Vasil, I.K. and Thorpe, T.A. (eds), *Plant Cell and Tissue Culture*. Kluwer Academic Publishers, Dordrecht, pp. 17–36.

Thorpe, T.A., Harry, I.S. and Kumar, P.P. (1991) Application of micropropagation to forestry. In: Debergh, P.C. and Zimmerman, R.H. (eds.), *Micropropagation*. Kluwer Academic Publishers, Dordrecht, pp. 311–336.

Toivonen, P.M.A. and Kartha, K.K. (1988) Regeneration of plantlets from *in vitro* cultured cotyledons of white spruce (*Picea glauca* (Moench) Voss). *Plant Cell Reports* 7, 318–321.

Torrey, J.G. (1966) The initiation of organized development in plants. *Advances in Morphogenesis* 5, 39–91.

Tranvan, H. (1979) *In vitro* adventitious bud formation on isolated seedling of *Pinus sylvestris* L. *Biologia Plantarum* 21, 230–233.

Trigiano, R.N. and May, R.A. (1994) Laboratory exercises illustrating organogenesis and transformation using chrysanthemum cultivars. *HortTechnology* 4, 325–327.

Van Aartrijk, J, Blom-Barnhoorn, G.J. and Bruinsma, J. (1986) A role for ethylene biosynthesis in adventitious bud formation on bulb-scale tissue of *Lilium speciosum*. In: Withers, L.A. and Alderson, P.G. (eds), *Plant Tissue Culture and its Agricultural Applications*. Butterworths, London, pp. 55–61.

Van der Linde, P.C.G. and Schipper, J.A. (1992) Micropropagation of iris with special reference to *Iris* × *hollandica* Tub. In: Bajaj, Y.P.S. (ed.), *Biotechnology in Agriculture and Forestry, Vol. 20. High-tech and Micropropagation IV*. Springer-Verlag, Berlin, pp. 173–197.

Van der Linde, P.C.G., Hol, G.M.G.M., Blom-Barnhoorn, G.J., Van Aartrijk, J. and De Klerk, G.J. (1988) In-vitro propagation of *Iris hollandica* Tub. cv. Prof. Blaauw. 1. Regeneration on bulb-scale explants. *Acta Horticulturae* 226, 121–128.

Vazquez, A.M., Davey, M.R. and Short, K.C. (1977) Organogenesis in cultures of *Saintpaulia ionantha*. *Acta Horticulturae* 78, 249–258.

Venverloo, C.J. (1973) The formation of adventitious organs. I. Cytokinin-induced formation of leaves and shoots in callus cultures of *Populus nigra* L. 'Italica'. *Acta Botanica Neerlandica* 22, 390–398.

Welander, M. (1988) Biochemical and anatomical studies of birch (*Betula pendula* Roth.) buds exposed to different climatic conditions in relation to growth *in vitro*. In:

Hanover, J.W. and Keathley, D.E. (eds), *Genetic Manipulation of Woody Plants*. Plenum, New York, pp. 79–99.

Welander, M. (1993) Micropropagation of birch. In: Ahuja. M.R. (ed.), *Micropropagation of Woody Plants*. Kluwer Academic Publishers, Dordrecht, pp. 223–246.

Wetzstein, H.Y. and Sommer, H.E. (1982) Leaf anatomy of tissue cultured *Liquidambar styraciflua* (Hamamelidaceae) during acclimatization. *American Journal of Botany* 69, 1579–1586.

Wilmink, A., Van de Ven, B.C.E., Custers, J.B.M., Nollen, Y. and Dons, J.J.M. (1995) Adventitious shoot formation in tulip: histological analysis and response to selective agents. *Plant Science* 110, 155–164.

Winton, L.L. (1970) Shoot and tree production from aspen tissue cultures. *American Journal of Botany* 57, 904–909.

Wolter, K.F. (1968) Root and shoot initiation in aspen callus culture. *Nature* 219, 509–510.

Zel, J. (1993) Micropropagation of *Pinus sylvestris*. In: Ahuja. M.R. (ed.), *Micropropagation of Woody Plants*. Kluwer Academic Publishers, Dordrecht, pp. 347–365.

Ziv, M., Halevy, A.H. and Shilo, R. (1970) Organs and plantlet regeneration of gladiolus through tissue culture. *Annals of Botany* 34, 671–676.

Approaches to Understanding Maturation or Phase Change

5

Wesley P. Hackett and John R. Murray
Department of Horticultural Science, University of Minnesota, St Paul, Minnesota 55108, USA

INTRODUCTION

Maturation or phase change has been intriguing but baffling to applied and basic scientists for many years (Goebel, 1900). The morphological and physiological characteristics associated with phase change have been described for many species, especially woody perennial species (Hackett and Murray, 1993), but little is known about the mechanisms involved. P.F. Wareing and his students (Wareing and Frydman, 1976) were some of the first scientists to begin to analyse the biological bases of phase change and as a result began to provide a conceptual framework for understanding the process. Subsequently, a relatively small number of scientists, interested mainly in woody species, have used a variety of approaches to try to manipulate and understand the process of maturation, but with limited success (Hackett, 1985). Recent advances in *in vitro* culture of plant tissues, plant biochemistry and recombinant DNA technology have provided new research tools to investigate phase change. The objectives of this chapter are to provide a current conceptual framework for investigation of maturation and to illustrate through research examples how biotechnological approaches have contributed to our understanding of this developmental process.

CONCEPT AND CHARACTERISTICS OF MATURATION

During the development of plants from seed there are changes in the characteristics of the structures produced by the shoot apical meristem. The most obvious of these is the transition from vegetative to reproductive structures. This change is a reflection of a change in competence to perceive, or respond to environmental or endogenous stimuli for flower induction. In addition, there

Table 5.1. Some characteristics that distinguish the juvenile and mature phases of *Hedera helix*.

Characteristics	Juvenile	Mature
Shoot apex width	*c.* 140 μm	*c.* 200 μm
Plastochron	4.2 days	3.2 days
Internode length	Long	Short
Leaf shape	Five-lobed, palmate	Entire, ovate
Phyllotaxis	Distichous (1,2)	Spiral (2,3)
Growth habit	Plagiotropic	Orthotropic
Anthocyanin stem pigmentation	Present	Absent
Photosynthetic capability	Low	High
Stem aerial roots	Present	Absent
Ability to flower	Absent	Present

are progressive changes in vegetative characteristics that involve morphological, anatomical, physiological and developmental differences. These include leaf shape, thickness and epidermal characteristics, phyllotaxis, thorniness, shoot orientation, shoot growth vigour, anthocyanin pigmentation, photosynthetic characteristics, disease and insect resistance, and competence to form adventitious buds and roots and somatic embryos (Hackett and Murray, 1993). Degree of change in such characteristics during development varies from species to species and most change gradually resulting in intermediate or transitional forms and competencies (Hackett, 1985). In addition, the various characteristics may change at different rates in a given species (Greenwood *et al.*, 1989; Steele *et al.*, 1989).

Those characteristics that appear early in development are referred to as juvenile and those that appear late in development are referred to as mature or adult characteristics. The change from juvenile to mature characteristics is referred to as maturation or phase change. Maturation-related differences in characteristics are most obvious in woody perennial species because of their prolonged juvenile and mature phases (Table 5.1) but are also observed in herbaceous species like maize (Poethig, 1990) (Table 5.2). In certain perennial species such as *Eucalyptus* sp. (Wiltshire *et al.*, 1992) or *Hedera helix* (ivy) (Rogler and Hackett, 1975), the differences are quite obvious and result in distinct phase-dependent phenotypes.

The apical or eumeristem and its cellular derivatives in primordial leaves and subapical meristem region of the shoot apex are the sites of origin of different paths of cellular, tissue and organ differentiation. This fact leads logically to the conclusion that these same sites are involved in the origin of maturation-related differences in physiological, anatomical and morphological characteristics. It is well documented, for example, that the apical and subapical meristems of juvenile and mature ivy are very different in size, configuration and cell division activity, and that these differences in meristem characteristics can be related to several phase-dependent, morphological

Table 5.2. Some characteristics that distinguish the juvenile and mature phases of maize.

Characteristics	Juvenile	Mature
Adventitious roots	Present	Absent
Lateral buds	Tiller-like	Ears or absent
Leaf shape	Narrow	Broad
Internode length	Compressed	Extended
Anthracnose resistance	Poor	Good
Epidermal characteristics		
Cuticle thickness	< 1 mm	> 1 mm
Epicuticular wax	Present	Absent
Epidermal hairs	Absent	Present
Bulliform cells	Absent	Present
Cell-wall invagination	Moderate	Extreme
Cell cross-section	Rounded	Rectangular
Toluidine blue O staining	Violet	Aqua

characteristics such as differences in phyllotaxis and internode lengths (Hackett *et al.*, 1987, Marc and Hackett, 1991, 1992).

Although the meristems are very different, and certain characteristics are related to these differences, much of the process of differentiation is common between the juvenile and mature phases of ivy. The anatomy of the cell types and the structural organization of the leaf laminae, petioles and stems of the distinct phase phenotypes are quite similar. This similarity in cell types and structural organization, but marked difference in the phenotypes of the two phases suggests there may be at least two levels of control of differentiation. The control of primary differentiation, that determines the cell types and structural organization of tissues, is fundamental and similar in both phases. Superimposed on primary differentiation is a level of control of differentiation which leads to the unique expression of the phase-dependent characteristics. This difference in expression of phase-dependent characteristics within an individual is referred to as epigenetic variation and is ultimately due to differential expression of genes (Goldberg, 1987). The epigenetic control of differential gene expression in a relatively small percentage of the cells in the plant body can result in rather large differences in phenotype. For example, red stems and leaf petioles in juvenile ivy plants as compared to green stems and petioles in mature plants is the result of differential expression of anthocyanin accumulation in four or five collenchyma cell layers in the hypodermis (Murray, 1988). However, it should be pointed out that it is not known whether this biochemical differentiation is related to changes in biochemical activities of shoot apical or subapical meristem.

The developmental behaviour of axillary buds reflects the phase-related characteristics of the primary apical meristem from which they arise. This means that the characteristics of lateral shoots depend on their position on the primary shoot and the maturation state of the shoot apical meristem at the time they were initiated (Hackett, 1985). As a result, the characteristics of plants derived from cuttings or grafting is dependent on the position on the primary or secondary shoot axis from which the cuttings or scions were taken. Plants derived from cuttings or scions from branches on the basal portion of shoot axes express juvenile characteristics while those from the more apical portions express mature characteristics, over long periods of time. For woody perennial plants, this stability is maintained at the shoot apical meristem level since excised meristems of ivy cultured *in vitro* maintain at least some of the characteristics of the phase of the shoot from which the meristem was excised (Hackett, 1985; James and Mantell, 1994). However, in maize, phase characteristics show less stability in that apical meristems in a mature phase of development will revert to the juvenile phase when excised and grown *in vitro* (Dudley and Poethig, 1993). Whether this observed difference in phase stability, as reflected in maize and ivy, is generalizable to a difference in stability in herbaceous and woody plants is not known. For woody perennial species, there is some evidence (Polito and Alliato, 1981; James and Mantell, 1994) that stability of phase-related characteristics resides at the cellular level.

THEORETICAL AND PRACTICAL SIGNIFICANCE

As can be seen from the discussion above, and as discussed by others (Poethig, 1990), maturation or phasic development is of considerable theoretical importance relative to morphogenetic control and differentiation in plant development. It also has practical significance for the following reasons:

1. The length of the juvenile period is inversely related to the breeding efficiency of woody perennials and to the selection of improved cultivars (Hansche and Beres, 1980).
2. The ease of adventitious root and bud initiation and somatic embryogenesis for many woody perennials is strongly affected by phase of maturation (Greenwood, 1987).
3. The quantity and quality of productivity of a forest tree species is related to its phase of maturation (Takemoto and Greenwood, 1993).

The ability to manipulate the maturation state would have great practical significance, particularly for long-lived perennial plants.

MECHANISMS OF PHASE CHANGE AND STABILITY

Wareing and Frydman (1976) pointed out that it is important to make a distinction between phase change and phase stability when thinking about mechanisms to explain phasic development. Relatively little is known about mechanisms for either phase change or phase stability but, as will be discussed later, biotechnological research approaches hold great promise and are beginning to give new insights into possible mechanisms. At this stage of our knowledge,

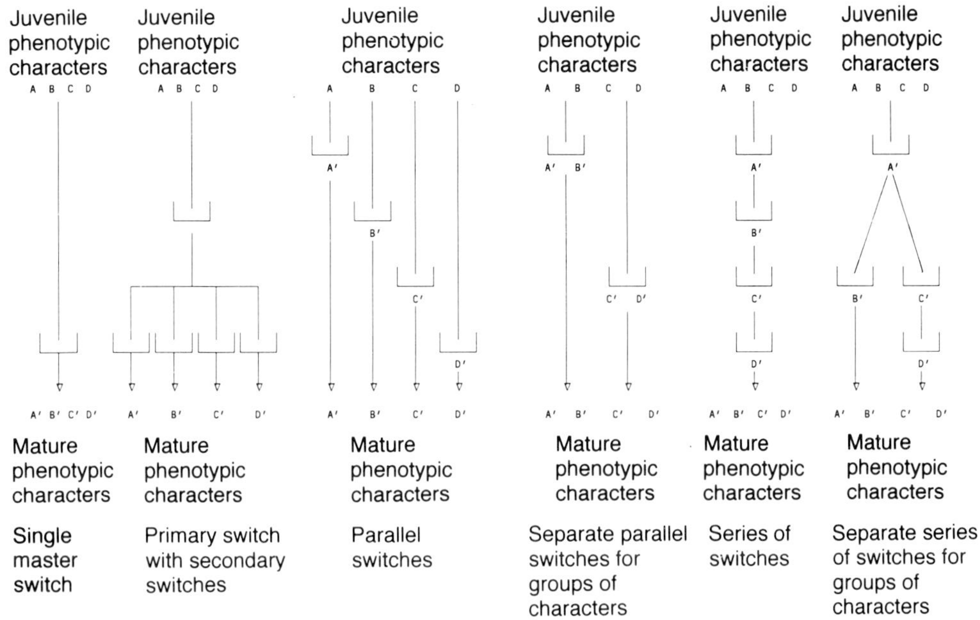

Fig. 5.1. Alternative models of phase change or maturation.

the best we can do is to put forth some general hypotheses or models that can form the conceptual framework for using biotechnological approaches to investigate mechanisms. For example, the differential time courses for changes of individual maturation-related characteristics (Greenwood *et al.*, 1989; Steele *et al.*, 1989; Browne, 1995), the coexistence of juvenile and mature characteristics in close developmental proximity on shoots of eucalyptus and ivy (Hackett and Murray, 1993), and the differential sensitivity of various morphological and physiological traits to rejuvenation (phase reversal) by gibberellic acid (GA_3) in ivy suggest that phase change is not a single step process but a composite of processes or programmes (Fig. 5.1, Hackett and Murray, 1993). Based on studies of maize, Poethig (1990) concluded that phasic development is specified by a series of independently regulated, overlapping programmes that modify the expression of a common set of processes required for shoot growth. Some evidence for this conclusion is provided by the observation that leaves produced during transition from juvenile to mature phase characteristics in maize have a combination of juvenile and mature cell types and express a variety of other traits in a quantitatively intermediate fashion.

No matter what the mechanism for phase change, mature characteristics accumulate during development. They accumulate because the characteristics are quite stable and are transmitted developmentally through cell division from one somatic cell generation to the next, even into branches of the main axis. However, such stability does not mean that maturation-related characteristics are permanent and non-reversible. They are universally reversible during sexual reproduction (formation of an embryo) and there are many examples to indicate they are somatically reversible or manipulable under specific conditions (Hackett and Murray, 1993). However, all maturation-related characteristics may not have the same ease of reversibility, and the ease of their reversibility may change over developmental time and may vary with species (maize as compared to ivy). Three general hypotheses have been presented by Wareing and Frydman (1976) to explain the stability of maturation-related characteristics:

1. The structure and properties of the apical meristem are controlled or stabilized by influences arising from pre-existing, differentiated tissue located in more or less remote parts of the plant.
2. The stable characteristic properties of the apical meristem are controlled by the structure and organization of the apex as a whole in terms of biophysical stability of the structure or diffusion–reaction schemes.
3. There are intrinsic differences in the meristematic cells of the juvenile and mature apices and their stable properties and organization arise from intrinsic differences in the cells (perhaps based on competence to express genes).

These three possible explanations can be termed the correlative, the structural and organizational, and the cellular or gene expression hypotheses. There is positive evidence from different species to support all three hypotheses (Hackett; 1985, Poethig, 1990).

APPROACHES TO UNDERSTANDING AND MANIPULATING MATURATION

The two general experimental approaches for study of maturation and phase change are epigenetic and genetic. The basis for these approaches is the knowledge that phenotypic variation during development can be due to genetic variation (difference in genotype) or epigenetic variation (differences in gene expression).

The epigenetic approach involves study of differences in maturation-related characteristics in tissue having the same genotype. This is the most common approach in studies with perennial species with prolonged juvenile phases, due to the difficulties of genetic studies involving long generation times. This approach takes advantage of the stability of differences in characteristics in many juvenile and mature tissues to make cellular, biochemical and molecular comparisons. An important outcome of such studies is the identification of markers (Table 5.3) that can be used to assess the progression of phase change or the degree of stability of maturation-related characteristics as a basis for understanding mechanisms of stability and devising protocols for manipulation. Study of the basis of stable, differing juvenile and mature characteristics, one at a time, provides the opportunity to determine the con-

Table 5.3. Some biochemical and molecular markers of phase change.

Marker	Method of analysis	Species	Reference
Chlorophyll *a/b* binding protein	Steady-state RNA	*Hedera helix* *Larix laricina*	Woo *et al.* (1994); Hutchison *et al.* (1990)
Ribulose bisphosphate carboxylase, large subunit	Steady-state RNA, protein blotting and immunodetection	*Prunus avium*	Besford *et al.* (1996)
Dihydroflavonol reductase	Nuclear run-on transcription assay, steady state RNA, enzyme activity	*H. helix*	Murray *et al.* (1994)
59.7 kDa protein	Immunoelectro-phoresis and immunochroma-tography	*Citrus* species	Snowball *et al.* (1991)
16 kDa protein	Protein blotting and immunodetection	*Sequoiadendron giganteum*	Bon (1988)
43.6 and 38.0 kDa proteins	SDS-PAGE	*Castanea sativa*	Amo-Marco *et al.* (1993)
Mitochondrial (mt) DNA	mt DNA restriction fragment length analysis	*Sequoia sempervirens*	Huang *et al.* (1995)
Phenylalanine-ammonia lyase	Enzyme activity	*H. helix*	Murray and Hackett (1991)
Caffeic and ferulic acids	HPLC with on-line spectrophotometry	*H. helix*	Murray and Hackett (1991)
Anthocyanin	Spectrophotometry	*H. helix*	Murray and Hackett (1991)
	Visual	*Malus*, *Betula papyrifera*, *Carya*	Stoutemeyer (1937) Brand and Lineberger (1992), Romberg (1994)
Abscisic acid	GLC	*H. helix*	Hillman *et al.* (1974)
4-ethyl (methyl)-1-hydroxy-4,8,8,10,10-pentamethyl 7,9-dioxo-2,3-dioxabicyclo (4.40) decane-5		*Eucalyptus grandis*	Nichols *et al.* (1972)
Myrcene and furanone (diterpenes)	GLC	*Dacrydium biforme*	Hayman *et al.* (1986)
Mono- and sesquiterpenes		*Juniperus scopulorum*	Adams and Hagerman (1978)

HPLC, high pressure liquid chromatography; GLC, gas-liquid chromatography.

trol mechanism of individual characteristics and to determine if or how control of one characteristic is related to another (Fig. 5.1).

Often the epigenetic approach makes use of excised tissues or organs cultured *in vitro*. This is done in order to isolate the experimental material from correlative influences of the remainder of the plant, to attain more precise environmental control, to make it easier to administer inductive agents and to take advantage of the greater ability to induce adventitious organogensis *in vitro*. However, it is important for studies aimed at identifying markers to verify the stability of juvenile and mature characteristics in excised systems.

The genetic approach utilizes a genotype or inbred line which has the expected progression of maturation-related characteristics for the species (wild genotype) and one or more single gene mutants of the wild genotype which alter the progression of phasic development or alter the expression of one of the maturation-related characteristics. The mutants may be natural variants or may be induced by utilization of mutagens. The wild type and mutant can be used for comparative studies as outlined for the epigenetic approach. In addition, if the sexual cycle is relatively short, it is feasible to do genetic analysis. Such analysis using genotypes that have various combinations of mutant genes is a powerful tool for understanding mechanisms (Fig. 5.1) of phase change (Poethig, 1990).

In the following paragraphs, examples of the epigenetic and genetic approaches to study maturation are described and discussed. The emphasis will be on use of biotechnological techniques.

EPIGENETIC APPROACHES

In vitro *adventitious bud formation from dedifferentiated tissues*

James and Mantell (1994) used *in vitro* cultured tissues of the woody perennial shrub *Solanum aviculare* to study the basis of stability of maturation-related characteristics. *S. aviculare* has several advantages for this kind of study including predictable heteroblastic leaf characters associated with progression through maturation, ease of culture of shoot apices *in vitro* and ease of induction of *de novo* adventitious shoot meristems from several excised tissues. By assessing the characteristics of plants regenerated from shoot apices and from adventitious shoots derived from various tissues including epidermal strips, pith and vascular stele, they concluded that some phase-specific characteristics are inherited even after dedifferentiation and subsequent *de novo* shoot meristem organogenesis. As phase-specific characters are imposed on all regenerates, regardless of tissue origin, they suggest this infers intrinsic differences in the meristem initial cells. However, there was some evidence of reversal (rejuvenation) of some mature characteristics in adventitiously derived plants. They interpret this to mean that the structure of the mature meristem may, at least partially, be a product of developmental history (distance from the roots for example). In general, juvenile and mature differences in the number of nodes to flower were more likely to be maintained in adventitiously derived plants than were juvenile and mature differences in leaf lobing and length. This suggests that some maturation-related characteristics are more intrinsically stable than others.

In vitro *multiplication of shoot tips from mature plants*

Brand and Lineberger (1992) assessed the stability of maturation-related characteristics of plants derived from *in vitro* cultured shoot tips from mature *Betula* plants. To do this they compared the characteristics of one-, four- and eight-month-old plants derived from: (i) seed, (ii) shoot tips from mature plants multiplied *in vitro* for two to three years, and (iii) buds from mature plants chip-grafted onto seedlings, with the characteristics of two-year-old cutting-derived plants. The cutting-derived plants had only mature characteristics including flowering competence. In contrast, plants derived from shoot tips multiplied *in vitro* for two to three years had several morphological and physiological characteristics similar to plants propagated from seed. There was some clone to clone variation in expression of juvenile characteristics, perhaps related to whether *in vitro* shoots were of axillary or adventitious origin, with less juvenile characteristics in the clone which was multiplied from axillary buds only. *In vitro* shoot-tip-derived plants that exhibited juvenile characteristics re-developed some mature characteristics more rapidly than did plants derived from seed. However, mature characteristics such as flowering and white bark did not re-develop in *in vitro* shoot-tip-derived plants during the eight-month duration of observation. Juvenile characteristics that were most transient in *in vitro* shoot-tip-derived plants were anthocyanin pigmentation and stem pubescence. Chip-graft-derived plants at four and eight months were nearly identical to cutting-propagated mature plants, but some juvenile characteristics were displayed during the first two months of growth. Brand and Lineberger concluded that some mature phase characteristics were unstable during long-term *in vitro* culture of mature phase shoot tips. However, complete reversal (rejuvenation) of maturation does not occur *in vitro* and individual maturation-related characteristics are not equally affected.

Biochemical analysis of juvenile and mature tissues

As noted earlier, the collenchyma cells of the hypodermis of juvenile stems and leaf petioles of ivy accumulate anthocyanin pigment while apparently identical collenchyma cells in mature phase stems and petioles do not. Under some environmental conditions, anthocyanin accumulates in the leaf laminae of juvenile phase shoots, whereas laminae of mature phase shoots do not accumulate anthocyanin. Murray and Hackett (1991) studied the biochemical basis for this difference in competence to accumulate anthocyanin in juvenile and mature laminae tissue. By measuring levels of phenylpropanoid and flavonoid compounds, which are precursors of anthocyanin, and by assaying the activities of enzymes of the phenylpropanoid and flavonoid pathways involved in anthocyanin biosynthesis, they found several differences that might be useful as markers of phase state in ivy. Using leaf discs cultured on sucrose in light, they found that leaf discs of both phases accumulate comparable levels of the flavonols kaempferol and quercetin, whereas only juvenile phase leaf discs accumulated anthocyanin. The accumulation of comparable levels of flavonols in both phases suggests that enzymes early in the flavonoid pathway, such as chalcone synthase (CHS), are not limiting in mature phase tissue. The accumulation of quercetin, but lack of accumulation of leucocyanidin or anthocyanin in mature phase discs, suggested that mature discs lack dihydroflavonol reductase (DFR) activity. Enzyme assays showed there was no detectable DFR activity in mature phase discs, whereas there was induction of DFR activity in juvenile phase discs in response to sucrose and light.

Phenylalanine ammonia-lyase (PAL), the first enzyme committed to phenylpropanoid metabolism and thus an enzyme early in the anthocyanin biosynthetic pathway, was induced above basal level by sucrose and light in discs of both phases of ivy, with greater activity in mature phase discs. Phenylpropanoids, a class of compounds that are precursors to flavanoids and other products, accumulated in leaf discs of both phases with greater levels of free or conjugated forms of caffeic and ferulic acid in mature phase discs. Murray and Hackett (1991) conclude that the lack of DFR activity limits the accumulation of anthocyanin in mature phase ivy tissue. Therefore, DFR activity is a definitive marker for juvenile phase leaves in ivy. Higher PAL activity and higher levels of free or conjugated forms of caffeic and ferulic acid may serve as quantitative markers for mature phase leaf tissue.

Differential gene expression in juvenile and mature tissues

Based on the work described in the previous section, Murray *et al.* (1994) determined the level of gene expression that accounts for the lack of DFR activity in mature phase lamina tissue of ivy. To do this, DNA clones of DFR and CHS were isolated from ivy using DNA clones from other species to probe an ivy genomic library. By use of nuclear 'run-on' transcription assays and RNA blot analyses, it was demonstrated that the lack of induction of DFR activity in mature phase lamina tissue is due to a lack of transcription of the DFR gene and the resultant lack of accumulation of detectable DFR mRNA. In contrast, both phases transcribe CHS and accumulate CHS mRNA in response to light and sucrose indicating that mature phase tissue is responsive to the inductive treatments. These responses are not unique to lamina tissue in that CHS mRNA accumulates in both juvenile and mature phase stem tissue, but DFR mRNA accumulates only in juvenile stem tissue. Murray *et al.* (1994) conclude that the lack of competence to induce anthocyanin pigmentation in lamina tissue of mature phase ivy is due to lack of transcription of the DFR gene late in the biosynthetic sequence. Further molecular genetic approaches are currently being used to determine whether the lack of transcription of the DFR locus in mature phase tissue is due to an altered characteristic of the

DFR locus, such as altered chromatin conformation, or due to the lack of expression of regulatory genes that control DFR expression.

GENETIC APPROACHES

Zea mays (maize) is the species with the best described maturation-related differences in characteristics (see Table 5.2) for which a genetic approach is feasible. This is because much is already known about the genetics of maize, many mutant genotypes are available, and the sexual generation time is relatively short so genetic analysis can be done relatively efficiently.

Poethig and co-workers, and more recently Moose and Sisco (1994), have made use of mutants to begin to analyse the mechanisms of phase change and maturation in maize. Poethig (1988a) characterized three maize genes, *Tp1*, *Tp2* and *Tp3*, that prolong the expression of juvenile characteristics. Because shoot growth in maize terminates with the initiation of a male inflorescence (tassel), the effect of these genes on reproductive development can be distinguished from their effect on maturation-related vegetative characteristics by assessing nodes to tassel formation. Using such an analysis, Poethig demonstrated that these genes prolong expression of juvenile vegetative characteristics but do not affect the rate or duration of shoot growth, or time of tassel initiation. He concludes from the phenotypes expressed as a result of these genes that the juvenile phase characteristics in maize are regulated, at least to some extent, independently from the mature phase characteristics. That is, extending juvenile phase characteristics does not necessarily change the stage of development when the plant becomes competent to flower.

Poethig (1988b) and Dudley and Poethig (1993) have done a clonal analysis of *Tp1* and *Tp2* using mosaic Teopod: wild-type plants. This analysis demonstrates that the functions of these genes are not expressed cell autonomously. Based on these results, they suggest that these genes are involved in the production or distribution of a diffusible factor that promotes the juvenile phase.

Recently, Evans *et al.* (1994) and Moose and Sisco (1994) have identified a recessive mutant, glossy 15 (*gl15*), which, in contrast to the *Tp* mutations, shortens the juvenile phase of maize and accelerates the appearance of mature characteristics. They have done an elegant phenotypic and genetic analysis of *gl15* which indicates that it conditions abbreviated expression of juvenile epidermal cell characteristics and the precocious expression of mature epidermal cell characteristics. Moose and Sisco (1994) isolated a transposon-induced mutable allele of Glossy 15 and used it to compare *Gl15* active and *Gl15* inactive sectors in leaves of variegated plants. They observed that the *Gl15* gene product acted in a cell-autonomous manner to direct juvenile epidermal cell differentiation but does not affect factors that regulate the overall process of phase change. Both Moose and Sisco (1994) and Evans *et al.* (1994) did phenotypic analysis of double mutants *gl15* and *Tp* and conclude that *gl15* acts downstream to *Tp1* and *Tp2* since it does not influence factors that regulate the overall process of phase change. Based on their results and those of Evans *et al.* (1994), Moose and Sisco (1994) conclude that the cell-autonomous Gl15 gene product responds to a juvenility programme that operates throughout the shoot to condition juvenile differentiation of maize leaf epidermal cells. This conclusion means that there are at least two levels of regulation of maturation-related characteristics in maize (Fig. 5.1).

Very recently Poethig (Pennsylvania, 1995, personal communication) has identified another maize mutant, referred to as glossy early flowering (*gef*), which shortens the juvenile phase and accelerates the appearance of mature characteristics. In contrast to *gl15, gef* has an effect on the expression of many phase characteristics. This recessive mutant has a rapid transition to mature vegetative characteristics (as early as leaf 1) and reduced total leaf number, suggesting that the mutation also accelerates competence to flower. Poethig concludes that *gef* plays a central role in shoot maturation and may be a component of the 'clock' that regulates the maturation process.

CONCLUSIONS

In the research discussed above, biotechnological approaches have been utilized to study the genetic and epigenetic bases of maturation. From this research the following conclusions can be drawn:

1. Maturation is not a single-step process and involves at least two levels of regulation in maize.
2. The expression of maturation-related characteristics can be regulated at the level of transcription, as observed for the *DFR* gene and anthocyanin pigmentation in ivy.
3. Degree of stability of expression of mature characteristics in excised tissues *in vitro* varies with the characteristic assessed.
4. Some mature characteristics are relatively stable even when the plants assessed are derived adventitiously from callus cultures.
5. Biochemical and molecular markers have been identified which are preferentially expressed in juvenile or mature tissues.
6. Genetic mutants for phase characteristics are a powerful tool for analysing the mechanisms of phase change.

REFERENCES

Adams, R.P. and Hagerman, A. (1976) A comparison of the volatile oils of mature versus young leaves of *Juniperus scopulorum*: chemosystematic significance. *Biochemical Systematic Ecology* 4, 75–79.

Amo-Marco, J.B., Vidal N., Vietiz, A.M. and Ballester, A. (1993). Polypeptide markers differentiating juvenile and adult tissues in chestnut. *Journal of Plant Physiology* 142:117–119.

Besford, R.T., Hand, P., Richardson, C.M. and Peppitt, S.D. (1996) Photoperiod effect on budburst in *Prunus* is phase dependent: significance for early photosynthetic development. *Tree Physiology* 16, 491-496.

Bon, M.-C. (1988) J16: an apex protein associated with the juvenility of *Sequoiadendron giganteum*. *Tree Physiology* 4, 381–387.

Brand, M.H. and Lineberger, R.D. (1992) *In vitro* rejuvenation of *Betula* (Betulaceae): morphological evaluation. *American Journal of Botany* 79, 618–625.

Browne, R.D. (1995) Maturation and rejuvenation in Jack pine (*Pinus banksiana* Lamb): aspects of shoot morphology, anatomy and adventitious rooting of cuttings. PhD thesis, University of Saskatchewan.

Dudley, M. and Poethig, R.S. (1993) The heterochronic *Teopod1* and *Teopod2* mutations of maize are expressed non-cell-autonomously. *Genetics* 133, 389–399.

Evans, M.M.S., Passas, H.J. and Poethig, R.S. (1994) Heterochronic effects of *glossy15* mutations on epidermal cell identity in maize. *Development* 120, 1971–1981.

Goebel, K. (1900) *Organography of Plants*. Clarendon Press, Oxford.

Goldberg, R.B. (1987) Emerging patterns of plant development. *Cell* 49, 298–300.

Greenwood, M.S. (1987) Rejuvenation of forest trees. *Plant Growth Regulation* 6, 1–12.

Greenwood, M.S., Hooper, C.A. and Hutchison, K.W. (1989) Maturation in larch. I. Effects of age on shoot growth, foliar characteristics and DNA methylation. *Plant Physiology* 90, 406–412.

Hackett, W.P. (1985) Juvenility, maturation and rejuvenation in woody plants. *Horticultural Reviews* 7, 109–155.

Hackett, W.P. and Murray, J.R. (1993) Maturation and rejuvenation in woody species. In: Ahuja, M.R. (ed.), *Micropropagation of Woody Plants*. Kluwer Academic Publishers, Dordrecht, pp. 93–105.

Hackett, W.P., Cordero, R.E. and Srinivasan, C. (1987) Apical meristem characteristics and activity in relation to juvenility in *Hedera*. In: Atherton, J.G. (ed.), *Manipulation of Flowering*. Butterworths, London, pp. 92–99.

Hansche, P.E. and Beres, W. (1980) Genetic remodeling of fruit and nut trees to facilitate cultivar development. *HortScience* 15, 710–715.

Hayman, A.R., Perry, N.B. and Evans, R.T. (1986) Juvenile–adult chemical dimorphism in foliage of *Dacrydium biforme*. *Phytochemistry* 25, 649–653.

Hillman, J.R., Young, I. and Knight, B.A. (1974) Abscisic acid in leaves of *Hedera helix* L. *Planta* 119, 263–266.

Huang, L.-C. and Lin, L.-Y., Chien, C.-M., Chen, L.-J., Huang, B.-L. and Murashige, T. (1995) Phase reversal in *Sequoia sempervirens* in relation to mtDNA. *Physiologia Plantarum* 94, 379–383.

Hutchison, K.W., Sherman, C.D., Smith, S.S., Singer, P.B. and Greenwood, M.S. (1990) Maturation in larch. II. Effects of age on photosynthesis and gene expression in developing foliage. *Plant Physiology* 94, 1308–1315.

James, A.C. and Mantell, S.H. (1994) Evidence for epigenetic inheritance of ontogenetic phase and tissue origin characteristics in the performance of plants derived through adventitious shoots. In: Lumsden, P.J. Nicholas, J.R. and Davies, W.J. (eds), *Physiology, Growth and Development of Plants in Culture*. Kluwer Academic Publishers. Dordrecht, pp. 339–345.

Marc, J. and Hackett, W.P. (1991) Gibberellin-induced reorganization of spatial relationships of emerging leaf primordia at the shoot apical meristem in *Hedera helix* L. *Planta* 185, 171–178.

Marc, J. and Hackett, W.P. (1992) Changes in the pattern of cell arrangement at the surface of the shoot apical meristem in *Hedera helix* L. following gibberellin treatment. *Planta* 186, 503–510.

Moose, S.P. and Sisco, P.H. (1994) *Glossy15* controls the epidermal juvenile-to-adult phase transition in maize. *Plant Cell* 6, 1343–1355.

Murray, J.R. (1988) Flavonoid metabolism in relation to differential anthocyanin accumulation in juvenile and mature phase *Hedera helix* L. PhD thesis, University of Minnesota.

Murray, J.R. and Hackett, W.P. (1991) Dihydroflavonol reductase activity in relation to differential anthocyanin accumulation in juvenile and mature phase *Hedera helix* L. *Plant Physiology* 97, 343–351.

Murray, J.R., Smith, A.G. and Hackett, W.P. (1994) Differential dihydroflavonol reductase transcription and anthocyanin pigmentation in the juvenile and mature phases of ivy (*Hedera helix* L.) *Planta* 194, 102–109.

Nichols, W., Crow, W.D. and Paton, D.M. (1972) Chemistry and physiology of rooting inhibitors in adult tissue of *Eucalyptus grandis*. In: Carr, D.J. (ed.), *Plant Growth Substances 1970*. Springer-Verlag, Berlin, pp. 324–329.

Poethig, R.S. (1988a) Heterochronic mutations affecting shoot development in maize. *Genetics* 119, 959–973.

Poethig, R.S. (1988b) A non-cell-autonomous mutation regulating juvenility in maize. *Nature* 336, 82–83.

Poethig, R.S. (1990) Phase change and regulation of shoot morphogenesis in plants. *Science* 250, 923–930.

Polito, V.S. and Alliato V. (1981) Growth of calluses derived from shoot apical meristems of adult and juvenile ivy (*Hedera helix* L.). *Plant Science Letters* 22, 387–393.

Rogler, E. and Hackett, W.P. (1975) Phase change in *Hedera helix*: induction of the mature to juvenile phase change by gibberellin A_3. *Physiologia Plantarum* 34, 141–147.

Romberg, L.D. (1944) Some characteristics of the juvenile and the bearing pecan tree. *Proceedings of the American Society for Horticultural Science* 44, 255–259.

Snowball, A.M., Zeman, A.M., Tchan, Y.T., Mullins, M.G. and Goodwin, P.B. (1991) Phase change in citrus: immunologically detectable differences between juvenile and mature plants. *Australian Journal of Plant Physiology* 18, 385–396.

Steele, M.J., Coutts, M.P. and Yeoman, M.M. (1989) Developmental changes in Sitka spruce as indices of physiological age. I. Changes in needle morphology. *New Phytologist* 113, 367–375.

Stoutemyer, V.T. (1937) Regeneration in various types of apple wood. *Research Bulletin of the Iowa Agricultural Experiment Station* 220, 308–352.

Takemoto, Y. and Greenwood, M.S. (1993) Maturation in larch: age-related changes in xylem development in the long-shoot foliage and main stem. *Tree Physiology* 13, 253–262.

Wareing, P.F. and Frydman, V.M. (1976) General aspects of phase change with special reference to *Hedera helix* L. *Acta Horticulturae* 56, 57–68.

Wiltshire, R.J.E., Potts, B.M. and Reid, J.B. (1992) A paedomorphocline in *Eucalyptus*. II. Variation in seedling morphology in the *E. risdonii/E. tenuiramis* complex. *Australian Journal of Botany* 40, 789–805.

Woo, H-H., Hackett, W.P. and Das, A. (1994) Differential expression of a chlorophyll a/b binding protein gene and a proline rich protein gene in juvenile and mature phase English ivy (*Hedera helix*). *Physiologia Plantarum* 92, 69–78.

Transformation Protocols for Ornamental Plants

6

S.C. Deroles, M.R. Boase and I. Konczak
New Zealand Institute for Crop & Food Research Ltd, Levin Research Centre, Private Bag 4005, Levin, New Zealand

INTRODUCTION

The genetic transformation of ornamental crops will considerably enhance the existing efforts of traditional and *in vitro* breeders in the generation of new crops. These new molecular techniques provide opportunities to alter specific characteristics such as plant form, colour, and pest and disease resistance.

The wholesale production of ornamentals passing through the Dutch auction system totalled over 5 billion Dutch guilders in 1993 (Table 6.1). The major proportion of this production was cut flowers at 70%, followed by house plants (26%) and garden plants (4%) (Table 6.1). Rose, chrysanthemum and carnation are the top three cut flowers and account for 48% of the total cut flower production, with the next three, tulip, lily and freesia totalling 18% (Table 6.2). The market for ornamentals has steadily increased over the last three years (Table 6.3) and a continued period of growth is expected with the opening up of new markets in Asia, the Eastern Block countries and China.

Traditional breeding continues to be the principal source of new cultivars and varieties for the ornamentals market. Its success is evident in the huge array of products available to the consumer. Breeders have selected for many phenotypes such as flower colour and shape, foliage colour, insect and pest resistance, leaf form, plant stature, plant vigour and scent. In spite of its success, traditional breeding has several limitations, the most obvious being the limited gene pool contained within a species. For example, it is not possible to breed for a blue rose or chrysanthemum because the gene(s) required for this colour is not present in the gene pools of these plants.

The techniques of *in vitro* breeding, such as embryo rescue, *in vitro* pollination and protoplast fusion, were developed to widen the available gene pool. Non-stigmatic pollination and embryo rescue avoid incompatibility systems

 Biotechnology of Ornamental Plants (eds R.L. Geneve, J.E. Preece and S.A. Merkle)

Table 6.1. Production of ornamentals in 1993 (millions of Dutch guilders).

Crop	Production	%
Cut flowers	3693	70
Garden plants	225	4
House plants	1347	26
Total	5265	100

Source: Vereniging van Bloemenveilingen (1993).

Table 6.2. Production of the top ten cut flowers (millions of Dutch guilders).

Crop	Production
Rose	902
Chrysanthemum	581
Carnation	276
Tulip	265
Lily	237
Freesia	145
Gerbera	133
Cymbidium	102
Gypsophila	82
Alstroemeria	78

Source: Vereniging van Bloemenveilingen (1993).

that cause premature abortion (Leduc *et al.*, 1992; Ozais-Akins *et al.*, 1992; Janson, 1993; Mont *et al.*, 1993; Kishi *et al.*, 1994). Protoplast fusion has been used to generate new hybrids from sexually incompatible species (Power *et al.*, 1986; Christey *et al.*, 1991). Using techniques such as these, breeders have generated several new ornamental cultivars, for example in *Gypsophila* and *Lilium*. An additional limitation of both traditional and *in vitro* breeding is the need for successive backcrosses to purify the desired trait away from undesirable traits in the original offspring. This process can take many years and often the final product is a compromise between desirable and undesirable characteristics.

Genetic transformation overcomes both the limited gene pool and the necessity for successive backcrossing. Since the DNA used in the transformation event can be isolated from any living organism, gene pool barriers are eliminated. Genetic transformation involves the transfer into the host plant of at most a few genes. The result is a progeny that is genetically identical to its parent with the addition of one or two additional characters. As a result

Table 6.3. Sale of ornamentals (millions of units).

Crop	1991	1992	1993
Cut flowers	9217	9619	9721
Garden plants	257	286	304
House plants	565	589	606

Source: Vereniging van Bloemenveilingen (1993).

the desirable horticultural properties of the host plant are preserved and no backcrossing is necessary to purify the desired phenotype.

The genetic transformation of a plant relies on the ability to transfer genetic material to a single plant cell followed by stable incorporation of the gene(s) into the plant genome and finally regeneration of an intact plant from that cell. The critical points in this process are the ability to deliver the DNA to a single cell and regeneration of a plant from that cell. *Agrobacterium*-mediated gene transfer remains the most popular technique for the generation of new ornamentals. Its major drawback, however, is its inability to deliver DNA to most monocotyledonous plants. Most of the other techniques covered in this review aim to provide a gene transfer system as simple and reliable as *Agrobacterium* but without the host range limitation.

There have been several recent reviews on the transformation of ornamentals (Dons *et al.*, 1991; Hutchinson *et al.*, 1992; Courtney-Gutterson, 1993; Robinson and Firoozabady, 1993). This review is structured differently in that it is written from the point of view of techniques. Table 6.4 lists the ornamental species that have been transformed to date and the methods used to transform them. Most of the ornamental crops that have been transformed are the cut flowers. This reflects the dominance that this sector has within the total ornamentals market. It is clear that *Agrobacterium*-mediated gene transfer is the preferred transformation technique in use for ornamental plants. The alternative technique is direct DNA uptake, which encompasses several different methods including electroporation and particle bombardment. In this review we will separate the technique of direct DNA uptake into its individual methods. We will review the following techniques: *Agrobacterium*-mediated gene transfer, chromosome-mediated gene transfer, electrophoresis, electroporation, imbibition, laser cell perforation, liposome-mediated gene transfer, microinjection, microprojectile-mediated gene transfer, polyethylene glycol(PEG)-induced DNA uptake, silicon carbide technique, and ultrasonication. Not all of the above techniques have been used on ornamentals, but all have potential where *Agrobacterium*-mediated transformation is not possible. The latest procedures and results will be discussed within each transformation technique.

Table 6.4. Ornamental species that have produced transgenic tissue.

Species	Common name	Method	Reference
Anthurium andraeanum	Anthurium	At	Kuehnle *et al.* (1993)
Antirrhinum majus	Snapdragon	At	Klemm *et al.* (1991)
		At	Godwin *et al.* (1991)
		Ar	Handa (1992)
		At	Diqui *et al.* (1994)
Cyclamen persicum	Cyclamen	At	Deroles *et al.* this chapter
Dendranthema grandiflora	Chrysanthemum	At	Lemieux *et al.* (1990)
		At	van Wordragen *et al.* (1991)
		At	Courtney-Guttersen (1993)
		At	Robinson and Firoozabady (1993)
		At	Renou *et al.* (1993)
		At	Dolgov *et al.* (1994)
		At	Pavingerová *et al.* (1994)
		At	Urban *et al.* (1992)
		At	Deroles *et al.* this chapter
Dendranthema indicum	Chrysanthemum	At	Ledger *et al.* (1991)
Dendrobium hybrid	Orchid	Bio	Kuehnle and Sugii (1992)
Dianthus caryophyllus	Carnation	At	Firoozabady *et al.* (1991)
		At	Lu *et al.* (1991)
		At	van Altvorst *et al.* (1995)
Eustoma grandiflorum	Lisianthus	At	Deroles *et al.* (1993)
		At	Semeria *et al.* (1994)
		Bio	Morikawa *et al.* (1994)
Forthysia intermedia	Forthysia	At	Duron (1994)
Gerbera hybrida	Gerbera	At	Elomaa *et al.* (1993)
		At	Robinson and Firoozabady (1993)
Gladiolus	Gladiolus	At	Graves and Goldman (1987)
Gypsophila paniculata	Gypsophila	At	Robinson and Firoozabady (1993)
Kalanchoe laciniata	Kalanchoe	At	Jia *et al.* (1989)
Lilium longiflorum	Lily	At	Cohen and Meridith (1992)
Pelargonium × *domesticum*	Pel (regal)	At	Deroles *et al.* this chapter
Pelargonium × *hortorum*	Pel (zonal)	At	Deroles *et al.* this chapter
Pelargonium sp.	Pelargonium	Ar	Pellegrineschi *et al.* (1994)
Petunia hybrida	Petunia	At	Fraley *et al.* (1985)
		Ar	Ondrej and Biskova (1986)
		C	Kruger-Lebus and Potrykus (1987)
Rosa hybrida	Rose	At/Ar	Firoozabady *et al.* (1994)
R. persica × *xanthina*	Rose	At	Robinson and Firoozabady (1993)
Salpiglossis sinuata		Ar/PF	Rietveld *et al.* (1994)
Tulipa gesneriana	Tulip	Bio	Wilmink *et al.* (1994)

At, *Agrobacterium tumefaciens*; Ar, *Agrobacterium rhizogenes*; Bio, biolistics; C, Chemical-mediated DNA transfer; PF, protoplast fusion.

AGROBACTERIUM-MEDIATED GENE TRANSFER

Agrobacterium tumefaciens and *A. rhizogenes* are the causative agents for crown gall and hairy root diseases. Genetic engineering via *Agrobacterium*-mediated transformation uses the natural gene transfer system present in these bacteria to generate transgenic plants.

The molecular process of *Agrobacterium*-mediated gene transfer has been extensively reviewed (Zambryski, 1988, 1992; Gelvin, 1990; Kado, 1991; Nester and Verma, 1991; Hookyaas and Schilperoort, 1992; Hookyaas and Mozo, 1994; Zupan and Zambryski, 1995). Most of the gene transfer functions of *Agrobacterium* reside on the Ti or Ri plasmid and are divided into two sections the T-region and the virulence region. The T-DNA region is bounded by a left and right border consisting of 25 bp direct repeats and is the segment of DNA integrated into the plant genome (Zambryski *et al.*, 1982; Hookyaas and Schilperoort, 1992). The second region on the Ti/Ri plasmid is called the Virulence (*vir*) region and contains genes that operate on the T-DNA region to control the production of the T-strand and its transport into the plant cell nucleus.

The principal limitation with *Agrobacterium*-mediated transformation is the host range of the bacteria. Many of the differences in plant host range have been attributed to differences in the *vir* genes, particularly *vir* A and G, which are responsible for the induction of the remaining *vir* loci (Leroux *et al.*, 1987; Gelvin and Liu, 1994). Enhanced expression of the *vir* G gene, for example in the Ti plasmid pTiBo542, has resulted in an extended plant host range (Jin *et al.*, 1987). This result has been further improved with the introduction of additional copies of the *vir* G gene and the development of a constitutively expressed *vir* G gene (Hansen *et al.*, 1994).

Researchers have developed two classes of *Agrobacterium* DNA vectors for plant transformation, co-integrate and binary (reviewed in Walkerpeach and Velten, 1994). Both systems result in the deletion of the genes responsible for gall formation thus allowing intact plants to develop. In the co-integrate vector, the wild-type T-DNA genes are replaced with the genes to be inserted by a single recombination step within the *Agrobacterium*. The co-integrate vector carries the genes of interest (including a bacterial selective marker) between two sequences homologous to regions just inside the wild-type T-DNA borders. After transfer to the host *Agrobacterium*, the recombination event is selected for via the transferred resistance marker, since the incoming vector is not able to replicate on its own in *Agrobacterium*. The binary vector was developed to bypass the recombination step (Hoekema *et al.*, 1983) and has a broad host range replicon (e.g. RK2) thus allowing it to be maintained in *Agrobacterium*. No recombination event is necessary and the binary vector is not limited to the copy number of the host Ti plasmid. The binary vector has its own T-DNA region bounded by the left and right border sequences. Within this region is a multiple cloning site, and a plant-selective marker gene (e.g. kanamycin resistance). Numerous binary vectors have been constructed, many with specialized functions such as promoter analysis and plant gene/promoter tagging (Walkerpeach and Velten, 1994).

There are four factors essential to a successful *Agrobacterium*-mediated transformation system:

1. A high frequency of regeneration from cells in the target area (e.g. the cut edge of an explant).
2. The *Agrobacterium* should be able to transform the majority of cells at the target site.
3. The method of selection (e.g. kanamycin resistance) must prevent the growth of non-transformed cells while allowing the recovery of the majority of transformed cells.
4. The transformed plants must have an intact copy of the gene(s) with stable expression and be able to transmit the gene(s) to their progeny.

The first step in establishing an *Agrobacterium*-mediated transformation system is to develop an efficient plant regeneration system that involves a minimum of callus growth to reduce somaclonal variation. In designing efficient *Agrobacterium* transformation methods, attention must also be paid to the plant- and bacteria-derived environmental factors that influence induction of the infection process. For example, compounds produced by the wounded plant cell, such as phenolic compounds (e.g. acetosyringone) and sugars (Bolton *et al.*, 1986; Ankenbauer and Nester, 1990; Godwin *et al.*, 1991), and compounds produced by *Agrobacterium* such as α-hydroxy-acetosyringone and the opines (Veluthambi *et al.*, 1989), all influence the transformation process. Osmoprotectants have also been implicated, for example glycine betaine (Vernade *et al.*, 1988) and proline (James *et al.*, 1993), as well as factors such as pH and temperature (Alt-Mörbe *et al.*, 1989; Godwin *et al.*, 1991).

The majority of gene transfer systems used in ornamental crops rely on *Agrobacterium*-mediated transformation. Recent advances in extending the host range now make *Agrobacterium*-mediated transformation a potential technique for monocotyledonous varieties. Bidney *et al.* (1992) found that using the particle gun as a wounding device prior to *Agrobacterium* infection enhanced the transformation frequency. Improvements are still being made in the virulence and host range of *Agrobacterium* strains (Wang *et al.*, 1990; Zyprian and Kado, 1990; Liu *et al.*, 1992; Hood *et al.*, 1993; Hansen *et al.*, 1994; Hiei *et al.*, 1994). In developing a new gene transfer system *Agrobacterium* is still the method of choice and should be tested before other systems.

PLANTS TRANSFORMED VIA *AGROBACTERIUM*-MEDIATED GENE TRANSFER

Anthurium andraeanum (anthurium)

There are two reports on the transformation of *Anthurium*. The first report demonstrated that *Agrobacterium* was able to stimulate tumour formation on *Anthurium* stems (Kuehnle and Sugii, 1991). This work was extended in the second report where the disarmed *Agrobacterium* strain LBA4404 was used to generate transgenic plants (Kuehnle *et al.*, 1993). The *Anthurium* cultivar Rudolph was transformed by co-cultivating etiolated internodes with LBA4404 carrying one of three binary vectors based on pBI121. Kanamycin was used as the selective agent. Several hundred transgenic plants were produced using this technique.

Antirrhinum majus (snapdragon)

There is much published work on the study of flower colour and form in *Antirrhinum* spp. (snapdragons) (Martin *et al.*, 1987, 1991; Jordan and Anthony, 1993; Martin and Gerats, 1993). The development of a reliable, high frequency transformation system would provide a valuable tool to elucidate these mechanisms further. To our knowledge no such system for snapdragon has been developed. However, there are several reports on the generation of transgenic snapdragon plants.

The generation of transgenic snapdragon plants was first reported by Klemm *et al.* (1991), who incubated etiolated hypocotyl segments with the *Agrobacterium* strain GV3101, a C58 type strain. Phosphonothricin was used as the selective agent. The addition of 100 μM acetosyringone (at pH 5.7) to the co-cultivation medium increased the transformation frequency 10- to 20-fold.

Holford *et al.* (1992) studied the effect of cultivar, *Agrobacterium* strain, elicitor compound and co-cultivation pH on the transformation of snapdragon. Although no transgenic plants were produced, this report shows clearly the large number of variables that need to be studied when establishing a new transformation system. There are also large differences in the efficiency of transformation between cultivars of snapdragon. This is similar to the situation with chrysanthemum.

Transgenic snapdragons have also been generated using an *Agrobacterium rhizogenes* strain A4 carrying the plasmid pRiA4 (Handa, 1992). Hairy roots were generated by inoculating the stems of three-week-old seedlings of *A. majus* 'Floral Carpet'. Tips (1 cm long) were excised from these adventitious hairy roots and cultured axenically. Adventitious buds that formed on the hairy root cultures were excised, acclimatized and grown to mature flowering plants. Three plants were recovered, all of which showed altered phenotypes because of the expression of the *rol* genes.

The most recent report describes the co-cultivation of hypocotyl explants with the *Agrobacterium* strain EHA101 harbouring the binary $pBYT_s$ (containing GUS and NPT II gene). Intact kanamycin-resistant, GUS positive plants were recovered (Diqui *et al.*, 1994).

Cyclamen persicum (cyclamen)

There have been several reports on the regeneration of plants from *Cyclamen persicum* Mill. (Geier, 1977; Wainwright and Harwood, 1985; Hoffmann and Preil, 1987). *Ex vitro* material is not suitable as an explant source due to high levels of contamination. Cotyledons and tuber tissue of cv. Rosamunde have shown the highest regeneration potential (Wainwright, and Harwood, 1985). We have found that petiole explants gave a higher regeneration frequency than leaf blade explants in the cultivar Sierra Rose (unpublished results).

We have inoculated leaf blades and petioles of *in vitro* seedlings with four *Agrobacterium* strains (A281, C58, A722 and LBA4404) carrying the binary vector pKIWI110 (Boase and Borst, 1995). By 20 days after inoculation, 90% of the wound sites inoculated with A281 showed tumour growth and positive GUS expression. At 48 days, some tumour growth was also present on sites inoculated with C58. To our knowledge this is the first occurrence of gene transfer into cyclamen. *Ex vitro* material infected with the same *Agrobacterium* strains showed no tumour growth or GUS-positive results.

Hypocotols and cotyledons from *in vitro* etiolated seedlings were then tested as an explant source (Boase and Borst, 1995). Three *Agrobacterium* strains were used, A218, C58 and LBA4404. The explants were co-cultivated for two or four days. When assayed for GUS expression after four days the best results were obtained from LBA4404 co-cultivated for two days (70–90% GUS positive). We are now combining our transformation and regeneration protocols to produce intact transgenic cyclamen plants. Our aim is to antisense the flavonoid 3′-hydroxylase gene to produce pelargonidin based flower colours such as orange and brick-red in cyclamen.

Dendranthema sp. (chrysanthemum)

Chrysanthemum is susceptible to infection by wild-type *Agrobacterium tumefaciens* (Miller, 1975; Miller *et al.*, 1975) but the efficiency of gene transfer is very cultivar specific (Bush and Pueppke, 1991; van Wordragen *et al.*, 1991). In addition, the conditions required for successful regeneration of plants via direct organogenesis or somatic embryogenesis are also specific to individual cultivars and vary depending on explant type and culture medium (Rademaker and de Jong, 1990; May and Trigiano, 1991; de Jong *et al.*, 1993; Lowe *et al.*, 1993; Pavingerová *et al.*, 1994).

Our group is developing transformation and regeneration systems for five chrysanthemum cultivars. We are using the β-glucuronidase (GUS) gene in the (pKIWI110) as a scoreable marker, to find which factors are critical to the gene transfer and cell selection process in the first 30 days, and the NPT II gene (kanamycin resistance) as a selectable marker. Factors being studied include bacterial-strain/plant-cultivar interactions, co-cultivation periods, length of preculture, and presence of silver thiosulphate and silver nitrate in the media. For regeneration and selection, we are examining the explant position on media, explant size and type, selectable markers and bacteriostatic antibiotics, and plant growth regulator concentrations and combinations in the medium by factorials.

We have made improvements to the published protocol for the Korean cultivar K4 (*Dendranthema zawadskii* (Herbich) Tzvelev × *D. grandiflora* (Ramat.) Kitam) (Ledger *et al.*, 1991; Soreng and Cope, 1991) by culturing explants in the presence of silver nitrate for up to 22 days after co-cultivation. The transformation efficiency has been increased from 1 to 13%, and preliminary data suggest further improvements can be gained by using *ex vitro* rather than *in vitro* material as an explant source and co-cultivating for four rather than two days (unpublished results).

The other four cultivars under study ('Yellow Lucondra', 'Peach Margaret', 'Horizon' and 'Nightingale') are all classified as *Dendranthema grandiflora* Tzvelev. (Anderson, 1987). We are conducting a similar range of GUS and regeneration experiments to increase the existing transformation efficiencies (1–3%) that we have already achieved with these cultivars. Our recent results indicate significant cultivar specific responses to *Agrobacterium* strain, effect of silver thiosulphate, leaf explant orientation, size and source (*ex vitro* cf. *in vitro*), and sodium hypochlorite concentration in the disinfestation protocol. In our most recent experiments 70 to 100% of leaf explants from the cultivars Horizon and Yellow Lucondra showed positive GUS loci 16 days post-inoculation, indicating integrative transfer of the GUS gene.

To date we have generated transgenic plants from the following cultivars: 'K4' (74 independent transgenics), 'Yellow Lucondra' (5), 'Peach Margaret' (16) and 'Horizon' (15) (unpublished results). We are presently transforming 'K4' with the Lc regulatory gene from maize. In addition we plan to inhibit the activity of the flavonoid 3′-hydroxylase gene in chrysanthemum via antisense to produce pelargonidin-based reds (Winefield *et al.*, 1994).

Other groups have transformed various cultivars of *Dendranthema grandiflora* Tzvelev. Four cultivars (Indianapolis White Giant #4, Indianapolis Dark Bronze #2, Lipstick and Puritan) have been transformed by inoculating peduncle explants with *A. tumefaciens* (Lemieux *et al.*, 1990).

An efficient transformation system has been developed for the cultivar 'Moneymaker' using leaf pieces as the explant. This protocol has been used to generate a white variant of 'Moneymaker' via co-suppression of the chalcone synthase gene (Lemieux *et al.*, 1990; Courtney-Gutterson, 1993). This transformation system has been tested across 15 cultivars with transformation efficiencies ranging from 1 to 10%, and over 800 transgenic plants have been produced to date (Robinson and Firoozabady, 1993; Lu, personal communication).

The cultivar White Hurricane has been transformed via co-cultivation of leaf discs with *Agrobacterium* strain A281 harbouring the plasmids pTiBo542 and pBTP-41 (*bt* gene). Kanamycin was used as the selective agent at 50 mg ml^{-1} to select for shoots and at 10 mg ml^{-1} for rooting of microshoots (Dolgov *et al.*, 1994).

Transgenic plants have been produced from the cultivar White Snowdon (Pavingerová *et al.*, 1994). Leaf segments were co-cultivated with the wild type *Agrobacterium* strain B6S3, and also with its disarmed equivalent carrying the helper plasmid pAL4404 and the binary vector p35SGUSint (GUS and NPT II). Plants recovered from experiments using the wild-type strain had significant morphological alterations, probably because of the presence of the phytohormone genes on the wild-type T-DNA.

The cultivar Pictor has been transformed using the disarmed binary

EHA101/pKHG3 (nos-NPT II, 35S-hygromycin^r) (Renou *et al.*, 1993). A variety of explants and transformation methods were tested. Leaf pieces, stem internodes and nodes all produced transgenic plants. In addition the complete elimination of the co-cultivation step produced a better transformation frequency when compared to co-cultivation for 24 hours. This was attributed to the sensitivity of the plant material to the inoculating *Agrobacterium* EHA101.

The cultivar 'Iridon' has also been transformed by infecting leaf explants using *Agrobacterium* EHA105/pBI121 (Urban *et al.*, 1992). Transgenic plants were recovered that expressed GUS activity and contained the nucleocapsid gene of tomato spotted wilt virus.

Transformed calluses have been generated from the cultivar 'Parliament' using strain A281/pTVK291 (harbouring the pTiBO542 supervirulence region) and A281/pCPO1-G (nos-NPT II, TR-2′-GUS) (van Wordragen *et al.*, 1991). Plants were not regenerated from these calluses.

De Jong and co-workers have published a useful report highlighting the effect of cultivar, ability to produce adventitious shoots, antibiotic selection procedure and type of binary vector on the generation of transformed chrysanthemums (de Jong *et al.*, 1994). Transgenic shoots were regenerated from the cultivars 8382 and 89100 using the binary vector pMOG410 in the *Agrobacterium* strain AGLO.

There is much interest in the development of gene transfer systems for chrysanthemum using *Agrobacterium*. From the results published to date, it is clear that different cultivars require individual approaches to their transformation. Significant differences have been reported in variables such as *Agrobacterium* strain, co-cultivation period, explant type, level of selection, preculture period, type of regeneration system and selectable marker. Analysis of the resultant transgenics by ourselves and Florigene (J. Mason, personal communication) indicates that there is a high frequency of shutdown of transgene expression, making it difficult to select for stable transformants.

Dianthus caryophyllus (carnation)

The first report of transformation in carnation via *Agrobacterium* was in 1989 (Woodson and Goldsbrough, 1989) when a single transgenic plant was generated. Since then two independent techniques have been developed for the efficient production of transgenic carnation plants. Firoozabady *et al.* (1991) have transformed the cultivars Improved White Sim Manon and Nathalie. They tested various explants such as, filaments, flower buds, leaf and petal bases, receptacles and styles. Transformation was most successful using leaf bases. The most suitable *Agrobacterium* strain was EHA101 carrying the binary vectors pSLJ1911 (p35SNPTII/p35SGUS) or pWTT2084(p35SALS/p35SGUS). Both chlorsulphuron and geneticin were successful as selection agents. The frequency of transformation was 3 to 10%. A number of the transformants were chimeric, indicating a multicellular origin of plant regeneration.

Lu *et al.* (1991) have reported the transformation of the cultivars Crowley, Red Sim and White Sim. This technique uses stem sections inoculated with the *Agrobacterium* wild type strain ICMP802 and the disarmed strain AGLO. Kanamycin was used as the selective agent and the transformation efficiency was approximately 3%. Using this technique, transgenic carnations have been produced that carry either an antisense copy of the EFE (ethylene forming enzyme) gene or the antisense ACC (1-aminocyclopropane-1-carboxylic acid) synthase gene. Some of these plants have a reduced senescence response leading to increased vase life (Michael *et al.*, 1993). This system has now been extended to 16 cultivars and over 2000 transgenic plants have been produced (Lu, personal communication).

Two other laboratories have also reported the successful transformation of carnation. Van Altvorst *et al.* (1995) co-cultivated leaf explants with the supervirulent *A. tumefaciens* strain AGLO. Kanamycin was used to select for transformation events. Thirty cultivars were examined and transgenic plants were obtained from five (CPRO clone numbers 89073, 89074, 89077, 89086 and 89172). Messeguer and Mele (1994) reported the development of a similar system. They also reported that a two-day cold (4°C) pretreatment enhanced explant growth, and the addition of 20 mM acetosyringone increased transformation frequency.

Vainstein *et al.* (1993) have described a technique for enhancing GUS expression in transgenic carnation cells. They reported that a phenolic compound specific to carnation inhibited the GUS enzyme activity. The inhibition was overcome by the addition of polyvinylpolypyrolidone (PVPP) as recommended by Jeffersen *et al.* (1987) for plant tissues containing high levels of polyphenols.

***Eustoma grandiflorum* (lisianthus)**

A reliable transformation system for lisianthus based on the leaf disc transformation protocol of Horsch *et al.* (1985) has been developed (Deroles *et al.*, 1993). Inoculation with the wild-type *Agrobacterium* A722 resulted in higher transformation efficiencies than LBA4404, C58 and A281. Transgenic plants were produced from two cultivars of lisianthus, #42 (white flowers) and #54 (purple flowers) using three binary vectors, pKIWI110, pGA643 and pLN6. The smaller vectors pGA643 and pLN6 gave significantly higher transformation frequencies than pKIWI110. In spite of the formation of tumorigenic tissue on the leaf edge, it has been possible to isolate consistent numbers of transgenic shoots that are morphologically identical to the parent plants. Southern and northern analysis of the transgenic plants has confirmed the presence and expression of the NPTII kanamycin resistance gene. Kanamycin at 70 mg ml^{-1} was used as the selective agent and ticarcillin was used to halt *Agrobacterium* growth. We found that the plant growth regulator effects of cefotaxime reduced the number of shoots produced per explant. The transformation frequency from these experiments was low at 1 to 5% but consistent and transgenic plants are routinely recovered. This technique is now being used to insert a range of flavonoid biosynthetic and regulatory genes into lisianthus to modify flower colour. Modifications to the established technique involving changes in explant source, preculture conditions and regeneration media, have boosted the transformation frequency to 15–20% across eight cultivars (unpublished results). Further experiments are in progress to confirm the cultivar range and critical parameters of transformation.

Transformation of *Eustoma* has also been attempted by Semeria *et al.* (1994). Using the strain EHA105 carrying pKIWI105 they have isolated several independent transgenic calluses. To our knowledge no transgenic plants have been produced.

***Forsythia* × *intermedia* (forsythia)**

Only a single transgenic plant has been regenerated from *Forsythia* × *intermedia* 'Spring Glory' (Duron, 1994). Internodes (0.5 cm long) from *in vitro* shoot cultures were co-cultivated with the disarmed strain EHA101 carrying the binary vector PFAJ3000. Kanamycin was used as the selection agent.

***Gerbera hybrida* (gerbera)**

Elomaa *et al.* (1993) have produced transgenic gerbera plants carrying the antisense chalcone synthase gene using the red-flowered cultivar 'Terra Regina'. In their system, petioles performed better than lamina pieces. The petioles were precultured for six to ten days prior to infection with *Agrobacterium* strain C58 carrying the disarmed Ti plasmid pGV2260. Explants were co-cultivated for two days, and then transferred to solid regeneration medium. Kanamycin was used as the selective marker at 25 mg ml^{-1}. Cefotaxime was used to control bacterial overgrowth, but it also significantly reduced the regeneration frequency. From 4800 explants, only four transgenic plants were recovered. Therefore this technique cannot be regarded as a reliable transformation system for *Gerbera*. Transgenic plants have also been generated using *Agrobacterium* LBA4404 carrying a binary vector with the genes for GUS and kanamycin resistance (Robinson and Firoozabady, 1993). No details are available for this technique.

***Gladiolus* (gladiolus)**

No transgenic gladiolus plants have been produced, but Graves and Goldman (1987) have shown that *Agrobacterium* can infect this plant. They cut discs from surface disinfested corms and inoculated them with either *Agrobacterium* strain B6 (pTiB6), C58 (pTiC58) or LBA4013 (pTiAch5). Tumour tissue formed from all three strains, and opine synthesis was detected in all cases. This indicates that *Agrobacterium*-mediated transformation is a good candidate for an efficient gene transfer system for gladiolus.

***Gypsophila paniculata* (gypsophila)**

This plant is susceptible to *Agrobacterium* (Clark *et al.*, 1989) and some transgenic plants have been generated that show stable expression of GUS and *nos*

(Robinson and Firoozabady, 1993). To our knowledge no repeatable transformation system has been published.

Kalanchoe laciniata (kalanchoe)

An efficient transformation system for kalanchoe using *Agrobacterium* has been published (Jia *et al.*, 1989). Three disarmed *Agrobacterium* strains were tested (A208SE, GV3111SE and EHA101), all containing a binary vector pROA93 (35SNPTII and 35SGUS). Leaf pieces were used as explants and kanamycin as the selective agent. Inoculation with any of the *Agrobacterium* strains resulted in transgenic plants with strain A208SE being the most efficient. Jia *et al.* (1989) also observed that kalanchoe transformation was significantly enhanced by a prolonged co-cultivation time of six to seven days.

Lathyrus odorata (sweet pea)

No transgenic plants have been reported for this species using *Agrobacterium*. However research is in progress using seedling tissue as the explant source (Fraser *et al.*, 1991).

Lilium longiflorum (lily)

There have been numerous reports on the tissue culture of lily from various explants, but to our knowledge no transgenic lily plants have been produced. There are two reports of transgenic tumorous lily tissue produced as a result of *Agrobacterium* infection. Cohen and Meridith (1992) inoculated bulb scale segments from *Lilium longiflorum* 'Nellie White' with *Agrobacterium* C58. The resulting tumour tissue was able to synthesize nopaline indicating its transformed state. Langveld *et al.* (1994) used both *Agrobacterium* C58 and the disarmed strain C58CI/pMP90. Both strains carried a binary vector with the intron-containing GUS gene. Inoculation of stems from the Asiatic hybrid cultivar 'Harmony' produced clusters of bright blue cells at the inoculation site after histochemical staining. These results show that *Agrobacterium* is capable of transforming lily, and further work should produce the appropriate combination of *Agrobacterium* strain, cultivar and regeneration protocol for the generation of transgenic lilies.

Pelargonium (pelargonium, geranium)

Pelargoniums are susceptible to a range of *Agrobacterium tumefaciens* strains (Boase and Smith, 1994). We have GUS positive results for eight regal cultivars (*Pelargonium* × *domesticum*) and four zonal cultivars (*Pelargonium* × *hortorum*) (Boase and Smith, 1994). The cultivars tested were the regals 'Alaska', 'Dubonnet', 'Hazel Cherry', 'Her Majesty', 'NC Faiss' (syn. 'NC Eatse'), 'Mrs James Rodgers', 'White Chiffon' and 'White Dawn'; and the zonals from F_1 hybrid seed: Orange Appeal, 'White Elite', 'White Multibloom' and 'White

Orbit'. Explants from cotyledons and primary leaves were inoculated with the *A. tumefaciens* strains A281, C58, A722 and LBA4404, each containing either the binary vector pKIWI110 (Janssen and Gardner, 1989), or pLN3 (pKIWI110 minus the acetolactate synthase gene). After two days of co-cultivation, explants were transferred to selection media containing 300 mg l^{-1} kanamycin sulphate. GUS expression was observed on explants at 4 to 6 days, 9 to 14 days, 35, 94 and 110 days after inoculation.

Kanamycin-resistant calluses were produced from cotyledons and primary leaves of *in vitro* seedlings of 'Orange Appeal', inoculated with pKIWI110/LBA4404 or pKIWI110/A722 (Boase, personal communication). These calluses were GUS positive at 35, 94 and 110 days post-inoculation. Leaf discs of 'NC Faiss' transformed by the supervirulent *Agrobacterium* strain A281, containing pKIWI110, produced kanamycin-resistant calluses that were GUS positive at 47 days post-inoculation.

We have developed leaf disc regeneration systems for the regal cultivars Dubonnet, Hazel Cherry and White Chiffon, and the zonal cultivars, Orange Appeal and White Multibloom (unpublished results). We have also developed a transformation and regeneration system for cv. Dubonnet. In four experiments with plasmids pGA643, pKIWI110 and pLN13, in the *Agrobacterium* strain LBA4404, a transformation efficiency of 67% was demonstrated. Analysis of transgene expression in 265 plants containing pKIWI110 has indicated a significant loss of gene expression. This could be due to the presence of rearranged, deleted or partially transferred T-DNA inserts or the loss of expression of intact transgenes.

Pellegrineschi *et al.* (1994) have transformed the scented pelargonium species referred to as 'lemon geranium' by inoculating stem and petiole fragments with a mixture of three agropine-type *Agrobacterium rhizogenes* strains, A4RSII, 15834 and LBA9402. Shoots with an altered phenotype regenerated spontaneously without exogenous plant growth regulators, from transformed roots. The altered phenotypes possessed improved ornamental characters and increased production of essential oils.

Rosa sp. (rose)

Regeneration and transformation techniques have only recently been developed for rose. For reviews on *in vitro* culture and regeneration of rose see Short and Roberts (1991) and Robinson and Firoozabady (1993). Most of the published regeneration systems are cultivar dependent and the frequency of regeneration is often low. As a result they are not suitable for use in a transformation protocol. To our knowledge no one has yet developed an efficient, cultivar-independent transformation system for rose. However, there are several reports of the transformation of individual rose cultivars. The somatic embryogenic callus line isolated by Noriega and Sondhal (1991) from the hybrid tea rose cultivar 'Royalty' has been transformed by Firoozabady *et al.* (1994). They co-cultivated the callus with *A. tumefaciens* LBA4404/pJJ3931 (pnosNPTII/p35SLUC) or *A. rhizogenes* 15834/pJJ3499 (pnosNPTII/p35SGUS).

Selection for transformed calluses was performed using 300 mg ml^{-1} kanamycin. Transformation efficiencies were high with between 40 and 60 kanamycin resistant calluses per gram of inoculated rose callus. Transgenic plants regenerated from the independent transformed calluses were shown by PCR to contain the transgenes. All plants produced were reported to be morphologically normal. This is fortunate, since there is an extended callus phase in the protocol, increasing the risk of somaclonal variation. This method is currently being used to introduce genes of interest into the rose cultivar Royalty.

R. persica × *xanthina* has also been transformed via co-cultivation of protoplasts with *A. tumefaciens* LBA4404 using hygromycin as the selective agent. Transformed callus that stably expresses the GUS gene has been produced (Robinson and Firoozabady, 1993).

Chin Yi Lu and co-workers have produced over 100 transgenic roses from six different cultivars (Lu, personal communication). This represents the most cultivar-independent rose transformation system to date.

Salpiglossis sinuata

Salpiglossis sinuata is a member of the *Solanaceae* that may have potential as a bedding plant because of its many coloured and campanulated flowers. Its principal drawback is its elongated and unbranched morphology. Rietveld *et al.* (1994) have developed an *Agrobacterium*-mediated transformation system for this plant, and have used it to generate transgenic plants carrying the *rol*C gene in an attempt to generate dwarf forms more suitable for the bedding plant industry.

Tulipa gesneriana (tulip)

Tulip is the top rated flower bulb on the international cut flower market. Breeding of the crop is significantly hampered by its long generation cycle. This is mainly because of a five-year juvenile period. As a result, the development of a transformation system for tulip is highly desirable. Two techniques have been used to transform tulip, *Agrobacterium* and particle bombardment (Wilmink *et al.*, 1992). The latter technique is discussed in the section on Microprojectile-mediated gene transfer. The following *Agrobacterium* strains have been used to transform tulip: *A. rhizogenes* LBA4902, *A. tumefaciens* A281, Ach5, C58C1, EHA101 and LBA4404. All the strains carried the 35S GUS-intron gene. Seven cultivars and seven *Tulipa* species were screened with these strains. Floral stem segments were co-cultivated with *Agrobacterium* in the presence of 1 mM acetosyringone. Strain LBA4902 showed the best results with the greatest number of blue staining cells produced at the inoculation site. However, the results do show that tulip is susceptible to *Agrobacterium*-mediated transformation. Results from similar experiments in both chrysanthemum and carnation show that successful transformation using *Agrobacterium* is dependent on a combination of *Agrobacterium* strain, cultivar, elicitors and co-cultivation pH, explant type, and regeneration system. The results in this report show that efficient *Agrobacterium*-mediated transformation of *Tulipa* is probable.

CHROMOSOME-MEDIATED GENE TRANSFER

Most of the transformation techniques described in this review are centred around the transfer of one or a few genes to the recipient plant. At the other end of the scale are whole genomic transformation techniques such as somatic hybridization (reviewed by Vasil *et al.*, 1979; Sybenga, 1983). Often a desirable phenotype is controlled by many genes, but somatic hybridization transfers too many genes and genetic transformation cannot transfer enough. Chromosome-mediated gene transfer bridges the gap between these two technologies and offers the possibility of transferring whole operons in a controlled fashion. This technology was originally developed for animal cells, but transformation of plants cells has also been reported (Griesbach, 1987, 1991). Szabados *et al.* (1987) have developed a PEG (polyethylene glycol)-based chemical procedure to transfer isolated chromosomes into plant protoplasts. Plant chromosomes have also been inserted into plants cells via microinjection (DeLaat and Blass, 1987; Griesbach, 1991).

ELECTROPHORESIS

Electrophoretic transformation has a significant advantage over *Agrobacterium* and other direct DNA transformation techniques in that it can be performed directly on intact meristematic tissue. Thus it requires little or no tissue culture procedure to obtain intact transgenic plants. In addition there are no host range limitations. Electrophoretic transfer of DNA into plants cells was first reported by Ahokas (1989). The negative electrode was placed in a pipette containing DNA coding for the GUS gene in agarose. The pipette was inserted into the top of a wheat seed to a point near the embryo's apical meristem. The positive electrode was placed at the base of the seed. After electrophoresis for one hour, DNA could be detected throughout the meristem. However no GUS expression was observed in the germinated plants.

Griesbach (1993) has improved this system for use in the generation of transformed orchid plants. Three-week-old protocorms (2 mm tall) were used as explants. At this stage protocorms have large single apical meristems that later give rise to shoot tips. pBI121 DNA (35SGUS) was dissolved in agarose made up in a pH buffer and poured into a 1 ml pipette tip to make the negative electrode. The positive electrode was a second 1 ml tip filled with agarose only. The negative electrode was placed over the meristem and the positive one at the protocorm base. Griesbach found that tris-borate-EDTA (TBE) buffer was the most suitable for the electrodes and that an electrophoresis time of 10 minutes at 0.5 mA and 10 volts gave the highest transformation frequency. Using this treatment, 55% of the protocorms survived the treatment and 57% of the resulting plants expressed the GUS marker gene in intact plant tissue. There were no chimeras reported, which indicates that all of the cells that gave rise to the intact plant were transformed at the same time.

The successful transfer and integration of DNA across a plant cell wall seems to conflict with the claims of Potrykus (1991) who maintains that DNA is not capable of passing through the cellulose of the plant cell wall. However, perhaps due to the rapid growth of the meristem surface the new cells may not have complete cell walls, thus allowing the DNA access to the plasma membrane. This technique holds much promise as a transformation technique for any crop that produces many meristematic loci in tissue culture. It also has an advantage over microprojectile bombardment in that it is capable of producing non-chimeral transformants from a multicellular origin such as the meristem.

ELECTROPORATION

Electroporation-mediated transformation is a technique originally developed for the transfer of DNA into mammalian cells (Neumann *et al.*, 1982). Electroporation is another technique for plants not amenable to transformation via *Agrobacterium*. It is species independent and so is an attractive technique for monocot species. Initially its principal limitation was the requirement for a regeneration system from protoplasts. Unfortunately such systems have not been developed for many of the important monocot species. In addition, regeneration from protoplasts results in a higher level of somaclonal variation. However, D'Halliun *et al.* (1992) have shown that electroporation of intact plant tissues is possible.

The mechanism of DNA transfer into the plant cell is not fully understood. Short pulses of electricity are thought to cause thinning of the membrane leading to the reversible formation of pores that allow large molecules such as DNA to pass through the plasma membrane (Fromm *et al.*, 1985, 1986; Lindelsy and Jones, 1990). There are several factors affecting the efficiency of electroporation, such as: cell cycle, DNA size, DNA conformation (linear or circular) and concentration, electroporation media and temperature, physiological condition of protoplasts, presence of carrier DNA, and strength and duration and wave form of the electric field. For recent reviews on electroporation see Potrykus (1991), Fisk and Dandekar (1993) and Lal and Lal (1993c). Electroporation has been used to successfully transform rice and corn, but to our knowledge has not been used on any ornamental species.

IMBIBITION

This technique represents the original effort to generate transgenic plants. Many attempts have been made to transform plants by imbibition of DNA into seedlings, organs, tissue explants and cell suspensions with DNA. No evidence is available that the generation of transgenic plants is possible using this technique. Indeed, Potrykus (1990, 1991) claimed that the transfer of DNA across an intact plant cell wall is highly unlikely.

LASER CELL PERFORATION

In this technique a nitrogen or CO_2 laser pulse of 3 to 15 ns is focused through a microscope to perforate the walls of individual cells (Weber *et al.*, 1990). The hole is sealed again in less than 5 ns and the cell continues to grow. Labelling DNA with the fluorescent dye bisbenzimide has demonstrated that DNA is taken into plant cells when they are perforated in a hypertonic buffer solution (Weber *et al.*, 1988a). It has been demonstrated that a focused laser is able to puncture individual organelles such as the chloroplast and that these organelles are able to import DNA (Weber *et al.*, 1988b). Pollen grains from *Brassica napus* transformed by this technique have been regenerated into plants (Weber *et al.*, 1990). Single cells perforated by laser treatment in the presence of a plasmid carrying a gene for hygromycin-resistance have been grown into stable hygromycin resistant colonies (Weber *et al.*, 1990). Pollen grains from the ornamental plant *Impatiens* have also been transformed in a hypertonic solution using a laser microbeam (Yamaguchi *et al.*, 1994). With a manual system, up to 10^3 individual cells can be irradiated in one hour. This rate can be raised to 10^6 cells with a semiautomatic system (Weber *et al.*, 1990).

LIPOSOME-MEDIATED GENE TRANSFER

Transformed plants have been generated by fusion of liposomes carrying DNA with plant protoplasts (Caboche, 1990; Sporlein and Koop, 1991). Attempts have also been made to fuse liposomes with pollen tubes and cell cultures, but no transgenic plants have been produced (Potrykus, 1991). An alternative technique is the injection of liposomes into the cell vacuole where they fuse with the tonoplast releasing the DNA into the cytoplasm (Lucas *et al.*, 1990). Stable transgenics have not yet been generated by this method.

MICROINJECTION

Microinjection involves the direct transfer of DNA into an individual plant cell nucleus (Neuhaus and Spandenberg, 1990; Lal and Lal, 1993d). The technique is very labour and equipment intensive, it requires a skilled operator and a regeneration system from a single plant cell. The large vacuole in many plant cells can make targeting of DNA via microinjection difficult. Microinjection can be used on a variety of explants such as protoplasts, cell cultures and embryos. It offers the potential to deliver DNA directly to pollen grains and microspores, thus generating transgenic haploid plants. To our knowledge this technique has not been used to transform ornamentals.

MICROPROJECTILE-MEDIATED GENE TRANSFER

Where *Agrobacterium*-mediated transformation is not effective, microprojectile bombardment is the method of choice for the generation of transgenic plants. This technique was developed by Sanford and co-workers (Klein *et al.*, 1987) and involves coating the microcarrier (1–5 μm-diameter tungsten or gold particles) with DNA. These particles are then accelerated by a variety of techniques to energies at which they can penetrate a plant cell. Once in the cell, the DNA is released from the particle and expressed transiently or stably incorporated into the plant genome. Such cells can then give rise to stable transgenic plants (for recent reviews see Potrykus, 1991; Gray and Finer 1993; Morikawa *et al.*, 1994).

Microprojectile bombardment has several advantages over *Agrobacterium*-mediated transformation. First, there is no host range limitation making microprojectile bombardment an attractive technique for the transformation of monocotyledonous species. DNA can be introduced directly into a range of organized morphogenic tissues such as embryos and meristems and into pollen. Such techniques have led to the generation of transgenic plants from maize (Gordan-Kamm *et al.*, 1990), rice (Christou *et al.*, 1991), soyabean (McCabe *et al.*, 1988) and wheat (Vasil *et al.*, 1992). Most recently haploid plants of *Nicotiana rustica* have been generated from bombarded pollen cells (Morikawa *et al.*, 1994). The principal limitation of microprojectile bombardment is the uncontrolled uptake and integration of the foreign DNA, leading to a high number of multiple and truncated inserts that can interfere with the stable expression of the desired phenotype. There are a variety of techniques and macrocarriers used to accelerate the particles, the most popular being some form of gas pressure device. Such devices fall into one of the following four categories.

ELECTRIC DISCHARGE PARTICLE ACCELERATION

This method uses an electric discharge as the propelling force. A high-voltage discharge is delivered to a small water droplet which instantly vaporizes, propelling a thin film (macrocarrier) carrying gold particles (microcarriers) against a fine screen. The microcarriers then continue on to the target cells (Lal and Lal, 1993b; McCabe and Christou, 1993). An advantage of this technique over the more commonly used gas discharge system is the ability to finely tune the impact velocity of the particles by altering the voltage to the electric discharge apparatus. This technique has been used to generate transgenic rice from scutellar tissue and transgenic soybean plants from meristems

(McCabe *et al.*, 1988; Christou *et al.*, 1991). Most of the transgenic soybean plants were chimeric.

HELIUM GAS DEVICES USING PLASTIC FILM AS THE MACROCARRIER

This type of device is the only particle delivery system that is sold commercially (Du Pont Biolistic[R] PDS1000/He device; Kikkert, 1993). Helium is used as it expands much faster than compressed air and thus for a given pressure imparts a higher velocity to the microcarriers. The gas pressure bursts a rupture disc allowing a pressure wave to accelerate the macrocarrier (a 2 mm-thick Kapton membrane disc) which then impacts on a stopper plate. The microcarriers, which were dried on to the macrocarrier, then continue on to the target tissue.

AIRGUN DEVICES USING A PLASTIC PROJECTILE AS A MACROCARRIER

This system is inexpensive to build and operate. It uses the gas propulsion generated in a commercial airgun to accelerate a macrocarrier against a stopping plate. The microcarriers then continue to the target tissue (Iida *et al.*, 1990; Oard, 1993).

GAS-STREAM DEVICES THAT HAVE NO MACROCARRIER

There have been two designs based on this principle. Vain *et al.* (1993a) constructed a simple device that has a pre-pressure chamber connected to a simple solenoid valve. Opening of the valve releases the gas charge in the pre-pressure chamber, which is directed through a syringe filter unit. The microcarriers are contained within a droplet of solution that has been placed on the screen of the filter unit. The pressure wave vaporizes the droplet and launches the particles toward the target tissue. This device has been further modified and now uses an inexpensive plastic desiccator as the vacuum chamber. The second system is designed to target small tissue samples accurately (Sautter, 1993) and is based on an air pistol. Helium gas is fed into the pressure chamber of a commercial air pistol. The pistol trigger is used as the release mechanism and the barrel and breech have been replaced with a microchamber in which the target tissue is placed. Upstream from the microchamber, there is a mount for a fine capillary tube. A small droplet containing the microcarriers is extruded from the capillary tube. The gas pressure wave vaporizes the droplet and accelerates the microcarrier into the target tissue.

Gas-stream devices are cheap to manufacture and produce similar transformation frequencies to the more expensive systems that use a macrocarrier.

We have constructed two helium gas-stream systems in our laboratory to develop transformation systems for *Sandersonia* (see below).

CRITICAL VARIABLES IN MICROPROJECTILE-MEDIATED GENE TRANSFER

In developing a particle delivery system that yields transgenic plants, a large number of variables must be considered. These include: amount of DNA per particle, cabinet type and size, DNA preparation and technique for binding to the particles, explant type and physiological condition, gas type and pressure, number of particles per shot, particle type and size, shot distance, shot duration and acceleration, and vacuum pressure (Klein *et al.*, 1988; Armaleo *et al.*, 1990; Vain *et al.*, 1993b). Morikawa *et al.* (1994) found, using lisianthus root tissue as a target explant, that the three most critical parameters were chamber vacuum level, particle size range and shot distance. In addition they showed that even a small layer of water covering the target explant (100 μm thick) can significantly reduce the transformation efficiency. Other workers have shown that preculture of the target explant can increase transformation efficiency (Seki *et al.*, 1991; Kodama *et al.*, 1993). Because of the large number of variables, the following summary of ornamental plants transformed via microprojectile bombardment will not include all the experimental parameters.

PLANTS TRANSFORMED VIA MICROPROJECTILE-MEDIATED GENE TRANSFER

Dendrobium hybrid (orchid)

Since orchids do not seem to be amenable to transformation by *Agrobacterium*, microprojectile bombardment is the preferred approach for this crop. In addition, no regeneration system suitable for *Agrobacterium* transformation has been developed for orchids. Kuehnle and Sugii (1992) have used microprojectile bombardment to generate transgenic orchids. Three-month-old protocorms were bombarded with tungsten particles (Sylvania M10) coated with DNA from the plasmid pGA482GG/cpPRV4 that contains the 35S GUS and NosNPT II genes. The particle gun used was that described by Klein *et al.* (1987).

Transformed tissue was selected via kanamycin resistance. After 17 months of culture, 13 plants were still resistant to kanamycin and contained copies of the kanamycin resistance gene as shown by Southern analysis. Kuehnle and Sugii (1992) found that the use of kanamycin as a selective marker did not completely halt the growth of non-transformed tissue, and recommended the use of hygromycin as the selective agent. In addition, some problems were experienced with chimerism in the transformed population. This may be overcome by using callus or embryogenic cell suspension cultures as the target tissue.

Eustoma grandiflorum (lisianthus)

GUS expression has been reported in lisianthus leaf and petal tissue after microprojectile bombardment using two devices modified from the designs of Sautter (1993) and Vain *et al.* (1993a) (J. Seelye and M. Morgan, personal communication, 1994). Morikawa *et al.* (1994) reported the generation of transgenic lisianthus plants via microprojectile bombardment. No details were provided of the technique or explant used.

Lilium longiflorum (lily)

There have been no reports of transgenic lilies produced via microprojectile bombardment. There have, however, been three reports of expression of the marker gene GUS in lily pollen after delivery via microprojectile bombardment (van der Leede-Plegt *et al.*, 1992; Nishihara *et al.*, 1993; Tabata *et al.*, 1993). Van der Leede-Plegt *et al.* (1992) showed that the *Agrobacterium* TR2' promoter was active in lily pollen. Nishihara *et al.* (1993) demonstrated the use of the tomato LAT52 promoter and Tabata *et al.* (1993) demonstrated the use of the yeast *mei2* promoter for expression in lily microspores. All reports used air- or helium-driven particle guns.

Sandersonia aurantiaca (sandersonia)

Sandersonia aurantiaca has distinctive orange, bell-shaped flowers and is a major export cut flower crop from New Zealand. *Sandersonia* is a monospecific member of the *Colchicaceae* with only limited phenotypic variation. Our goal is to introduce new flower colours into this species via direct gene transfer. In order to transform sandersonia, we have constructed two helium-driven particle guns (J. Seelye, and E. Morgan, personal communication). The first was constructed using a polycarbonate vacuum desiccator based on the design of Finer *et al.* (1992), the second gun uses a modified air pistol based on the design of Sautter *et al.* (1991). *In vitro* leaves, meristematic tissue and petal pieces were used as the target tissue. The plasmids pRT99 and pLN36 (both contain the GUS gene) were precipitated onto gold and tungsten particles. Transient GUS expression was observed in all tissues. Evidence of stable expression was found in the leaf tissue 10 to 14 days after shooting. Gold particles were more effective than tungsten although their shot-to-shot consistency was more variable. The smaller plasmid (pRT99) was also more effective (unpublished results).

Tulipa gesneriana (tulip)

Wilmink *et al.* (1992) have shown that tulip cells (*Tulip gesneriana*) can be transformed via microprojectile bombardment. The helium driven Du Pont PDS1000 system was used to shoot tungsten particles (average diameter 1.1 μm) into nodal stem segments. Particle bombardment and *Agrobacterium*-mediated gene delivery gave similar numbers of blue staining cells after delivery of vectors containing the GUS gene. Seven *Tulipa* species and seven cultivars were tested. All seven cultivars gave a similar response with a significant number of blue spots being generated. In contrast the response of the botanical species was much lower. Wilmink *et al.* (1994) reported further development of this technique. The selective agent was changed to phosphinothricin (the herbicide, BASTA) and immature flower stem segments were used as the target tissue. Transgenic shoots that expressed the GUS gene were produced.

POLYETHYLENE GLYCOL (PEG)-INDUCED DNA UPTAKE

This method is similar to electroporation except that PEG rather than an electric pulse is used to permeabilize the cell membrane (Shillito *et al.*, 1985). As a result, PEG-induced DNA uptake shares the advantages and disadvantages of electroporation (Lal and Lal, 1993a). In addition, PEG-induced DNA uptake has not been reported across an intact cell wall, thus it requires the ability to regenerate plants from protoplasts. The technique exposes protoplasts to foreign DNA in the presence of PEG, Ca^{2+} ions and high pH. The efficiency of stable transformation is much lower than for *Agrobacterium*-mediated gene transfer when using the Ti plasmid (Draper, 1982). There was also a higher frequency of truncated and rearranged inserts present in the transgenic plants. The use of smaller plasmids and linearized DNA has increased the transformation frequency. The technique has produced stable transformed plants (Paszkowski *et al.*, 1984; Zhang and Wu, 1988). Apart from the model plant petunia, there have been no reports of this technique being used to generate transgenics from ornamental species. Direct DNA transfer to protoplasts may have potential for plants recalcitrant to *Agrobacterium* and that have an established protoplast-to-plant regeneration system.

SILICON CARBIDE TECHNIQUE

Silicon-fibre mediated gene transfer to plant cells is another new technique that holds much promise. The technique is simple, quick and inexpensive. Suspension cultures of cells are mixed with plasmid DNA and silicon carbide fibres (0.6 μm average diameter, 10–80 μm long). The mixture is vortexed and cultured in the presence of selectable markers (Asano *et al.*, 1991; Kaeppler *et al.*, 1992). Stable transgenic cell colonies of both tobacco ('Xanthi') and maize ('Black Mexican Sweet') have been generated using this method. The only disadvantage of this method is the requirement for a system to regenerate plants from single cells.

ULTRASONICATION

Ultrasonication is a relatively new technique that shows promise as a simple, cheap method for transforming plants (Joersbo and Brunstedt, 1992; Fisk and Dandekar, 1993). It has been used to transform plant protoplasts (Joersbo and Brunstedt, 1992) and intact plant tissues (Zhang *et al.*, 1991). In the latter example, tobacco leaf pieces were ultrasonicated in a buffer solution containing plasmid DNA. The final transformation frequency was 22%. If this high transformation frequency can be repeated in several plant species, ultrasonication could become the most effective direct DNA transfer system for plants.

CONCLUDING REMARKS

Much effort is being focused on the development of gene transfer techniques for ornamental crops. The new vectors and *Agrobacterium* strains being developed indicate that *Agrobacterium*-mediated transformation will continue to be the most popular gene transfer technique for ornamental crops. Microprojectile bombardment is the most common alternative. However, the most effective method of particle delivery for this technique has not been determined. Several of the newer gene transfer techniques, such as electrophoresis, silicon carbide mediated gene transfer, and ultrasonication, offer considerable promise as viable alternative techniques.

REFERENCES

Ahokas, H. (1989) Transfection of germinating barley seed electrophoretically with exogenous DNA. *Theoretical and Applied Genetics* 77, 469–472.

Alt-Mörbe, J., Kühlmann, P. and Schröder, J. (1989) Differences in induction of Ti plasmid virulence genes *vir*G and *vir*D, and continued control of *vir*D expression by four external factors. *Molecular Plant–Microbe Interactions* 2, 301–308.

Anderson, N.O. (1987) Reclassification of the genus *Chrysanthemum* L. *HortScience* 22, 313.

Ankenbauer, R.G. and Nester, E.W. (1990) Sugar-mediated induction of *Agrobacterium tumefaciens* virulence genes: structural specificity and activities of monosaccharides. *Journal of Bacteriology* 172, 6442–6446.

Armaleo, D., Ye, C.N., Klein, T.M., Shark, K.B. and Sanford, J.C. (1990) Biolistic nuclear transformation of *Saccharomyces cerevisiae* and other fungi. *Current Genetics* 17, 97–103.

Asano, Y., Otsuki, Y. and Ugaki, M. (1991) Electroporation-mediated and silicon carbon fibre-mediated DNA delivery in *Agrostis alba* L. (Redtop). *Plant Science* 79, 247–252.

Bidney, D., Scelonge, C., Martich, J., Burrus, M., Sims, L. and Huffman, G. (1992) Microprojectile bombardment of plant tissues increases transformation frequency by *Agrobacterium tumefaciens*. *Plant Molecular Biology* 18, 301–313.

Boase, M.R. and Borst, N.K. (1995) Tumour production and GUS expression in *Cyclamen persicum* transformed by *Agrobacterium tumefaciens*. *XIth Biennial Conference of the NZ Branch IAPTC,* Auckland, Abstract.

Boase, M.R. and Smith, J.L. (1994) Expression of the GUS gene in 11 *Pelargonium* cultivars transformed by *Agrobacterium tumefaciens*. *IVth International Congress of Plant Molecular Biology* Amsterdam, Abstract 1974.

Bolton, G.W., Nester, E.W. and Gordon, M.P. (1986) Plant phenolic compounds induce expression of *A. tumefaciens* loci needed for virulence. *Science* 232, 983–985.

Bush, A.L. and Pueppke, S.K. (1991) Cultivar–strain specificity between *Chrysanthemum morifolium* and *Agrobacterium tumefaciens*. *Physiological and Molecular Plant Pathology* 39, 309–323.

Caboche, M. (1990) Liposome-mediated transfer of nucleic acids in plant protoplasts. *Physiologia Plantarum* 79, 173–176.

Christey, M.C., Makaroff, C.A. and Earle, E.D. (1991) Atrazine-resistant cytoplasmic male-sterile-*nigra* broccoli obtained by protoplast fusion between cytoplasmic male-sterile *Brassica oleracea* and atrazine-resistant *Brassica campestris*. *Theoretical and Applied Genetics* 83, 210–218.

Christou, P., Ford, T.L. and Kofron, M. (1991) Production of transgenic rice (*Oryza sativa* L.) plants from agronomically important *indica* and *japonica* varieties via electric discharge particle acceleration of exogenous DNA into immature zygotic embryos. *Bio/Technology* 9, 957–962.

Clark, E., Vigodsky-Haas, H. and Gafni, Y. (1989) Characteristics in tissue culture of hyperplasias induced by *Erwinia herbicola* pathovar *gypsophilae*. *Physiological and Molecular Plant Pathology* 35, 383–390.

Cohen, A. and Meridith, C.P. (1992) *Agrobacterium*-mediated transformation of *Lilium*. *Acta Horticulturae* 325, 611–618.

Courtney-Gutterson, N. (1993) Molecular breeding for colour, flavour and fragrance. *Scientia Horticulturae* 55, 141–160.

de Jong, J., Rademaker, W. and van Wordragen, M.F. (1993) Restoring adventitious shoot formation on chrysanthemum leaf explants following co-cultivation with *Agrobacterium tumefaciens*. *Plant Cell, Tissue and Organ Culture* 32, 263–270.

de Jong, J., Mertens M.M.J. and Rademaker, W. (1994) Stable expression of the GUS reporter gene in chrysanthemum depends on binary plasmid DNA. *Plant Cell Reports* 14, 59–64.

DeLaat, A. and Blass, J. (1987) An improved method for protoplast microinjection suitable for transfer of entire plant chromosomes. *Plant Science* 50, 161–169.

Deroles, S.C., Ledger, S.E., Miller, R.M., Davies, K.M. and Given, N.K. (1993) Transformation in *Eustoma grandiflorum* (Lisianthus). In: Bajaj, Y.P.S. (ed.), *Biotechnology in Agriculture and Forestry. Vol. 22. Plant Protoplasts and Genetic Engineering III.* Springer-Verlag, Berlin, pp. 202–212.

D'Halliun, K., Bonne, E., Bossut, M., De Beuckeleer, M. and Leemans, J. (1992) Transgenic maize plants by tissue electroporation. *Plant Cell* 4, 1495–1505.

Diqui, Y., Pingge, H., Yisnian, S. and Baojian, L. (1994) Tissue culture and gene transformation of *Antirrhinum majus*. *VIIIth International Congress of Plant Tissue and Cell Culture.* Abstract S7–103.

Dolgov, S.V., Mityushkina, T.Yu., Rukavtsova, E.B. and Buryanov, Ya.I. (1994) Genetic transformation of *Chrysanthemum morifolium* Ramat. *IVth International Congress of Plant Molecular Biology,* Amsterdam, Abstract.

Dons, J.J.M., Mollema, C., Stiekema, W.J. and Visser, B. (1991) Routes to the development of disease resistant ornamentals. In: Harding, J., Singh, F. and Mol, J.N.M.

(eds), *Genetics and Breeding of Ornamental Species*. Kluwer Academic Publishers, Dordrecht, pp. 387–417.

Draper, J. (1982) Transformation of plant protoplasts by *Agrobacterium* and isolated Ti plasmid. Unpublished PhD thesis, University of Nottingham, UK.

Duron, M. (1994) Genetic transformation of *Forsythia* × *intermedia* 'Spring Glory' via *Agrobacterium tumefaciens*. *VIIIth International Congress of Plant Tissue and Cell Culture,* Abstract S7–90.

Elomaa, P., Honkanen, J., Puske, R., Seppönen, P., Helariutta, Y., Mehto, M., Kotilainen, M., Nevalainen, L. and Teeri, T.H. (1993) *Agrobacterium*-mediated transfer of antisense chalcone synthase cDNA to *Gerbera hybrida* inhibits flower pigmentation. *Bio/Technology* 11, 508–511.

Finer, J.J., Vain, P., Jones, M.W. and McMullem, M.D. (1992) Development of the particle inflow gun for DNA delivery to plant cells. *Plant Cell Reports* 11, 323–328.

Firoozabady, E., Lemieux, C.S., Moy, Y.S., Moll, B., Nicholas, J.A. and Robinson, K.E.P. (1991) Genetic engineering of ornamental crops. *In Vitro* 27, 96A.

Firoozabady, E., Moy, Y., Courtney-Gutterson, N. and Robinson, K.E.P. (1994) Regeneration of transgenic rose (*Rosa hybrida*) plants from embryogenic tissue. *Bio/Technology* 12, 609–613.

Fisk, H.J. and Dandekar, A.M. (1993) The introduction and expression of transgenes in plants. *Scientia Horticulturae* 55, 5–36.

Fraley, R.T., Rogers, S.G., Horsch, R.B., Eichholtz, D.A., Flick, J.S., Fink, C.L., Hoffmann, N.L. and Sanders, P.R. (1985) The SEV system: a new disarmed Ti plasmid vector system for plant transformation. *Bio/Technology* 3, 629–635.

Fraser, L., Kent, J. and Harvey, C. (1991) Transformation studies on *Lathyrus odoratus* L. (sweet pea). *Journal of Cellular Biochemistry* Suppl 15A.

Fromm, M.E., Taylor, M.P. and Walbot, V. (1985) Expression of genes transferred into monocot and dicot plant cells by electroporation. *Proceedings of the National Academy of Sciences, USA* 82, 5824.

Fromm, M.E., Taylor, M.P. and Walbot, V. (1986) Stable transformation of maize after gene transfer by electroporation. *Nature* 319, 791.

Geier, T. (1977) Morphogenesis and plant regeneration from cultured organ fragments of *Cyclamen persicum*. *Acta Horticulturae* 78, 167–174.

Gelvin, S.B. (1990) Crown gall disease and hairy root disease. *Plant Physiology* 92, 281–285.

Gelvin, S.B. and Liu, C.-N. (1994) Genetic manipulation of *Agrobacterium tumefaciens* strains to improve transformation of recalcitrant plant species. In: Gelvin, S.B. and Schilperoert, R.A. (eds), *Plant Molecular Biology Manual.* Kluwer Academic Publishers, Belgium, B4: 1–13.

Godwin, I., Todd, G., Ford-Lloyd, B. and Newbury, H.J. (1991) The effects of acetosyringone and pH on *Agrobacterium*-mediated transformation vary according to plant species. *Plant Cell Reports* 9, 671–675.

Gordan-Kamm, W.J., Spencer, T.M., Mangano, M.L., Adams, T.R., Daines, R.J., Start, W.G., O'Brien, J.V., Chambers, S.A., Adams, W.R., Willetts, N.G., Rice, T.B., Mackey, C.J., Krueger, R.W., Kausch, A.P. and Lemaux, P.G. (1990) Transformation of maize cells and regeneration of fertile transgenic plants. *Plant Cell* 2, 603–618.

Graves, A.C.F. and Goldman, S.L. (1987) *Agrobacterium tumefaciens*-mediated transformation of the monocot genus *Gladiolus*: detection of expression of T-DNA encoded genes. *Journal of Bacteriology* 169, 1745–1746.

Gray, D.J. and Finer, J.J. (eds) (1993) Development and operation of five particle guns for introduction of DNA into plant cells. Special Section on Particle Bombardment: *Plant Cell, Tissue and Organ Culture* 33, 219–257.

Griesbach, R.J. (1987) Chromosome mediated transformation via microinjection. *Plant Science Letters* 30, 297–301.

Griesbach, R.J. (1991) Techniques for chromosomal transformation. In: Harding, J., Singh, F. and Mol, J.N.M. (eds), *Genetics and Breeding of Ornamental Species*. Kluwer Academic Publishers, Dordrecht, pp. 69–78.

Griesbach, R.J. (1993) Genetic transformation of ornamentals through *in vivo* electrophoresis. In: Schiva, T. and Mercuri, A. (eds), *Creating Genetic Variation in Ornamentals. Proceedings of the XVIIth Eucarpia Symposium,* Sanremo, Italy, pp. 33–36.

Handa, T. (1992) Genetic transformation of *Antirrhinum majus* L. and inheritance of altered phenotype induced by Ri T-DNA. *Plant Science* 81, 199–206.

Hansen, G., Das, A. and Chilton, M.-D. (1994) Constitutive expression of the virulence gene improves the efficiency of plant transformation by *Agrobacterium. Proceedings of the National Academy of Sciences, USA* 91, 7603–7607.

Hiei, Y., Ohta, S., Komari, T. and Kumashiro, T. (1994) Efficient transformation of rice (*Oryza sativa* L.) mediated by *Agrobacterium* and sequence analysis of the boundaries of the T-DNA. *Plant Journal* 6, 271–282.

Hoekema, A., Hirsch, P.R., Hooykaas, P.J.J. and Schilperoort, R.A. (1983) A binary plant vector strategy based on the separation of the *vir*- and T-region of the *Agrobacterium tumefaciens* Ti-plasmid. *Nature* 303, 179–181.

Hoffmann, M. and Preil, W. (1987) *In vitro* culture of leaf explants of various *Cyclamen persicum* idiotypes. *Gartenbauwissenschaft* 52, 145–148.

Holford, P., Hernandez, N. and Newbury, H.J. (1992) Factors influencing the efficiency of T-DNA transfer during co-cultivation of *Antirrhinum majus* with *Agrobacterium tumefaciens. Plant Cell Reports* 11, 196–199.

Hood, E.E., Gelvin, S.B., Melchers, L.S. and Hoekema, A. (1993) New *Agrobacterium* helper plasmids for gene transfer to plants. *Transgenic Research* 2, 208–218.

Hooykaas, P.J.J. and Mozo, T. (1994) *Agrobacterium* molecular genetics. In: Gelvin, S.B. and Schilperoort, R.A. (eds), *Plant Molecular Biology Manual.* Kluwer Academic Publishers, Belgium, B3: 1–9.

Hooykaas, P.J.J. and Schilperoort, R.A. (1992) *Agrobacterium* and plant genetic engineering. *Plant Molecular Biology* 19, 15–38.

Horsch, R.B., Fry, J.E., Hoffmann, N.L., Eicholtz, D., Rogers, S.G. and Fraley. R.T. (1985) A simple and general method for transferring genes into plants. *Science* 227, 1229–1231.

Hutchinson, J.F., Kaul, V., Maheswaran, G., Moran, J.R., Graham, M.W. and Richards, D. (1992) Genetic improvement of floriculture crops using biotechnology. *Australian Journal of Botany* 40, 765–787.

Iida, A., Seki, M., Kamada, M., Yamada, Y. and Morikawa, H. (1990) Gene delivery into cultured plant cells by DNA-coated gold particles accelerated by a pneumatic particle gun. *Theoretical and Applied Genetics* 80, 813–816.

James, D.J., Uratsu, S.L., Cheng, J., Negri, P., Viss, P. and Dandekar, A.M. (1993) Acetosyringone and osmoprotectants like betaine and proline synergistically enhance *Agrobacterium*-mediated transformation of apple. *Plant Cell Reports* 12, 559–563.

Janson, J. (1993) Placental pollination in *Lilium longiflorum* Thumb. *Plant Science* 90, 105–115.

Janssen, B. and Gardner, R.C. (1989) Localized transient expression of GUS in leaf discs following co-cultivation with *Agrobacterium. Plant Molecular Biology* 14, 61–72.

Jeffersen, R.A., Kavanagh, T.A. and Bevan, M.W. (1987) GUS fusion: β- glucuronidase as a sensitive and versatile gene fusion marker in higher plants. *EMBO Journal* 6, 3901–3907.

Jia, S.-R., Yang, M,-Z., Ott, R. and Chua, N.-H. (1989) High frequency transformation of *Kalanchoe laciniata. Plant Cell Reports* 8, 336–340.

Jin, S., Komari, T., Gordon, M.P. and Nester, E.W. (1987) Genes responsible for the super virulence phenotype of *Agrobacterium tumefaciens* A281. *Journal of Bacteriology* 169, 4417–4425.

Joersbo, M. and Brunstedt, J. (1992) Direct gene transfer to plant protoplasts by mild sonication. *Plant Cell Reports* 9, 207–210.

Jordan, B.R. and Anthony, R.G. (1993) Floral homeotic genes: isolation, characterization and expression during floral development. In: Jordan, B.R. (ed.), *The Molecular Biology of Flowering*. CAB International, Wallingford, UK, pp. 93–116.

Kado, C.I. (1991) Molecular mechanisms of crown gall tumorigenesis. *Critical Reviews in Plant Science* 10, 1–32.

Kaeppler, H.F., Somers, D.A., Rines, H.W. and Cockburn, A.F. (1992) Silicon carbide fibre-mediated stable transformation of plant cells. *Theoretical and Applied Genetics* 84, 560–566.

Kikkert, J.R. (1993) The Biolistic PDS-1000/He device. *Plant Cell, Tissue and Organ Culture* 33, 221–226.

Kishi, F., Kagami, Y., Shinohara, M., Hatano, S. and Tsurushima, H. (1994) Production of an interspecific hybrid in *Gypsophila* by ovule-embryo culture. *Euphytica* 74, 85–90.

Klein, T.M., Wolf, E.D., Wu, R. and Sanford, J.C. (1987) High velocity microprojectiles for delivering nucleic acids into living cells. *Nature* 327, 70–73.

Klein, T.M., Gradziel, T., Fromm, M.E. and Sandorf, J.C. (1988) Factors influencing gene delivery into *Zea mays* cells by high velocity microprojectiles. *Bio/Technology* 6, 559–563.

Klemm, M., Scharz-Sommer, Zs. and Saedler, H. (1991) Transformation of *Antirrhinum majus*: Regeneration of transgenic plants and development of a transient expression system. *IIIrd International Congress of Plant Molecular Biology,* Tuscon, Arizona. Abstract No. 885.

Kodama, H., Irifune, K., Kamada, H. and Morikawa, H. (1993) Transgenic roots produced by introducing Ri-*rol* genes into cucumber cotyledons by particle bombardment. *Transgenic Research* 2, 147–152.

Kruger-Lebus, S. and Potrykus, I. (1987) Direct gene transfer to *Petunia hybrida* without electroporation. *Plant Molecular Biology Reporter* 5, 289–294

Kuehnle, A.R. and Sugii, N. (1991) Induction of tumours in *Anthurium andraeanum* by *Agrobacterium tumefaciens*. *HortScience* 26, 1325–1328.

Kuehnle, R. and Sugii, N. (1992) Transformation of *Dendrobium* orchid using particle bombardment of protocorms. *Plant Cell Reports* 11, 484–488.

Kuehnle, A.R., Chen, F.C. and Jaynes, J.M. (1993) Engineering bacterial blight resistance into *Anthurium*. In: Schiva, T. and Mercuri, A. (eds), *Creating Genetic Variation in Ornamentals. Proceedings XVIIth Eucarpia Symposium,* Sanremo, Italy, pp. 127–129.

Lal, R. and Lal, S. (1993a) Chemically mediated DNA uptake. In: *Genetic Engineering of Plants for Crop Improvement*. CRC Press, Boca Raton, pp. 12–13.

Lal, R. and Lal, S. (1993b) Electric discharge particle acceleration method. In: *Genetic Engineering of Plants for Crop Improvement*. CRC Press, Boca Raton, 34.

Lal, R. and Lal, S. (1993c) Electroporation. In: *Genetic Engineering of Plants for Crop Improvement*. CRC Press, Boca Raton, pp. 14–23.

Lal, R. and Lal, S. (1993d) Microinjection. In: *Genetic Engineering of Plants for Crop Improvement*. CRC Press, Boca Raton, pp. 27–28.

Langeveld, S.A., Derks, A.F.L.M., Gerrits, M.M., Boonekamp, P.M. and Bol, J.F. (1994) Transformation of lily by *Agrobacterium. VIIIth International Congress of Plant Tissue and Cell Culture*, Abstract M–23.

Ledger, S.E., Deroles, S.C. and Given, N.K. (1991) Regeneration and *Agrobacterium*-mediated transformation of chrysanthemum. *Plant Cell Reports* 10, 195–199.

Leduc, N., Douglas, G.C. and Monnier, M. (1992) Methods for non- stigmatic pollination in *Trifolium repens* (*Papilionaceae*): seed set with self- and cross-pollinations *in vitro*. *Theoretical and Applied Genetics* 83, 912–918.

Lemieux, C. S., Firoozabady, E. and Robinson, K.E.P. (1990) *Agrobacterium*-mediated transformation of chrysanthemum. In: *Integration of* In Vitro *Techniques in Ornamental Plant Breeding. Proceedings of the Eucarpia Symposium*, Wageningen, pp. 150–155.

Leroux, B., Yanofsky, M.F., Winans, S.C., Ward, J.E., Ziegler, S.F. and Nester, E.W. (1987) Characterization of the *vir*A locus of *Agrobacterium tumefaciens*: a transcriptional regulator and host range determinant. *EMBO Journal* 6, 849–846.

Lindelsy, K. and Jones, M.G.K. (1990) Electroporation of cells. *Physiologia Plantarum* 79, 168.

Liu, C.-N., Li, X.-Q. and Gelvin, S.B. (1992) Multiple copies of *vir*G enhance the transient transformation of celery, carrot and rice tissues by *Agrobacterium tumefaciens*. *Plant Molecular Biology* 20, 1071–1087.

Lowe, J.M., Davey, M.R., Power, J.B. and Blundy, K.S. (1993) A study of some factors affecting *Agrobacterium* transformation and plant regeneration of *Dendranthema grandiflora* Tzvelev (syn. *Chrysanthemum morifolium* Ramat.). *Plant Cell, Tissue and Organ Culture* 33, 171–180.

Lu, C., Nugent, G., Wardley-Richardson, T., Chandler, S.F., Young, R. and Dalling, M.J. (1991) *Agrobacterium*-mediated transformation of carnation (*Dianthus caryophyllus* L.). *Bio/Technology* 9, 864–868.

Lucas, W.J., Lansing, A., de Wet, J.R. and Walbot, V. (1990) Introduction of foreign DNA into walled plant cells via liposomes injected into the vacuole; a preliminary study. *Physiologia Plantarum* 79, 184–189.

Martin, C. And Gerats, T. (1993) The control of flower coloration. In: Jordan, B.R. (ed.), *The Molecular Biology of Flowering*. CAB International, Wallingford, UK, pp. 219–256.

Martin, C., Carpenter, R., Coen, E.S. and Gerats, A.G.M. (1987) The control of floral pigmentation in *Antirrhinum majus*. In: Thomas, H. and Grierson, D. (eds), *Developmental Mutants in Plants*, SEB Seminar Series 32. Cambridge University Press, Cambridge, pp. 19–53.

Martin, C., Prescott, A., Machay, S., Bartlett, J. and Vrijlandt, E. (1991) Control of anthocyanin biosynthesis in flowers of *Antirrhinum majus*. *The Plant Journal* 1, 37–49.

May, R.A. and Trigiano R.N. (1991) Somatic embryogenesis and plant regeneration from leaves of *Dendranthema grandiflora*. *Journal of the American Society of Horticultural Science* 116, 366–371.

McCabe, D. and Christou, P. (1993) Direct DNA transfer using electric discharge particle acceleration (ACCELL™ technology). *Plant Cell, Tissue and Organ Culture* 33, 227–236.

McCabe, D.E., Swain, W.F., Martinell, B.J. and Christou, P. (1988) Stable transformation of soybean (*Glycine max*) by particle acceleration. *Bio/Technology* 6, 923–926.

Messeguer, J. and Mele, E. (1994) Factors influencing genetic transformation in carnation. *VIIIth International Congress of Plant Tissue and Cell Culture,* Abstract S7–77.

Michael, M.Z., Savin, K.W., Baudinette, S.C., Graham, M.W., Chandler, S.F., Lu, C-Y., Caesar, C., Gautrais, R., Young, R., Nugent, G.D., Stevenson, K.R., O'Conner, E.L.-J., Cobbett, C.S. and Cornish, E.C. (1993) Cloning of ethylene biosynthetic genes involved in petal senescence of carnation and petunia, and their antisense expression in transgenic plants. In: Pech J.C. *et al.* (eds), *Cellular and Molecular Aspects of the Plant Hormone Ethylene.* Kluwer Academic Publishers, Dordrecht, pp. 298–303.

Miller, H.N. (1975) Leaf, stem, crown, and root galls induced in *Chrysanthemum* by *Agrobacterium tumefaciens. Phytopathology* 65, 805–811.

Miller, H.N., Miller, J.W. and Crane, G.L. (1975) Relative susceptibility of chrysanthemum cultivars to *Agrobacterium tumefaciens. Plant Disease Reporter* 59, 576–581.

Mont, J., Iwanaga, M., Orjeda, G. and Watanabe, K. (1993) Abortion and determination of stages for embryo rescue in crosses between sweet potato, *Ipomoea batatas* Lam. ($2n = 6x = 90$) and its wild relative, *I. trifida* (H.B.K.) G.Don. ($2n=2x=30$). *Sexual Plant Reproduction* 6, 176–182.

Morikawa, H., Nishihara, M., Seki, M. and Irifune, K. (1994) Bombardment-mediated transformation of plant cells. *Journal of Plant Research* 107, 117–123.

Nester, E.W. and Verma D.P.S. (eds) (1991) *Agrobacterium*–plant interactions. In: *Advances of Molecular Genetics of Plant–Microbe Interactions*, Vol. 2. Kluwer Academic Publishers, Dordrecht, pp. 37–125.

Neuhaus, G. and Spandenberg, G. (1990) Plant transformation by microinjection techniques. *Physiologia Plantarum* 79, 213–217.

Neumann, E., Shaefer-Ridder, M., Wang, Y. and Hofschneider, P.H. (1982) Gene transfer into mouse lyoma cells by electroporation in high electric fields. *EMBO Journal* 1, 841.

Nishihara, M., Ito, M., Tanaka, I., Kyo, M., Ono, K., Irifune, K. and Morikawa, H. (1993) Expression of the β-glucuronidase gene in pollen of lily (*Lilium longiflorum*), tobacco (*Nicotiana tabacum*), *Nicotiana rustica*, and peony (*Paeonia lactiflora*) by particle bombardment. *Plant Physiology* 102, 357–361.

Noriega, C. and Sondhal, M.R. (1991) Somatic embryogenesis in hydrid tea roses. *Bio/Technology* 9, 991–993.

Oard, J. (1993) Development of an airgun device for particle bombardment. *Plant Cell, Tissue and Organ Culture* 33, 247–250.

Ondrej, M. and Biskova, R. (1986) Differentiation of *Petunia hybrida* tissues transformed by *Agrobacterium rhizogenes* and *Agrobacterium tumefaciens. Biologica Plantarum* 28, 152–155.

Ozais-Akins, P., Singsit, C. and Brabch, W.D. (1992) Interspecific hybrid inviability in crosses of *Arachis hypogea* × *A. stenosperma* can be overcome by *in vitro* embryo maturation of somatic embryogenesis. *Journal of Plant Physiology* 140, 207–212.

Paszkowski, J., Shillito, R.D., Saul, M., Mandak, V., Hohn, T., Hohn, B. and Potrykus, I. (1984) Direct gene transfer to plants. *EMBO Journal* 3, 2717–2722.

Pavingerová, D., Dostál, J., Bísková, R. and Benetka, V. (1994) Somatic embryogenesis and *Agrobacterium*-mediated transformation of chrysanthemum. *Plant Science* 97, 95–101.

Pellegrineschi, A., Damon, J.P., Valtorta, N., Paillard, N. and Tepfer, D. (1994) Improvement of ornamental characters and fragrance production in lemon scented geranium through genetic transformation by *Agrobacterium rhizogenes. Bio/Technology* 12, 64–68.

Potrykus, I. (1990) Gene transfer to plants: assessment and perspectives. *Physiologia Plantarum* 79, 125–134.

Potrykus, I. (1991) Gene transfer to plants: assessment of published approaches and results. *Annual Review of Plant Physiology and Plant Molecular Biology* 42, 205–225.

Power, J.B., Davey, M.R., Freeman, J.P., Mulligan, B.J. and Cocking, E.C. (1986) Fusion and transformation of plant protoplasts. *Methods in Enzymology* 118, 578–594.

Rademaker, W. and de Jong, J. (1990) Genetic variation on adventitious shoot formation in *Dendranthema grandiflora* (*Chrysanthemum morifolium*) explants. In: de Jong (ed.), *Integration of In Vitro Techniques in Ornamental Plant Breeding Symposium Proceedings*, pp. 34–38.

Renou, J.P., Brochard, P. and Jalouzot, R. (1993) Recovery of transgenic chrysanthemum (*Drendranthema grandiflora* Tzvelev) after hygromycin resistance selection. *Plant Science* 89, 185–197.

Rietveld, E.M., Gielen, J.J.L., Westerbroek, A., Tan, M.M.C. and Schram, A. (1994) Introduction of the *rol*C gene of *Agrobacterium rhizogenes* into *Salpiglossis sinuata* by transformation and cell fusion. *VIIIth International Congress of Plant Tissue and Cell Culture* Abstract S7–126.

Robinson, K.E.P. and Firoozabady, E. (1993) Transformation of floriculture crops. *Scientia Horticulturae* 55, 83–99.

Sautter, C. (1993) Development of a microtargeting device for particle bombardment of plant meristems. *Plant Cell, Tissue and Organ Culture* 33, 251–258.

Sautter, C., Waldner, H., Neuhaus-Url, G., Galli, A., Neuhaus, G. and Potrykus, I. (1991) Micro-targeting: high efficiency gene targeting using a novel approach for the acceleration of microprojectiles. *Bio/Technology* 9, 1080–1085.

Seki, M., Komeda, Y., Iida, A., Yamada, Y. and Morikawa, H. (1991) Transient expression of β-glucuronidase in *Arabidopsis thaliana* leaves and roots and *Brassica napus* using a pneumatic particle gun. *Plant Molecular Biology* 17, 259–263.

Semeria, L., Vaira, A.M., Accotto, G.P., Ruffoni, B. and Allavena, A. (1994) Genetic transformation of *Eustoma grandiflorum* Griseb. by *Agrobacterium tumefaciens*. *VIIIth International Congress of Plant Tissue and Cell Culture,* Abstract S7–78.

Shillito, R.D., Saul, M.W., Paszkowski, J., Muller, M. and Potrykus, I. (1985) High frequency direct gene transfer to plants. *Bio/Technology* 4, 1099–1103.

Short, K.C. and Roberts, A.V. (1991) *Rosa* spp. (roses): *In vitro* culture, micropropagation and the production of secondary products. In: Bajaj, Y.P.S. (ed.), *Biotechnology in Agriculture and Forestry. Vol. 15. Medicinal and Aromatic Plants III.* Springer-Verlag, Berlin, pp. 376–397.

Soreng, R.J. and Cope, E.C (1991) On the taxonomy of cultivated species of the *Chrysanthemum* genus-complex (Anthemideae; Compositae). *Baileya* 23, 145–165.

Sporlein, B. and Koop, H.-U. (1991) Lipofection: direct gene transfer to higher plants using cationic liposomes. *Theoretical and Applied Genetics,* 83, 1–5.

Sybenga, J. (1983) Genetic manipulation in plant breeding: somatic virus generative. *Theoretical and Applied Genetics* 66, 179–201.

Szabados, L.G., Hadlaczky, G. and Dudits, D. (1987) Uptake of isolated plant chromosomes by plant protoplasts. *Planta* 151, 141–145.

Tabata, S., Sato, S., Watanabe, Y., Yamamoto, M. and Hata, Y. (1993) Evidence of meiosis-specific regulation of gene expression in lily microsporocytes. *Plant Science* 89, 31–41.

Urban, L.A., Sherman, J.M., Moyer, J.W. and Daub, M.E. (1994) High frequency shoot regeneration and *Agrobacterium*-mediated transformation of chrysanthemum (*Dendranthema grandiflora*). *Plant Science* 98, 69–70.

Vain, P., Keen, N., Murillo, J., Rathus, C., Nemes, C. and Finer, J.J. (1993a) Development of the particle inflow gun. *Plant Cell, Tissue and Organ Culture* 33, 237–246.

Vain, P., McMullen, M.D. and Finer, J.J. (1993b) Osmotic treatment enhances particle bombardment-mediated transient and stable transformation of maize. *Plant Cell Reports* 12, 84–88.

Vainstein A., Fisher, M. and Ziv, M. (1993) Applicability of reporter gene to carnation transformation. *HortScience* 28, 1122–1124.

Van Altvorst, A.-C., Risken T., Koehorst H. and Dons, H.J.M. (1995) Transgenic carnations obtained by *Agrobacterium tumefaciens*-mediated transformation of leaf explants. *Transgenic Research* 4, 105–113

van der Leede-Plegt, L.M., van de Ven, B.C.E., Bino, R.J., ven der Salm, T.P.M. and van Tunen, A.J. (1992) Introduction and differential use of various promoters in pollen grains of *Nicotiana glutinosa* and *Lilium longiflorum*. *Plant Cell Reports* 11, 20–24.

van Wordragen, M.F., de Jong, J., Huitema, H.B.M. and Dons, H.J.M. (1991) Genetic transformation of *Chrysanthemum* using wild type *Agrobacterium* strains: strain and cultivar specificity. *Plant Cell Reports* 9, 505–508.

Vasil, I., Ahuja, M. and Vasil, V. (1979) Plant tissue cultures in genetics and plant breeding. *Advances in Genetics* 20, 127–210.

Vasil, V., Castillo, A.M., Fromme, M.E. and Vasil, I.K. (1992) Herbicide resistant, fertile transgenic wheat plants obtained by microprojectile bombardment of regenerable embryogenic callus. *Bio/Technology* 10, 667–674.

Veluthambi, K., Krishnan, M., Gould, J.H., Smith, R.H. and Gelvin, S.B. (1989) Opines stimulate the induction of the *vir* genes of the *Agrobacterium tumefaciens* Ti plasmid. *Journal of Bacteriology* 171, 3969–3703.

Vereniging van Bloemenveilingen (VBN) (1993), *Statistiekboek* Leiden, The Netherlands.

Vernade, D., Herrera-Estrella, A., Wang, K. and Van Montagu, M. (1988) Glycine betaine allows enhanced induction of the *Agrobacterium tumefaciens vir* genes by acetosyringone at low pH. *Journal of Bacteriology* 170, 5822–5829.

Wainwright, H. and Harwood, A.C. (1985) *In vitro* organogenesis and plant regeneration of *Cyclamen persicum* Mill. using seedling tissue. *Journal of Horticultural Science* 60, 397–403.

Walkerpeach, C.R. and Velten, J. (1994) *Agrobacterium*-mediated gene transfer to plant cells: co-integrate and binary vector systems. *Plant Molecular Biology Manual*. Kluwer Academic Publishers, Belgium, B1: 1–19.

Wang, K., Herrera-Estrella, A. and Van Montagu, M. (1990) Over expression of *vir*D1 and *vir*D2 gene in *Agrobacterium tumefaciens* enhancers T-complex formation and plant transformation. *Journal of Bacteriology* 172, 4432–4440.

Weber, G., Monajembashi, S., Greulich, K.-O. and Wolfrum, J. (1988a) Injection of DNA into plant cells with a UV-laser microbeam. *Naturwissenschaften* 75, 35–36.

Weber, G., Monajembashi, S., Greulich, K.-O. and Wolfrum, J. (1988b) Genetic manipulation of plant cells' and organelles with a laser microbeam. *Plant Cell, Tissue and Organ Culture* 12, 219–222.

Weber, G., Monajembashi, S., Wolfrum, J. and Greulich, K.-O. (1990) Genetic changes induced in higher plants cells by a laser microbeam. *Physiologia Plantarum* 79, 190–193.

Wilmink, A., van de Ven, B.C.E. and Dons, J.J.M. (1992) Expression of the GUS gene in the monocot tulip after introduction by particle bombardment and *Agrobacterium*. *Plant Cell Reports* 11, 76–80.

Wilmink, A., van de Ven, B.C.E and Dons, H.J.M. (1994) A transformation procedure for tulip. *IVth International Congress of Plant Molecular Biology* 19–24 June, Amsterdam, The Netherlands, Abstract 2038.

Winefield, C., Davies, K., Schwinn, K., Deroles, S., Boulton, G., Miller, R., Boase, M. and Konczac-Islam, I. (1994) Genetic engineering for pelargonidin production in chrysanthemum. *IVth Annual Queenstown Molecular Biology Meeting*, Queenstown, New Zealand, Abstract P72.

Woodson, W.R. and Goldsbrough, P.B. (1989) Genetic transformation of carnation using *Agrobacterium tumefaciens*. *HortScience* 24, 80.

Yamaguchi, T., Wakizuka, T., Osaki, T. and Withers, L.A. (1994) Gene introduction into pollen by using N_2-laser microbeam. *VIIIth International Congress of Plant Tissue and Cell Culture,* Abstract S7–51.

Zambryski, P.C. (1988) Basic processes underlying *Agrobacterium*-mediated DNA transfer to plant cells. *Annual Review of Genetics* 22, 1–30.

Zambryski, P.C. (1992) Chronicles from the *Agrobacterium*-plant cell DNA transfer story. *Annual Review of Plant Physiology and Plant Molecular Biology* 43, 465–490.

Zambryski, P.C., Holsters, M., Kruger, K. and Goodman, H. (1982) Tumour induction by *Agrobacterium tumefaciens*: analysis of the boundaries of T-DNA. *Journal of Molecular and Applied Genetics* 1, 361–370.

Zhang, L.-J., Cheng, L.-M., Xu, N., Zhao, N.-M., Li, C.-G., Yuan, J. and Jia, S.-R. (1991) Efficient transformation of tobacco by ultrasonication. *Bio/Technology* 9, 996–997.

Zhang, W. and Wu, R. (1988) Efficient regeneration of transgenic plants from rice protoplasts and correctly regulated expression of the foreign gene in the plants. *Theoretical and Applied Genetics* 76, 835–840.

Zupan, J.R. and Zambryski, P. (1995) Transfer of T-DNA from *Agrobacterium* to the plant cell. *Plant Physiology* 107, 1041–1047.

Zyprian, E. and Kado, C.I. (1990) *Agrobacterium*-mediated plant transformation by novel mini-T vectors in conjunction with a high-copy *vir* region helper plasmid. *Plant Molecular Biology* 15, 245–256.

Reproductive Barriers in Flowering Plants

7

Norbert Nass, Adrienne E. Clarke and Ed Newbigin
Plant Cell Biology Research Centre, School of Botany, University of Melbourne, Parkville, Victoria 3052, Australia

INTRODUCTION

The flower is a striking and attractive feature of many ornamental plants. The flower is also the site of sexual reproduction in plants and the female pistil is generally readily accessible to pollen from other flowers, borne by the wind or by vectors such as insects. Many species produce both male and female gametes on the same plant (monoecious) and it is common for both types of gametes to develop within the same flower (hermaphrodite). A flower can thus receive pollen from a wide variety of sources, including itself; however, there are many mechanisms that prevent or restrict the ability of either self pollen, or pollen from related species, to fertilize the ovules. Together, the various restrictions on fertilization are termed incompatibility barriers. Interspecific incompatibility is used to describe floral adaptations that prevent the formation of interspecific hybrids and, in hermaphroditic flowers, adaptations that prevent fertilization by the plants' own pollen are referred to as self-incompatibility. Interspecific incompatibility restricts gene flow between species and consequently favours speciation, whereas self-incompatibility establishes limits on the degree of inbreeding that is possible within a species or population.

Of the two types of incompatibility, self-incompatibility has received the greater attention and is better understood. According to Darwin (1877), it was Kölreuter in the eighteenth century who first described self-incompatibility when he reported that *Verbascum* did not set seed when pollinated with its own fertile pollen but readily set seed when cross-pollinated. However, Darwin published the best and most comprehensive early descriptions of self-incompatibility (for example, see Darwin, 1877). Since Darwin, a number of excellent reviews of self-incompatibility have been written, including the classic monograph *Incompatibility in Angiosperms* by de Nettancourt (1977). More recent reviews that emphasize progress on the molecular biology of self-

 Biotechnology of Ornamental Plants
(eds R.L. Geneve, J.E. Preece and S.A. Merkle)

incompatibility include Nasrallah and Nasrallah (1993), Newbigin *et al.* (1993), and Kao and Huang (1994). Less is known about interspecific incompatibility although good general descriptions of a number of systems are found in Frankel and Galun (1977) and Liedl and Anderson (1993). In this review, we will focus on the molecular and genetic mechanisms involved in incompatibility and discuss the role of self-incompatibility in interspecific incompatibility.

SELF-INCOMPATIBILITY

Self-incompatibility, the inability of a flower that is otherwise fully fertile to set seed following self-pollination, is widespread throughout angiosperms. Some authors estimate that it is present in half of all species (East, 1940; Whitehouse, 1950). Despite its ubiquity in flowering plants, there are many different forms of self-incompatibility. Two types of self-incompatibility are defined on the basis of floral architecture as heteromorphic and homomorphic. As the name implies, heteromorphic self-incompatibility refers to the different forms of flowers in individuals from the same species; a well-known example is distyly in *Primula* in which flowers have either a long style and short anthers ('pin'), or a short style and long anthers ('thrum'). In species with homomorphic self-incompatibility, on the other hand, there is no obvious physical barrier to self-pollination. The arrangement of floral organs in species with heteromorphic self-incompatibility provides a physical barrier to self-pollination, although self-incompatibility in these plants is basically independent of morphological variation; an effective genetic barrier prevents self pollen from fertilizing the ovules (Gibbs, 1986).

Two types of self-incompatibility can be defined genetically as sporophytic and gametophytic (Fig. 7.1). In most, but not all cases, self-incompatibility in both gametophytic and sporophytic systems is controlled by a single genetic locus (the S locus) with multiple alleles. Because self-incompatibility requires the exchange of information between the pollen and the pistil, the S-locus must be expressed in both these tissues. Figure 7.1 shows the two types of self-incompatibility and how they can be distinguished through appropriate pollinations. In both systems, pollen is rejected and fertilization does not occur if the products of the S alleles present on pollen and pistil match. In gametophytic self-incompatibility, this rejection is determined by the haploid genotype of the individual pollen grain, whereas in sporophytic self-incompatibility, the incompatibility phenotype is conferred by the two S-locus alleles of the diploid pollen parent. The sporophytic and gametophytic incompatibility systems are believed to differ in the timing and site of S-locus expression in the anther; in gametophytic self-incompatible plants, the S locus must be expressed in the haploid pollen grain; in contrast, in sporophytic self-incompatible plants, the S locus can be expressed either in the diploid pollen-mother cell before meiosis, or contributed to maturing pollen grains by the diploid tissues of the anther tapetum (Fig. 7.2). Thus, although genetically distinct, many authors have speculated that some gametophytic and sporophytic systems may be closely related with a change from one system to the

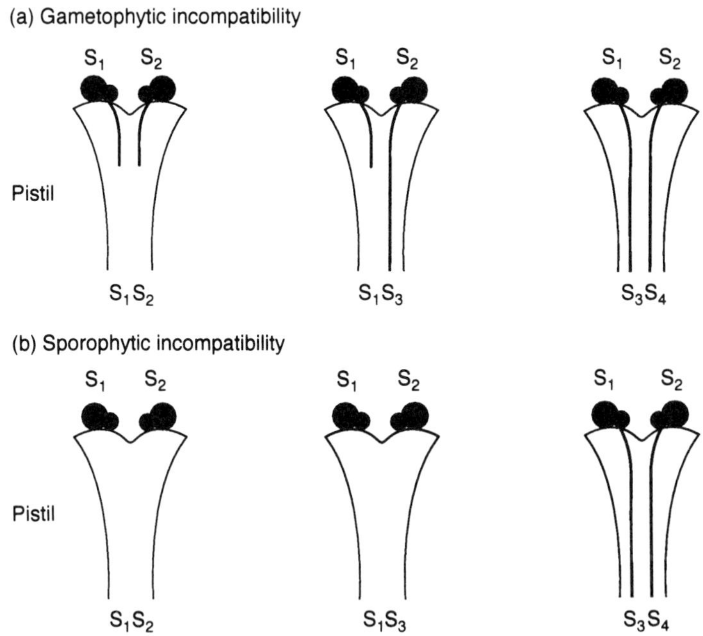

Fig. 7.1. Behaviour of pollen in the two major self-incompatibility systems. (a) Behaviour of pollen in a single-locus gametophytic system. The pollen parent genotype is S_1S_2. When an allele in the individual haploid pollen grain matches either allele in the diploid style tissues, growth of the pollen tube is arrested, usually in the style. When there is no match of alleles (e.g. pollen grains from an S_1S_2 plant on an S_3S_4 pistil), the pollen tubes grow through the style to the embryo sac. (b) Behaviour of pollen in a single-locus sporophytic system. The pollen parent genotype is S_1S_2. When an allele in the pollen parent matches that of the pistil (e.g. S_1S_2 or S_1S_3), pollen germination is arrested at the stigma surface. When there is no match (S_3S_4), the pollen may germinate and grow through the style to the embryo sac. The central panel applies only if the S_1 allele is dominant to or co-dominant with S_2 in the pollen and if S_1 is dominant to or co-dominant with S_3 in the style. If S_3 is dominant to S_1 in the style, or if S_2 is dominant to S_1 in the pollen, pollen from the S_1S_2 parent will be compatible. (a and b, modified from Anderson *et al.*, 1983; with permission of John Wiley and Sons.)

other requiring only an alteration in either the timing or site of S-locus expression (de Nettancourt, 1977; Pandey, 1980). This point and its implications for the evolution of self-incompatibility are discussed later in the review.

PRODUCTS OF THE S-LOCUS IN PLANTS WITH GAMETOPHYTIC SELF-INCOMPATIBILITY

There were a number of early attempts to characterize products of the S locus in different plants, but significant advances in this research were not possible until the techniques of molecular biology were applied to the problem. The

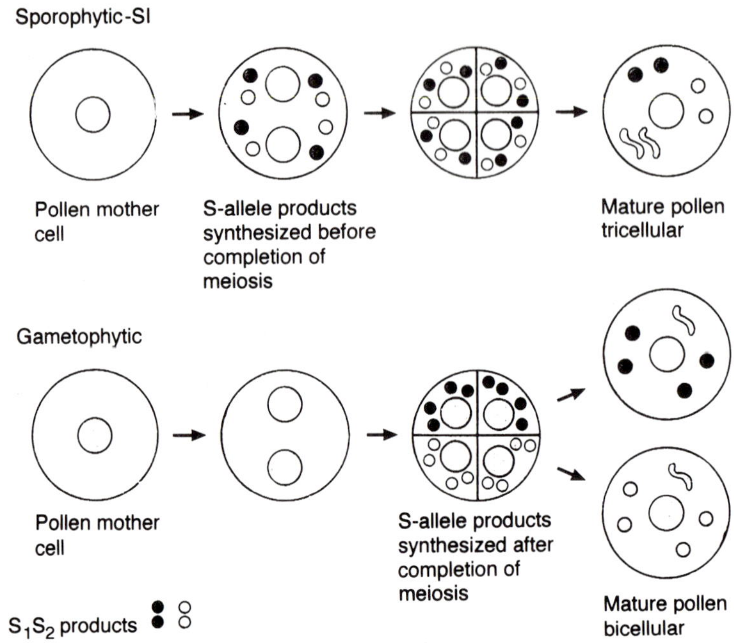

Fig. 7.2. Possible developmental basis for the difference between sporophytic and gametophytic self-incompatibility. One major difference in the two systems is believed to be the timing of expression of the S locus in the anther. In the sporophytic system, the S-allele products are thought to be synthesized before completion of meiosis (upper panel), whereas in gametophytic systems, the S-allele products are thought to be synthesized after completion of meiosis. The difference in timing of expression is one explanation for the fact that pollen in the sporophytic system displays its maternal genotype, whereas pollen in the gametophytic system displays its own (haploid) genotype. (Modified from Newbigin *et al.*, 1993; with permission of The American Society of Plant Physiologists.)

products of the S-locus have been characterized in the *Solanaceae* (including petunias, ornamental tobacco, potatoes and tomatoes), *Papaveraceae* (poppies), *Scrophulariaceae* (snapdragons) and *Rosaceae* (cherries and apples), and most thoroughly studied in the *Solanaceae*.

In the *Solanaceae*, both incompatible and compatible pollen grains swell on the stigma surface and produce a pollen tube that penetrates the stigma, and grows into the style. There, the tubes grow between the longitudinal files of cells of the central transmitting tract, where, at some point, usually in the upper third of the style, pollen tube growth toward the ovary is arrested (Fig. 7.1). This is in contrast to the *Brassica* sporophytic system, in which pollen tube growth in an incompatible mating is usually arrested on the surface of the stigma.

The molecular biology of self-incompatibility in solanaceous plants was opened up by the observation that extracts of pistils of certain species contained glycoproteins that segregated with particular S genotypes. The first of

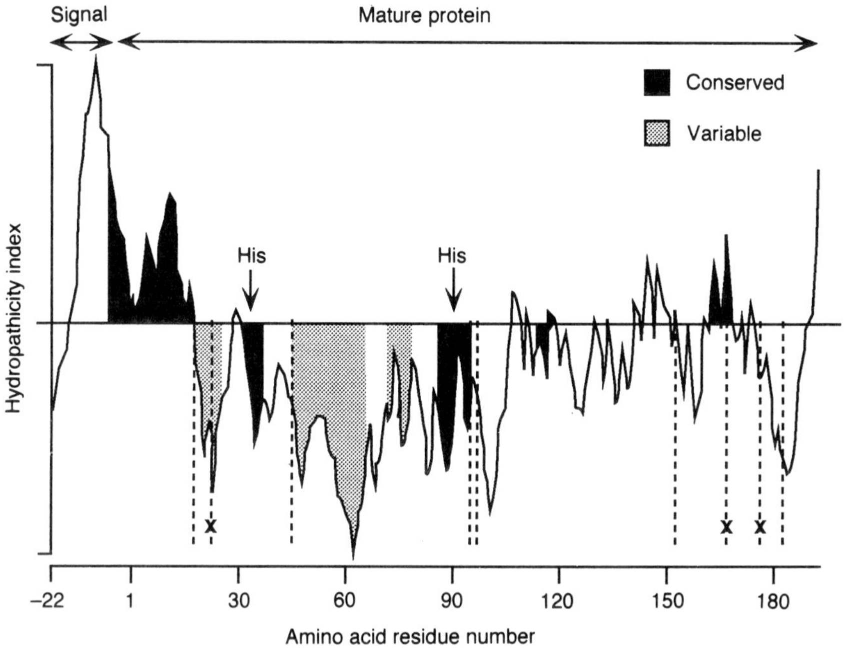

Fig. 7.3. Hydropathy plot based on the *Nicotiana alata* S_2-glycoprotein. The *N. alata* sequence was aligned with cDNAs encoding other S glycoproteins from *N. alata*, *Petunia inflata*, and potato. The variable regions are shown in grey, and the conserved regions are shown in black. Conserved cysteine residues are shown as dashed lines; cysteine residues that are not conserved in alignments with fungal ribonucleases are marked with a cross. The signal peptide was not included in the alignment. (Modified from Newbigin *et al.*, 1993; with permission of The American Society of Plant Physiologists.)

these glycoproteins to be isolated was the glycoprotein that co-segregates with the S_2 allele of *Nicotiana alata* (Anderson *et al.*, 1986). The sequence of amino acids at the N-terminus of this protein was used to identify a cDNA encoding the protein component of this molecule; subsequently, cDNAs corresponding to the glycoproteins of the S_3 and S_6 alleles of *N. alata* were obtained (Anderson *et al.*, 1989). Comparison of these cDNAs showed that the derived proteins included a putative signal sequence and a conserved 15-amino acid hydrophobic region at the N-terminus of the mature protein as well as several other conserved sequences that are depicted in Fig. 7.3. This information allowed the cloning of cDNAs corresponding to further S alleles of *N. alata* (Kheyr-Pour *et al.*, 1990) and S alleles from *Petunia inflata* (Ai *et al.*, 1990), *P. hybrida* (petunia, Clark *et al.*, 1990), *Solanum chacoense* (Xu *et al.*, 1990), *S. tuberosum* (potato, Kaufmann *et al.*, 1991), and *Lycopersicon peruvianum* (Tsai *et al.*, 1992; Rivers *et al.*, 1993; Royo *et al.*, 1994a,b). Alignment of a number of these sequences showed that, overall, approximately 16% of the amino acids are identical, including eight cysteine residues. Apart from the cysteine residues, there are five short conserved regions, two of which contain a histidine residue (Fig. 7.3).

These cDNAs are the allelic products of a single locus that co-segregates with the S locus. This was shown by DNA gel blot analysis of genomic DNA from *N. alata* (Anderson *et al.*, 1989). Genomic DNA from plants homozygous for the S_1, S_2, S_3, S_6 or S_7 alleles was digested with restriction enzymes and probed with the S_2, S_3 or S_6 cDNAs. The hybridizing fragments from the different S genotypes were dissimilar in size and constituted a series of restriction fragment length polymorphisms (RFLPs). The behaviour of these RFLPs was examined in a population of plants segregating for four alleles of the S locus. The S genotype of the plants was also determined by conventional breeding experiments. The DNA gel blot analysis of an individual plant with its breeding behaviour and each RFLP was associated with only one S allele. Together, these data show that the cDNAs encoding the style-specific glycoproteins are the allelic products of the S locus in *N. alata*. Similar results have been obtained in other solanaceous species (Bredemeijer and Blass, 1981; Kamboj and Jackson, 1986; Bernatzky, 1993; Kowyama *et al.*, 1994). Collectively, the proteins that segregate with the S locus are referred to as S glycoproteins, with individual proteins named for the S allele with which they are associated (e.g. S_2 glycoprotein is the product of the S_2 allele).

After the initial cloning of a cDNA for an S glycoprotein, the next major breakthrough was built on an observation by Sakiyama and co-workers of two short stretches of amino acid similarity between two extracellular ribonucleases of fungi and the S_2-glycoprotein of *N. alata*. These short stretches corresponded to the sequences surrounding the catalytic domains of fungal ribonucleases and included two histidine residues that formed the putative active site of the molecule (Kawata *et al.*, 1990). These short regions were examined in an allelic series of S glycoproteins from *N. alata* and were found to be conserved in all cases (McClure *et al.*, 1989; Fig. 7.3). This led to the testing of purified S glycoproteins for ribonuclease activity (McClure *et al.*, 1989). Style extracts from *N. alata* were fractionated using cation-exchange chromatography (Jahnen *et al.*, 1989) and individual fractions tested for ribonuclease activity. Ribonuclease activity co-eluted with individual S glycoproteins and, overall, the activity of these fractions accounted for 40–80% of the total activity in the style extracts. Subsequently, S glycoproteins from other solanaceous species have been shown to have ribonuclease activity (Broothaerts *et al.*, 1991; Kaufmann *et al.*, 1991; Singh *et al.*, 1991; Ai *et al.*, 1992; Kowyama *et al.*, 1994) and, as a result, these proteins are now referred to as S ribonucleases (S-RNases).

The latest results on the molecular basis of self-incompatibility in solanaceous plants come from studies with either self-compatible variants of self-incompatible species (Kowyama *et al.*, 1994; Royo *et al.*, 1994b) or plants in which the expression of S-RNases has been modified by transgenic approaches (Huang *et al.*, 1994; Lee *et al.*, 1994; Murfett *et al.*, 1994). Both approaches confirm that S-RNases determine the self-incompatibility phenotype of the pistil and that the ribonuclease activity of these glycoproteins is required for rejection of incompatible pollen tubes. These findings strengthen the view that, if S-RNases enter incompatible pollen tubes, then they would act as cytotoxins by degrading RNA, including ribosomal (r) RNA (McClure *et al.*, 1989). As rRNA genes do not appear to be expressed in mature pollen (Mascarenhas, 1990), degrading the fixed amount of RNA synthesized during early pollen development would be an effective way of arresting pollen-tube growth. This model of self-incompatibility is consistent with observations made earlier by Gray and co-workers (1991) who found that S-RNases were taken up without detectable alteration by *N. alata* pollen tubes grown *in vitro*, and with the degradation of rRNA in pollen following incompatible but not compatible pollinations (McClure *et al.*, 1990). Although there is some evidence as to the site of action (cytoplasm of incompatible pollen tubes) and the mechanism of cell death (RNA degradation), the nature of the allelic interaction between the products of the S locus in pollen and pistil is not established. A major quest in self-incompatibility research is the identification of the S locus product in pollen. For reasons that are not entirely clear, the experimental approaches that led to the identification of S-RNases in the pistil have not proved useful in characterizing the pollen products of the S locus, although one report shows that low amounts of S-RNases are produced during post-meiotic pollen development in *N. alata* (Dodds *et al.*, 1993). Our current thinking is that S-RNases are specifically taken up by incompatible pollen tubes and this may involve domains on the surface of the protein that are very different in sequence between different S-RNase alleles (possibly, the variable domains shown in Fig. 7.3; Ai *et al.*, 1990). In this context, it is worth noting that some S-RNases isolated from the same species are nearly identical in sequence (Royo *et al.*, 1994a; Saba-El-Leil *et al.*, 1994) yet still behave as S alleles in crosses. Thus, the product of the S-locus in pollen must be capable of distinguishing molecules that can be very similar.

Although gametophytic self-incompatibility arose independently a number of times during the evolution of the angiosperms (for review, see Matton *et al.*, 1994), self-incompatibility in the *Scrophulariaceae* and the *Rosaceae* appears to be controlled by stylar ribonucleases similar to the solanaceous S-RNases (Xue *et al.*, 1996). The *Solanaceae* and the *Scrophulariaceae* are closely-related families and only distantly related to the *Rosaceae*. This result suggests that one and possibly two families other than the *Solanaceae* have acquired gametophytic self-incompatibility mechanisms that involve ribonucleases, although it is not known if other aspects of the pathway that leads to rejection of incompatible pollen are similar. Gametophytic self-incompatibility in at least one family, the *Papaveraceae*, involves molecules that are unrelated to the S-RNases and lack ribonuclease activity (Foote *et al.*, 1994).

PRODUCTS OF THE S LOCUS IN PLANTS WITH SPOROPHYTIC SELF-INCOMPATIBILITY

The sporophytic system of self-incompatibility has been most thoroughly studied in the cabbage family, *Brassicaceae*. Self-incompatibility in vegetable crops of this family is used in seed production and there are several practical problems, involving, for example, the periodic loss of alleles which could be solved with a better understanding of the system. Over 50 different S alleles have been identified in *Brassica oleracea* (Ockendon, 1974) and the genetics of self-incompatibility in this group is more complex than the gametophytic self-incompatibility systems of the *Solanaceae*, as it involves dominance and co-dominance of certain alleles.

The early work on the molecular biology of this sporophytic self-incompatibility system was in *B. oleracea* (Nasrallah and Nasrallah, 1984; Nasrallah *et al.*, 1987) and *B. campestris* (Hinata and Nishio, 1978; Isogai *et al.*, 1987). The first indication of components associated with S genotype was the identification of a number of abundant proteins of approximately 60 kDa that accumulated in the stigma one to two days prior to anthesis, coinciding with the onset of self-incompatibility in the flower. Information about the abundance and tissue specificity of these proteins was used to devise a strategy that led to the cloning of a cDNA that encoded a protein of the predicted molecular weight and sequence (Nasrallah *et al.*, 1985). Later, direct sequencing of the purified protein confirmed the identity of this cDNA (Takayama *et al.*, 1987). DNA blot analysis indicated that the genome of *Brassica* contains several copies of the sequence present in the cDNA. Currently, these sequences are classified as SLGs (S locus glycoproteins), that encode the proteins previously identified in the stigma; SLRs (S locus-related sequences), which are genomic sequences, some which are expressed, related in sequence to the SLGs but inherited independently of the S locus; and SRKs (S-receptor kinases), that encode membrane-associated proteins able to phosphorylate serine and threonine residues (for more details, see Nasrallah and Nasrallah, 1993). The genes for both the SLGs and the SRKs are found within the *Brassica* S locus and are within a few hundred kilobases of each other in the genome (Boyes and Nasrallah, 1993). The genes for SLGs and SRKs are expressed both in the pistil and in the pollen grain and anther, although the level of expression found in these latter two tissues is much lower than that found in the pistil, and SRKs are expressed at a much lower level than SLGs (Nasrallah *et al.*, 1985; Sato *et al.*, 1991; Stein *et al.*, 1991). SLGs are particularly abundant in stigmatic papillae, the cells of the pistil where the growth of incompatible pollen grains is arrested. The predicted sequence of the SRK contains a potentially glycosylated, extracellular domain with extensive similarity to SLG at the N-terminus (Stein *et al.*, 1991); indeed, on the basis of sequence comparisons of a number of SLG/SRK gene pairs from the same S allele, it appears that the SLGs may be derived by duplication from the SRKs (Tantikanjana *et al.*, 1993). The C-terminus of the protein has similarity with serine/threonine kinases and it is thought this region is located within the cytoplasm and joined to the extracellular SLG-like domain by a short, membrane-spanning domain (Stein *et al.*, 1991).

A number of lines of evidence implicate both SLGs and SRKs in sporophytic self-incompatibility (Toriyama *et al.*, 1991; Nasrallah *et al.*, 1992; Tantikanjana *et al.*, 1993; Nasrallah *et al.*, 1994) and several models have been proposed to describe the interaction between these two molecules (for example, see Nasrallah and Nasrallah, 1993). Unlike the gametophytic system outlined above in which the cells of the stigma and style play no other known role in pollen rejection other than to secrete S-RNases, in *Brassica*, the cells of the stigmatic papillae prevent penetration of the style by the emerging pollen tube and most models used to describe this system concentrate on molecular events occurring in these cells. Most models assume that the extracellular SLG binds to the SLG-like domain on the membrane-associated SRK found on the stigmatic papillae. The binding of the SLG to the SRK is thought to initiate a signal transduction pathway within the stigmatic cells that begins with the phosphorylation of intracellular substrates by the kinase domain on the SRK and ends with changes to the papillar cell wall and pollen rejection. The specificity of the binding of the SLG to the SRK is presumed to be mediated by an as-yet-unidentified extracellular molecule that is encoded by the S locus and delivered to the stigma by the pollen. This molecule is thought to either modify the SLG to allow binding to the SRK or bind with the SLG to the SRK. In both cases, the molecule acts in an allele-specific manner (Nasrallah and Nasrallah, 1993).

EVOLUTION OF SELF-INCOMPATIBILITY

A major interest of many early writers was the evolutionary implications of the broad taxonomic distribution of self-incompatibility among different lineages of angiosperms. For example, some workers held that self-incompatibility played a significant role in the evolutionary success of flowering plants and indeed, after considering a number of theoretical and phylogenetic arguments, Whitehouse (1950) and Stebbins (1957) concluded that the ancestors of the angiosperms were self-incompatible. According to this view, the different forms of self-incompatibility found in existing angiosperms are derived from the form present in the protoangiosperms and hence, all self-incompatibility systems are related. Other authors viewed the different types of self-incompatibility as unrelated to each other with the transition from self-fertility to self-incompatibility having occurred a number of times during the evolution of the angiosperms (East, 1940; Bateman, 1952). The molecular evidence outlined above supports the latter view and raises the possibility that each system arose by adaptation of pre-existing signalling pathways in the plant. These pathways may not necessarily have been involved in fertilization or pollen–stigma interactions; indeed, it is widely thought that both the *Brassica* and solanaceous self-incompatibility systems are derived from systems involved in defence against pathogens (Dickinson, 1994; Matton *et al.*, 1994). One interesting possibility is that some closely related families may have evolved different self-incompatibility systems based on the same molecules. This is most likely to be true for the *Convolvulaceae* (sweet potato) and the *Solanaceae*, which have sporophytic and gametophytic self-incompatibility, respectively, and is supported by the observation that allelism at the S locus appears to predate speciation of the *Solanaceae* (Ioerger *et al.*, 1990). Thus, the alleles present at the solanaceous S locus could have arisen in a species that was ancestral to both the *Convolvulaceae* and *Solanaceae*; the subsequent evolution of self-incompatibility in the two families was presumably accompanied by a change in the timing or site of expression of the S locus as outlined in Fig. 7.2.

INTERSPECIFIC INCOMPATIBILITY

Plants often form viable seed following pollination with pollen from closely related species. Humans have exploited this feature to introduce agriculturally important traits such as disease resistance from wild relatives into domesticated crops. Often, however, pollinations of this type fail because of reproductive barriers that genetically isolate the species participating in the cross. There are many types of these reproductive barriers that have been variously classified by different authors (see Hogenboom, 1975, for one example). Here, we discuss interspecific barriers characterized by the arrest of pollen tube growth within the style, and therefore superficially resembling self-incompatibility. Typically, incompatibility of this type occurs between a self-compatible species and a self-incompatible relative. Crosses between such species are often successful (i.e. result in the production of viable seed) when the self-incompatible plant is the pollen donor, but not when pollen from the self-compatible plant is used to pollinate the self-incompatible species. This phenomenon was first described by Harrison and Darby (1955), who intercrossed in all possible combinations, two self-compatible and six self-incompatible species of *Antirrhinum*. It is also well documented in many important taxonomic groups such as the *Brassicaceae* (Sampson, 1962; Hiscock and Dickinson, 1993), *Lycopersicon* (Martin, 1967) and *Solanum* (Grun and Aubertin, 1966). Numerous terms have been used to describe this reproductive barrier of which the most common is unilateral incompatibility.

Although incompatibility generally occurs between the pollen of a self-compatible species and the style of a self-incompatible species, there are exceptions; for example, there are reports of unilateral incompatibility between self-compatible species, between self-incompatible species, and even rejection of pollen from a self-incompatible species by the style of a self-compatible species (Martin, 1967, 1968; Pandey, 1969). The relationship between unilateral and self-incompatibility is not clear although many authors favour models in which the S locus plays a major role in determining the outcome of interspecific crosses (Lewis and Crowe, 1958; Pandey 1979). Consistent with this, Chetelat and DeVerna (1991) showed that a locus at or near the S locus has an essential role in unilateral incompatibility in *Lycopersicon* and, in the *Brassicaceae*, Hiscock and Dickinson (1993) showed that treatments used to overcome self-incompatibility also overcome unilateral incompatibility. Other authors, notably Hogenboom (1973, 1975), have argued that there is no relationship between unilateral and self-incompatibility, and this may be true in some cases involving crosses between more distantly related species. The observation that loci unlinked to the S locus in *Lycopersicon* play an essential role in unilateral incompatibility argues against hypotheses that the S locus is the sole cause of unilateral incompatibility (Chetelat and DeVerna, 1991). It is worth mentioning in this context that self-incompatibility in the *Solanaceae* is also influenced by loci unlinked to the S locus (Ai *et al.*, 1991; Thompson *et al.*, 1991) although any relationship between these 'modifier loci' and the loci involved in unilateral incompatibility is untested. Further work on the genetics of unilateral incompatibility is required before the importance of self-incompatibility to its function can be understood. A practical outcome of research on mapping the loci involved in unilateral incompatibility would be the ability to identify particular interspecific hybrid plants likely to be cross-compatible with both wild and cultivated relatives. These plants could be used as bridging lines in the transfer of wild germplasm to cultivated species and hence of agricultural benefit in the continued development of new crop varieties. Since the S locus undoubtedly plays a role in unilateral incompatibility in many plants, rapid assays to detect the S-RNases in solanaceous plants and the SLGs/SRKs of the *Brassicaceae*, could be useful in breeding programmes involving species from these families and could potentially be extended to other families with self-incompatibility systems based on related molecules (see above).

CONCLUSIONS AND PERSPECTIVES

In this review, we have discussed unilateral and self-incompatibility, two reproductive barriers found in flowering plants. Both are examples of cell–cell communication in higher plants; in self-incompatibility, the pollen and the pistil perceive information about the S alleles present in each with a match of S alleles leading to rejection of the pollen grain (see Fig. 7.1). Some of the molecules involved in self-incompatibility are known and models describing this phenomenon at the molecular level are available.

Self and unilateral incompatibility are both a boon and a hindrance to plant breeders. Self-incompatibility between inbred lines has been used for a number of years in the production of F_1-hybrid seed since it eliminates the need for costly hand emasculation and pollination (Frankel and Galun, 1977; Liedl and Anderson, 1993). However, self-incompatibility can be a disadvantage; for example, self-incompatibility in fruit trees necessitates the interplanting of cultivars with different self-incompatibility alleles and breeding programmes can be severely restricted when the number of S alleles is limited. Currently, sporophytic self-incompatibility is widely used in the production of hybrid seeds from a variety of *Brassica* species including *B. oleracea* (Brussels sprouts, cabbages and cauliflowers) and *B. chinensis* (Chinese cabbage). Because many important horticultural crops such as tomato (*L. esculentum*) have wild relatives that are self-incompatible, the utility of self-incompatibility to hybrid seed production could potentially be extended by introgressing this genetic trait into the cultivated species. The reason this has not been done reflects the complex nature of reproductive barriers within the *Lycopersicon* genus (see above). Knowledge of the molecular biology of self-incompatibility in the *Solanaceae* and recent success with transformation experiments involving S-RNases (Huang *et al.*, 1994; Lee *et al.*, 1994; Murfett *et al.*, 1994) opens up the possibility that self-incompatibility could be introduced into a wide range of self-compatible and transformable crops. The S locus in solanaceous plants is probably complex and a number of genes in the locus, particularly those involved in the allelic interaction between pollen and pistil, remain to be discovered. Even with a complete description of all the genes involved in self-incompatibility, the task of transferring and correctly regulating these genes in target species is far from straightforward.

Many authors have compared self-incompatibility to other processes in plants that involve cell–cell signalling, such as host–pathogen interactions (for example, see Pryor, 1987; Hodgkin *et al.*, 1988). Like self-incompatibility, the infection of a host by certain pathogens can be either compatible, resulting in a systemic colonization of the plant by the pathogen, or incompatible, in which further colonization of the plant by the pathogen is prevented. The underlying genetic mechanisms that control this response are also similar to self-incompatibility in that they involve interactions between genes for resistance in the plant and genes for avirulence in the pathogen. Like the S locus that controls self-incompatibility, there are many alleles of both the plant's resistance gene and the avirulence gene of the pathogen with certain combinations of resistance and avirulence alleles leading to an incompatible interaction. Recent molecular data on the sequence of a tomato gene encoding resistance to bacterial speck (*Pseudomonas syringae*) reinforces the similarities between host–pathogen interactions and self-incompatibility. The gene, known as *Pto*, encodes a protein with similarity to plant serine/threonine kinase, including the SRKs from *Brassica* involved in sporophytic self-incompatibility (Martin *et al.*, 1993). However, unlike SRKs, the putative *Pto* protein contains no obvious membrane-spanning or extracellular domains and may be localized to the cytoplasm. Thus both sporophytic self-incompatibility in *Brassica* and resistance to bacterial pathogens probably involve phosphorylation, although the nature of the substrates and the subsequent steps in the signal-transduction pathway are not known in either case.

Apart from some reproductive barriers and host–pathogen interactions, do any other processes in plants discriminate between compatible and incompatible cells, and do any of these processes also involve cell–cell communication? One possible example occurs during grafting and some authors think this may involve specific cell signalling between the understock and the scion (Clarke and Knox, 1978; Yeoman, 1984). There are three distinct stages in the formation of a graft: cohesion of the stock to the scion, proliferation of callus at the graft interface, and vascular redifferentiation across the graft interface. Obviously, grafts can fail for a variety of reasons, such as environmental stress and infection. However, grafts between certain species always fail, a phenomenon known as graft incompatibility. In these cases, rejection of an incompatible scion occurs late in development of the graft and probably results from the failure to achieve vascular continuity with the stock (Yeoman, 1984). As a rule, graft incompatibility increases as the taxonomical distance between the species whose tissues are being grafted increases, although there are notable exceptions; for example, it is possible to graft *Helianthus annus* and *Vicia faba* (Kollmann and Glockmann, 1985), but grafts between some cultivars of apple do not take (Herrero, 1951). Graft incompatibility is complex and in many instances can be accounted for by the failure of long-distance hormonal interactions or the release of toxic products by the stock (Moore, 1984; Sachs, 1991). Some examples of graft incompatibility appear to involve short-distance interactions that are overcome by an interstock, a small piece of tissue taken from a compatible genotype and used to graft between two incompatible genotypes (Yeoman, 1984). The nature of the local interactions between cells at the graft interface is unknown, but of considerable importance given the impact of graft incompatibility on horticulture and forestry. Further research in this area is warranted, but requires the development of a suitable model; a strong candidate for such a model is *Lycopersicon esculentum* which combines a well-defined genetic system (Tanksley *et al.*, 1992) with a rapid and reliable assay for graft incompatibility (Parkinson *et al.*, 1987). The combination of classical genetics, cell biology and molecular biology led to the beginnings of a molecular understanding of self-incompatibility, and a similar approach could be applied to the problem of graft incompatibility. Notwithstanding our current lack of molecular information on the mechanism of graft incompatibility, it is clear that plant cells have elaborate recognition and discrimination systems that are expressed in both vegetative and reproductive tissues. The ways in which these systems evolved and contributed to the success of flowering ornamental plants will remain challenging research questions for the future.

ACKNOWLEDGEMENTS

The authors thank their colleagues at the PCBRC and the School of Botany at Melbourne University for contributions to various parts of this work. N.N. is supported by a BASF research fellowship granted by the Studienstifung des Deutschen Volkes. Research at the PCBRC is supported by a Special Research Centre grant from the Australian Research Council.

REFERENCES

Ai, Y., Singh, A., Coleman, C.E., Ioerger, T.R., Kheyr-Pour, A. and Kao, T.-H. (1990) Self-incompatibility in *Petunia inflata*: isolation and characterization of cDNAs encoding three S allele-associated proteins. *Sexual Plant Reproduction* 3, 130–138.

Ai, Y., Kron, E. and Kao, T.-H. (1991) S alleles are retained and expressed in self-compatible cultivar of *Petunia hybrida*. *Molecular and General Genetics* 230, 353–359.

Ai, Y., Tsai, D.-S. and Kao, T.-H. (1992) Cloning and sequencing of cDNAs encoding two S proteins of a self-compatible cultivar of *Petunia hybrida*. *Plant Molecular Biology* 19, 523–528.

Anderson, M.A., Hoggart, R.D. and Clarke, A.E. (1983) The possible role of lectins in mediating plant cell–cell interactions. In: Goldstein, I.J. and Etzler, M.E. (eds), *Progress in Clinical and Biological Research, Vol. 138: Chemical Taxonomy, Molecular Biology and Function of Plant Lectins.* Alan R. Liss, New York, pp. 143–161.

Anderson, M.A., Cornish, E.C., Mau, S.-L., Williams, E.G., Hoggart, R., Atkinson, A., Bönig, I., Grego, B., Simpson, R., Roche, P.J., Haley, J.D., Penschow, J.D., Niall, H.D., Tregear, G.W., Coghlan, J.P., Crawford, R.J. and Clarke, A.E. (1986) Cloning of a cDNA for a stylar glycoprotein associated with expression of self-incompatibility in *Nicotiana alata. Nature* 321, 38–44.

Anderson, M.A., McFadden, G.I., Bernatzky, R., Atkinson, A., Orpin, T., Dedman, H., Tregear, G., Fernley, R. and Clarke, A.E. (1989) Sequence variability of three alleles of the self-incompatibility gene of *Nicotiana alata. Plant Cell* 1, 483–491.

Bateman, A.J. (1952) Self-incompatibility systems in angiosperms. I. Theory. *Heredity* 6, 285–310.

Bernatzky, R. (1993) Genetic mapping and protein product diversity of the self-incompatibility locus in wild tomato (*Lycopersicon peruvianum*). *Biochemical Genetics* 31, 173–184.

Boyes, D.C. and Nasrallah, J.B. (1993) Physical linkage of the SLG and SRK genes at the self-incompatibility locus of *Brassica oleracea. Molecular and General Genetics* 236, 369–373.

Bredemeijer, G.M.M. and Blass, J. (1981) S-specific proteins in styles of self-incompatible *Nicotiana alata. Theoretical and Applied Genetics* 59, 185–190.

Broothaerts, W., Vanvinckenroye, P., Decock, B., Van Damme, J. and Vendrig, J.C. (1991) *Petunia hybrida* S-proteins: ribonuclease activity and the role of their glycan side chains in self-incompatibility. *Sexual Plant Reproduction* 4, 258–266.

Chetelat, R.T. and DeVerna, J.W. (1991) Expression of unilateral incompatibility in pollen of *Lycopersicon penellii* is determined by major loci on chromosomes 1, 6 and 10. *Theoretical and Applied Genetics* 82, 704–712.

Clark, K.R., Okuley, J.J., Collins, P.D. and Sims, T.L. (1990) Sequence variability and development expression of S alleles in self-incompatible and pseudo-self-incompatible *Petunia. Plant Cell* 2, 815–826.

Clarke, A.E. and Knox, R.B. (1978) Cell recognition in flowering plants. *Quarterly Reviews of Biology* 53, 3–28.

Darwin, C. (1877) *The Different Forms of Flowers and Plants of the Same Species.* John Murray, London.

de Nettancourt, D. (1977) *Incompatibility in Angiosperms.* Springer-Verlag, New York.

Dickinson, H. (1994) Simply a social disease? *Nature* 367, 517–518.

Dodds, P.N., Bönig, I., Du, H., Rödin, J., Anderson, M.A., Newbigin, E. and Clarke, A.E. (1993) S-RNase gene of *Nicotiana alata* is expressed in developing pollen. *Plant Cell* 5, 1771–1782.

East, E.M. (1940) The distribution of self-sterility in flowering plants. *Proceedings of the American Philosophical Society* 82, 449–518.

Foote, H.C.C., Ride, J.P., Franklin-Tong, V.E., Walker, E.A., Lawrence, M.J. and Franklin, F.C.H. (1994) Cloning and expression of a distinctive class of self-incompatibility (S) gene from *Papaver rhoeas* L. *Proceedings of the National Academy of Sciences, USA* 91, 2265–2269.

Frankel, R. and Galun, E. (1977) *Pollination Mechanisms, Reproduction and Plant Breeding.* Springer-Verlag, Berlin.

Gibbs, P.E. (1986) Do homomorphic and heteromorphic self-incompatibility systems have the same sporophytic mechanisms? *Plant Systematics and Evolution* 154, 285–323.

Gray, J.E., McClure, B.A., Bönig, I., Anderson, M.A. and Clarke, A.E. (1991) Action of the style product of the self-incompatibility gene of *Nicotiana alata* (S-RNase) on *in vitro*-grown pollen tubes. *Plant Cell* 3, 271–283.

Grun, P. and Aubertin, M. (1966) The inheritance and expression of unilateral incompatibility in *Solanum*. *Heredity* 21, 131–138.

Harrison, B.J. and Darby, L.A. (1955) Unilateral hybridization. *Nature* 176, 982

Herrero, J. (1951) Studies of compatible and incompatible graft combinations with special reference to hardy fruit trees. *Journal of Horticultural Science* 26, 186–237.

Hinata, K. and Nishio, T. (1978) S allele specificity of stigma proteins in *Brassica oleracea* and *Brassica campestris*. *Heredity* 41, 93–100.

Hiscock, S.J. and Dickinson, H.G. (1993) Unilateral incompatibility within the *Brassicaceae*: further evidence for the involvement of the self-incompatibility (S)-locus. *Theoretical and Applied Genetics* 86, 744–753.

Hodgkin, T., Lyon, G.D. and Dickinson, H.G. (1988) Recognition in flowering plants: a comparison of the *Brassica* self-incompatibility system and plant pathogen interactions. *New Phytologist* 110, 557–569.

Hogenboom, N.G. (1973) A model for incongruity in intimate partner relationships. *Euphytica* 22, 219–233.

Hogenboom, N.G. (1975) Incompatibility and incongruity. Two different mechanisms for the non-functioning of intimate partner relationships. *Proceedings of the Royal Society of London, Series B* 188, 361–375.

Huang, S., Lee, H.-Y., Karunanandaa, B. and Kao, T.-H. (1994) Ribonuclease activity of *Petunia inflata* S proteins is essential for rejection of self-pollen. *Plant Cell* 6, 1021–1028.

Ioerger, T.R., Clark, A.G. and Kao, T.-H. (1990) Polymorphism at the self-incompatibility locus of Solanaceae predates speciation. *Proceedings of the National Academy of Sciences, USA* 87, 9732–9735.

Isogai, A., Takayama, S., Tsukamoto, C., Ueda, Y., Shiozawa, H., Hinata, K., Okazaki, K. and Suzuki, A. (1987) S-locus-specific glycoproteins associated with self-incompatibility in *Brassica campestris*. *Plant and Cell Physiology* 28, 1279–1291.

Jahnen, W., Batterham, M.P., Clarke, A.E., Moritz, R.L. and Simpson, R.J. (1989) Identification, isolation and N-terminal sequencing of style glycoproteins associated with self-incompatibility in *Nicotiana alata*. *Plant Cell* 1, 493–499.

Kamboj, R.K. and Jackson, J.F. (1986) Self-incompatibility alleles control a low molecular weight, basic protein in pistils of *Petunia hybrida*. *Theoretical and Applied Genetics* 71, 815–819.

Kao, T.-H. and Huang, S. (1994) Gametophytic self-incompatibility: a mechanism for self/nonself discrimination during sexual reproduction. *Plant Physiology* 105, 461–466.

Kaufmann, H., Salamini, F. and Thompson, R.D. (1991) Sequence variability and gene structure at the self-incompatibility locus of *Solanum tuberosum*. *Molecular and General Genetics* 226, 457–466.

Kawata, Y., Sakiyama, F., Hayashi, F. and Kyogoku, Y. (1990) Identification of two essential histidine residues of ribonuclease T_2 from *Aspergillus oryzae*. *European Journal of Biochemistry* 176, 683–697.

Kheyr-Pour, A., Bintrim, S.B., Ioerger, T.R., Remy, R., Hammond, S.A. and Kao, T.-H. (1990) Sequence diversity of pistil S proteins associated with gametophytic self-incompatibility in *Nicotiana alata*. *Sexual Plant Reproduction* 3, 88–97.

Kollmann, R. and Glockmann, C. (1985) Studies on graft union. I. Plasmodesmata between cells of plants belonging to plants of different, unrelated taxa. *Protoplasma* 124, 224–235.

Kowyama, Y., Kunz, C., Lewis, I., Newbigin, E., Clarke, A.E. and Anderson, M.A. (1994) Self-compatibility in a *Lycopersicon peruvianum* variant (LA2157) is associated with a lack of style S-RNase activity. *Theoretical and Applied Genetics* 88, 859–864.

Lee, H.-S., Huang, S. and Kao, T.-H. (1994) S proteins control rejection of incompatible pollen in *Petunia inflata*. *Nature* 367, 560–563.

Lewis, D. and Crowe, L.K. (1958) Unilateral interspecific incompatibility in flowering plants. *Heredity* 12, 233–256.

Liedl, B.E. and Anderson, N.O. (1993) Reproductive barriers: identification, uses and circumvention. In: Janick, J. (ed.), *Plant Breeding Reviews*. John Wiley, New York, vol. II, pp. 11–154.

Martin, F.W. (1967) The genetic control of unilateral incompatibility between two tomato species. *Genetics* 56, 391–398.

Martin, F.W. (1968) The behaviour of *Lycopersicon* incompatibility alleles in an alien genetic milieu. *Genetics* 60, 101–109.

Martin, G.B., Brommonschenkel, S.H., Chunwongse, J., Frary, A., Ganal, M.W., Spivey, R., Wu, T., Earle, E.D. and Tanksley, S.D. (1993) Map-based cloning of a protein kinase gene conferring disease resistance in tomato. *Science* 262, 1432–1436.

Mascarenhas, J.P. (1990) Gene activity during pollen development. *Annual Reviews of Plant Physiology and Plant Molecular Biology* 41, 317–338.

Matton, D.P.M., Nass, N., Clarke, A.E. and Newbigin, E. (1994) Self-incompatibility: how plants avoid illegitimate offspring. *Proceedings of the National Academy of Sciences, USA* 91, 1992–1997.

McClure, B.A., Haring, V., Ebert, P.R., Anderson, M.A., Simpson, R.J., Sakiyama, F. and Clarke A.E. (1989) Style self-incompatibility gene products of *Nicotiana alata* are ribonucleases. *Nature* 342, 955–957.

McClure, B.A., Gray, J.E., Anderson, M.A. and Clarke, A.E. (1990) Self-incompatibility in *Nicotiana alata* involves degradation of rRNA. *Nature* 347, 757–760.

Moore, R. (1984) A model for graft compatibility–incompatibility in higher plants. *American Journal of Botany* 71, 752–758.

Murfett, J., Atherton, T.L., Mou, B., Gasser, C.S. and McClure, B.A. (1994) S-RNases expressed in transgenic *Nicotiana* cause S allele-specific pollen rejection. *Nature* 367, 563–566.

Nasrallah, J.B. and Nasrallah, M.E. (1984) Electrophoretic heterogeneity exhibited by the S allele specific glycoproteins of *Brassica*. *Experientia* 40, 279–280.

Nasrallah, J.B. and Nasrallah, M.E. (1993) Pollen–stigma interactions in the sporophytic self-incompatibility response. *Plant Cell* 5, 1325–1335.

Nasrallah, J.B., Kao, T.-H., Goldberg, M.L. and Nasrallah, M.E. (1985) A cDNA clone encoding an S locus specific glycoprotein from *Brassica oleracea*. *Nature* 318, 617–618.

Nasrallah, J.B., Kao, T.-H., Chen, C.-H., Goldberg, M.L. and Nasrallah, M.E. (1987) Amino-acid sequence of glycoproteins encoded by three alleles of the S locus of *Brassica oleracea*. *Nature* 326, 617–619.

Nasrallah, J.B., Rundle, S.J. and Nasrallah, M.E. (1994) Genetic evidence for the requirement of the *Brassica* S locus receptor kinase gene in the self-incompatibility response. *Plant Journal* 5, 373–384.

Nasrallah, M.E., Kandasamy, M.K. and Nasrallah, J.B. (1992) A genetically defined *trans* acting locus regulates S locus function in *Brassica*. *Plant Journal* 2, 497–506.

Newbigin, E., Anderson, M.A. and Clarke, A.E. (1993) Gametophytic self-incompatibility systems. *Plant Cell* 5, 1315–1324.

Ockendon, D.J. (1974) Distribution of self-incompatibility alleles and breeding structure in open-pollinated cultivars of Brussels sprouts. *Heredity* 33, 159–171.

Pandey, K.K. (1969) Elements of the S gene complex. V. Interspecific cross-compatibility relationships and theory of the evolution of the S complex. *Genetica* 40, 447–474

Pandey, K.K. (1979) The genus *Nicotiana*: evolution of incompatibility in flowering plants. In: Hawkes, J.G., Lester, R.N. and Skelding, A.D. (eds), *Biology and Taxonomy of the Solanaceae*. Academic Press, London, pp. 421–433.

Pandey, K.K. (1980) Evolution of incompatibility systems in plants: origins of 'independent' and 'complementary' control of incompatibility in angiosperms. *New Phytologist* 84, 381–400.

Parkinson, M., Jeffree, C.E. and Yeoman, M.M. (1987) Incompatibility in cultured explant grafts between members of the Solanaceae. *New Phytologist* 107, 489–498.

Pryor, A. (1987) The origin and structure of fungal disease resistance genes in plants. *Trends in Genetics* 3, 157–161.

Rivers, B.A., Bernatzky, R., Robinson, S.J. and Jahnen-Dechent, W. (1993) Molecular diversity at the self-incompatibility locus is a salient feature in natural populations of wild tomato (*Lycopersicon peruvianum*). *Molecular and General Genetics* 238, 419–427.

Royo, J., Kowyama, K. and Clarke, A.E. (1994a) Cloning and nucleotide sequence of two S-RNases from *Lycopersicon peruvianum* (L.) Mill. *Plant Physiology* 105, 751–752.

Royo, J., Kunz, C., Kowyama, Y., Anderson, M.A., Clarke, A.E. and Newbigin, E. (1994b) Loss of a histidine at the active site of S-RNases is associated with self-compatibility in *Lycopersicon peruvianum*. *Proceedings of the National Academy of Sciences, USA* 91, 6511–6514.

Saba-El-Leil, M.K., Rivard, S., Morse, D. and Cappadocia, M. (1994) The S11 and S13 self incompatibility alleles of *Solanum chacoense* Bitt. are remarkably similar. *Plant Molecular Biology* 24, 571–583,

Sachs, T. (1991) *Pattern Formation in Plant Tissues.* Cambridge University Press, Cambridge.

Sampson, D.R. (1962) Intergenic pollen–stigma incompatibility in Cruciferae. *Canadian Journal of Genetics and Cytology* 4, 38–49.

Sassa, H., Hirano, H. and Ikehashi, H. (1993) Identification and characterization of stylar glycoproteins associated with self-incompatibility genes of Japanese pear, *Pyrus serotina* Rehd. *Molecular and General Genetics* 241, 17–25.

Sato, T., Thorsness, M.K., Kandasamy, M.K., Nishio, T., Hirai, M., Nasrallah, J.B. and Nasrallah, M.E. (1991) Activity of an S locus gene promoter in pistils and anthers of transgenic *Brassica*. *Plant Cell* 3, 867–876.

Singh, A., Ai, Y. and Kao, T.-H. (1991) Characterization of ribonuclease activity of three S allele-associated proteins of *Petunia inflata*. *Plant Physiology* 96, 61–68.

Stebbins, G.L. (1957) Self-fertilization and population variability in the higher plants. *American Naturalist* 91, 337–354.

Stein, J.C., Howlett, B., Boyes, D.C., Nasrallah, M.E. and Nasrallah, J.B. (1991) Molecular cloning of a putative receptor protein kinase gene encoded by the self-incompatibility locus of *Brassica oleracea*. *Proceedings of the National Academy of Sciences, USA* 88, 8816–8820.

Takayama, S., Isogai, A., Tsukamoto, C., Ueda, Y., Hinata, K., Okazaki, K. and Suzuki, A. (1987) Sequences of S glycoproteins, products of *Brassica campestris* self-incompatibility locus. *Nature* 326, 102–104.

Tanksley, S.D., Ganal, M.W., Prince, J.P., de Vicente, M.C., Bonierbale, M.W., Broun, P., Fulton, T.M., Giovannoni, J.J., Grandillo, S., Martin, G.B., Messeguer, R., Miller, J.C., Miller, L., Paterson, A.H., Pineda, O., Röder, M.S., Wing, R.A., Wu, W. and Young, N.D. (1992) High density molecular linkage maps of the tomato and potato genomes. *Genetics* 132, 1141–1160.

Tantikanjana, T., Nasrallah, M.E., Stein, J.C., Chen, C.-H. and Nasrallah, J.B. (1993) An alternative transcript of the S locus glycoprotein gene in a class II pollen-recessive self-incompatibility haplotype of *Brassica oleracea* encodes a membrane-anchored protein. *Plant Cell* 5, 657–666.

Thompson, R.D., Uhrig, H., Hermsen, J.G.Th., Salamini, F. and Kaufmann, H. (1991) Investigation of a self-compatible mutation in *Solanum tuberosum* clones inhibiting S allele activity in pollen differentially. *Molecular and General Genetics* 226, 283–288.

Toriyama, K., Stein, J.C., Nasrallah, M.E. and Nasrallah, J.B. (1991) Transformation of *Brassica oleracea* with an S locus gene from *B. campestris* changes the self-incompatibility phenotype. *Theoretical and Applied Genetics* 81, 769–776.

Tsai, D.-S., Lee, H.S., Post, L.C., Kreiling, K.M. and Kao, T.-H. (1992) Sequence of an S protein of *Lycopersicon peruvianum* and comparison with other solanaceous S proteins. *Sexual Plant Reproduction* 5, 256–263.

Whitehouse, H.L.K. (1950) Multiple-allelomorph incompatibility of pollen and style in the evolution of the angiosperms. *Annals of Botany* 14, 198–216.

Xu, B., Mu, J., Nevins, D.L., Grun, P. and Kao, T.-H. (1990) Cloning and sequencing of cDNAs encoding two self-incompatibility associated proteins in *Solanum chacoense. Molecular and General Genetics* 224, 341–346.

Xue, Y., Carpenter, R., Dickinson, H.G. and Coen, E.S. (1996) Origin of allelic diversity in *Antirrhinum* S locus RNases. *Plant Cell* 8, 805–814.

Yeoman, M.M. (1984) Cellular recognition systems in grafting. In: Linskens, H.F. and Heslop-Harrison, J. (eds), *Cellular Interactions, Encyclopedia of Plant Physiology.* Springer-Verlag, Berlin, pp. 453–472.

Genome Mapping 8

Ross Whetten
Department of Forestry, North Carolina State University, Raleigh, North Carolina 27695, USA

INTRODUCTION

The development of techniques for identifying genetic variation at a molecular level has changed the nature of genetic research. This chapter briefly describes several methods of detecting genetic variation, and discusses the relative strengths and weaknesses of each method. Strategies for implementation of mapping tools are outlined, and the chapter concludes with a discussion of the applications of genetic mapping to plant breeding.

TWO APPROACHES TO GENOMIC MAPPING

Two methods of mapping genomes are available: physical mapping and genetic linkage mapping. A complete characterization of the genome of an organism would use both methods, as the information gained is complementary rather than redundant. The cost of physical mapping is high, however, and, in many cases, physical map information is less useful to plant breeders than genetic linkage map information. This chapter will focus on genetic mapping; interested readers are referred to Davies and Tilghman (1990) and Beckmann and Osborn (1992) for further discussion of both genetic and physical mapping.

GENETIC MAPPING

Genetic linkage maps are made by analysing the segregation of genetic markers in progeny derived from a known parent or parents. The genetic markers can be functional genes that affect the phenotype of the progeny, or randomly selected regions of DNA. The only requirement is that there must

 Biotechnology of Ornamental Plants
(eds R.L. Geneve, J.E. Preece and S.A. Merkle)

be reproducible and heritable differences between the alleles of the marker loci, so that different alleles of the markers are detectable in different progeny. Alleles of markers on the same chromosome tend to segregate together, allowing the markers to be classified into linkage groups. The amount of recombination that occurs between two markers in the same linkage group is taken as a measure of the genetic distance between the markers.

The nature of the pedigree used in a mapping experiment depends on the mating system and the degree of heterozygosity found in the species of interest. Many ornamental plant species are outbred and highly heterozygous, and the strategies used for mapping in inbred crop species are not directly applicable. The primary requirement for mapping is production of 50 to 100 offspring from a highly heterozygous individual or set of genetically identical individuals, and this requirement can usually be met.

The traditional approach to genetic mapping with inbred species has been to make a cross between two inbred lines to form an F_1 hybrid with maximal heterozygosity. The F_1 hybrids are then selfed or intercrossed to give an F_2 segregating population, or backcrossed to one parent to give a BC_1 segregating population. In either case, because the original inbred lines are homozygous at all loci, only two alleles are present at any locus in the segregating population (Fig. 8.1A). This is frequently not the case with populations derived from crossing two highly heterozygous individuals. As many as four alleles can be present at any one locus in the segregating population, presenting calculation problems that software designed for use with inbred species cannot solve (e.g. Lander *et al.*, 1987). New strategies have been developed for genetic mapping in outbred species, notably in forest tree species (Grattapaglia *et al.*, 1992). New software is available to analyse segregation data in which up to four different alleles per locus are present in the population (B.-H. Liu, Department of Forestry and Statistics, North Carolina State University, 1994, personal communication). Loci at which four different alleles are segregating become highly informative when the data are properly analysed (Fig. 8.1B).

METHODS OF DETECTING MOLECULAR GENETIC MARKERS

DETECTION OF PROTEIN VARIATION

Variation in functional genes can be manifested as alterations in the size or charge of the protein products encoded by the genes. Gel electrophoresis can separate proteins on the basis of net charge, total mass, or charge-to-mass ratio. Two types of molecular markers rely on detecting variation at the level of proteins.

Isozymes

Variation in genes that encode enzymes can be detected in some cases as differences in electrophoretic mobility of the enzymes. More than 30 enzymes can be assayed after separation by electrophoresis on either starch or acrylam-

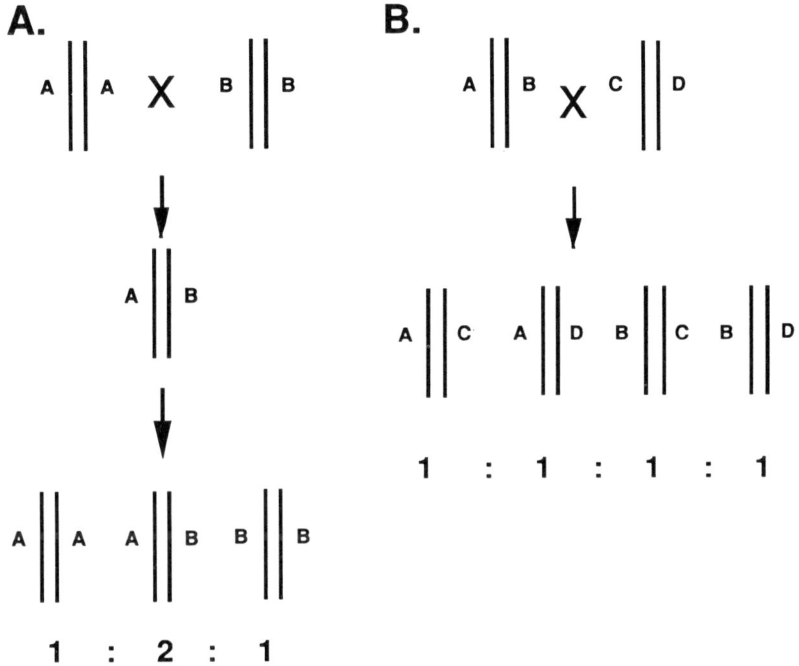

Fig. 8.1. Comparison of pedigrees available in inbred species and outbred species. (A) Most mapping pedigrees used in inbred species segregate only two alleles per locus, whether analysed in F_2 populations derived from F_1 selfing or intercrossing, or in backcross populations derived from crosses of F_1 hybrids back to each parent. Separate backcross populations can be used to analyse recombination rates in pollen parent and seed parent independently. (B) Crosses between highly heterozygous outbred species can segregate up to four alleles per locus. Recombination rates for pollen parent and seed parent can be measured simultaneously in the F_1 hybrid population, and all progeny are informative.

ide gels (Brewer, 1970; Acquaah, 1992). These enzymes can be used as genetic markers, but are limited in number. Isozymes are still used in conjunction with other molecular markers, however, because of the extensive experience that has accumulated using isozymes to analyse genetic variation in plants (e.g. Hemmat *et al.*, 1994).

Anonymous proteins

Two-dimensional gel electrophoresis, separating proteins first on the basis of net charge then on the basis of molecular mass, can be used to resolve mixtures of thousands of proteins (O'Farrell, 1975). The ability to examine thousands of gene products simultaneously for variation in size or charge has been used as a means of genetic mapping (Leonardi *et al.*, 1988). The number of potential genetic markers available through this technique is large. The technical difficulty of carrying out two-dimensional protein electrophoresis as a means of identifying genetic markers is likely to limit the general utility of this approach, although it has found application in some studies (Damerval *et al.*, 1994).

DETECTION OF DNA SEQUENCE VARIATION

Detection of DNA sequence variation can be based on cloned DNA fragments, on amplification of arbitrary sequences, or on analysis of total genomic DNA on two-dimensional gels. Markers detected on the basis of cloned DNA fragments include restriction fragment length polymorphism (RFLP) markers and microsatellite or variable number of tandem repeat (VNTR) markers. Markers detected by arbitrary amplification include random amplified polymorphic DNA (RAPD) markers and amplified fragment length polymorphism (AFLP) markers. Techniques for detecting markers by two-dimensional analysis of genomic DNA include restriction landmark genome scanning (RLGS) and a two-dimensional DNA typing technique that has not yet been given an acronym.

Markers detected on the basis of cloned DNA fragments usually require a greater investment of time and effort before genetic mapping experiments can begin, but yield more information per marker during the mapping process. Markers detected by arbitrary amplification require little if any preliminary work to identify useful markers, but yield somewhat less information per marker during the mapping experiments. Markers based on two-dimensional gel analysis of genomic DNA are as yet highly specialized techniques that have not been widely applied, so little information is available regarding the costs and benefits relative to other marker systems. The balance between the initial investment required to identify markers and the information yield per marker must be decided on a case-by-case basis, as too many variables enter into the cost–benefit analysis to allow any generalization.

Markers detected on the basis of cloned DNA fragments

Restriction fragment length polymorphism (RFLP) markers

The detection of restriction fragment length polymorphisms involves several steps. DNA from each individual to be analysed is digested with a restriction enzyme previously determined to detect variation, and the fragments are separated by size in an agarose gel. The DNA is then immobilized on a hybridization membrane, and the membrane is placed in a solution of denatured labelled probe DNA. After hybridization of probe sequences to complementary strands of DNA immobilized on the membrane, the remaining probe is washed away, and the retained probe is detected by virtue of the label it contains. This label can be radioactive, fluorescent, a hapten that can be detected immunologically, or a specific ligand for which a binding protein is available (e.g. biotin and avidin). Different alleles are represented by different locations of probe retention in the different progeny DNA samples on the membrane.

RFLP markers are typically co-dominant, providing full information on the genotype of the individuals tested. Some difficulties can arise if the probe used hybridizes to DNA sequences found at multiple locations in the genome, because it can be difficult to determine which bands are allelic and which are genetically distinct loci. This difficulty becomes acute if multiple families of different parentages are to be tested with the same set of probes. Different alleles may be present at corresponding loci in different pedigrees, and different loci may be polymorphic in different pedigrees. If individual probes detect multiple loci in both pedigrees, determining the correspondence of a locus in one pedigree to a locus in another pedigree requires additional information.

Identifying probes that detect single polymorphic loci in pedigrees of interest is an essential preliminary step to using RFLPs as genetic markers. This process can be carried out by screening random cDNA or genomic DNA clones for the ability to detect different alleles in the parents of the pedigrees of interest, although this is a labour-intensive and time-consuming process. A new strategy has recently been developed that avoids the necessity of constructing genomic or cDNA libraries, and selectively identifies polymorphic single-copy probes (Rosenberg *et al.*, 1994). This advance considerably simplifies the process of identifying suitable RFLP probes, particularly in species with relatively low levels of genetic variation.

Short repeated sequences or microsatellite markers

The speed and sensitivity of the polymerase chain reaction (PCR) (Saiki *et al.*, 1988) has contributed to development of several new methods of identifying polymorphic genetic markers for linkage mapping, as well as methods for physical mapping. PCR methods for detection of variability at known loci include amplification of short repeated sequences, known variously as microsatellites, STR (simple tandem repeats), or SSR (simple sequence repeats). These acronyms all refer to short tandem repeats of simple dinucleotide, trinucleotide or tetranucleotide sequences. Such repeated sequences are highly polymorphic in many genomes, often having several distinguishable alleles per locus and a high frequency of heterozygosity in the population. A large number of distinguishable alleles is desirable for analysis of multiple pedigrees within a larger population, because it means that more information is available regarding gene flow and linkage disequilibrium in the population. The polymorphisms are detected by amplification of known microsatellite loci with flanking PCR primers, then separation of the reaction products by electrophoresis to detect variation in length.

Identification of microsatellite loci suitable for use as markers requires specific procedures for cloning genomic DNA fragments containing microsatellites, and determination of flanking unique sequences from which PCR primers can be made. A substantial investment must be made to obtain the markers, but the utility of the markers is very high and the effort involved in genotyping an individual is relatively low.

Other markers based on cloned DNA sequences

Other marker systems based on cloned loci include sequence tagged sites (STSs), a general term for genetic loci identified on the basis of sequence and scored by PCR, and expressed sequence tags (ESTs), which are that subset of STSs derived from cDNA clones. EST and STS markers are used in the Human Genome Project primarily for physical mapping purposes (Olson *et al.*, 1989; Palazzolo *et al.*, 1991). A method for mapping mutations in *Arabidopsis*, using PCR to amplify portions of cloned genes and restriction enzyme digestions to detect polymorphism in the amplified DNA, has been named cleaved amplified polymorphic sequences, or CAPSs (Konieczny and Ausubel, 1993). An analogous approach has also been used to map loci corresponding to cDNAs that have been partially sequenced to allow design of PCR primers from 5′ and 3′ end sequences (Harry and Neale, 1994). These methods share with microsatellites the requirement for preliminary DNA sequence data on which to base synthesis of PCR primers before marker polymorphism can be assayed.

One problem with using restriction enzymes to detect polymorphism is that such enzymes only sample a small fraction of the potential sequence variation, because cleavage is based solely on the presence or absence of a four- to eight-nucleotide sequence recognized by the restriction enzyme used. An alternative means of detecting sequence variation in amplified DNA fragments is single-strand conformational polymorphism (SSCP) (Orita *et al.*, 1989). This method exploits the tendency of single-stranded DNA to form intramolecular base pairs, resulting in a sequence-dependent conformation. Changes in DNA sequence can cause alterations in the conformation of single-stranded DNA molecules, and result in changes in migration during gel electrophoresis (Orita *et al.*, 1989).

Markers detected by arbitrary amplification

Random amplified polymorphic DNA (RAPD) markers

This method uses single arbitrarily chosen decanucleotide primers of known sequence to amplify segments of genomic DNA that happen to lie between two appropriately oriented primer-binding sites (Williams *et al.*, 1990). The occurrence of appropriate primer-binding sites is dependent on the sequence complexity of the genome, but is rarely limiting. Genotyping individual progeny is a two-step process of carrying out a PCR reaction using a single primer and a small amount of genomic DNA as template, then separating the reaction products on an agarose gel and detecting them by staining with ethidium bromide. The different alleles are distinguished by presence or absence of a product of a particular size, and multiple loci can be scored in a single reaction provided the products are distinguishable. RAPD markers behave as dominant factors, present in homozygous or in heterozygous progeny containing the allele that amplifies, and absent in progeny homozygous for the allele that does not amplify. This feature renders RAPD inconvenient for analysis of F_2 intercross pedigrees (Fig. 8.1A). Selection of markers heterozygous in one parent and homozygous for the non-amplifying allele in the other parent of an F_1 hybrid pedigree allows analysis of 1:1 segregation of markers from each parent independently. This has been referred to as a 'pseudo-testcross strategy' (Grattapaglia and Sederoff, 1994; Hemmat *et al.*, 1994).

Amplified fragment length polymorphism (AFLP) markers

AFLP is intermediate between RFLP and RAPD in some respects: it relies on the presence or absence of restriction sites as a source of detectable polymorphisms (as does the RFLP technique), but it uses PCR amplification to detect the polymorphisms, which behave as dominant markers that are present or absent (similar to RAPD). The technique is based on digestion of genomic DNA samples with two restriction enzymes, ligation of adaptor oligonucleotides that provide binding sites for PCR primers, and then amplification of subsets of the genomic DNA fragments by PCR with primers that extend up to four nucleotides past the known restriction site into the genomic DNA (Zabeau and Vos, 1992). The additional nucleotides on the PCR primers reduce the number of fragments that amplify, because only the subset of DNA fragments that contain nucleotides complementary to the extensions on the primers will be amplified. A three-nucleotide extension on each of two primers reduces the number of fragments amplified by a factor of 4^6, or 4096. The choice of nucleotides in the extension dictates which subfraction of the genomic DNA will amplify. The technique requires that the amplification products be labelled by using one radioactive or fluorescent primer, and that the reaction products be separated on high-resolution polyacrylamide gels for maximum resolution of the many amplified DNA fragments. As a result, a relatively high level of technical skill is required to carry out the technique successfully.

Markers based on two-dimensional analysis of DNA

Two different methods have been described for two-dimensional analysis of DNA. Restriction landmark genome scanning (RLGS) involves restriction enzyme digestion of genomic DNA isolated from individual progeny, followed by labelling of the ends of the fragments with radioactive nucleotides. The labelled DNA fragments are separated by size using agarose gel electrophoresis, and a strip of the agarose containing the DNA fragments separated by size is cut from the gel. The entire strip of agarose gel is placed into a second restriction endonuclease digestion reaction. After the second digestion is complete, the agarose gel strip is laid across the top of a high-resolution polyacrylamide gel and the DNA fragments are separated according to size again (Hatada *et al.*, 1991). The intensity of the signal from each DNA fragment on the gel can be used to determine whether it is present in two copies per genome (i.e. the individual offspring was homozygous) or in a single copy per genome (i.e. the offspring was heterozygous) (Asakawa *et al.*, 1994).

The two-dimensional DNA typing technique uses polyacrylamide gel electrophoresis to separate DNA restriction fragments on the basis of size, then uses a length-independent electrophoresis system to separate the fragments in the second dimension (Uitterlinden and Vijg, 1994). The separated DNA is immobilized on a membrane and hybridized to labelled repetitive DNA sequences. Differences in the distribution of hybridizing fragments is reported to reflect underlying genetic variation, although homozygotes are not distinguishable from heterozygotes (te Meerman *et al.*, 1993).

STRATEGIES FOR GENETIC MAPPING EXPERIMENTS

CHARACTERISTICS OF DIFFERENT MARKER DETECTION METHODS

The best marker detection method for any given genetic mapping experiment depends on the objectives and scope of the experiment. The length of the committment to genetic mapping, the kinds of pedigrees available and the resources to be invested in identifying markers are all important variables. The number of progeny to be genotyped and the number of markers needed are also relevant. A short-term experiment to identify a few genetic markers linked to loci controlling qualitative traits would be best served by anonymous amplification-based markers, because of the low initial investment required. A long-term project to build saturated genetic maps, identify loci controlling quantitative traits and use the identified loci in a large multi-generation breeding programme might be better served by markers based on cloned DNA fragments, because the utility of co-dominant markers would repay the investment required to identify them. PCR-based markers are very effective at genotyping many progeny for just a few markers, while two-dimensional gels (either of protein or DNA fragments) and RFLP analysis are more efficient methods of genotyping a few individuals for a very large number of markers. Multiplexing, or the ability to genotype an individual for several markers simultaneously, is more important in projects involving many markers *and* many progeny than in projects involving either few markers *or* few progeny.

MAXIMIZING RETURN ON EFFORT

The nature of the traits to be analysed will dictate the strategy to be used in any mapping project. Several techniques have been described that increase the power of analysis or reduce the effort involved in carrying out mapping experiments. Two kinds of techniques will be described here: identification of markers linked to a locus of interest by screening pooled DNA samples; and identification of individuals bearing recombinant chromosomes of interest for further analysis. The best marker for the first stage of a mapping project (such as identifying markers linked to particular loci) may not be the best marker for the next stage (such as using markers for breeding). Strategies for converting between types of markers have been described.

Identification of markers linked to a specific locus: bulk segregant analysis

Identification of markers linked to individual loci of interest relies on differential presence of marker alleles in the individuals in different phenotypic classes (Arnheim *et al.*, 1985). Application of this technique to quantitative traits was suggested by Lander and Botstein (1989), and reported by investigators working with qualitative traits in plants such as morphological characteristics (Giovannoni *et al.*, 1991) and disease resistance (Michelmore *et al.*, 1991). Individual progeny from a segregating population are scored for the phenotype conferred by the locus of interest, and DNA samples from individuals of like phenotypes are pooled in groups of 5 to 20 samples per pool. The pooled DNA samples are then screened for molecular markers that appear polymorphic between the two pools. Portions of the genome not linked to the locus conferring the phenotype scored in the progeny segregate randomly between the two pooled DNA samples, and so are not polymorphic between the pools. Markers linked to the locus of interest are polymorphic between the pools because one pool contains a much higher fraction of one marker allele than the other pool.

Michelmore and co-workers (1991) reported reconstruction experiments indicating that the RAPD technique allows detection only of alleles making up more than about 20% of the alleles in the pool, so that the window of detection of linked markers is about 20 to 25 cM on either side of the locus of interest. This value may depend greatly on the complexity of the genome on which mapping is carried out, and the conditions under which the RAPD reactions are done. Application of the pooling technique to analysis of quantitative traits uses pools of individuals from the extremes of the trait distribution (Lander and Botstein, 1989). This approach has recently been tested in a study of quantitative trait inheritance in a half-sib family of *Eucalyptus* (Grattapaglia, 1994). Individuals of largest circumference at breast height (CBH) were pooled, the individuals of smallest CBH were pooled, and RAPD markers providing coverage of most of the genome were screened for polymorphism between bulks. Two linked markers were identified that were later shown by co-segregation analysis in a randomly chosen set of progeny to be linked to a locus that accounts for 5.8% of the phenotypic variation in CBH (Grattapaglia, 1994).

Pooling should also work with more sensitive methods, such as RFLP subtraction (Rosenberg *et al.*, 1994) or PCR methods that use unique sequence primers (e.g. the microsatellite and AFLP methods). These methods are likely to detect only those linked markers for which little or no recombination has occurred to break the linkage of a marker allele with the locus of interest, resulting in a smaller genetic window around the locus of interest.

A pooling strategy to detect rare recombination events

Separation of linked loci by recombination is desirable for some breeding applications as well as for fine-scale linkage mapping. Pooling of progeny before screening for recombination improves the efficiency of the screening process, and can provide information on gene order as well (Churchill *et al.*, 1993). This technique is likely to play a central role in experiments to isolate the genes that control quantitative traits, as large numbers of recombinant progeny must be analysed in order to precisely map the genes of interest (Tanksley, 1993).

Conversion of markers

Conversion of markers from one type to another can help to overcome the disadvantages of both marker types. RAPD and AFLP markers based on PCR amplification of anonymous sequences are very cost-effective in early stages of mapping experiments, but lack the co-dominant inheritance patterns of markers based on cloned DNA sequences. Anonymous DNA fragments can be cloned and sequenced, and specific PCR primers made for specific amplification of DNA fragments (Paran and Michelmore, 1993). Polymorphism may exist in the length of the amplified fragments, or may be detectable by restriction digestion or SSCP analysis.

APPLICATIONS OF GENETIC MAPPING

The ability to construct a dense genetic map with molecular genetic markers makes possible detection and segregation analysis of loci controlling quantitative traits (Lander and Botstein, 1989). This approach allows investigation of the genetic basis of traits that vary continuously, and separates the many genes that contribute to these traits into a major effect class and a minor effect class. Prior to the development of quantitative trait analysis by molecular mapping, quantitative traits were assumed to be governed by many genes, all of small effect. The involvement of many genes in quantitative traits has not been questioned. It is now clear, however, based on many experiments in both plants and animals, that some genes affect a given trait much more than other genes, and that the chromosomal locations of genes with major effects can be identified (Tanksley, 1993). The precision and accuracy of the mapping process depends on the number of progeny scored for phenotype and genotype, and the reproducibility of the phenotypic measurements.

A second use of genetic mapping has been to identify regions of the genome associated with inheritance of particular qualitative traits that are difficult or expensive to analyse. Molecular markers can then be identified that are linked in coupling to a desired allele, such as an allele conferring resistance to a fungal pathogen, and the inheritance of the marker used to infer the inheritance of the allele of interest (e.g. Kelly, 1994). This technique can be used to advantage with traits that are expressed only in mature plants. Identification of individuals of the desired genotype at the seedling stage reduces the number of plants that must be grown to maturity for further selection on other characteristics.

Marker-assisted selection, or the use of genetic markers to identify progeny of interest for direct deployment, has been criticized as unnecessary when heritability is high, and unlikely to work when heritability is low (Strauss *et al.*, 1992). This criticism does not apply to markers that predict genotypes that are not expressed until maturity, or to the selection of progeny for use in advanced generation breeding strategies. The ability to identify commercially important recessive alleles in the heterozygous state and to use that knowledge in designing advanced generation breeding strategies will be of great value to plant breeders. Transgressive segregation and overdominance effects can be more valuable in clonally propagated ornamental species than in inbred crops or outbred forest trees, because a single individual of superior quality can be captured through propagation. Detailed knowledge of the genetic basis of commercially important characteristics such as length of juvenile phase, chilling response, disease resistance, flower number or plant size could provide ornamental plant breeders with a host of new options for improvement of current varieties and development of novel traits (de Jong, 1991).

REFERENCES

Acquaah, G. (1992) *Practical Protein Electrophoresis for Genetic Research.* Discorides Press, Portland, Oregon.

Arnheim, N., Strange, C. and Erlich, H. (1985) Use of pooled DNA samples to detect linkage disequilibrium of polymorphic restriction fragments and human disease: studies of the HLA class II loci. *Proceedings of the National Academy of Sciences, USA* 82, 6970–6974.

Asakawa, J.-I., Kuick, R., Neel, J.V., Kodaira, M., Satoh, C. and Hanash, S. M. (1994) Genetic variation detected by quantitative analysis of end-labeled genomic DNA fragments. *Proceedings of the National Academy of Sciences, USA* 91, 9052–9056.

Beckmann, J.S. and Osborn, T.C. (eds) (1992) *Plant Genomes: Methods for Genetic and Physical Mapping.* Kluwer Academic Publishers, Dordrecht.

Botstein, D., White, R.L., Skolnik, M. and Davis, R.W. (1980) Construction of a genetic linkage map in man using restriction fragment length polymorphisms. *American Journal of Human Genetics* 32, 314–331.

Brewer, G.J. (1970) *An Introduction to Isozyme Techniques.* Academic Press, New York.

Churchill, G., Giovannoni, J. J. and Tanksley, S.D. (1993) Pooled-sampling makes high-resolution mapping practical with DNA markers. *Proceedings of the National Academy of Sciences, USA* 90, 16–20.

Damerval, C., Maurice, A., Josse, J.M. and de Vienne, D. (1994) Quantitative trait loci underlying gene product variation: a novel perspective for analyzing regulation of genome expression. *Genetics* 137, 289–301.

Davies, K.E. and Tilghman, S.M. (1990) *Genetic and Physical Mapping.* Cold Spring Harbor Laboratory Press, Cold Spring Harbor, New York.

De Jong, J. (1991) Selection for physiological traits. In: Harding, J., Singh, F., and Mol, J.N.M. (eds), *Genetics and Breeding of Ornamental Species.* Kluwer Academic Publishers, Dordrecht, pp. 109–133.

Giovannoni, J.J., Wing, R.A., Ganal, M.W. and Tanksley, S.D. (1991) Isolation of molecular markers from specific chromosomal intervals using DNA pools from existing mapping populations. *Nucleic Acids Research* 19, 6553–6558.

Grattapaglia, D. (1994) Genetic mapping of quantitatively inherited economically important traits in *Eucalyptus*. Unpublished PhD Thesis, Department of Genetics, North Carolina State University, Raleigh.

Grattapaglia, D. and Sederoff, R. (1994) Genetic linkage maps of *Eucalyptus grandis* and *Eucalyptus urophylla* using a pseudo-testcross mapping strategy and RAPD markers. *Genetics* 137, 1121–1137.

Grattapaglia, D., Chaparro, J.X., Wilcox, P., McCord, S., Werner, D., Amerson, H., McKeand, S., Bridgwater, F., Whetten, R., O'Malley, D. and Sederoff, R. (1992) Mapping in woody plants with RAPD markers: applications to breeding in forestry and horticulture. In: *Proceedings of the Symposium on Applications of RAPD Technology to Plant Breeding*, Crop Science Society/American Society for Horticultural Science/ American Genetic Association, Minneapolis, MN, USA, 1 November 1992, pp. 37–40.

Harry, D.E. and Neale, D.B. (1994) PCR-based genetic markers for loblolly pine developed from cloned cDNAs. Abstract for Plant Genome II, available on WWW at http://probe.nalusda.gov:8000/plant/index.html.

Hatada, I., Hayashizaki, Y., Hirotsune, S., Komatsubara, H. and Mukai, T. (1991) A genomic scanning method for higher organisms using restriction sites as landmarks. *Proceedings of the National Academy of Sciences*, USA 88, 9523–9527.

Hemmat, M., Weeden, N.F., Manganaris, A.G. and Lawson, D.M. (1994) Molecular marker linkage map for apple. *Journal of Heredity 85, 4–11.*

Kelly, J.D. (1994) Use of molecular markers in breeding for resistance to fungal pathogens. *HortScience* 29(5). Abstract number 997, 91st Annual Meeting of the American Society for Horticultural Science, Corvallis, Oregon, 7–10 August 1994.

Konieczny, A. and Ausubel, F. (1993) A procedure for mapping *Arabidopsis* mutations using co-dominant ecotype-specific PCR-based markers. *Plant Journal* 4, 403–410.

Lander, E.S. and Botstein, D. (1989) Mapping Mendelian factors underlying quantitative traits using RFLP linkage maps. *Genetics* 121, 185–199.

Lander, E.S., Green, P., Abrahamson, I., Barlow, A., Daly, M.J., Lincoln, S.L. and Newbers, L. (1987) MAPMAKER: an interactive computer package for constructing primary genetic linkage maps of experimental populations. *Genomics* 1, 182–195.

Leonardi, A., Damerval, C. and De Vienne, D. (1988) Organ-specific variability and inheritance of maize proteins revealed by two-dimensional electrophoresis. *Genetical Research* 52, 97–103.

Michelmore, R.W., Paran, I. and Kesseli, R.V. (1991) Identification of markers linked to disease-resistance genes by bulked segregant analysis: a rapid method to detect markers in specific genomic regions by using segregating populations. *Proceedings of the National Academy of Sciences, USA,* 88, 9828–9832.

O'Farrell, P.H. (1975) High-resolution two-dimensional electrophoresis of proteins. *Journal of Biological Chemistry* 250, 4007–4021.

Olson, M., Hood, L., Cantor, C. and Botstein, D. (1989) A common language for physical mapping of the human genome. *Science* 245, 1434–1435.

Orita, M., Iwahana, H., Kanazawa, H., Hayashi, K. and Sekiya, T. (1989) Detection of polymorphisms of human DNA by gel electrophoresis as single-strand conformation polymorphisms. *Proceedings of the National Academy of Sciences, USA* 86, 2766–2770.

Palazzolo, M.J., Sawyer, S.A., Martin, C.H., Smoller, D.A. and Hartl, D.L. (1991)

Optimized strategies for sequence-tagged-site selection in genome mapping. *Proceedings of the National Academy of Sciences, USA* 88, 8034–8038.

Paran, I. and Michelmore, R.W. (1993) Development of reliable PCR-based markers linked to downy mildew resistance genes in lettuce. *Theoretical and Applied Genetics* 85, 985–993.

Rosenberg, M., Przybylska, M. and Straus, D. (1994) RFLP subtraction: a method for making libraries of polymorphic markers. *Proceedings of the National Academy of Sciences, USA* 91, 6113–6117.

Saiki, R.K., Gelfand, D.H., Stoffel, S., Scharf, S.J., Higuchi, R., Horn, G.T., Mullis, K.B. and Ehrlich, H.A. (1988) Primer-directed enzymatic amplification of DNA with a thermostable DNA polymerase. *Science* 239, 487–491.

Strauss, S., Lande, R. and Namkoong, G. (1992) Limitations of molecular marker-aided selection in forest tree breeding. *Canadian Journal of Forest Research* 22, 1050–1061.

Tanksley, S.D. (1993) Mapping polygenes. *Annual Reviews of Genetics* 27, 205–233.

te Meerman, G.J., Mullaart, E., van der Meulen, M.A., den Daas, J.H.G., Morolli, B., Uitterlinden, A.G. and Vijg, J. (1993) Linkage analysis by two-dimensional DNA typing. *American Journal of Human Genetics* 53, 1289–1297.

Uitterlinden, A.G. and Vijg, J. (1994) *Two-Dimensional DNA Typing. A Parallel Approach to Genome Analysis.* Ellis Horwood, London.

Weber, J.L. and May, P.E. (1987) Abundant class of human DNA polymorphisms which can be typed using the polymerase chain reaction. *American Journal of Human Genetics* 44, 388–396.

Williams, J., Kubelik, A.R., Livak, D.J., Rafalski, J.A. and Tingey, S.V. (1990) DNA polymorphisms amplified by arbitrary primers are useful as genetic markers. *Nucleic Acids Research* 18, 6531–6535.

Zabeau, M. and Vos, P. (1992) Selective restriction fragment amplification: a general method for DNA fingerprinting. European patent application 92402629.7, publication number 0 534 858 A1, European Patent Office.

Cultivar Identification Using Molecular Methods

9

Sriyani Rajapakse and Robert E. Ballard
Department of Biological Sciences, Clemson University, Clemson, South Carolina 29634, USA

Molecular biological technology developed over the past 20 years has provided new methods for observing genetic differences between cultivars. These techniques offer many advantages over the conventional methods of germplasm characterization. The four main types of molecular methods available for use in cultivar identification are restriction fragment length polymorphism (RFLP), random amplified polymorphic DNA (RAPD), DNA amplification fingerprinting (DAF), and simple sequence repeat (SSR) polymorphism. In ornamentals, cultivar identification based on molecular techniques is used primarily for two very different purposes; first, to develop DNA fingerprints which can be used in patent protection, and, second, to study the genetics of cultivars.

DNA FINGERPRINTING OF CULTIVARS FOR PATENT PROTECTION

Under the US Plant Patent Act of 1930, cultivars of plants that are only propagated asexually, but excluding tuber-forming plants, are protected from unauthorized propagation for 17 years. In addition, the Plant Variety Protection (PVP) Act of 1970 provides patent-like protection for either open pollinated cultivars or inbred lines of seed propagated crops, but not for hybrids. Protection is also available under the general utility patent and trademark law which was extended in 1985 to protect all plant material (Barton, 1982; Davis and Luby, 1988). Most ornamental cultivars are protected, but not exclusively, through the Plant Patent Act. Ornamental plants that have been protected include rose (mostly Hybrid Tea), chrysanthemum, turf grass species, azalea, rhododendron, heather, holly, juniper, and many others. In a situation of suspected patent infringement, owners of the patented cultivars are faced with the problem of providing proof of infringement for litigation. It is necessary to show that the cultivars in dispute are identical in order to prove

 Biotechnology of Ornamental Plants
(eds R.L. Geneve, J.E. Preece and S.A. Merkle)

an infringement has actually occurred. Therefore, accurate identification of plant cultivars is essential in enforcing the property rights of breeders who develop novel plant cultivars.

Traditionally, cultivars have been identified by morphological, physiological or horticultural descriptions. These descriptions are largely subject to environmental conditions and the human judgement. Development of new cultivars that differ little from the existing cultivars also required more reliable methods to discriminate between plant cultivars. Comparative analysis of isozyme patterns (Scandalios, 1969; Pierce and Brewbaker, 1973) has been introduced as a supplementary method to identify cultivars and has been used to identify cultivars of rose (Kuhns and Fretz, 1978; Millan *et al.*, 1995), camellia, (Wendel and Parks, 1983) and carnation (Messeguer and Arus, 1985). Although isozymes can be very useful in cultivar identification, they can be tissue specific, and the plant tissue must also be sampled at comparable physiological states to obtain uniform and repeatable banding patterns. Furthermore, the limited number of isozyme loci available for examination may be insufficient for identification of many cultivars (Murphy *et al.*, 1990). Another method used to distinguish ornamental plant cultivars is based on differences in secondary metabolites such as flavonoids (Asen, 1979, 1984). Unfortunately, at the cultivar level, most differences are quantitative. Quantitative distinctions between cultivars are significant only if they are greater than those differences that result from experimental error or the environment (Asen, 1979).

Molecular genetic techniques have been applied to plant cultivar identification within the past decade. By developing molecular markers that detect differences in DNA sequences between cultivars, highly specific marker profiles, commonly known as DNA fingerprints, can be developed for each cultivar and used in its identification. Compared to isozyme and flavonoid markers, molecular markers have many advantages; they are detectable in all tissues at all ages of the plant, thereby enabling early identification. They are developmentally stable and not affected by the environment. Furthermore, the number of DNA markers available for examination is unlimited because genomes are composed of billions of base pairs and many different types of molecular marker systems are available to compare these differences.

In addition to providing proof of infringement or misappropriation in plant intellectual property protection, DNA fingerprints can assist in demonstrating the patentability of a new cultivar, obtaining a Plant Variety Protection (PVP) certificate, and in proving the ownership (Jondle, 1992). Two requirements for patentability are novelty and non-obviousness, which can be demonstrated by evidence that shows that the claims in an application are distinguished from prior art (Jondle, 1992).

CULTIVAR IDENTIFICATION FOR GENETIC STUDIES

Apart from the potential use in plant breeders' protection, DNA profiles are also used to study genetic relatedness/diversity among cultivars. In this type of research, the DNA markers developed are used in constructing phylogenies and in calculating genetic distances which help understand evolutionary relationships among cultivars, species and other taxa. Molecular phylogenies are useful in verification of existing taxonomy based mostly on morphology and in elucidating ambiguous areas in plant systematics. DNA profiles are also used to assess the origin of hybrids, to construct genetic linkage maps, and to check for mutations (somaclonal variations) after lengthy tissue culture propagation.

TYPES OF MOLECULAR MARKERS USED IN CULTIVAR IDENTIFICATION

RESTRICTION FRAGMENT LENGTH POLYMORPHISM (RFLP)

RFLP is the first and the most widely used molecular method in cultivar identification. Base substitutions in a restriction endonuclease site or insertions or deletions between sites can give rise to detectable differences in the length of DNA when digested with restriction enzymes. These differences constitute a RFLP.

To develop RFLP profiles, DNA extracted from leaves is cut with a restriction enzyme and subjected to size-separation by agarose gel electrophoresis. The DNA in the gel is transferred to a nylon membrane and hybridized with a segment of DNA known as a probe. To visualize the hybridized DNA profile, probes are labelled using either radioactive or non-radioactive methods. The DNA fingerprint obtained is specific to the restriction enzyme and probe combination used. Probes can be selected from a variety of sources including a random genomic DNA library or cDNA library constructed from the same species, libraries made from heterologous species, or minisatellite probes derived from other organisms.

When using clones from a random genomic library, clones are first screened to find ones that give rise to few (about 3–20) bands upon hybridization. Prescreening the library to select clones with moderate numbers of copies can reduce the number of probes that need to be actually hybridized to find DNA fingerprint probes. The reader is referred to Eldredge *et al.* (1992) for details in prescreening libraries. RFLPs produced using a prescreened genomic library of the same species have been used to identify cultivars of Hybrid Tea and miniature roses (Hubbard *et al.*, 1992; Rajapakse *et al.*, 1992, 1993). Fifteen probes that display RFLPs useful in rose cultivar identification were identified (Rajapakse *et al.*, 1992). One of the probes can distinguish 12 cultivars. By combining the RFLP profile of this probe with that of another probe, 16 cultivars have been characterized. Several dogwood (*Cornus*) species and four cultivars of *C. florida* have been fingerprinted using clones from a random genomic library from the same genus (Culpepper *et al.*, 1991). These authors found that polymorphisms were more common among the species and less among the cultivars. This was expected due to the close taxonomic relationship among the cultivars. Despite the common origin, they were able to identify the four cultivars using RFLPs.

RFLP analysis in chrysanthemum was successful using *Pst*I and *Hin*dIII genomic libraries of the same species (Wolff *et al.*, 1994). Of the probes from the *Pst*I and *Hin*dIII libraries, 91% and 35% appeared to have low-copy sequences, respectively. *Pst*I is a methylation sensitive enzyme and, because many repeated sequences are methylated, they are underrepresented in the genomic library. Because of the polyploid nature of chrysanthemum, complex patterns consisting of 6–12 bands were observed with these low copy clones. Wolff *et al.* (1994) reported a high level of genetic variation among chrysanthemum cultivars. Some of the RFLPs obtained in this study were later converted to PCR markers for use in marker-assisted selection.

In general, cDNA probes which represent the gene coding regions are less useful in cultivar identification than genomic clones, which also include non-coding sequences, since most of the differences among cultivars are found in non-coding regions of DNA. Genomic clones were indeed found to be more useful when compared to cDNA clones in detecting RFLPs in chrysanthemum (Wolff *et al.*, 1994), pine (Devey *et al.*, 1991), and in many other non-ornamental plant species such as peach (S. Rajapakse, unpublished results), maize (Helentjaris *et al.*, 1986), and citrus (Durham *et al.*, 1992).

Interspecific and intraspecific variation in poplar (*Populus*) has been examined using heterologous probes of cloned petunia chloroplast DNA (cpDNA) with success (Rajora and Dancik, 1995a,b). They were able to separate *P. maximowiczii* clones of Chinese origin from those of Japanese origin by the change of one restriction site. Based on findings from these studies, they have suggested a revision to the existing classification that was based on morphology and reproductive characters. A detailed phylogenetic and taxonomic study has been conducted in bamboo (*Phyllostachys*) species and clones using RFLPs with bamboo genomic and rice cDNA probes (Friar and Kochert, 1994). Genetic markers are especially useful in bamboo systematics as this plant's unique life cycle severely limits the use of floral characters. Tobe (1993) used cpDNA to determine phylogenetic relationships among ten *Magnolia* taxa and to demonstrate the maternal inheritance of cpDNA in *Magnolia*.

DNA fingerprinting in ornamentals has also used probes derived from other organisms. Cultivars of rose, gerbera daisy, and carnation have been fingerprinted using human derived minisatellite probes, bacteriophage derived M13 minisatellite probe, and cattle derived probes (Tzuri *et al.*, 1991; Vainstein *et al.*, 1992). The successful use of probes derived from other organisms, such as humans and cattle, indicates the presence of similar 'core' sequences among the different taxonomic groups, suggesting that 'core' sequences did not change significantly during evolution (Jeffreys *et al.*, 1985). The M13 probe (Vassart *et al.*, 1987) has also been used to find RFLPs in box elder (*Acer negundo*) (Nybom and Rogstad, 1990), quaking aspen (*Populus tremuloides*) (Rogstad *et al.*, 1991), *Pinus*, *Asimina*, *Populus* and *Polyalthia* (Rogstad *et al.*, 1988). This probe has been used widely to detect RFLPs in other plants, animals and microorganisms (Ryskov *et al.*, 1988; Nybom & Hall, 1991).

RANDOM AMPLIFIED POLYMORPHIC DNA (RAPD)

The discovery of polymerase chain reaction (PCRs) (Mullis *et al.*, 1986; Mullis and Faloona, 1987) has led to the development of another genetic marker system for detecting DNA polymorphisms (Welsh and McClelland, 1990; Williams *et al.*, 1990). Random amplified polymorphic DNA (RAPD) involves PCR amplification of total genomic DNA using a single random primer of about ten bases and separating amplified products by agarose gel electrophoresis.

Using RAPD markers offers several advantages over RFLPs; most important of all is the simplicity of the method and hence the fast rate at which the markers can be developed. Other advantages include generally high levels of polymorphism obtained, lower cost, and the adaptability of the procedure to automation. Because of these attractions, many studies have relied on RAPDs for cultivar fingerprinting (Table 9.1). However, amplification of genomic DNA using such random short primers is very sensitive to reaction conditions and amplification protocol. Concentrations of template DNA, oligonucleotide primer, Mg^{2+}, and DNA polymerase in the reaction mix, as well as quality of

Table 9.1. Ornamental cultivar identification using RAPDs.

Ornamental	Reference	Purpose of study or remarks
Rose	Torres *et al.* (1993)	Patent protection
	Deberner (1995)	Genetic mapping
	Millan *et al.* (1995)	
	Reynders-Aloisi and Bollereau (1995)	Genetic diversity
	Rajapakse *et al.* (1995a)	Genetic mapping
Chrysanthemum	Wolff and Peters-Van Rijn (1993)	Genetic variability
Sunflower	Teulat *et al.* (1994)	Characterization of inbred lines
Rhododendron	Rayburn *et al.* (1993)	Broad general purpose
Lilac	Marsolais *et al.* (1993)	Determine origin of hybrids
Buffalograss	Huff *et al.* (1993)	Population genetic variation
	Wu and Lin (1994)	Patent protection
Ryegrass	Sweeney and Danneberger (1994)	Comparison of bulked samples vs. individuals
Poplar	Lin *et al.* (1994)	Random primers and M13 universal primer
	Castiglione *et al.* (1993)	Taxonomy/patent protection
Aspen	Liu and Furnier (1993)	Genetic variation
Willow	Lin *et al.* (1994)	Random primers and M13 universal primer
	Chong *et al.* (1995)	PCR using M13 universal primer
Juniper	Adams *et al.* (1993)	Taxonomy/evolution
Eucalyptus	Keil and Griffin (1994)	Verification of genotypes

template DNA and thermal cycler parameters such as annealing temperature, can affect the amplification of DNA. Therefore, levels of these critical parameters should be optimized initially to get RAPD reactions to work, and thereafter, only very consistent laboratory techniques can guarantee successful amplification.

More worrisome than the problems in developing RAPDs are the questions regarding the reliability of RAPDs as genetic markers in discriminating cultivars. These questions have arisen due to non-reproducible bands, faint or spurious bands that cannot be scored confidently, and non-genetic variation. Even if one only uses reproducible major bands to distinguish cultivars, there is no easy way to confirm that the variations are from the cultivars' genetic origin and not from contaminating microorganisms or due to some dubious aspect in the amplification process not yet understood (Weeden *et al.*, 1992). Genetic or non-genetic origin of RAPD bands can be verified by testing both parents (if known) of the cultivar. However, in confirmed hybrids used in genetic mapping experiments, some RAPD bands which are not present in either parent have been observed (Rajapakse *et al.*, 1995b; M. Pooler, West Virginia, 1993, personal communication). These types of bands, though likely to be heteroduplex formations, should not be used as markers in cultivar identification unless verified by parent–offspring template mixing experiments (Davis *et al.*, 1995).

DNA AMPLIFICATION FINGERPRINTING (DAF)

Another spin-off method of PCR uses very short (five to eight nucleotides) random primers for amplification of genomic DNA and visualizes the products by polyacrylamide gel electrophoresis coupled with silver staining (Caetano-Anolles *et al.*, 1991). The DAF method produces much more variation among different taxa than RAPDs, due to the shortness of primers. As many as 100 DAF bands could be amplified with a single primer (Caetano-Anolles *et al.*, 1991). Therefore, the main advantage of this method resides in its ability to distinguish genetically similar cultivars.

DAF markers can distinguish a series of closely related chrysanthemum cultivars (Scott *et al.*, 1995). DAF markers also have been used to examine the genetic variation among bermudagrass (*Cynodon*) species and cultivars (Caetano-Anolles *et al.*, 1995). Digesting template DNA with restriction enzymes prior to amplification or use of arbitrary primers that have mini-hairpin structures further increased the level of polymorphism and simplified the task of distinguishing closely related cultivars of bermudagrass.

Although DAFs have a special role in cultivar fingerprinting and discrimination of related cultivars, they are less suitable for genetic studies at or above the species level, mainly due to very high level of variation which may even bring out variations among individual plants. This was observed in petunia, where different plants within a species produced distinctive fingerprints (T.A. Cerney, South Carolina, 1995, personal communication). In non-ornamental *Ipomoea* species, intraspecific variation was found to be as high as interspecific variation using DAF markers (Rajapakse *et al.*, unpublished results). In the above mentioned petunia DAF study, ten cultivars were fingerprinted, and five species of wild petunia were used in constructing a molecular phylogeny. As with the RAPDs, concerns involving reproducibility and non-genetic origin of bands also apply to DAFs.

SIMPLE SEQUENCE REPEATS (SSR)

A new class of molecular markers based on simple sequence repeats is very useful in observing genetic diversity in animals and plants (Weber and May, 1989; Love *et al.*, 1990; Beyermann *et al.*, 1992). Tandem repeats of two to five nucleotides, e.g. $(GT)_n$ or $(GATA)_n$, are ubiquitous in regions known as microsatellites in eukaryotic genomes. The copy number (*n*) of these repeats can vary among individuals and this variation provides the basis for the polymorphism. Since SSRs are abundant throughout plant genomes and are associated with high levels of polymorphism, they are well suited for cultivar DNA fingerprinting.

The DNA sequences flanking microsatellite regions are generally conserved. Therefore, SSR polymorphisms are developed by designing primers specific for the flanking sequences of repeat regions and PCR amplifying genomic DNA with these primers. We are unaware of SSR polymorphism observed in ornamental plants in this classical way, for example as carried out in fingerprinting soybean genotypes (Rongwen *et al.*, 1995). However, synthetic oligonucleotide probes having simple repeat sequences have been used as RFLP probes in ornamental cultivar DNA fingerprinting. Sunflower (*Helianthus annuus*) lines and hybrids were analysed using synthetic oligonucleotides $(GACA)_4$, $(GATA)_4$ and $(GGAT)_4$ as hybridization probes (Mosges and Friedt, 1994). In addition to these repeat probes, they have tested the same sunflower lines using isozymes and PCR using random primers. They concluded that oligonucleotide repeats showed the greatest variability among the lines and are most appropriate for generating fingerprints.

Markers developed from different methods discussed above could be combined to obtain fingerprints. Cumulative markers obtained by hybridizing with human, M13 and oligonucleotide probes have been used to fingerprint rose cultivars (Vainstein and Ben-Meir, 1994).

CONCLUSION

Molecular markers provide a reliable and powerful means of identity testing, whether it is a human, animal, plant or a microbe that is under test. Human DNA fingerprints have been successfully used as evidence for criminal cases in US courts since 1987 (Jondle, 1992). However, plant DNA fingerprints have not been tested in court, but that may soon change. Isozyme markers are being used in some cases involving legal protection (R. Jondle, Nebraska, 1995, personal communication). DNA fingerprints are now being submitted as a part of PVP application for some plants including ornamentals (A. Ashley, Virginia, 1995, personal communication). The mere power of DNA based methods, and the readiness of the technology to be used if necessary, tends to discourage illegal propagations. Some disputes of plant ownership are settled out of court because of the cost involved in litigation. DNA fingerprinting in plants has been an academic interest for many years, and only recently has it emerged as a commercial and legal interest.

Of the molecular methods available today for DNA fingerprinting, RFLPs and SSRs provide the most reliable markers, but these techniques are laborious. Although not yet exploited fully in ornamentals, SSRs have great potential as DNA fingerprints because of the high level of polymorphisms observed in plants. The RAPD method is simple and yields more polymorphisms than RFLP, but less than DAF which can distinguish closely related cultivars. If using RAPDs or DAF to file for patent protection or as evidence for infringement of a legal protection, reproducibility and other problems associated with these methods should be addressed.

REFERENCES

Adams, R.P., Demeke, T. and Abulfaith, H.A. (1993) RAPD DNA fingerprints and terpenoids: clues to past migrations of *Juniperus* in Arabia and east Africa. *Theoretical and Applied Genetics* 87, 22–26.

Asen, S. (1979) Flavonoid chemical markers in *Poinsettia* bracts. *Journal of the American Society for Horticultural Science* 104, 223–226.

Asen, S. (1984) High pressure liquid chromatographic analysis of flavonoid chemical markers in petals from *Gerbera* flowers as an adjunct for cultivar and germplasm identification. *Phytochemistry* 23, 2523–2526.

Barton, J.H. (1982) The international breeders rights system and crop plant innovation. *Science* 26, 1071–1075.

Beyermann, B., Nurnberg, P., Weihe, A., Meixner, M., Epplen, J.T. and Borner, T. (1992) Fingerprinting plant genomes with oligonucleotide probes specific for simple repetitive DNA sequences. *Theoretical and Applied Genetics* 83, 691–694.

Caetano-Anolles, G., Bassam, B.J. and Gresshoff, P.M. (1991) DNA amplification fingerprinting using very short arbitrary oligonucleotide primers. *Bio/Technology* 9, 553–557.

Caetano-Anolles, G., Callahan, L.M., Williams, P.E., Weaver, K.R. and Gresshoff, P.M. (1995) DNA amplification fingerprinting analysis of bermudagrass (*Cynodon*): genetic relationships between species and interspecific crosses. *Theoretical and Applied Genetics* 91, 228–235.

Castiglione, S., Wang, G., Damiani, G., Bandi, C., Bisoffi, S. and Sala, F. (1993) RAPD fingerprints for identification and for taxonomic studies of elite poplar (*Populus* spp.) clones. *Theoretical and Applied Genetics* 87, 54–59.

Chong, D.K.X., Yeh, F.C., Aravanopoulos, F.A. and Zsuffa, L. (1995) DNA fingerprinting of willows (*Salix* L.) using polymerase chain reaction with the M13 universal primer. *Scandinavian Journal of Forest Research* 10(1), 27–31.

Culpepper, J.H., Sayavedra-Soto, L.A., Bassam, B. J. and Gresshoff, P.M. (1991) Characterization of *Cornus* (dogwood) genotypes using DNA fingerprinting. *Journal of the American Society for Horticultural Science* 116, 1103–1107.

Davis, D.W. and Luby, J.J. (1988) Some current options in the use of plant variety protection in horticulture. *HortScience* 23, 15–18.

Davis, T.M., Haigis, K., McGowan, P. and Yu, H. (1995) Template mixing: a means of enhancing detection of codominant RAPD markers in mapping populations. In: *Final Program and Abstract Guide, Plant Genome III: The International Conference on the Status of Plant Genome Research,* San Diego, p. 32.

Deberner, T. (1995) Genetic and molecular analysis in wild and cultivated roses. In: *Abstracts of Contributed Papers: 2nd International Rose Symposium,* Antibes, France, p. IV-C1.

Devey, M.E., Jermstad, KD., Tauer C.G. and Neale, D.B. (1991) Inheritance of RFLP loci in a loblolly pine three-generation pedigree. *Theoretical and Applied Genetics* 83, 238–242.

Durham, R.E., Liou, P.C., Gmitter, F.G. Jr and Moore, G.A. (1992) Linkage of restriction fragment length polymorphisms and isozymes in *Citrus*. *Theoretical and Applied Genetics* 84, 39–48.

Eldredge, L., Ballard, R., Baird, W.V., Abbott, A.G., Morgens, P., Callahan, A., Scorza, R. and Monet, R. (1992) Application of RFLP analysis to genetic linkage mapping in peaches. *HortScience* 27(2), 160–163.

Friar, E. and Kochert. G. (1994) A study of genetic variation and evolution of *Phyllostachys* (Bambusoideae: Poaceae) using nuclear restriction fragment length polymorphisms. *Theoretical and Applied Genetics* 89, 265–270.

Helentjaris, T., Slocum, M., Schaefer, S. and Nienhuis, J. (1986) Construction of genetic linkage maps in maize and tomato using restriction fragment length polymorphisms. *Theoretical and Applied Genetics* 72, 761–769.

Hubbard, M., Kelly, J., Rajapakse, S., Abbott, A.G. and Ballard, R.E. (1992) Restriction fragment length polymorphisms in rose and their use for cultivar identification. *HortScience* 27, 172–173.

Huff, D.R., Peakall, R. and Smouse, P.E. (1993) RAPD variation within and among natural populations of outcrossing buffalograss [*Buchloe dactyloides* (Nutt.) Engelm.]. *Theoretical and Applied Genetics* 86, 927–934.

Jefferys, A.J., Wilson, V. and Thein, S.L. (1985) Individual specific ‘fingerprints’ of human DNA. *Nature* 316, 76–79.

Jondle, R.J. (1992) Legal aspects of varietal protection using molecular markers. In: *Applications of RAPD Technology to Plant Breeding Symposium Proceedings*, Minneapolis, pp. 50–52.

Keil, M. and Griffin, A.R. (1994) Use of random amplified polymorphic DNA (RAPD) markers in the discrimination and verification of genotypes in *Eucalyptus*. *Theoretical and Applied Genetics* 89, 442–450.

Kuhns, L.J. and Fretz, T.A. (1978) A study of potential use of electrophoresis in distinguishing rose cultivars. *Research Bulletin 1094, Ohio Agricultural and Developmental Center*, Wooster, pp. 1–33.

Lin, D., Hubbes, M. and Zsuffa, L. (1994) Differentiation of poplar and willow clones using RAPD fingerprints. *Tree Physiology* 14, 1097–1105.

Liu, Z, and Furnier, G.R. (1993) Comparison of allozyme, RFLP, and RAPD markers for revealing genetic variation within and between trembing aspen and bigtooth aspen. *Theoretical and Applied Genetics* 87, 97–105.

Love, J.M., Knight, A.M., McAleer, M.A. and Todd, J.A. (1990) Towards construction of a high resolution map of the mouse genome using PCR-analysed microsatellites. *Nucleic Acids Research* 18(14), 4123–4130.

Marsolais, J.V., Pringle, J.S. and White, B.N. (1993) Assessment of random amplified polymorphic DNA (RAPD) as genetic markers for determining the origin of interspecific lilac hybrids. *Taxon* 42, 531–537.

Messeguer, R. and Arus, P. (1985) Electrophoretic identification of carnation cultivars. *HortScience* 20, 372–373.

Millan, T., Torres, A.M. and Cubero, J.I. (1995) Varietal identification in *Rosa* by using isozyme and RAPD markers. In: *Abstracts of Contributed Papers: 2nd International Rose Symposium,* Antibes, France, pp. IV-C3.

Mosges, G. and Friedt, W. (1994) Genetic 'fingerprinting' of sunflower lines and F1 hybrids using isozymes, simple and repetitive sequences as hybridization probes, and random primers for PCR. *Plant Breeding* 113, 114–124.

Mullis, K.B. and Faloona, F.A. (1987) Specific synthesis of DNA *in vitro* via a polymerase catalysed reaction. *Methods in Enzymology* 255, 335–350.

Mullis, K.B., Faloona, F.A., Scharf, S., Saiki, R., Horn, G. and Erlich, H. (1986) Specific enzymatic amplification of DNA *in vitro*: the polymerase chain reaction. Cold Spring Harbor Symp. *Quantitative Biology* 51, 263–273.

Murphy, R.W., Sites, J.W. Jr, Buth, D.G. and Haufler, C.H. (1990) Proteins I: Isozyme electrophoresis. In: Hillis, D.M. and Moritz, C. (eds), *Molecular Systematics*. Sinauer Associates, Sunderland, Massachusetts, pp. 45–126.

Nybom, H. and Hall, H.K. (1991) Minisatellite DNA 'fingerprints' can distinguish *Rubus* cultivars and estimate their degree of relatedness. *Euphytica* 53, 107–114.

Nybom, H. and Rogstad, S.H. (1990) DNA 'fingerprints' detect genetic variation in *Acer negundo* (Aceraceae). *Plant Systematics and Evolution* 173, 49–56.

Pierce, L.C. and Brewbaker, J.L. (1973) Application of isozyme analysis in horticultural science. *HortScience* 8, 17–22.

Rajapakse, S., Hubbard, M., Kelly, J., Abbott, A.G. and Ballard, R.E. (1992) Identification of rose cultivars by restriction fragment length polymorphism. *Scientia Horticulturae* 52, 237–245.

Rajapakse, S., Abbott, A.G. and Ballard, R.E. (1993) Fingerprinting roses. *The American Rose*, July 1993, 20–21.

Rajapakse, S., Abbott, A.G. and Ballard, R.E. (1995a) DNA markers in rose and their use for cultivar identification and genome mapping. In: *Abstracts of Contributed Papers: 2nd International Rose Symposium*, Antibes, France. p. IV-C4.

Rajapakse, S., Belthoff, L.E., He, G., Scorza, R., Ballard, R.E., Baird, W.V., Callahan, A., Monet, R. and Abbott, A.G. (1995b) Genetic linkage mapping in peach using morphological, RFLP, and RAPD markers. *Theoretical and Applied Genetics* 90, 503–510.

Rajora, O.P. and Dancik, B.P. (1995a) Chloroplast DNA variation in *Populus*. I. Intraspecific restriction fragment diversity within *Populus deltoides*, *P. nigra* and *P. maximowiczii*. *Theoretical and Applied Genetics* 90, 317–323.

Rajora, O.P. and Dancik, B.P. (1995b) Chloroplast DNA variation in *Populus*. II. Interspecific restriction fragment polymorphisms and genetic relationships among *Populus deltoides*, *P. nigra*, *P. maximowiczii*, and *P.* × *canadensis*. *Theoretical and Applied Genetics* 90, 324–330.

Rayburn, A.L., Iqbal, M.J. and Paden, D.W. (1993) Positive identification of rhododendron through DNA fingerprinting. *Journal of American Rhododendron Society*, Summer 1993, 137–138.

Reynders-Aloisi, S. and Bollereau, P. (1995) Characterisation of genetic diversity in genus *Rosa* by RAPD. In: *Abstracts of Contributed Papers: 2nd International Rose Symposium*, Antibes, France, p. IV–C2.

Rogstad, S.H., Patton II, J.C. and Schaal, B.A. (1988) M13 repeat probe detects DNA minisatellite-like sequences in gymnosperms and angiosperms. *Proceedings of the National Academy of Sciences, USA* 85, 9176–9178.

Rogstad, S.H., Nybom, H. and Schaal, B.A. (1991). The tetrapod 'DNA fingerprinting' M13 repeat probe reveals genetic diversity and clonal growth in quaking aspen (*Populus tremuloides*, Salicaceae). *Plant Systematics and Evolution* 175, 115–123.

Rongwen, J., Akkaya, M.S., Bhagwat, A.A., Lavi, U. and Cregan, P.B. (1995) The use of microsatellite DNA markers for soybean genotype identification. *Theoretical and Applied Genetics* 90, 43–48.

Ryskov, A.P., Jincharadze, A.G., Prosnyak, M.I., Ivanov, P.L. and Limborska, S.A. (1988) M13 phage DNA as a universal marker for DNA fingerprinting of animals, plants and microorganisms. *Federation of European Biochemical Societies Letters* 233, 388–392.

Scandalios, J.G. (1969) Genetic control of multiple molecular forms of enzymes in plants: a review. *Biochemical Genetics* 3, 37–79.

Scott, M.C., Caetano-Anolles, G. and Trigiano, R.N. (1995) DNA amplification fingerprinting (DAF) identifies closely related cultivars in three series of chrysanthemums. *HortScience* 30(3), 427.

Sweeney, P.M. and Danneberger, T.K. (1994) Random amplified polymorphic DNA in perennial ryegrass: a comparison of bulk samples vs. individuals. *HortScience* 29(6), 624–626.

Teulat, B., Zhang, Y.X. and Nicolas, P. (1994) Characteristics of random amplified polymorphic DNA markers discriminating *Helianthus annuus* inbred lines. *Agronomie* 14, 497–502.

Tobe, J.D. (1993). A molecular systematic study of Eastern North American species of *Magnolia* L. PhD dissertation, Clemson University, USA.

Torres, A.M., Millan, T. and Cubero, J.I. (1993) Identifying rose cultivars using random amplified polymorphic DNA markers. *HortScience* 28, 333–334.

Tzuri, G., Hillel, J., Lavi, U., Haberfield, A. and Vainstein, A. (1991) DNA fingerprint analysis of ornamental plants. *Plant Science* 76, 91–97.

Vainstein, A. and Ben-Meir, H. (1994) DNA fingerprint analysis of roses. *Journal of the American Society for Horticultural Science* 119(5), 1099–1103.

Vainstein, A., Hillel, J., Lavi, U. and Tzuri G. (1992) Genetic variation detected by DNA fingerprinting in flowers. *Acta Horticulturae* 314, 345–351.

Vassart, G., Georges, M., Monsieur, R., Brocas, H., Lequarre, A.S. and Christophe, D. (1987) A sequence in M13 phage detects hypervariable minisatellites in human and animal DNA. *Science* 235, 683.

Weber, J.L. and May, P.E. (1989) Abundant class of human DNA polymorphisms which can be typed using the polymerase chain reaction. *American Journal of Human Genetics* 44, 388–396.

Weeden, N.F., Timmerman, G.M., Hemmat, M., Kneen, B.E. and Lodhi, M.A. (1992) Inheritance and reliability of RAPD markers. In: *Applications of RAPD Technology to Plant Breeding Symposium Proceedings*, Minneapolis, pp. 12–17.

Welsh, J. and McClelland, M. (1991) Fingerprinting genomes using PCR with arbitrary primers. *Nucleic Acids Research* 18, 7213–7218.

Wendel, J.F. and Parks, C.R. (1983) Cultivar identification in *Camellia japonica* L. using allozyme polymorphisms. *Journal of the American Society for Horticultural Science* 108, 290–295.

Williams, J.G.K., Kubelik, A.R., Livak, K.J., Rafalski, J.A. and Tingey, S.V. (1990) DNA polymorphism amplified by arbitrary primers as useful genetic markers. *Nucleic Acid Research* 18, 6531–6535.

Wolff, K. and Peters-Van Rijn, J. (1993) Rapid detection of genetic variability in chrysanthemum (*Dendranthema grandiflora* Tzvelev) using random primers. *Heredity* 71, 335–341.

Wolff, K., Peters-Van Rijn, J. and Hofstra, H. (1994) RFLP analysis in chrysanthemum. I. Probe and primer development. *Theoretical and Applied Genetics* 88, 472–478.

Wu, L. and Lin, H. (1994) Identifying buffalograss [*Buchloe dactyloides* (Nutt.) Engelm.] cultivar breeding lines using random amplified polymorphic DNA (RAPD) markers. *Journal of the American Society for Horticultural Science* 119(1), 126–130.

Somaclonal Variation 10

Han Bouman and Geert-Jan de Klerk
Centre for Plant Tissue Culture Research, PO Box 85, 2160 AB Lisse, The Netherlands

INTRODUCTION

The term 'somaclonal variation' (SV) was introduced to describe the genetic variation in plants regenerated from tissue culture (Larkin and Scowcroft, 1981). The term has been adopted widely, but is used in various senses, especially in practical discussions (Soh, 1987). In all organisms, spontaneous mutations occur from one generation to the next (e.g. Stebbins, 1982). In this chapter, SV indicates the additional mutations in plants produced in tissue culture. Variation that originates from differential expression of the genome (and not from alterations of the genome itself) is called 'epigenetic' and is discussed later (see section on 'Nature').

Many reviews have appeared on SV in general (see D'Amato, 1990; Karp, 1991) or on specific aspects (see De Klerk, 1990; Brown, 1991; Geier, 1991; Van den Bulk, 1991; Phillips *et al.*, 1994). The present article deals with SV in relation to ornamental plants. We first discuss general aspects of SV. After that, we deal with SV as a source of desirable mutations for the breeder and unwanted mutations in various applications of ornamental plant biotechnology. An assay to determine the extent of SV is indispensable for both practical applications and research on the backgrounds of SV. Therefore, we discuss at length various assays.

ORIGINS AND NATURE

ORIGINS

Plants are generated by a series of cell divisions in meristematic tissues. The apical meristem is formed during the early stages of embryogenesis from the

 Biotechnology of Ornamental Plants (eds R.L. Geneve, J.E. Preece and S.A. Merkle)

zygote and consists of two outer cell layers (L_1 and L_2, or tunica) and the inner body (L_3 or corpus). Axillary meristems originate from the apical meristem and have the same histogenic arrangement (Reeve, 1948). New apical meristems may also originate from non-zygotic cells, in particular from somatic cells and from cells in callus or cell suspensions. Plants generated from these adventitious meristems are often genetically different from the mother plant. Although quantitative data are scarce (De Klerk, 1990), it appears that only plants generated from adventitious meristems display SV, whereas tissue-cultured plants produced by the outgrowth of pre-existing apical or axillary meristems are typically true-to-type. This indicates that mutations leading to the aberrant phenotypes occur in the period that begins when cells do not form part of an apical or axillary meristem and ends with the formation of a new meristem. This period includes (i) the time during which the cells form part of the mother plant but are located outside the meristems, (ii) a period of unorganized callus growth in tissue culture and (iii) the regeneration process. With direct regeneration, the second period is absent. The third period, regeneration, can be divided into three phases (cf. Christianson and Warnick, 1983), namely, dedifferentiation, induction and differentiation. Since the new meristem essentially has been formed at the end of the induction phase, SV is induced before. From this, several possible origins of SV can be deduced.

1. Programmed DNA changes in the mother plant, in particular endopolyploidy, which is not an exception but rather an integral part of the morphogenesis of plants (Smulders *et al.*, 1994). Plants regenerated from these cells will be polyploids. Endopolyploidy may also lead to an increase of other types of mutations. Apart from endopolyploidy, other programmed quantitative changes in DNA may occur, especially in the repetitive sequences (Bassi, 1990). This depends on the type of tissue and developmental stage.
2. Increased rate of mutations occurring in the mother plant in tissues other than the apical or axillary meristems.
3. 'Programmed' DNA changes in tissue culture (endopolyploidy and other quantitative DNA changes similar to the changes mentioned under 1).
4. Increased rate of mutations during cell divisions occurring after the initial start of tissue culture until the formation of a new meristem.

Because plants produced by direct regeneration are mostly true-to-type, it is generally believed that SV usually has the latter origin. A general explanation for the increase of the mutation rate (De Klerk, 1990) is that in cell divisions occurring outside meristems, the mechanisms for a careful duplication of the genome are absent. Only in meristems is accurate duplication of the genome required, because meristems are the germ line (from which the pollen and egg cells originate) and transfer the genome from one generation to the next. Another explanation is that the tissue culture procedure itself is mutagenic. In an assay for mutagenesis, colour mutants in stamen hairs of *Tradescantia*, medium components including 2,4-D (2,4-dichlorophenoxyacetic acid) appeared not to be mutagenic (Dolezel and Novak, 1984). Pavlica *et al.* (1991, 1992), however, reported that both 2,4-D (45 μM) and benzyladenine

(0.125 μM) resulted in cell division abnormalities in shallot root tips. It should be noted that the rapidly growing root tips in liquid medium may be very sensitive to added plant growth regulators. It has been observed that tissue culture conditions alter the level of DNA methylation (Lo Schiavo *et al.*, 1988). This, in turn, may induce mutations (Phillips *et al.*, 1994). The various explanations for the occurrence of SV are not mutually exclusive.

NATURE

In plants regenerated from tissue culture, all possible types of genetic variation have been observed, viz., changes in DNA sequence (point mutations, activation of transposons), chromosome structure (duplications, translocations) and chromosome number (leading to polyploidy or aneuploidy). All these types of genetic variation are transferred through meiosis from one generation to the next. However, the difference between genetic and epigenetic variation is not a black-and-white phenomenon, but a matter of degree. DNA methylation and other chromatin states that influence the plant phenotype are transmitted through meiosis but they are also reversible (Jorgensen, 1994; Wolffe, 1994). In this chapter, we classify these types of genomic changes as somaclonal.

The occurrence of differential methylation of the genome has aroused major interest because the increased rate of mutations in tissue culture possibly results from this (Phillips *et al.*, 1994). It is believed that differences in methylation are caused by the stress imposed by tissue culture conditions. Results from our research with begonia also strongly suggest the occurrence of alterations in methylation. Regeneration of begonia plants after a long callus phase resulted in a large percentage of plants that remained in a juvenile state: they had round leaves with a juvenile texture and did not flower, even after three years. They kept these characteristics after propagation via direct regeneration from leaf fragments in tissue culture. However, plants regenerated from leaf explants after a short treatment of the explants with 5-azacytidine (this compound results in hypomethylation) frequently returned to the normal state. Another indication of methylation in tissue culture of begonia will be discussed later in this chapter.

SV IN BREEDING

In their influential review, Larkin and Scowcroft (1981) advocate the view that SV is a new source of variability for breeders. Thus, in an existing cultivar SV might result in an additional, advantageous trait by mutation of a single gene. Selection of cells with a desirable mutation (caused by SV or by chemical mutagenesis) in cell suspension, followed by regeneration of plants from the selected cells also aroused much interest. On these principles, many groups have carried out research (Dixon and Gonzales, 1994). Although various new cultivars have been created, it is now clear that SV has not lived up to its promise.

USE OF SV IN BREEDING OF ORNAMENTALS: SELECTED EXAMPLES

Chrysanthemum

Khalid *et al.* (1989) obtained two variants from a white genotype of *Chrysanthemum* with possible commercial value. It should be noted that in *Chrysanthemum* the spontaneous mutation rate ('sports') is high.

Paulownia

Jagannathan and Marcotrigiano (1986) described a variegated form of the ornamental tree *Paulownia*. In other crops, variegation is regularly found after adventitious plant induction from callus. It is not stable in most plants. It could be a sectoral chimera, or result from activation of a transposable element (Evans, 1989).

Begonia

In begonia, we frequently found leaf variegation in combination with sectoral chimeras after a callus phase (Fig. 10.1). At adventitious regeneration from the various types of leaf tissue, only white leaf-explants regenerated some chimeric plants but at a very low frequency (*c.* 5%), whereas all other regenerants were normal green plants. This suggests reversible changes in the chromatin state. In our studies, we also found many other variants, some with potential ornamental value, for example dwarfs, mini's, plants that cover soil like moss, etc. The aberrant plants were propagated vegetatively by direct regeneration from leaf explants for several cycles and did not revert to normal plants. This indicates that the aberrations are genetic.

Lisianthus

Griesbach and Semeniuk (1987) reported new somaclonal variants of lisianthus (*Eustoma grandiflorum*) after regeneration from callus. They found dwarfs, mini's and bushy phenotypes. They conclude 'that it offers hope of obtaining new genes for improving the crop'.

Daylily

For daylily (*Hemerocallis*) Griesbach (1989) described a dwarf mutant obtained through tissue culture, but in this case the commercial value was not great.

Fig. 10.1. Begonia (*Begonia* × *hiemalis* cv. Schwabenland Red) regenerated after a long callus phase on 5 μM 2,4-D. Note the different states of albinism (sectoral chimera) and the variegated leaf.

EVALUATION OF THE USE OF SV IN BREEDING

The following conclusions can be drawn about the role of SV as a new source of variation in breeding:

1. It is unlikely that SV results in mutations different from those obtained by other mutagenic treatments (Gavazzi *et al.*, 1987; Pius *et al.*, 1994). Since the success of mutation breeding (Broertjes and Van Harten, 1978) is very limited, it is not surprising that the number of new genotypes produced with SV is only small.

2. As with all mutations, most effects of SV on the phenotypic level are disadvantageous for the performance of plants under natural conditions (Stebbins, 1982). Thus, many advantageous mutants that have been described, e.g. dwarf mutants in cereals and rice, are good for agriculture but not for performance under natural conditions. It should be noted that particularly in ornamentals, mutations may be found that are advantageous for horticulture because the ornamental value is different from the survival value in nature.

3. When an individual plant with advantageous mutations has been found, in particular when cell selection has been applied, it is likely that in addition many deleterious mutations have occurred.
4. For ornamental crops, there are large, unexploited possibilities to obtain new genes for breeding programmes from natural populations.

SV IN OTHER BIOTECHNOLOGICAL BREEDING METHODS

As mentioned before, all biotechnological breeding methods involve a regeneration step. In the most promising technique, genetic engineering with *Agrobacterium*, selected genes are incorporated into existing cultivars. After that, transgenic ornamental plants are usually vegetatively propagated and not propagated generatively by selfing. Moreover, ornamental plants are often polyploid. Thus, in ornamental plants deleterious mutations caused by SV will usually not become apparent at the level of the phenotype.

Other biotechnological methods, such as cell selection and methods involving protoplasts, will result in plants with high levels of SV. Plader *et al.* (1994) reported percentages of aberrant cucumber plants after selfing to be 0, 0, 8 or 90 for axillary branching, direct adventitious regeneration, indirect regeneration from callus or indirect regeneration from protoplasts. For rose, direct regeneration gave no SV, but plants from callus, especially induced on zygotic embryonic tissue, showed variability (Arene *et al.*, 1993). Thus, in techniques that include a callus phase, high numbers of plants should be produced, after which plants are selected with little or no unwanted variation. Alternatively, unwanted traits can be removed by repeated crossings. Since ornamentals have often long generation cycles, and as they are 'small' crops not allowing large investments in terms of time and money, these techniques are unlikely to be used at a large scale.

SV IN MICROPROPAGATION

For the production of new shoots in micropropagation, two pathways may be chosen: (i) forced outgrowth of axillary buds by addition of a high level of cytokinin or by decapitation and (ii) regeneration of new meristems from, for example, leaf explants or cell suspensions. As noted before, SV is only observed if adventitious meristems are induced. Other types of aberration from the mother plant, epigenetic variation and the disappearance of the chimeric nature, may occur in both types of propagation. The disappearance of the chimeric nature is the rule, rather than the exception, in regeneration of adventitious meristems.

Even in the years that SV was considered predominantly as a new, positive phenomenon, commercial micropropagation laboratories using adventitious methods for various ornamentals, viz., *Saintpaulia*, *Lilium*, *Anthurium* and *Cymbidium*, frequently observed unwanted variation during the first growing season

Table 10.1. Rules to avoid variation.

I. Starting material:	• select stable genotype • induce regeneration from meristematic tissues • induce regeneration from young tissues (?)
II. Tissue culture:	• avoid repeated regeneration steps • keep callus phase as short as possible • avoid unorganized callus • avoid protoplasts
III. Planting:	• check material between times by planting in soil and observing plantlet performance

in soil. No publications appeared, probably to protect commercial interests. It is now clear that the observed variation was occasionally epigenetic and not mutational, and was related to high cytokinin concentrations during propagation. Many aberrations though were probably caused by mutations, e.g. defect flowers and colour changes. Rules-of-thumb to avoid SV are summarized in Table 10.1.

In commercial applications, planting in soil of a small sample of plants shortly after the start of tissue culture to test the genetic and epigenetic stability of the genotype and the tissue culture procedure has become routine for 'difficult' crops. A typical example is *Anthurium*. With *Anthurium*, some genotypes regenerate very few plants. This plant often requires a long period in the callus stage (a typical example would be that 6 out of 200 leaf pieces regenerate one plantlet each after only 10 months). These plantlets are propagated individually as a subclone by axillary branching. To test their conformity, plantlets from these subclones are routinely planted in soil. Often, only half the subclones are true-to-type. An early performance test is, of course, only feasible for plants that require short times to reach maturity.

HIDDEN MUTATIONS

From a practical point of view, the occurrence of hidden and difficult-to-detect mutations has to be kept in mind. There are three types:

1. Recessive mutations in plants that become apparent when these plants are propagated by selfing (Malepszy and Nadolska-Orczyk, 1989). As noted before, this is only a minor problem for ornamental plants, because the regenerated plants are usually not selfed but propagated vegetatively.

2. Mutations that are expressed only when the regenerated plants meet specific environmental conditions or reach a certain developmental stage. A mutation in the gene for alcoholdehydrogenase in maize, for example, will affect the resistance to flooding and will be detected only under this circumstance (Schwartz, 1969). For carrot, mutants are described that only have an effect at high temperature (Lo Schiavo *et al.*, 1988). Of course, in ornamentals such hidden mutations will also occur. In laboratories, greenhouses and nurseries, the plants will grow normally, but under the conditions of living room and garden, severe problems may arise. We observed among regenerants of begonia normal-looking plants that developed wilted, yellowish leaves only if exposed to high light levels in summer.

3. Mutations that have a minor effect on quantitative traits, in particular on yield. For growers a decrease in flower yield of only a few per cent will have large economic effects, as will small decreases in other quantitative characters, e.g. vase-life, number of flowers per plant/branch, etc. It is, however, very difficult to detect this in small-scale experiments.

MEASUREMENT OF SV

De Klerk (1990) argued the need for an assay to assess the extent of SV in a population of regenerated plants. Such an assay is indispensable to select cultivars and tissue culture procedures that result in little or no SV. Moreover, such an assay is necessary for studies on the mechanisms underlying SV. An assay to measure the level of SV in a population of regenerated plants should be fast, simple, objective, inexpensive and applicable shortly after regeneration. The assays that are now available are summarized in Table 10.2.

ASSAYS ON THE LEVEL OF THE PHENOTYPE

Determination of the percentage of aberrant plants

The most obvious method to measure the extent of SV is determination of the percentage of aberrant plants in a large sample (100 or more) of somaclonal plants. However, this method requires much space (usually in the glasshouse) and time (both for raising plants to maturity and for evaluation), and is subjective. Moreover, aberrations may be caused by epigenetic variation. Thus, the determination should be carried out in the generation after selfing.

A phenotypic, quantitative method

De Klerk *et al.* (1990) described a quantitative, phenotypic method for measuring SV in a small group of adult plants after regeneration. This method uses the coefficient of variation of a quantitative trait and has the following rationale.

> As stated before, most mutations are deleterious. Thus, quantitative traits that are related to performance, e.g. the length of a plant, will be affected unidirectionally (negatively). Quantitative traits that are not related to survival-value, are likely affected in a random manner, so bidirectionally.

Table 10.2. Methods to measure somaclonal variation.

	Interference by epigenetic variation	Labour-intensive[a]	Early/late[b]	Objective/ subjective[c]	Correlation with phenotype[d]
Phenotype: % aberrant plants	++[e]	– – –	– – –	–	+++
Phenotype: variation coefficient	+	++	–	+	+
Chromosomes: number	++	– –	+	+	–
Chromosomes: structure	++	– – – –	+	±	±
Isozyme/protein pattern	±	±	±	+	–
DNA sequence: RFLP	++	–	+	+	–
DNA sequence: RAPD	++	+	+	+	–
DNA sequence: AFLP	++	–	+	+	–
DNA sequence: STS	++	–	+	+	–

[a]The amount of work involved in the assay itself and in the production of (adult) plants or their progeny.
[b]The time at which the assay can be performed: shortly after regeneration, in adult R_0 plants or in the progeny of R_0 plants after selfing.
[c]The observation of a parameter should not depend upon the eye of the beholder.
[d]From a practical point of view, it is the phenotype that counts. It has been found that the extent of variation on, e.g. the molecular level is not the same as the extent of variation at the level of the phenotype (see De Klerk, 1990).
[e]Measured after selfing of the regenerated plants.

Leaf-shape might be a quantitative trait that is 'performance-neutral'. Thus, if the mutation-frequency has increased, an increased variation in the leaf-shape might occur: because of mutations some plants will have wide, short leaves and other plants narrow, long leaves. We have taken the natural logarithm of the ratio of two ribs (ln a/c) as a measure of the leaf-shape (Fig. 10.2). From a sample of somaclonal plants (usually, we take a sample of 20 to 30 plants for each tissue culture condition), we calculate from each plant the leaf-shape (= ln a/c). For the sample, the mean, the standard deviation (SD) and the coefficient of variation (SD as a percentage of the mean value) are calculated. The difference of the coefficients of variation of two samples is assessed in an F-test.

We observed an increase of the coefficient of variation in regenerated plants originating from callus induced by 5 μM 2,4-D relative to directly regenerated plants from 10 to 20% (Table 10.3). The mean of the ratio was not affected (0.46 and 0.47, respectively) demonstrating that the effect of the callus phase was bidirectionally. The method was validated by measuring populations of plants that had been regenerated from leaf fragments treated with the chemical mutagen nitrosomethylurea (NMU, Table 10.3). Measurements were only made for plants that looked normal (so the phenotypically aberrant plants were not included). There was a dose-response reaction for

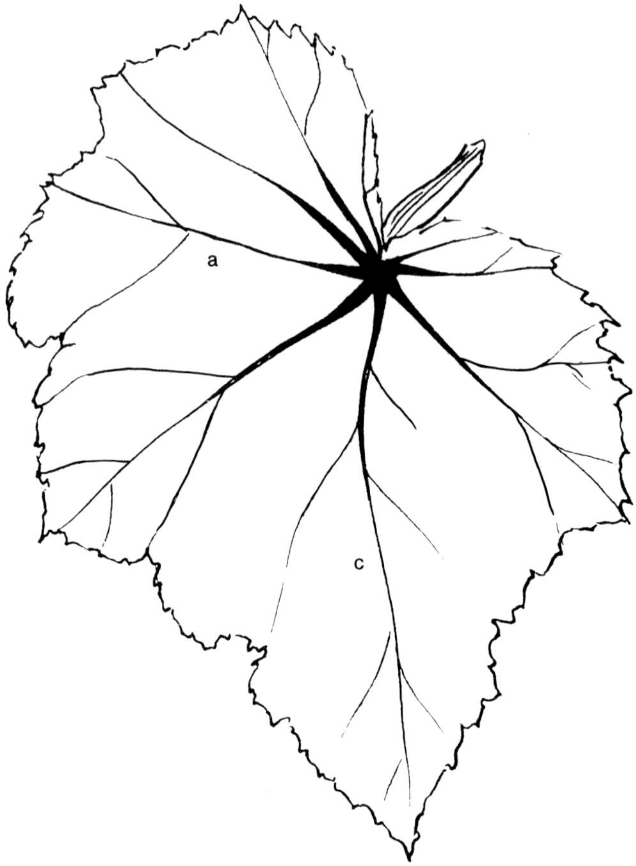

Fig. 10.2. A schematic drawing of a begonia leaf. The ribs *a* and *c* are indicated.

the coefficient of variation, but the mean of ln *a*/*c* was the same for all NMU concentrations.

We examined the effects of different callus regimes on SV using this assay. A short callus phase with 0.5 μM 2,4-D followed by regeneration from the callus on medium with 0.5 μM NAA (naphthyleneacetic acid) and 1 μM BA (benzyladenine) resulted in low SV. A long period of callus proliferation at the same 2,4-D concentration, followed by regeneration from the callus, resulted in maximal variation, similar to that observed from a short period on 5 μM 2,4-D. It should be noted that under the latter condition, regeneration took much longer than after the short period on 0.5 μM 2,4-D, and involved a long period of additional callus growth. There was a slight increase of aberrant plants as dwarfs, juvenile plants, chlorophyll mutants, venation mutants, etc. (Fig. 10.3) only after a long period of callus proliferation. The phenotypic variation of the plants shown in Fig. 10.3 was still present after a second and a third cycle of direct regeneration and was therefore most probably genetic.

Table 10.3. Coefficient of variation of the leaf shape in a somaclonal population of begonia plants regenerated under various conditions. Values are means of 2–7 experiments of 25–60 plants. The experiments with picloram and 2,4,5,-T were done only once. The values of the coefficient of variation were tested against the control. *: $P<0.05$; **: $P<0.01$; ***: $P<0.001$.

			Coefficient of variation	
			leaf 1	leaf 2
Direct regeneration, control				
0.5 μM NAA + 1 μM BAP			9.9	10.1
Direct regeneration + mutagen				
0.2 mM NMU			13.1*	15.3**
2 mM NMU			20.8***	21.1***
10 mM NMU			19.1***	22.5***
Regeneration after callus interphase				
	Duration of callus phase	Extent of organization		
0.5 μM 2,4-D	short	–	13.7***	13.2**
0.5 μM 2,4-D	long	– – –	18.2***	17.5***
5.0 μM 2,4-D	long	– – –	18.9***	20.0***
0.5 μM 2,4,5-T	short	+	8.8	8.3
2.5 μM 2,4,5-T	long	+	10.7	9.9
5.0 μM 2,4,5-T	long	–	12.5	13.4*
0.5 μM picloram	short	+	11.4	11.5
2.5 μM picloram	long	–	12.5	15.5*
5.0 μM picloram	long	–	11.6	17.5***

Other auxins, such as 2,4,5-T (2,4,5-trichlorophenoxyacetic acid) and picloram were not as potent in inducing variation (Table 10.3). Low concentrations of these auxins induced organized callus in contrast to the genuine, unorganized callus induced by 2,4-D. The callus looked like it consisted of clusters of embryonic structures. Geier (1991) reported that regeneration from organogenic callus did not induce variation in ploidy levels whereas callus without organ-like structures regenerated plants with an aberrant ploidy level.

The begonia cultivar used in our experiments is sterile. To examine the stability of the high value of the coefficient of variation, we regenerated plantlets directly from leaves of plants that originally were cultured on the medium with high 2,4-D. Although the coefficient of variation was still significantly higher than the control, it had dropped strongly (Table 10.4). However, such a drop did not occur in plants that had a high coefficient of variation after mutagenesis with NMU (Table 10.4). Thus, assuming that the assay measured genetic variation only, we observed reversible mutations after a 2,4-D treatment and stable mutations after NMU treatment. The most likely

Fig. 10.3. (*and opposite*) Some examples of abnormal begonia plants regenerated from callus. (A) Left, normal plant; right, dwarf plant with abnormal leaf (note different texture and shape). (B) Left, normal plant; middle and right, two juvenile plants with different leaf size. (C) Left, normal leaf; right, leaf with abnormal venation.

Table 10.4. Change of the coefficient of variation of the leaf shape after an additional step of direct regeneration. Plants were regenerated directly (R_0) with addition of various concentrations of the mutagens NMU, or indirectly from callus. The coefficients of variation of the leaf shape were determined. After that, from each R_0 plant, leaf explants were taken and a second generation was produced by direct regeneration. For each R_0 plant, a regenerated plant was taken at random. So for each somaclonal population a corresponding population was produced via direct regeneration. From the new populations, the coefficients of variation were determined. The values of the coefficient of variation were tested against the control. *: $P<0.05$; **: $P<0.01$; ***: $P<0.001$.

	Regenerated plants (R_0)	Plants directly regenerated from leaf explants of R_0
Direct (control)	11.5	10.5
Direct, 0.1 mM NMU	14.2*	13.4*
Direct, 2 mM NMU	19.4**	18.1**
Direct, 10 mM NMU	22.1***	21.8***
Via callus, 5 μM 2,4-D	21.2**	13.9*

explanation for this is that the reversible mutations had occurred because of altered levels of methylation of DNA or to altered chromatin states.

An increase in the coefficient of variation of the leaf shape of regenerated plants has been found in *Dianthus* (Cassels *et al.*, 1993), *Saintpaulia* and begonia (Jain, 1993), *Pelargonium* (Cassels *et al.*, 1995) and potato (P.C.G. van der Linde, Metslawier, 1991, personal communication).

ASSAYS ON THE LEVEL OF PROTEIN PATTERNS

Many authors claim the conformity of regenerated plants by reference to isozyme patterns or two-dimensional (total) protein fingerprints. There are various pitfalls. First, the portion of the genome that is examined is only small. In a critical study, Orton (1985) examined callus of celery plants that were heterozygous for five isozymes. After 12 months of culture, he did not observe a single change in isozyme patterns in 50 callus lines originating from these plants. For chromosome structure and number, though, there were gross shifts in all but one of 40 callus lines examined. This finding casts doubts on the sensitivity of this isozyme analysis. Second, when changes are observed, they may be related to other physiological or developmental states of the tissue. Thus, in the case of isozymes, developmentally and physiologically stable enzymes should be examined, such as alcohol dehydrogenase, malate dehydrogenase, phosphoglucomutase and phosphoisomerase, and not peroxidases and esterases that are very variable.

ASSAYS ON THE LEVEL OF THE GENOME

Variation at the chromosomal level

Many authors determine the ploidy level to establish genetic conformity. Obviously this will only give a very rough and unreliable indication of the occurrence of genetic changes. This assay is particularly unreliable since various developmental processes result in changes in the ploidy level (Smulders *et al.*, 1994) More sophisticated determinations consider not only the number of chromosomes, but also their structure, and are extremely time-consuming.

Molecular methods to measure SV

SV is often a genetic event, and is therefore detectable at a very early stage after the regeneration event. For such detection, various techniques are available: analysis by RAPD (random amplified polymorphic DNA), RFLP (restriction fragment length polymorphism) and, more recently, AFLP (amplified fragment length polymorphism) and analysis of differences in SSR (short sequence repeats) and microsatellites.

Brown *et al.* (1991,1993) detected differences in RAPD patterns in almost

Table 10.5. Calculation of the efficiacy of the RAPD analysis in screening the genome for mutations.

Assumptions:
- haploid plant genome: 10^9 b.p.
- primer: 10 b.p.
- RAPD pattern with a mean of ten polymorphic DNA bands with a mean length of 500 bp
- no overlap and no mispairing in the primer/DNA pairing

Deletions/insertions
DNA length of $10 \times 500 = 5.10^3$ bp, i.e. $5 \times 10^3 / 10^9 = 0.0005\%$ of the genome

Point mutations
Primer pairing in $2 \times 10 \times 10 = 200$ bp, i.e. $200/10^9 = 0.00002\%$ of the genome

all regenerants from maize and wheat callus; however, others failed to find any variation in RAPD patterns from callus regenerants when heritable variation had been found (tomato, Rus-Kortekaas *et al.*, 1994). In begonia, we occasionally found regenerants that had an altered RAPD pattern, but there was no correlation with the severeness of the phenotypic aberration. The plants shown in Fig. 10.3, for example, had all the same RAPD patterns using ten primers. We also examined RAPD patterns in plants after chemical mutagenesis (the same plants as used for validation of the phenotypic assay shown in Table 10.3). Using ten different primers, we did not find any differences in the populations from the different NMU treatments, also not in the very aberrant plants (mostly from the highest concentration, 10 mM NMU). It should be remembered that in the quantitative, phenotypic assay a large effect of the mutagen had been observed (Table 10.3). Mulcahy *et al.* (1993) calculated that two primers analyse only a very small part of a plant genome (less than 0.002%). According to our calculation (Table 10.5), the portion of the genome that is screened is somewhat lower. Using 100 primers and presuming there is no overlap in the DNA fragments analysed, the percentage of the genome that has been examined is only 0.05%. Like RAPD analysis, RFLP analysis also only screens a very small portion of the genome. The conclusion of various authors drawn from identical RFLP and RAPD patterns that the regenerated plants are genetically stable (Valles *et al.*, 1993; Isabel *et al.*, 1993) is therefore not justified.

RAPD patterns are often not reproducible. It should also be noted that the mutations probably (but not always, Nuti Ronchi, 1995) occur on one or two (or more, because many ornamentals are polyploid) chromosomes. RFLP patterns are more reproducible but require carefully purified DNA. The AFLP technique that can detect many more bands (30–60) is semiquantitative, but has other disadvantages. As well as the necessity to purify DNA extensively for the restriction enzymes used in this assay (otherwise partial digestion may occur), even a smaller part of the base sequence is investigated by a restriction enzyme in comparison with the 10-mer primers used in the RAPD method.

Of course a longer stretch of DNA is represented by the higher number of bands. The other DNA techniques mentioned above are already used in DNA fingerprinting for identification and estimation of genetic distances, but have the same disadvantages as RAPD and RFLP, analysing only part of the genome.

In conclusion, the problem with techniques on the DNA level is that they are not reproducible or require extensively purified DNA. Moreover, these techniques are not sufficiently sensitive, because they screen only a very small part of the genome. However, mutation rates vary within genomes (Lenski and Miller, 1993). Thus, if some hypervariable part of the DNA can be assayed specifically, and if this part is representative for the remainder of the genome, DNA analysing methods may be used for measuring the level of SV.

CONCLUSIONS

It is clear that the use of SV in breeding is only limited. As a matter of fact, in most applications of biotechnology SV is very undesirable. However, at present it is not clear whether SV constitutes a major problem in the various applications of biotechnology since there are only few quantitative data about its occurrence.

SV has been described carefully on the level of DNA, chromosome structure, number of chromosomes and phenotypic traits. Experimental manipulation to examine which factors influence the level of SV has only occasionally been carried out. Thus it is, for example, still not known whether various auxins result in different levels of SV or why various genotypes respond to the same tissue culture conditions with very different levels of SV. We believe that an assay to measure SV is indispensable in this type of research.

ACKNOWLEDGEMENTS

We thank Anne-Marie Kuijpers for the valuable assistance in the experiments with begonia and Merel Langens-Gerrits for critical reading of the manuscript.

REFERENCES

Arene, L., Pellegrino, C. and Gudin, S. (1993) A comparison of the somaclonal variation level of *Rosa hybrida* L. cv Meirutral plants regenerated from callus and direct induction from different vegetative and embryonic tissues. *Euphytica* 71, 83–90.

Bassi, P. (1990) Quantitative variations of nuclear DNA during plant development: a critical analysis. *Biological Reviews* 65, 185–225.

Broertjes, C. and Van Harten, A.M. (1978) *Application of Mutation Breeding Methods in the Improvement of Vegetatively Propagated Crops*. Elsevier, Amsterdam.

Brown, P.T.H. (1991) The spectrum of molecular changes associated with somaclonal variation. *IAPTC Newsletter* 66, 14–25.

Brown, P.T.H, Gobel, E. and Lörz, H. (1991) RFLP analysis of *Zea mays* callus cultures and their regenerated plants. *Theoretical and Applied Genetics* 81, 227–232.

Brown, P.T.H., Lange, F.D., Kranz, E. and Lörz, H. (1993) Analysis of single protoplasts and regenerated plants by PCR and RAPD technology. *Molecular and General Genetics* 237, 311–317.

Cassels, A.C., Walsh, C. and Periapurram, C. (1993) Diplontic selection as a positive factor in determining the fitness of mutants of *Dianthus* 'Mystere', derived from x-irradiation of nodes in *in vitro* culture. *Euphytica* 70, 167–174.

Cassels, A.C., Barrett, C. and Carney, B.F. (1995) Creation of variability in *Pelargonium* × *domesticum* by hybridisation of stable with unstable genotypes and enhancement of expression of instability in the hybrids by adventitious regeneration *in vitro*. *Acta Horticulturae* 420, 81–84.

Christianson, M.L. and Warnick, D.A. (1983) Competence and determination in the process of *in vitro* shoot organogenesis. *Developmental Biology* 95, 288–293.

D'Amato, F. (1990) Somatic nuclear mutations *in vivo* and *in vitro* in higher plants. *Caryologia* 43, 191–204.

De Klerk, G.-J. (1990) How to measure somaclonal variation. *Acta Botanica Neerlandica* 39, 129–144.

De Klerk, G.-J., ter Brugge, J. and Bouman, H. (1990) An assay to measure the extent of somaclonal variation in micropropagated plants of *Begonia* × *hiemalis*. *Acta Botanica Neerlandica* 39, 145–151.

Dixon, R.A. and Gonzales, R.A. (1994) *Plant Cell Culture. A Practical Approach.* 2nd edition. Oxford University Press, Oxford.

Dolezel, J. and Novak, F.J. (1984) Effect of plant tissue culture media on the frequency of somatic mutations in *Tradescantia* stamen hairs. *Zeitschrift für Pflanzenphysiologie* 114, 48–51.

Evans, D.E. (1989) Somaclonal variaton: genetic basis and breeding applications. *Trends in Genetics* 5, 46–50.

Gavazzi, G., Tonelli, C., Todesco, G., Arreghini, E., Raffalid, F., Vecchio, F., Barbuzzi, G., Biasini, M.G. and Sala, F. (1987) Somaclonal variation versus chemically induced mutagenesis in tomato (*Lycopersicon esculentum* L.). *Theoretical and Applied Genetics* 74, 733–738.

Geier, Th. (1991) Chromosome variability in callus produced plants. In: Harding, J., Singh, F. and Mol, J.N. (eds), *Genetics and Breeding of Ornamental Species*. Kluwer Academic Publishers, Dordrecht, pp. 79–106.

Giorgetti, L., Vergara, M.R., Evangelista, M., Lo Schiavo, F., Terzi, M. and Nuti Ronchi, V. (1995) On the occurrence of somatic meiosis in embryogenic carrot cell structures. *Molecular and General Genetics* 246, 657–662.

Griesbach, R.J. (1989) Selection of a dwarf *Hemerocallis* through tissue culture. *HortScience* 24, 1027–1028.

Griesbach, R.J. and Semeniuk, P. (1987) Use of somaclonal variation in the improvement of *Eustoma grandiflorum*. *Journal of Heredity* 78, 114–116.

Isabel, N., Tremblay, L., Michaud, M., Tremblay, F.M. and Bousquet, J. (1993) RAPDs as an aid to evaluate the genetic integrity of somatic embryogenesis-derived populations of *Picea mariana* (Mill.) B.S.P. *Theoretical and Applied Genetics* 86, 81–87.

Jagannathan, L. and Marcotrigiano, M. (1986) Phenotypic and ploidy status of *Paulownia tomentosa* trees regenerated from cultured hypocotyls. *Plant Cell, Tissue and Organ Culture* 7, 227–236.

Jain, S.M. (1993) Somaclonal variation in *Begonia* × *elatior* and *Saintpaulia ionantha* L. *Scientia Horticulturae* 54, 221–231.

Jorgensen, R. (1994) Developmental significance of epigenetic impositions of the plant genome: a paragenetic function for chromosomes. *Developmental Genetics* 15, 523–532.

Karp, A. (1991) On the current understanding of somaclonal variation. In: Miflin, B.J. (ed.), *Oxford Surveys of Plant Molecular and Cell Biology,* Vol. 7, 1–58

Khalid, N., Davey, M.R. and Power, J.B. (1989) An assessment of somaclonal variation in *Chrysanthemum morifolium*: the generation of plants of potential commercial value. *Scientia Horticulturae* 38, 287–294.

Larkin, P.J. and Scowcroft, W.R. (1981) Somaclonal variation – a novel source of variability from cell cultures for plant improvement. *Theoretical and Applied Genetics* 60, 197–214.

Lenski, R.E. and Miller, J.E. (1993) The directed mutation controversy and neo-darwinism. *Science* 259, 188–194.

Lo Schiavo, F., Giuliano, G. and Sung, Z.R. (1988) Characterization of a temperature-sensitive carrot cell mutant impaired in somatic embryogenesis. *Plant Science* 54, 157–164.

Malepszy, S. and Nadolska-Orczyk, A. (1989) *In vitro* culture of *Cucumis sativus*. VIII Variation in the progeny of phenotypically not altered R_0 plants. *Plant Breeding* 102, 66–72.

Mulcahy, D.L., Cresti, M., Sansavini, S., Douglas, G.C., Linskens, H.F., Bergamini Mulcahy, G., Vignani, R. and Pancaldi, M. (1993) The use of random amplified polymorphic DNAs to fingerprint apple genotypes. *Scientia Horticulturae* 54, 89–96.

Orton, T.J. (1985) Genetic instability during embryogenic cloning of celery. *Plant Cell, Tissue and Organ Culture* 4, 159–169.

Pavlica, M., Papes, D. and Nagy, B. (1991) 2,4-Dichlorophenoxyacetic acid causes chromatin and chromosome abnormalities in plant cells and mutation in cultured mammalian cells. *Mutation Research* 263, 77–81.

Pavlica, M., Papes, D., Franekic, J. and Nagy, B. (1992) Effects of benzyladenine on prokaryotic and eukaryotic cells. *Mutation Research* 281, 277–282.

Phillips, R.L., Kaeppler, S.M. and Olhoft, P. (1994) Genetic instability of plant tissue cultures: breakdown of normal controls. *Proceedings of the National Academy of Sciences USA* 91, 5222–5226.

Pius, J., George, L., Eapen, S. and Rao, P.S. (1994) Evaluation of somaclonal and mutagen induced variation in fingermillet. *Plant Breeding* 112, 239–243.

Plader, W., Burza, W. and Malepszy, W. (1994) Relations between various regeneration systems and somaclonal variation by cucumber (*Cucumis sativus* L.). *Abstracts VIIIth International Congress of Plant Tissue and Cell Culture*, Florence. p. 116.

Reeve, R.M. (1948) The 'tunica-corpus' concept and development of shoot apices in certain dicotyledons. *American Journal of Botany* 35, 65–75.

Rus-Kortekaas, W., Smulders, M.J.M., Arens, P. and Vosman, B. (1994) Direct comparison of levels of genetic variation in tomato detected by a GACA-containing microsatellite probe and by random amplified polymorphic DNA. *Genome* 37, 375–381.

Schwartz, D. (1969) An advantage of gene fixation resulting from selective advantage in suboptimal conditions. *American Naturalist* 103, 479–481.

Smulders, M.J.M., Rus-Kortekaas, W. and Gilissen, L.J.W. (1994) Development of polysomaty during differentiation in diploid and tetraploid tomato (*Lycopersicon esculentum*) plants. *Plant Science* 97, 53–60.

Soh, A.C. (1987) Abnormal oil palm clones. Possible causes and implications: further discussions. *Planter* 63, 59–65.

Stebbins, G.L. (1982) *Darwin to DNA, Molecules to Humanity.* W.H. Freeman, San Francisco.

Valles, M.P., Wang, Z.Y., Montavon, P., Potrykus, I. and Spangenberg, G. (1993) Analysis of genetic stability of plants regenerated from suspension cultures and protoplasts of meadow fescue (*Festuca pratensis* Huds.). *Plant Cell Reports* 12, 101–106.

Van den Bulk, R.W. (1991) Application of cell and tissue culture and *in vitro* selection for disease resistance breeding – a review. *Euphytica* 56, 269–285.

Wolffe, A.P. (1994) Inheritance of chromatin states. *Developmental Genetics* 15, 463–470.

T-DNA Insertion Mutagenesis for Improvement of Ornamentals

11

Matthew A. Jenks and Kenneth A. Feldmann
Department of Plant Sciences, University of Arizona, Tucson, Arizona 85721, USA

While T-DNA insertion mutagenesis has found wide popularity among plant molecular biologists as a tool for identifying and cloning genes, the potential for using T-DNA gene tagging technologies in the improvement of ornamental crops has yet to be realized. Gene cloning from T-DNA mutagenized populations of ornamental species, as will be shown in this review, would probably be extremely inefficient. Instead, T-DNA insertion and cloning of genes using the well-developed model plant system of *Arabidopsis thaliana* has been established as a relatively efficient method for acquiring useful gene sequences. Recent reviews have shown that a variety of plant species can be modified by genetic engineering with genes identified and cloned from *Arabidopsis* (Feldmann and Meyerowitz, 1991; Forsthoefel *et al.*, 1992; Feldmann *et al.*, 1994). Therefore, T-DNA insertion mutagenesis to acquire genes from *Arabidopsis* may become a valuable method for research programmes that employ genetic engineering to improve the aesthetic value or environmental stress resistance of important ornamental crops.

GENERAL REVIEW OF T-DNA TECHNOLOGY

MUTATION INDUCTION

Induction of plant mutations by a variety of means has been used to produce lines that differ in only one gene (Koornneef *et al.*, 1982; Feldmann and Marks, 1987; Jenks *et al.*, 1992). Comparisons between wild types and induced mutants have proven to be effective for analysing how single genes determine plant traits (Koornneef *et al.*, 1989; Feldmann, 1991; Jenks *et al.*, 1992) and the role of these genes in plant biochemistry (Jenks *et al.*, 1995), development (Meinke,

 Biotechnology of Ornamental Plants (eds R.L. Geneve, J.E. Preece and S.A. Merkle)

1986; Aarts, *et al.*, 1993) and physiology (Su and Howell, 1992; Jenks *et al.*, 1994a,b).

The three most commonly used plant mutagens are ethylmethanesulphonate (EMS), transposons (transposable elements) and T-DNAs (transferred DNA). EMS has proved to be among the most efficient of all mutagens producing the highest frequency of mutations under optimum conditions. EMS is a DNA alkylating agent that predominantly induces point mutations. Genes altered by EMS can be isolated using positional cloning in species whose chromosomes have been saturated with genetic markers, however, the task can be labour intensive (Arondel *et al.*, 1992). 'Chromosome walking' will become an increasingly efficient method for isolating genes in *Arabidopsis* as its physical map is completed (Chang *et al.*, 1988; Nam *et al.*, 1989). Another advantage of EMS as a mutagen is that it can cause both gain-of-function and loss-of-function mutations, whereas, insertion elements cause predominantly the latter type.

Transposons are endogenous mobile pieces of DNA that are excised from the chromosome during cell divisions and then reinserted at another, usually linked, position on the chromosome (Dooner and Blachew, 1989; Dooner *et al.*, 1991; Bancroft and Dean, 1993). Transposons cause mutations by inserting into and thus altering or inactivating genes at their target sites. Isolation of plant DNA flanking the inserted transposable element is recovered using part of the transposon sequence as a probe. Cloning the transposon tagged gene therefore requires knowledge of the transposon sequence. Once the plant sequence is isolated, it is used as a probe to recover the wild-type gene from a genomic library made from the parent line. Most plant species lack an active and well-characterized transposon. However, transposon systems from one species, e.g. the *Ac/Ds* elements from maize, have been transformed into other species where they are active (Aarts *et al.*, 1993; Bancroft and Dean, 1993). Future research will determine whether transposon tagging will become an efficient means of gene identification and cloning.

T-DNA insertion mutagenesis has proved to be a highly effective means to both identify and clone important plant genes (Feldmann *et al.*, 1989; Feldmann, 1991, 1992; Van Lijsebettens *et al.*, 1991). Like transposons, T-DNA induced mutations result from the insertion of DNA segments into a coding or regulating sequence of a gene to alter or inactivate gene function. The T-DNA is carried on an *Agrobacterium tumefaciens Ti* (tumor inducing) plasmid that is transferred to plant cells during transformation (Koncz *et al.*, 1989, 1992). Unlike transposons, T-DNAs insert randomly within the genome of the host plant (Feldmann, 1991). The T-DNAs used for insertion mutagenesis are engineered with a gene coding for resistance to a selectable marker, e.g. kanamycin, which provides an efficient way to select for transformed plants within a treated population (e.g. Feldmann and Marks, 1987). Like transposon tagging, isolation of the inserted T-DNA recovers the associated plant flanking DNA which can be used as a probe to isolate the wild-type gene from a genomic library made from the parent line.

METHODS OF T-DNA MUTAGENESIS AND GENE CLONING

The best developed T-DNA insertion mutagenesis protocols have been developed for *Arabidopsis*. The first gene cloned using T-DNA tagging protocols, *chlorata-42*, came from a population of *Arabidopsis* transformants generated using tissue culture techniques (Koncz *et al.*, 1990). This mutant had a yellow–green phenotype and was viable because the insert was in the 3′ end of the gene and as such did not cause a loss-of-function mutation. Other mutants identified from *Arabidopsis* tissues co-cultivated with *A. tumefaciens* did not have T-DNA insertions. These mutations probably arose due to normal somaclonal variation caused by *in vitro* culture (Larkin and Scowcroft, 1981).

In 1987, Feldmann and Marks developed a non-tissue culture T-DNA transformation protocol for *Arabidopsis* which avoids the somaclonal variation associated with tissue culture regenerates. This approach involved transformation of sterile seeds in batches of 3000 which had been pre-imbibed for 10–18 hours in liquid cocultivation medium. Seeds were infected by adding an overnight culture of *A. tumefaciens* and then cocultivated for 24 hours. The *A. tumefaciens* was a C58C1Rif strain containing the co-integrate plasmid p3850:1003 (Velton and Schell, 1985). The neomycin phosphotransferase II gene, which confers kanamycin resistance (Kan^r) to the plant, is located between the two T-DNA borders. The resulting T_1 seeds were grown to maturity and allowed to self-pollinate to produce T_2 seeds. T_2 seeds were plated onto kanamycin containing media and each Kan^r seedling transferred to vermiculite and grown to maturity. See Feldmann and Marks (1987) and Feldmann (1991) for more details of the procedure. More than 14,000 transformants have been generated so far (Feldmann and Marks, 1987; Feldmann, 1991; Forsthoefel *et al.*, 1992; and unpublished). To generate these transformants, more than 1.1×10^6 infected plants (T1) have been grown to maturity, and more than 10^8 seed (T2) from these plants have been tested on kanamycin. Segregation analysis of the Kan^r marker has established that the average number of inserts per resistant segregant is 1.5. Therefore, in the population of 14,000 transformants there are an estimated 21,000 independent inserts.

When seed infection was first proposed, it was thought that *A. tumefaciens* would move through the micropyle and into Layer II of the embryonic meristem. Fortuitously, transformation does not occur until very late in the development of the infected plant. Apparently, premeiotic cells, meiotic products or zygotes are transformed in the developing flower bud. While seed infection/transformation was an improvement over tissue culture methods, seed transformation has remained labour intensive and undependable.

Recognizing that T-DNA insertion occurs late in development, Chang *et al.* (1994) developed another whole-plant transformation system. In this procedure, the newly emerging inflorescence is removed near the base and *A. tumefaciens* are inoculated around the wounded area of the growing plant. After adventitious floral buds break, the newly emerging inflorescences are again removed. The wounded area is again inoculated with *A. tumefaciens* and this time the plants are allowed to bolt and selfed progeny are collected. This procedure is more labour intensive and results in fewer transformants, however, it has the advantage of being more reproducible.

An even better strategy for routine T-DNA transformation for insertion mutagenesis is vacuum infiltration of *A. tumefaciens* solution into young whole *Arabidopsis* plants (Bechtold *et al.*, 1993). After infiltration, the plants are allowed to recover in the greenhouse and after several weeks progeny are collected. Vacuum infiltration circumvents the need to get the *A. tumefaciens* into the seed and is thus more reliable than seed infection/transformation. Moreover, screening of the progeny from vacuum-infiltrated plants on selective medium results in a higher frequency of transformants than was realized for either of the previous seed or wounding transformation procedures.

Interestingly, all seed and whole-plant T-DNA transformation procedures apparently produce transformants by the same mechanism. The *A. tumefaciens* get into the interstitial spaces as the plant bolts and are carried along until flowering. At some stage late in flowering the *A. tumefaciens* are induced to attach to a competent plant cell and transfer the T-DNA.

Three common techniques used to isolate plant DNA flanking T-DNA inserts in transformed lines are (i) screening lambda libraries made from genomic DNA of the T-DNA tagged mutants, (ii) polymerase chain reaction (PCR) amplification of plant flanking DNA, and (iii) plasmid rescue of self-ligated DNA fragments from the genomic DNA of T-DNA transformed lines. In the first approach, the lambda library is screened for clones which contain the T-DNA insert using parts of the T-DNA borders as probes. Positive genomic clones are purified and analysed using standard molecular techniques (Sambrook *et al.*, 1989; Schultz *et al.*, 1994). Plant DNA flanking a T-DNA insert can also be amplified using inverse PCR. For inverse PCR, DNA from the transformed plant is restricted with a number of restriction enzymes that cut the T-DNA border regions within a few hundred base pairs of the borders. Alternatively, the genomic DNA from the transformant can be analysed in some detail to ascertain which restriction enzyme cuts near enough to the border that it will be possible to amplify the intervening plant flanking DNA. In addition, it is possible to size select DNA that contains putative plant flanking DNA and use it in subsequent steps (Deng *et al.*, 1992). The DNA is ligated and primers made from the border region and pointing outward from each other are used for PCR (Earp *et al.*, 1990; Deng *et al.*, 1992). Another round of PCR with primers made for the linear PCR products and pointing inward will amplify a band that can be cloned into a sequencing vector for sequence analysis. Standard molecular techniques can be used to clone the full length gene.

The most successfully used cloning technique for T-DNA tagged genes has been plasmid rescue. This technique utilizes the antibiotic resistance markers and the ColE1 origin of replication of the two copies of pBR322 in the T-DNA vector, pGV3850:1003. Genomic DNA from a T-DNA transformed line is digested by restriction enzymes, self-ligated, and transformed into *E. coli*. Transformed colonies are isolated that survive after plating on antibiotic containing media. The plant flanking DNA is used as a probe to screen a non-transformed genomic library for an intact homologous wild-type gene.

To confirm that an isolated gene is responsible for the mutant phenotype, the mutant is transformed with the wild-type gene to restore the wild-type phenotype. Transcriptionally active regions of these clones can be restriction mapped by probing wild-type poly(A)$^+$ RNA (Deng *et al.*, 1992). Standard protocols to perform these and other methods associated with T-DNA gene cloning are presented in Sambrook *et al.* (1989) and Schulz *et al.* (1994).

USING ARABIDOPSIS *IN AN ORNAMENTALS IMPROVEMENT PROGRAMME*

Genetic and cultural characteristics of *Arabidopsis* makes it one of the finest plant species for development of an efficient plant system for T-DNA insertional mutagenesis. *Arabidopsis* is a flowering plant with typical cellular physiology and development such that genes obtained can generally be assumed to have similar functions in most other higher plants. The small size of *Arabidopsis* (25 to 40 cm height) means that less growth space is required for *Arabidopsis* than for most other plant species to grow the large T-DNA mutagenized populations that must be screened to identify new mutants. The short life cycle (six to eight weeks), self pollination, and large seed production (5000–10,000 seeds per plant) of *Arabidopsis* mean that less time will be spent in genetic analysis than might have been required for another plant species. The small haploid genome of *Arabidopsis* (100,000 Mbp; Arumuganathan and Earle, 1991) and near absence of interspersed, repetitive DNA (Leutwiler *et al.*, 1994) increases the likelihood for a T-DNA insert to tag functional and important genes and also reduces the work required in gene isolation. For these reasons, *Arabidopsis* has become the premier plant model for T-DNA insertion mutagenesis to identify and clone plant genes.

RECENT T-DNA GENE ISOLATION RESEARCH AND POTENTIAL APPLICATIONS TO THE IMPROVEMENT OF ORNAMENTALS

Current work in our laboratory has the potential to make significant contributions to the ornamentals industry through genetic engineering of ornamental crops. Our laboratory is currently investigating three classes of T-DNA tagged genes that, when mutated, cause dwarfism, alter the epicuticular waxes and change leaf shape. Examples of T-DNA tagged *Arabidopsis* genes being investigated in other laboratories also have utility in the improvement of ornamentals.

As communities move increasingly from rural to urban settings, the potential to produce smaller plants for smaller living spaces may be realized through plant genetic engineering with dwarfism genes. We have identified five T-DNA tagged genes which, when mutated, cause *Arabidopsis* plants to become dwarfed (*dwf*) (Fig. 11.1A, B.). These mutants appear blocked in their ability to undergo cell elongation. Industry attempts at inducing mutations in ornamental plant genes, that are orthologous to *dwf* genes, have potential for increasing the number of dwarf plants for use in the urban environment. However, the use of transformation techniques to control expression of *dwf* genes has the potential for increased control, through genetic engineering, of ornamental plant size and shape.

Other research that could benefit the ornamentals industry is T-DNA tagging of genes that affect flower structure and development. Since the aesthetic value of many ornamentals arises from their floral structures, modification of flowers could greatly increase the aesthetic value of many economically important ornamentals. Flowers of *Arabidopsis* are divided into four major whorls of organs, the outer being sepals, then petals, then stamens, and then the inner whorl composed of a bilocular gynecium with a papillate stigma. A range of different flower structure mutants in *Arabidopsis* induced by T-DNA insertion have been identified (Feldmann and Meyerowitz, 1991). These mutants alter flower organs of individual whorls. For example, *flw2* has flower parts that resemble leaves and is similar to the EMS-induced mutant designated *leafy* (Haughn and Somerville, 1988). Other flower mutants convert inner whorl structures into an outer whorl-type morphology (Feldmann and Meyerowitz, 1991), greatly increasing the aesthetic value of even the lowly *Arabidopsis* flower (Fig. 11.1C). The flower gene *agamous* (*ag*) induced by EMS was allelic to the first floral mutant identified by T-DNA tagging (Yanofsky *et al.*, 1990). After the agamous allele was shown to co-segregate with the Kan^r marker, plant DNA flanking the T-DNA insert was used to screen a wild-type genomic library and the wild-type gene then sequenced (Yanofsky *et al.*, 1990). Transformation of the wild-type allele into the *ag* mutant restored the wild-type phenotype, thus verifying that the proper gene (that which when mutated caused the mutant phenotype) had been sequenced. This method for verifying gene identity is termed complementation.

The whitish appearance of many plant surfaces is due to the deposition of epicuticular wax (EW) crystalline microstructures. Genetic engineering to increase or decrease EW microstructures on the surface could modify surface glaucousness and create novel ornamental phenotypes. Moreover, EW has been shown to play a role in resistance to both biotic and abiotic environmental stresses (Jordan *et al.*, 1984; Jenks *et al.*, 1994a; Eigenbrode and Espelie, 1995). Therefore, changing the surface glaucousness and/or improving environmental stress resistance by genetic engineering to alter EW on ornamentals could greatly benefit the ornamentals industry.

Visible screening of an EMS-induced mutant population of *Arabidopsis* uncovered unique lines with altered visible glaucousness over the stem surface. These mutants were called *eceriferum* (*cer*) meaning 'wax bare' (Fig 11.1D) and

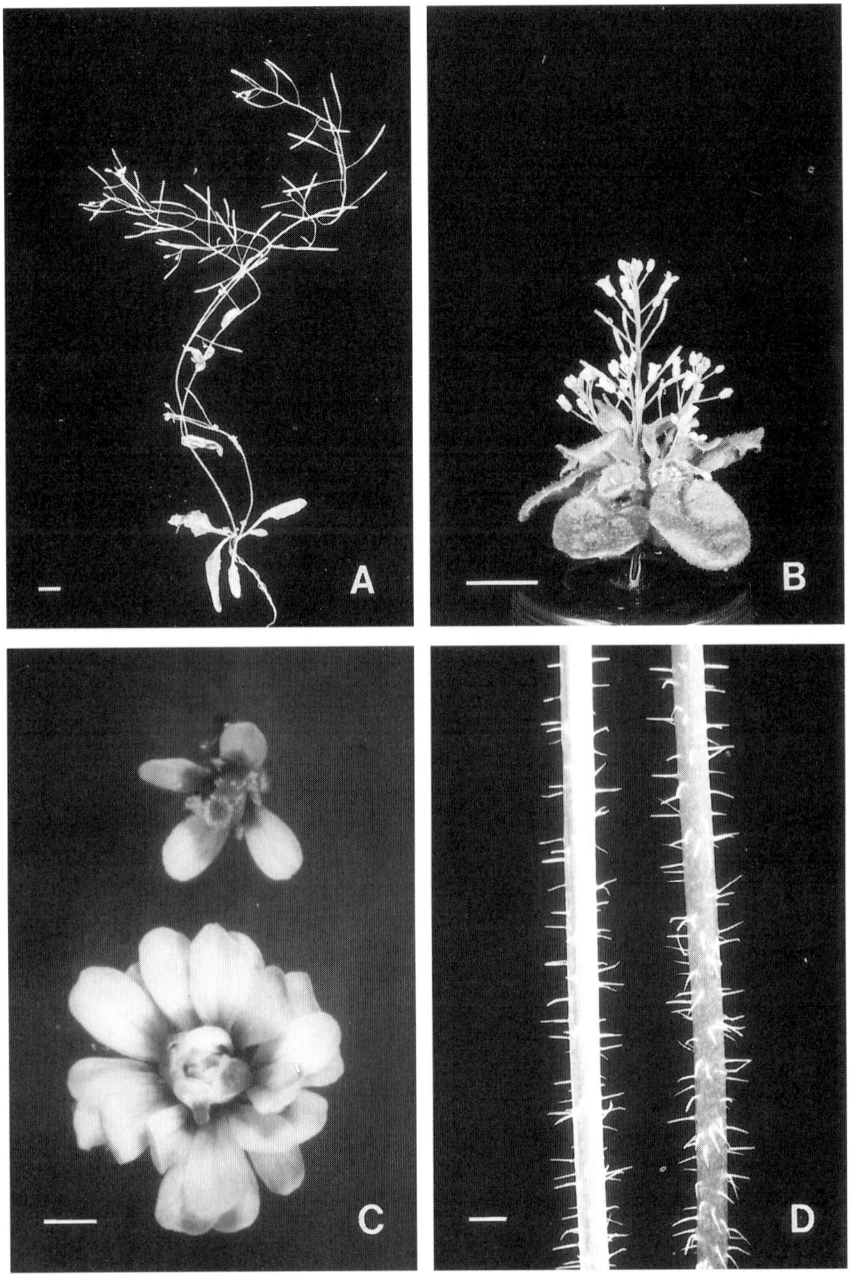

Fig. 11.1 (A–D). Wild-type *Arabidopsis* ecotype Wassilewskija and isogenic mutants isolated from a T-DNA mutagenized population (approximately 25 days old). (A) Wild type. Bar = 1.5 cm. (B) *dwarf4* (*dwf4*). Bar = 2.0 cm. (C) *apetals2* (*ap2*). Bar = 1.1 mm. (D) *eceriferum4* (*cer4*). Bar = 1.4 mm.

it was shown that these lines had altered EW crystallization patterns (Koornneef *et al.*, 1982) and EW chemical constituent profiles (Hannoufa *et al.*, 1993; Jenks *et al.*, 1995). So far, mutations in 20 loci cause *cer* phenotypes. We screened a T-DNA mutagenized population and discovered four *cer* mutants whose genes corresponded to *cer1*, *cer2*, *cer3* and *cer4* and which also segregated with Kan^r (McNevin *et al.*, 1993). The *CER2* and *CER3* gene coding sequences have recently been cloned and *CER1* and *CER4* are expected soon. Based on detailed chemical analysis, each of the *CER* gene products appears to play a role in the EW biosynthetic pathway. The *CER1* gene product appears to function in the conversion of aldehydes to both alkane and primary alcohol EW constituents. The *CER2* gene product appears to regulate elongation of fatty acid precursors (acyl-CoA) beyond 26 carbons in length. We suspect that *CER3* gene product is involved in the release or transport of acyl-CoA away from the elongase complexes whereas the *CER4* gene product appears to function in aldehyde reduction to primary alcohols. Once the role of EW crystallization patterns and EW chemistry in ornamental plant aesthetics and environmental stress resistance has been determined, novel genetic approaches using CER genes can be developed to create superior ornamental plants.

Genes that control leaf size and shape were also identified from a T-DNA mutagenized population of *Arabidopsis*. One of these mutant genes caused the lanceolate leaves of wild-type *Arabidopsis* (Fig. 11.2A) to become large, oval and multilobed (Fig. 11.2B) while another mutation caused leaves to become small and circular with a nearly cordate base (Fig. 11.2C). These genes might be used to genetically engineer the textural quality of ornamentals used in landscape design. Likewise, the foliage plant industry might benefit by increasing the selection of novel and attractive plants.

Other genes identified from T-DNA mutagenized *Arabidopsis* populations that have potential value to the ornamentals industry include those that control flowering time (Feldmann laboratory), trichome development (Herman and Marks, 1989) and flavonoid production (Shirely *et al.*, 1992). In the future, many other *Arabidopsis* genes will be discovered that can be used for genetic engineering to modify ornamental plant characteristics.

FUTURE APPLICATIONS OF T-DNA TECHNOLOGY FOR ORNAMENTALS

EMPLOYING ARABIDOPSIS *T-DNA TAGGING PROTOCOLS FOR THE IMPROVEMENT OF ORNAMENTALS*

The basic requirements for identifying genes that control specific traits include the availability of an efficient screen for selection of the trait, knowledge of the gene's potential for application to an improvement programme and an efficient transformation system for the ornamental of interest. Thousands of individual T2 plants from an *Arabidopsis* T-DNA transformed population will need to be screened to identify mutant genes that are associated with

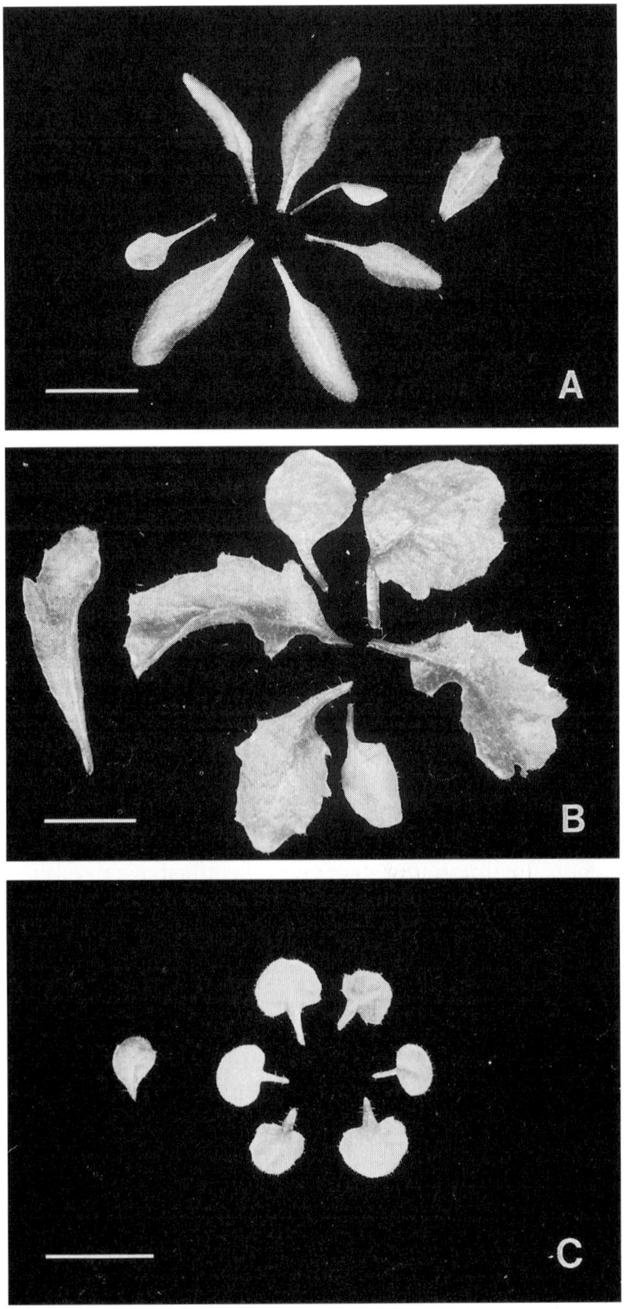

Fig. 11.2. Leaves of wild-type *Arabidopsis* ecotype Wassilewskija and isogenic mutants isolated from a T-DNA mutagenized population (approximately 25 days old). (A) Wild-type leaves. (B) Mutant with large, oval and multi-lobed leaves. (C) Mutant with small and circular leaves. Bars = 2.5 cm.

the desired trait. Therefore, a very efficient screening method for the expected mutant phenotype must be available or developed.

Once a set of genes affecting the trait are shown to possess a T-DNA insertion, it must be determined which genes have the best potential for use in genetic engineering of the trait in the ornamental of interest. Therefore, knowledge about the gene function in plant aesthetics or environmental interactions is needed. For certain traits, the T-DNA tagged lines themselves make an excellent system to answer these questions. Since the lines are isogenic, physiological and biochemical studies can be performed in the light of proposed gene function. Once a gene has been sequenced, its function can often be determined from a gene sequence data base by matching its sequence with that of a known gene sequence with known function. Studies to predict or verify the function of individual genes will help predict the kind of genetic manipulations which are possible.

Once it has been determined which newly cloned gene to use in transformation of the ornamental, a transformation protocol must be developed for the ornamental if one is not already in place. Success in transformation of certain ornamentals could be difficult since tissue culture and transformation protocols for many have not been developed.

POTENTIAL FOR TAGGING GENES THAT ARE UNIQUE TO ORNAMENTALS

One drawback to utilization of the *Arabidopsis* T-DNA tagging protocol is that *Arabidopsis* lacks many of the most valuable traits, and corresponding genes that would be useful for the improvement of ornamentals. For example, the genes responsible for regulated expression in the flower of most colour pigments do not occur in the *Arabidopsis* species. The identification and cloning of genes unique to ornamentals might be facilitated using a T-DNA tagging protocol developed for the ornamental carrying the trait. However, this approach would certainly be labour intensive and the cost might not be justifiable. For instance, an efficient T-DNA infection protocol would be needed for the generation of large numbers of T1 seeds. Inbred lines would be needed to reduce phenotypic variability for selecting the mutant trait in the T2 generation and also to reduce the amount of work required for cloning the gene once it is shown to contain a T-DNA insert. The ornamental must have a short life cycle (or the researcher a long one) in order to produce the number of generations needed for efficient genetic analysis. The ornamental should also have efficient cultural requirements since a tremendous number of plants would probably need to be screened to find the desired mutants. Moreover, the ornamental should have a relatively small genome and a small amount of non-expressed DNA to increase the frequency of T-DNA insertions into expressed genes and reduce the number of plants that need to be screened. These and other requirements currently make a T-DNA tagging approach to clone genes in ornamentals very inefficient.

Perhaps the method of choice for identifying valuable genes found only in ornamentals is transposon insertion mutagenesis. Mobile genetic elements in ornamental crops like *Antirrhium majus* (Lister *et al.*, 1993; Noda *et al.*, 1994; Simon *et al.*, 1994) and *Petunia hybrida* (Chuck *et al.*, 1993) have been used to tag a number of flower-related genes that could prove valuable to the ornamentals industry. A review covering recent advances in transposon tagging techniques using *Arabidopsis* systems was recently published (Feldmann *et al.*, 1994).

SUMMARY

The *Arabidopsis* T-DNA tagging protocol provides a powerful tool for identifying and cloning genes in *Arabidopsis* that have value for genetic engineering of ornamental plants. T-DNA tagged genes from *Arabidopsis* that control plant development, morphology and physiology have great potential for improving the aesthetic value and stress resistance of ornamental plants.

REFERENCES

Aarts, M.G.M., Dirkse, W.G., Stiekema, W.J. and Pereira, A. (1993) Transposon tagging of a male sterility gene in *Arabidopsis. Nature* 363, 715–717.

Arondel, V., Lemieux, B., Hwang, I., Gibson, S., Goodman, H.M. and Somerville, C.R. (1992) Map-based cloning of a gene controlling *omega-3* fatty acid desaturation in *Arabidopsis. Science* 258, 1453–1355.

Arumuganathan, K. and Earle, E.D. (1991) Nuclear DNA content of some important plant species. *Plant Molecular Biology Reporter* 9, 208–218.

Bancroft, I. and Dean, C. (1993) Transposition pattern of the maize element *Ds* in *Arabidopsis thaliana. Genetics* 134, 1221–1229.

Bechtold, N., Ellis, J. and Pelletier, G. (1993) In planta *Agrobacterium* mediated gene transfer by infiltration of adult *Arabidopsis thaliana* plants. *C.R. Academy of Science* 316, 1194–1199.

Chang, C., Bowman, J.L., DeJohn, A.W., Lander, E.S. and Meyerowitz, E.M. (1988) Restriction fragment length polymorphism linkage map of *Arabidopsis thaliana. Proceedings of the National Academy of Sciences, USA* 85, 6856–6860.

Chang, S.S., Park, S.K., Kim, B.C., Kang, B.J., Kim, D.U. and Nam, H.G. (1994) Stable genetic transformation of *Arabidopsis thaliana* by *Agrobacterium* inoculation in planta. *Plant Journal* 5, 551–558.

Chuck, G., Robbins, T., Nijjar, C., Ralston, E., Courtney-Gutterson, N. and Dooner, H.K. (1993) Tagging and cloning of a petunia flower color gene with the maize transposable element Activator. *Plant Cell* 5, 371–378.

Deng, X.-W., Matsui, M., Wei, N., Wagner, D., Chu, A.M., Feldmann, K.A. and Quail, P.H. (1992) *COP1*, an *Arabidopsis* regulatory gene, encodes a protein with both a zinc-binding motif and a Gβ homologous domain. *Cell* 71, 791–801.

Dooner, H.K. and Belachew, A. (1989) Transposition pattern of maize element *Ac* from the *bz-m2* (*Ac*) allele. *Genetics* 122, 447–457.

Dooner, H.K., Keller, J., Harper, E. and Ralston, E. (1991) Variable patterns of transposition of the maize element Activator in tobacco. *Plant Cell* 3, 473–482.

Earp, D., Lowe, B. and Baker, B. (1990) Amplification of genomic sequences flanking transposable elements in host and heterologous plants: a tool for transposon tagging and genome characterization. *Nucleic Acids Research* 18, 3271–3279.

Eigenbrode, S.D. and Espelie, K.E. (1995) Effects of plant epicuticular lipids on insect herbivores. *Annual Review of Entomology* 40, 117–142.

Feldmann, K.A. (1991) T-DNA insertional mutagenesis in *Arabidopsis*: mutational spectrum. *Plant Journal* 1, 71–82.

Feldmann, K.A. (1992) T-DNA insertion mutagenesis in *Arabidopsis*: seed infection/transformation. In: Conez, C. (ed.), *Methods in Arabidopsis Research*. World Scientific, London, pp. 274–289.

Feldmann, K.A. and Marks, M.D. (1987) *Agrobacterium*-mediated transformation of germinating seeds of *Arabidopsis thaliana*: a non-tissue culture approach. *Molecular and General Genetics* 208, 1–9.

Feldmann, K.A. and Meyerowitz, E.M. (1991) Tagging floral structure genes. In: Harding, J. (ed.), *Genetics and Breeding of Ornamental Species*. Kluwer Academic Publishers, Dordrecht, pp. 271–283.

Feldmann, K.A., Marks, M.D., Christianson, L. and Quatrano, R.S. (1989) A dwarf mutant of *Arabidopsis* generated by T-DNA insertion mutagenesis. *Science* 243, 1351–1354.

Feldmann, K.A., Malmberg, R.G. and Dean, C. (1994) Mutagenesis in *Arabidopsis*. In: Meyerowitz, E.M. and Somerville, C.R. (eds), *Arabidopsis*. Cold Springs Harbor Press, New York, pp. 137–172.

Forsthoefel, N.R., Wu, Y., Schulz, B., Bennett, M.J. and Feldmann, K.A. (1992) T-DNA insertion mutagenesis in *Arabidopsis*: prospects and perspectives. *Australian Journal of Plant Physiology* 19, 353–366.

Hannoufa, A., McNevin, J.P. and Lemieux, B. (1993) Epicuticular waxes of *eceriferum* mutants of *Arabidopsis thaliana*. *Phytochemistry* 33, 851–855.

Haughn, G.W. and Somerville, C.R. (1988) Genetic control of morphogenesis in *Arabidopsis*. *Developmental Genetics* 9, 73–89.

Herman, P.L. and Marks, M.D. (1989) Trichome development in *Arabidopsis thaliana*. II. Isolation and complementation of the GLABROUS1 gene. *Plant Cell* 1, 1051–1055.

Jenks, M.A., Rich, P.J., Peters, P.J., Axtell, J.D. and Ashworth, E.N. (1992) Epicuticular wax morphology of bloomless (*bm*) mutants in *Sorghum bicolor*. *International Journal of Plant Science* 153, 311–319.

Jenks, M.A., Joly, R.J., Peters, P.J., Rich, P.J., Axtell, J.D. and Ashworth, E.A. (1994a) Chemically induced cuticle mutation affecting epidermal conductance to water vapor and disease susceptibility in *Sorghum bicolor* (L.) Moench. *Plant Physiology* 105, 1239–1245.

Jenks, M.A., Rich, P.J. and Ashworth, E.N. (1994b) Involvement of cork cells in the secretion of filaments on *Sorghum bicolor* (L.) Moench. *International Journal of Plant Science* 155, 506–518.

Jenks, M.A., Tuttle, H.A., Eigenbrode, S.D. and Feldmann, K.A. (1995) Leaf epicuticular waxes of the *eceriferum* mutants in *Arabidopsis*. *Plant Physiology* 108, 367–377.

Jordan, W.R., Shouse, P.J., Blum, A., Miller, F.R. and Monk, R.C. (1984) Environmental physiology of sorghum. II. Epicuticular wax load and cuticular transpiration. *Crop Science* 24, 1168–1173.

Koncz, C., Martini, N., Mayerhofer, R., Kalman, Z., Korber, H., Redei, G.P. and Schell, J. (1989) High-frequency T-DNA mediated gene tagging in plants. *Proceedings of the National Academy of Sciences*, USA 86, 8467–8471.

Koncz, C., Mayerhofer, R., Kalman, Z., Nawrath, C., Reiss, B., Redei, G.P. and Schell, J. (1990) Isolation of a gene encoding a novel chloroplast protein by T-DNA tagging in *Arabidopsis thaliana. EMBO Journal* 5, 1137–1146.

Koncz, C., Schell, J. and Redei, G.P. (1992) T-DNA transformation and insertion mutagenesis. In: Koncz, C., Chua, H-H. and Schell, J. (eds), *Methods in Arabidopsis Research*. World Scientific Publishing, New Jersey, pp. 224–273.

Koornneef, M., Dellaert, L.W.M. and van der Veen, J.H. (1982) EMS- and radiation-induced mutation frequencies at individual loci in *Arabidopsis thaliana* (L.) Heynh. *Mutation Research* 93, 109–123.

Koornneef, M., Hanhart, C.J. and Thiel, F. (1989) A genetic and phenotypic description of *eceriferum* (*cer*) mutants in *Arabidopsis thaliana. Journal of Heredity* 80, 118–122.

Larkin, P.J. and Scowcroft, W.R. (1981) Somaclonal variation-A novel source of variability from cell cultures for plant improvement. *Theoretical and Applied Genetics* 60, 197–214.

Leutwiler, L.S., Hough-Evans, B.R. and Meyerowitz, E.M. (1984) The DNA of *Arabidopsis thaliana. Molecular and General Genetics* 219, 225–234.

Lister, C., Jackson, D. and Martin, C. (1993) Transposon-induced inversion in *Antirrhinum* modifies nivea gene expression to give a novel flower color pattern under the control of cycloidea-radicalis. *Plant Cell* 5, 1541–1553.

McNevin, J.P., Woodward, W., Hannoufa, A., Feldmann, K.A. and Lemieux, B. (1993) Isolation and characterization of *eceriferum* (*cer*) mutants induced by T-DNA insertions in *Arabidopsis thaliana. Genome* 36, 610–618.

Meinke, D.W. (1986) Embryo-lethal mutants and the study of plant embryo development. *Oxford Survey of Plant Molecular Biology* 3, 122–165.

Nam, H.-G., Giraudat, J., Boer, B. den, Moonan, F., Loos, W.D.B., Hauge, B.M. and Goodman, H.M. (1989) Restriction fragment length polymorphism map of *Arabidopsis thaliana. Plant Cell* 1, 989–1003.

Noda, K.I., Glover, B.J., Linstead, P. and Martin, C. (1994) Flower color intensity depends on specialized cell shape controlled by a Myb-related transcription factor. *Nature* 369, 661–664.

Sambrook, J., Fritsch, E.F. and Maniatus, T. (1989) *Molecular cloning: a laboratory manual.* Cold Springs Harbor Laboratory Press, Cold Springs Harbor, New York.

Schultz, B., Bennet, M.J., Dilkes, B.P. and Feldmann, K.A. (1994) T-DNA tagging in *Arabidopsis thaliana*: cloning by gene disruption. In: Gelvin, S.B. and Schilperoort, R.A. (eds), *Plant Molecular Biology Manual*. Kluwer Academic Publishers, Boston.

Shirley, B.W., Hanley, S. and Goodman, H.M. (1992) Effects of ionizing radiation on a plant genome: analysis of two Arabidopsis *transparent testa* mutations. *Plant Cell* 4, 333–347.

Simon, R., Carpenter, R., Doyle, S. and Coen, E. (1994) Fimbriata controls flower development by mediating between meristem and organ identity genes. *Cell* 78, 99–107.

Su, W. and Howell, S.H. (1992) A single genetic locus, *Ckr1*, defines *Arabidopsis* mutants in which root growth is resistant to low concentrations of cytokinin. *Plant Physiology* 99, 605–613.

Van Lijsebettens, M., Vanderhaeghen, R. and Van Montagu, M. (1991) Insertion mutagenesis in *Arabidopsis thaliana. Plant Science* 80, 27–37.

Velton, J. and Schell, J. (1985) Selection-expression plasmid vectors for use in genetic transformation of higher plants. *Nucleic Acids Research* 13, 6981–6998.

Yanofsky, M.F., Ma, H., Bowman, J.L., Drews, G.N., Feldmann, K.A. and Meyerowitz, E.M. (1990) The protein encoded by the *Arabidopsis* homeotic gene agamous resembles transcription factors. *Nature* 346, 35–39.

Biotechnological Approaches to Modifying Plant Form

12

Brent H. McCown
Department of Horticulture, University of Wisconsin-Madison, Madison, Winsconsin 53706, USA

Ornamental horticulturists must be holistic in perspective. In evaluating plants, all aspects of plant growth and development are of concern for it is the performance of the total plant over its lifetime that is critical. The characteristic of plant form exemplifies this perspective. The ultimate form of an ornamental plant is of major importance and is necessarily the final integration of many genetic, physiological and cultural influences. As such, then, the development of the form of a plant is a very complex process. The goal of being able, through genetics, to more predictably predetermine the form of a plant to suit special or general needs is an emerging goal of plant biotechnology.

This chapter will discuss the use of various biotechnological tools that might enable the ornamental plant breeder to have better control over the ultimate plant form. The scope of the discussion is limited by several constraints. First, very little work of this type has been conducted using ornamental plants. Thus much of the discussion will centre on the knowledge base and examples drawn from work on model plant species. Second, by choice, the emphasis (but not exclusively) will be on woody perennial ornamentals. By nature of their size and long life, these plants dominate the ornamental setting but have only brief flower displays; thus the form of the plant becomes a dominant and critically important trait. Also significant, plant breeders have had much greater success controlling plant form of herbaceous plants because of the considerably greater ease with which herbaceous species can be bred. Biotechnological approaches have immense potential for improving woody perennial crops where such tools may be the only feasible approach for predictably changing the genetics of plant form.

 Biotechnology of Ornamental Plants (eds R.L. Geneve, J.E. Preece and S.A. Merkle)

THE GENETIC BASIS OF PLANT FORM

That plant form has a specific and identifiable genetic basis is well established and is a common selection criterion in breeding programmes. However, for biotechnological purposes, the important question is whether plant form can be controlled, at least in part, by single genes, since, using present biotechnology, single gene characteristics are the most manageable. The answer to the question in many cases is 'yes'. For woody perennials, some of the most detailed information comes from the forestry and pomological sectors. For example, Timmis and Trotter (1989) discussed a number of examples where major changes in the architecture of tree crowns of forest trees (*Picea abies*, *Pinus sylvestris*, and *Pseudotsuga menzeisii*) are controlled by single genes.

Of particular note is the marked columnar habit conferred to MacIntosh apples by the *Co* gene. Micropropagation of such columnar cultivars showed that these plants had an unusually high tolerance of the levels of the cytokinin benzyladenine in the medium (Marin *et al*., 1993). Although circumstantial, this unusual response to a plant growth regulator may indicate that the columnar habit may be related to an alteration in the cytokinin dynamics of the plants.

Little is known about the precise mode of action of genes that markedly alter plant form. However, as noted in the columnar apple example, most hypotheses associate changes in plant form with alterations in the hormonal balance in the affected plants. Indeed, this theme of modifying plant form by altering hormonal balances is predominant in our current understanding of the determinants of plant form.

MICROPROPAGULE MANIPULATION AND PLANT FORM

The most advanced application of plant biotechnology is the micropropagation (true-to-type cloning) of elite germplasm. Although one use of micropropagation is the cloning of selections of unique form (such as the columnar apples), the question to be addressed here is 'Can micropropation be used to alter plant form?'

Considerable work detailing the form developed in the field by micropropagules has been done in forestry (e.g. *Pinus taeda*, Frampton, 1992; *Pseudotsuga menzeisii*, Ritchie and Long, 1986; *Betula*, McCown, 1989; Jokinen and Tormala, 1991) and pomology (particularly with *Malus*; Zimmerman and Miller, 1991). The general theme of most of the work is that micropropagules often exhibit growth forms similar to seedlings or intermediate between seedlings and cuttings/grafts of plants with more mature phase state. Thus, for example, loblolly pine micropropagules displayed a more mature growth form (heavier needles, fewer branches, fewer flushes per season) than seedlings (Anderson *et al*., 1992). This effect was determined to be mostly the result of the shoot system and not the root system of the micropropagule. Interestingly, the original source of the micropropagules (derived from axillary/fascicular buds or cotyledons) resulted in different initial plant form in the field. The differences between various types of propagules generally lessens and disappears after multiple growing seasons in the field. Thus any form differences of trees as a result of micropropagation *per se* appear to be only temporary and are probably mostly related to juvenility or other epigenetic differences.

With shrubs, however, early differences in plant form can ramify to marked differences in the mature plant. With blueberry (Grout *et al.*, 1986; Morrison and Smagula, 1986), micropropagules had significantly greater basal branching, lateral growth of shoots, and a more stoloniferous habit than conventional cuttings. Similar increased basal branching of micropropagules is a common characteristic of ornamental *Ericaceae* [*Kalmia, Rhododendron* (Preece, 1992)] and tropical foliage plants (*Dieffenbachia, Syngonium*). The early establishment of a well-branched habit leads to the formation of a more bushy and well-formed mature plant. The response is probably related in large part to a rejuvenation of the shrubs in culture (McCown, 1988) as a number of nurseries have noted that seedlings of closely related genotypes respond similarly to micropropagules (McCown, 1989).

How the micropropagules are handled can as strongly influence the ultimate plant form as the micropropagation process itself. For example, rooted birch micropropagules stored under refrigerated conditions for 3.5 months had more than sevenfold greater branching after the first season's growth as compared to those micropropagules planted directly to the field from the greenhouse (McCown, 1989). Such data indicate that at least some aspects of plant form can be somewhat 'predetermined' by planned (or unplanned) manipulations of the young micropropagules.

CHANGES IN FORM AS A RESULT OF SOMACLONAL VARIATION

Unlike the purely epigenetic differences noted above, the literature is resplendent with examples of 'mutant' changes in plant form derived from microculture of plants. The general topic of somaclonal variation is covered elsewhere in this book and thus I will only refer to it here as a method of modifying plant form.

For ornamentals, a predominant variant offtype observed in commercial production involves some type of dwarfing. Where this topic is researched, comparable results are obtained. For example, with begonia, 25% of regenerants obtained from callus culture were more bushy than the control and at least 30% were shorter (Jain, 1993). While taller offtypes were also observed with begonia, this is seldom the case with trees and shrubs where offtypes are generally of smaller stature than controls (Saieed *et al.*, 1994).

Why are dwarf offtypes so common in somaclonal variation of ornamental plants? One explanation is that there may be a selection pressure for such aberrant types in the microculture process. Lane *et al.* (1982) found some evidence that this is indeed the case by determining that among three strains of

MacIntosh apples grown as shoot cultures, the dwarf strain was the most tolerant of high levels of cytokinin in the culture medium. Thus by using high concentrations of cytokinin during micropropagation to stimulate rapid multiplication, one may simultaneously be preferentially selecting for any dwarf offtypes that may arise.

The experience concerning the stability of somaclonal variants in form is mixed. Nurseries report many of the variants grow out of the dwarfed habit while others see plants remaining dwarfed for many years. Some dwarfed rhododendrons have been introduced as new cultivars. Detailed observation of a large number of such aberrants from a wide range of species over a decade or more of seasons would be required to answer fully the stability question.

GENETIC ENGINEERING OF PLANT FORM

As was alluded to at the beginning of this discussion, plant form and hormone balance in the plant are usually interrelated. Thus it is no surprise that one of the most effective biotechnological approaches to changing plant form is by genetically modifying the hormonal dynamics of a plant through insertion of genes coding for proteins involved in hormonal regulation. There are four general approaches to endogenously altering hormone dynamics via genetics:

1. *Overproduction of a hormone.* This is usually accomplished by inserting genes that (a) complement a biosynthetic hormonal pathway common in all plants thus increasing synthesis, or (b) change an inactive form of hormone into an active state, thus effectively increasing the physiologically active levels of the hormone.
2. *Underproduction of a hormone.* Reduced levels of a hormone in a plant have been achieved through lowering the levels of a key enzyme or of a key intermediate in the biosynthetic pathway of a hormone.
3. *Inactivation of a hormone.* This can be accomplished by direct modification of a hormone itself or by producing an analogue that competes with a hormone at active sites.
4. *Change in the sensitivity of tissues to a hormone.* Changes in the capacity to inactivate a hormone or in the signal transduction pathway may be involved.

Over the last decade, biotechnology has contributed considerably to our understanding of hormone dynamics in plants and their influence on plant form and other characteristics (Hamill, 1993; Reid, 1993; Cline, 1994). Much of this work has been through two major experimental approaches: (i) detecting phenotypic mutants in form and determining the genetic basis of these changes and the subsequent altered hormonal physiology, and (ii) transferring genes known to code for plant hormonal components from pathogenic bacteria into plants and observing the subsequent altered plant phenotype. Although the study of plant mutants has identified hundreds of single genes specifically coding for steps in the biosynthetic pathway of many plant hormones (Reid, 1993), few of these genes have been cloned and transferred into test plants. Thus the usefulness of these genes in plant biotechnology to alter plant form remains unrealized (except as utilized in conventional breeding within that species). In the near future, plant-derived genes may be particularly important sources for genes involved in gibberellin (Bensen and Briggs, 1994) and abscisic acid (Giraudat *et al.*, 1992) metabolism, since genes involving these hormones have not been as readily available from microbial sources as have genes altering auxin, cytokinin or ethylene metabolism.

One area where plants have been important as a direct source of genes influencing plant form is phytochrome. Molecular studies using isolated and cloned phytochrome genes have advanced phytochrome research immensely. Some of this work has shown profound influences on plant form; this work will be discussed more thoroughly later. Since the changes in form induced by transformation with phytochrome genes may involve concomitant changes in hormone dynamics and tissue sensitivity, they have often been grouped under alteration in hormonal regulation. However, the work on this aspect is far from conclusive and thus I will discuss the phytochrome work separately.

ALTERING PLANT FORM THROUGH THE USE OF HORMONE GENES FROM MICROBIAL SOURCES

iaa*M (tms1) and* iaa*H (tms2) genes – overproduction of IAA*

The *iaa* genes isolated from either *Agrobacterium tumefaciens* or *A. rhizogenes* have been among the most utilized plant-hormone-altering genes in plant biotechnology. *iaa*M and *iaa*H can act in concert to overproduce IAA in transformed plants by coding for tryptophan monooxygenase and indoleacetamide hydrolase respectively:

$$\text{Tryptophan} \xrightarrow[\text{monooxygenase}]{(iaa\text{H})} \text{indole-3-acetamide} \xrightarrow[\text{hydrolase}]{(iaa\text{M})} \text{indole-3-acetate}$$

However, transformation with the *iaa*H gene alone often shows strong effects on IAA dynamics. Analyses of tissues from plants transformed with such genes under the control of CaMV35S or heat shock promoters can show very elevated levels of endogenous IAA, often approaching ten times normal levels.

The effects of overproducing IAA are remarkably similar among the diversity of plants that have been transformed with these genes. In addition to the model species tobacco (Sitbon *et al.*, 1992; Romano *et al.*, 1993), herbaceous genera including *Petunia* (Klee *et al.*, 1987) and *Solanum* (McInnes *et al.*, 1991) have been transformed. Commonly observed changes in phenotype include:

- Increased apical dominance, resulting in less branching.
- Reduced stature (shorter internodes).
- Increased secondary xylem and lignin.

- Smaller and narrower leaves, often accompanied by epinasty.
- Increase in adventitious rooting.

Changes in auxin dynamics usually have marked effects on the dynamics of ethylene; increased auxin levels can be expected to lead to the stimulation of ethylene production (McKeon and Yang, 1987). The effect is direct as the enzyme 1-aminocyclopropane-1-carboxylate (ACC) synthetase, an essential enzyme in ethylene biosynthesis, can be induced by auxin (Yang and Hoffman, 1984). Romano *et al.* (1993) elegantly separated the auxin and ethylene overproduction effects by overproducing auxin via the *iaa*M gene in plants that had reduced capacity to synthesize ethylene (tobacco) or were ethylene insensitive mutants (*Arabidopsis*, see Bleeker, 1991). A lowered capacity to synthesize ethylene was conferred by engineering the tobacco plant with the bacterial ACC-deaminase gene, thus limiting the key intermediate ACC, which is necessary for ethylene biosynthesis. These plants showed all the classical changes in form noted above for plants overproducing auxin except the reduction in internode length leading to reduced stature. Thus combining auxin overproduction and the inhibition of ethylene synthesis/action may yield a columnar plant that is taller than if only auxin overproduction is utilized.

Thus incorporating the *iaa*H gene alone, or *iaa*H and *iaa*M genes combined, possibly coupled with the ACC-deaminase gene, may be useful in providing more of a columnar habit to ornamentals. How effective these genes will be for modifying woody perennial ornamentals is an intriguing question awaiting future study.

ipt *gene (tmr; gene 4) – overproduction of cytokinins*

A number of studies have used the *ipt* gene isolated from *Agrobacterium tumefaciens* to determine how overproduction of endogenous cytokinins will affect plant development, including plant form. The *ipt* gene encodes for isopentenyl transferase, which catalyses a condensation reaction:

isopentenyl pyrophosphate + adenosine monophosphate ——> isopentenyl AMP

Isopentenyl AMP is a precursor for synthesis of cytokinins (Letham and Palni, 1983; Akiyoshi *et al.*, 1984) and increasing the amounts of AMP in plants often leads to increased levels (to 25 times normal concentrations) of naturally synthesized cytokinins such as zeatin riboside (Letham and Palni, 1983). The phenotypic effects have been well studied in tobacco using either the native, heat shock (hsp 70), CaMV, nopaline synthase, or SAUR promoters (Schmülling *et al.*, 1988, Smigocki and Owens, 1988; Medford *et al.*, 1989; Klee and Estelle, 1991; Smart *et al.*, 1991; Li *et al.*, 1992). Again, there are some general changes in phenotypic charactersicis to be expected when overexpressing this gene in transformed plants:

- Reduced plant height accompanied by a reduced apical dominance and increased lateral branching.
- Smaller stems often with reduced xylem development.

- Decrease in root system development and possibly a marked loss of ability to form adventitious roots.
- Increased chlorophyll content and delayed leaf senescence which may give the appearance of darker leaves.

Interestingly, many of these traits are opposite to those observed in plants where auxins were overproduced. Thus one can readily surmise the importance of the balance of endogenous hormone dynamics in the plant in the phenotypic expression of plant form. Likewise, altering this balance by genetic engineering can be used to alter simultaneously many aspects of plant form.

In contrast to the *iaa*H and *iaa*M genes, the *ipt* gene may be useful to produce a more compact plant form in ornamentals. The gene has been used in the transformation of the woody genera, *Populus* (Schwartzenberg *et al.*, 1994) and *Prunus* (Smigocki and Hammerschlag, 1991). In *Populus*, the gene was driven by a strong promoter and the recovered plants were highly distorted and unable to form roots; however, many of the same phenotypic traits as described above for herbaceous plants were observed. In peach, plants rooted with difficulty and showed branching in early growth in the greenhouse but were probably not stably transformed.

iaa*L gene (gene 5) – reduced levels of endogenous free IAA*

Unlike the above approaches where hormones are increased in transformed plants, use of the *iaa*L gene leads to reduced hormone levels. This gene was isolated from *Pseudomonas syringae* subsp. *savastonoi* and encodes for IAA–lysine synthetase, which catalyses the condensation of lysine and IAA. The resultant product, indole-3-lactate, has little auxin activity and may compete with IAA for binding sites (Körber *et al.*, 1991). Thus use of this gene in transformed plants may reduce IAA levels by both reducing free active IAA up to 20 fold (Romano *et al.*, 1991) and competing with the remaining free IAA.

Romano *et al.* (1991) transformed tobacco with the *iaa*L gene and found a marked reduction in apical dominance and an increase in lateral branch outgrowth. Alterations in development of tobacco and potato engineered with *iaa*L were also noted by Spena *et al.* (1991). Thus this approach may be another avenue in addition to the *ipt* gene to acquire plants that are dwarfed and have improved bushiness.

rolA, rolB, *and* rolC *genes – increase in sensitivity and altered hormone levels*

The *rol* ('root locus') genes were isolated from *Agrobacterium rhizogenes* and have been widely used in plant biotechnology. Together they appear to act synergistically in modifying hormone dynamics in plants and may all increase the sensitivity of tissues to hormones (Shen *et al.*, 1990; Maurel *et al.*, 1991). The biochemical basis of the action induced by *rolA* is not clear but appears to involve both alteration in auxin and cytokinin dynamics. Some have hypothesized that *rolA* may be involved in the signal transduction pathway for auxin in the plasma membrane (Julliard *et al.*, 1992; Vansuyt *et al.*, 1992). *RolB* may code for an indole glucosidase (Estruch *et al.*, 1991b) that may increase free auxin by hydrolyzing indole-3-acetyl-β-glucosides. The product of the *rolC* gene may catalyse the release of cytokinins from glucoside conjugates (Estruch *et al.*, 1991a, b). Thus these genes, if used separately or together, can have profound affects on the auxin/cytokinin balance in transformed plants by increasing the effectiveness of a given level of hormone. This differs markedly from the mode of action of the previously discussed *iaa* and *ipt* genes.

As would be expected from their proposed modes of action, transforming plants with the *rol* genes generates phenotypes that may resemble, although not exactly (Nilsson *et al.*, 1993), plants produced with both auxin and cytokinin overproduction:

- Decreased apical dominance with increased branching.
- Shorter internodes.
- Smaller leaves that are often wrinkled.
- Smaller flowers with reduced pollen/seed production.
- Modified root system, often more abundant.

Scorza *et al.* (1994) engineered tobacco with the *rolC* gene to determine the horticultural value of such an approach. The transgenic clones and their first generation sexual offspring varied in the degree of expression of the growth alterations, offering the opportunity to select plants with the most desirable phenotype. Of the four transgenic clones detailed, height at flowering was 53 to 64% of the control. The number of leaves per plant and photosynthesis rates were not altered; however, the leaf area of all the transgenics was reduced by about 35%. Total above-ground dry weight of the transgenics was less than half that of the controls while specific root lengths did not vary from the controls. Such data indicate that the total productivity of the transgenics was much reduced.

Handa (1992) transformed snapdragon (*Antirrhinum majus*) with a plasmid containing all of the *rol* genes. Dwarfed phenotypes were obtained that had heavily increased branching, shorter internodes and smaller leaves and flowers. However, because of the increased number of branches, the number of total flowers per plant was dramatically increased by threefold.

The *rol* genes thus present yet another approach to modifying plant form. Although more complex in their physiology, the *rol* genes used in combination offer the opportunity to modify plant form through changes in both auxin and cytokinin dynamics. This may possibly overcome some of the limitations (such as difficulty in rooting) of using hormonal overproduction strategies.

ALTERING PLANT FORM THROUGH THE USE OF PHYTOCHROME GENES FROM PLANT SOURCES

Phytochrome through the signalling between environment and plant development can be a major determinant of plant form (photomorphogenesis). The most obvious example is the marked difference in phenotype of a plant grown in shade as compared to full sun. Thus a modification in the phytochrome dynamics of a plant would be expected to also affect the ultimate plant form.

Recent evidence indicates that a multitude of phytochromes appear to exist in plants (Vierstra, 1993). Phytochrome is synthesized as the P_r form that is biologically relatively inactive. After photoconversion to P_{fr}, a vast array of responses can be triggered.

Phytochrome genes (*phy*) have been isolated from a diversity of phylogenetic divisions (Quail, 1991; Thümmler *et al.*, 1992). Work with *Arabidopsis thaliana* has been especially thorough and five *phy* genes have been identified. *phyA* encodes the most abundant phytochrome (phytochrome A) in dark-grown plants; in light, unlike the other phytochromes, phytochrome A transcription is repressed; since the P_{fr} form of phytochrome A is also metabolically unstable, levels of phytochrome A are low in light.

Phytochrome dynamics can be modified by engineering plants with *phy* genes so that *phy* is overexpressed (Kay *et al.*, 1989; Keller *et al.*, 1989; Boylan and Quail, 1991; Cherry *et al.*, 1991; Wagner *et al.*, 1991). Interestingly, there is broad functionality between phylogenetic divisions as, for example, *phyA* from monocots will express and produce biologically active chromophores in dicots. When the phytochromes accumulate from such overexpression, a phenotype results that has a strongly modified plant form:

- Dwarfing through reduction of internode length.
- Increased leaf chlorophyll, giving a darker leaf coloration.
- Decreased apical dominance, resulting in increased branching.
- Delayed foliar senescence.

Such a phenotype has been interpreted as an exaggeration of the normal light-grown phenotype; that is, overexpression of *phy*A results in an exaggeration of the differences normally observed between a shade-grown plant and a full sun-grown plant.

Cherry *et al.* (1992, 1993) showed that there was a very strong non-linear and negative relationship between plant height and the amount of full length phytochrome A in a transgenic plant. Transgenic tobacco that had only a two fold increase in phytochrome A over the wild type was only one quarter as tall; further increases in phytochrome A had no additional effect. Although there appears to be a narrow range between no and full response, this does suggest that one may be able to obtain a range of plant dwarfing depending on the degree of expression of *phyA* in a particular transformant.

The approach to modifying plant form by engineering with the *phy* genes has been utilized only in herbaceous test species, particularly tobacco and arabidopsis. However, since phytochrome is such a universal photoreceptor, similar responses to overexpression of *phyA* would be expected in all plants, including ornamentals. Interestingly, after observing a strong *phyA* overexpressing plant, one is struck by the similarity to a plant overtreated with growth retardants. That *phyA* overexpressing plants so closely resemble growth-retardant treated plants conjures up a number of considerations:

1. Does *phyA* overexpression involve suppression of gibberellin synthesis, activity, or tissue sensitivity? Certainly some plant phytochrome mutants appear to act through alteration of gibberellin dynamics (Reid, 1993). Whether alteration in gibberellin status is directly involved in *phyA* overexpression is unclear.
2. The predicted phenotype for a species being considered for genetic engineering with *phyA*, for all practical purposes, would be expected to resemble a plant treated with growth retardants. Thus one should be able to experimentally predict with some degree of certainty the value of this approach for any particular ornamental.
3. The 'growth retardant' phenotype produced by *phyA* overexpression also suggests a major application for such an approach: replacing the use of growth retardants in production of ornamental plants.

CONCLUSIONS

We now have enough examples of the application of plant biotechnology to modify those characteristics of a plant that contribute to its form to conclude that such goals are now reasonable and approachable for ornamentals. Single-gene-driven determinism of many aspects of plant form have been identified and, in some instances, suitable genes have been isolated and cloned. However, there are some major precautionary considerations:

1. Pleiotropic effects are common. The principal approach that is currently available for modifying plant form, namely altering hormonal balances, is non-specific in the traits it affects. Not all the changes may be beneficial, for example poor root system development with cytokinin overexpression, smaller flowers with the use of the *rol* genes, distorted leaves with auxin overproduction, or reduced plant productivity with any of the approaches. Similar criticisms hold for the somaclonal approach where multiple changes (Ostry *et al.*, 1994) are common and for phytochrome where the effect of changes in phytochrome A content on plant hardiness, dormancy and flowering are of concern.
2. Dwarfed plant form is the principal phenotype to be expected through the use of the tools of biotechnology. Although this plant form is certainly of major importance in horticulture and ornamentals, a vast array of other plant form modifications (such as canopy architecture) await more refined and varied approaches.

One intriguing approach that has not been well developed is genetically modifying rootstocks so that the form of grafted plants can be controlled. Modifying rootstocks would lessen one of the severe limitations in working with ornamentals – the great number of individual cultivars of each species that are economically important. If graft compatibility among a large number of genotypes could be achieved (as with apple rootstocks), only a number of rootstocks need be engineered and not every scion cultivar. Some characteristics may be transmissible through a graft union (Murfet and Reid, 1993), while others such as the *rol* gene effects are not (Spena *et al.*, 1989).

Of the approaches cited in this chapter, genetic engineering for altered plant form holds the most promise. Gene transfer for ornamentals is rapidly becoming less of an obstacle, particularly with the success of the such broadly-applicable technologies as particle bombardment (McCown *et al.*, 1991; Serres *et al.*, 1992; Ellis *et al.*, 1993). Most encouraging, but with one precaution, is the likely prospect that what is learned with the herbaceous models and heavily researched agronomic and horticultural food and fibre crops can be transferred to ornamentals. The precaution is that we still do not have enough experience with perennials, especially woody perennials, to be confident that major complications might not arise in the more complex life cycle of perennials. Such complications, if they occur, may become a limitation to this transferability between plants. For genetic engineering approaches, the truly exciting period will come when promoter technology has advanced enough to be able to precisely and predictably alter plant traits on specific temporal and tissue agendas. That time is not far away!

REFERENCES

Akiyoshi, D., Klee, H., Amasino, R., Nester, E.W. and Gordon, M.P. (1984) T-DNA of *Agrobacterium tumefaciens* encodes an enzyme of cytokin biosynthesis. *Proceedings of the National Academy of Sciences, USA.* 81, 5994–5998.

Anderson, A.B., Frampton, L.J., McKeand, S.E. and Hodges, J.F. (1992) Tissue-culture shoot and root system effects on field performance of loblolly pine. *Canadian Journal of Forest Research* 22, 56–61.

Bensen, R.J. and Briggs, S.P. (1994) Cloning gibberellin biosynthetic genes: a review of recent progress. In: Fritz, C.D. (ed.), *Proceedings of the Plant Growth Regulator Society of America*, 21st Annual Meeting, Portland, Oregon. Plant Growth Regulator Society of America, Research Triangle Park, North Carolina, pp. 1–14.

Bleeker, A.B. (1991) Genetic analysis of ethylene reponse in *Arabidopsis thaliana*. In: Jenkins, G.I. and Schuch, W. (eds), *Molecular Biology of Plant Development*. The Company of Biologists Limited, Cambridge, pp. 149–158.

Boylan, M.T. and Quail, P.H. (1991) Phytochrome a overexpression inhibits hypocotyl elongation in transgenic *Arabidopsis. Proceedings of the National Academy of Sciences, USA* 88, 10806–10810.

Cherry, J.R., Hershey, H.P. and Vierstra, R.D. (1991) Characterization of tobacco expressing functional oat phytochrome. *Plant Physiology* 96, 775–785.

Cherry, J.R., Hondred, D., Walker, J.M. and Vierstra, R.D. (1992) Phytochrome requires the 6-kDa N-terminal domain for full biological activity. *Proceedings of the National Academy of Sciences, USA* 89, 5039–5043.

Cherry, J.R., Hondred, D., Walker, J.M., Keller, J.M., Hershey, H.P. and Vierstra, R.D. (1993) Carboxyl-terminal deletion analysis of oat phytochrome A reveals the presence of separate domains required for structure and biological activity. *Plant Cell* 5, 565–575.

Cline, M.G. (1994) The role of hormones in apical dominance. New approaches to an old problem in plant development. *Physiologia Plantarum* 90, 230–237.

Ellis, D.D., McCabe, D.E., McInnis, S., Ramachandran, R., Russell, D.R., Wallace, K.M., Martinell, B.J., Roberts, D.R. and McCown, B.H. (1993) Stable transformation of *Picea glauca* by particle acceleration-A model system for conifer transformation. *Bio/Technology* 11(1), 84–89.

Estruch, J.J., Chriqui, D., Grossmann, K., Schell, J. and Spena, A. (1991a) The plant oncogene *rolC* is responsible for the release of cytokinins from glucosidic conjugates. *EMBO Journal* 10, 2889–2895.

Estruch, J.J., Chriqui, D., Grossmann, K., Schell, J. and Spena, A. (1991b) The protein encoded by the *rolB* plant oncogene hydrolyses indole glucosides. *EMBO Journal* 10, 3125–3128.

Frampton, L.J. (1992) Field performance comparisons between vegetative propagules and seedlings of loblolly and slash pines. In: Foster, G.S. and Diner, A.M. (eds), *Applications of Vegetative Propagation in Forestry. Proceedings of the Southern Regional Information Exchange Group Biennial Symposium on Forest Genetics*. Southern Forest Experiment Station, New Orleans, pp. 115–121.

Giraudat, J., Hauge, B.M., Valon, C., Smalle, J., Parcy, F. and Goodman, H.M. (1992) Isolation of the *Arabidopsis AB13* gene by positional cloning. *Plant Cell* 4, 1251–1261.

Grout, J.M., Read, P.E. and Wildung, D.K. (1986) Influence of tissue culture and leaf bud propagation on the growth habit of 'Northblue' blueberry. *Journal of the American Society of Horticultural Science* 111, 372–375.

Hamill, J.D. (1993) Alterations in auxin and cytokinin metabolism of higher plants due to expression of specific genes from pathogenic bacteria: a review. *Australian Journal of Plant Physiology* 20, 405–423.

Handa, T. (1992) Genetic transformation of *Antirrhinum majus* L. and inheritance of altered phenotype induced by T-DNA. *Plant Science* 81, 199–206.

Jain, S.M. (1993) Studies on somaclonal variation in ornamental plants. In: Hammerschlag, F.A., Zimmeramn, R.H. and Owens, L.D. (eds), *Second International Symposium on In Vitro Culture and Horticultural Breeding. Acta Horticulturae* number 336, Wageningen, pp. 365–372.

Jokinen, K. and Tormala, T. (1991) Micropropagation of silver birch (*Betula pendula* Roth.) and clonal fidelity of mass propagated birch plants. In: Ahuja, M.R. (ed.), *Woody Plant Biotechnology*, Plenum, New York, pp. 31–36.

Julliard, J., Sotta, B., Pelletier, G. and Miginiac, E. (1992) Enhancement of naphthalene acid-induced rhizogenesis in T_i-DNA-trnsformed *Brassica napus* without significant modification of auxin levels and auxin sensitivity. *Plant Physiology* 100, 1277–1282.

Kay, S.A., Nagatani, A., Keith, B., Deak, M., Furuya, M. and Chua, N.-H. (1989) Rice phytochrome is biologically active in transgenic tobacco. *Plant Cell* 1, 775–782.

Keller, J.M., Shanklin, J., Vierstra, R.D. and Hershey, H.P. (1989) Expression of a functional monocotyledonous phytochrome in transgenic tobacco. *EMBO Journal* 8, 1005–1012.

Klee, H. and Estelle, M. (1991) Molecular genetic approaches to plant hormone biology. *Annual Review of Plant Physiology and Plant Molecular Biology* 42, 529–551.

Klee, H.J., Horsch, R.B., Hinchee, M.A., Hein, M.B. and Hoffmann, N.L. (1987) The effects of overproduction of two *Agrobacterium tumefaciens* T-DNA auxin biosynthetic gene products in transgenic petunia plants. *Genes and Development* 1, 86–96.

Körber, H., Strizhov, N., Straiger, D., Feldwisch, J., Olsson, O., Sandberg, G., Palme, K., Schell, J. and Koncz, C. (1991) T-DNA gene 5 of *Agrobacterium* modulates auxin response by autoregulated synthesis of a growth hormone antagonist in plants. *EMBO Journal* 10, 3983–3991.

Lane, W.D., Looney, N.E. and Mage, F. (1982) A selective tissue culture medium for growth of compact (dwarf) mutants of apple. *Theoretical and Applied Genetics* 61, 219–223.

Letham, D.S. and Palni, L.M.S. (1983) The biosynthesis and metabolism of cytokinins. *Annual Review of Plant Physiology* 24, 163–197.

Li, Y., Hagen, G. and Guilfoyle, T.J. (1992) Altered morphology in transgenic tobacco plants that overproduce cytokinins in specfic tissues and organs. *Developmental Biology* 153, 386–395.

Marin, J.A., Jones, O.P. and Hadlow, W.C.C. (1993) Micropropation of columnar apple trees. *Journal of Horticultural Science* 68, 289–297.

Maurel, C., Barbier-Brygoo, H., Spena, A., Tempe, J. and Guern, J. (1991) Single *rol* genes from *Agrobacterium rhizogenes* Ti-DNA alter some of the cellular responses to auxin in *Nicotiana tabacum*. *Plant Physiology* 97, 212–216.

McCown, B.H. (1988) Adventitious rooting of tissue-cultured plants. In: Davis, T.D. and Haissig, B. (eds), *Adventitious Root Formation on Cuttings*. Dioscorides Press, Portland, Oregon, pp. 289–302.

McCown, B.H. (1989) Birch (*Betula* spp.). In: Bajaj, Y.P.S. (ed.), *Biotechnology in Agricuture and Forestry. Vol. 5. Trees II*. Springer-Verlag, Berlin, pp. 324–341.

McCown, B.H., McCabe, D.E., Russell, D.R., Robison, D.J., Barton, K.A. and Raffa, K.R. (1991) Stable transformation of *Populus* and incorporation of pest resistance by electrical discharge particle acceleration. *Plant Cell Reports* 9, 590–594.

McInnes, E., Morgan, A.J., Muligan, B.J. and Davey, M.R. (1991) Phenotypic effects of isolated $pRiA_4$ TL-DNA *rol* genes an the presence of intact TR-DNA in transgenic plants of *Solanum dulcamara* L. *Journal of Experimental Botany* 42, 1279–1286.

McKeon, T. and Yang, S.F. (1987) Biosynthesis and metabolism of ethylene. In: Davies, P.J. (ed.), *Plant Hormones and Their Role in Plant Growth and Development*. Martinus Nijhoff, Boston, pp. 94–112.

Medford, J.I., Horgan, R., El-Sawi, Z. and Klee, H.J. (1989) Alterations of endogenous cytokinins in transgenic plants using a chimeric isopentenyl transferase gene. *Plant Cell* 1, 403–413.

Morrison, S.E. and Smagula, J.M. (1986) Morphology, growth and rhizome development of lowbush blueberry tissue culture plants, seedlings, and rooted softwood cuttings. *HortScience* 21, 734 Abstract.

Murfet, I.C. and Reid, J.B. (1993) Developmental mutants. In: Casey, R. and Davies, D.R. (eds), *Peas – Genetics, Molecular Biology and Biotechnology*. CAB International, Wallingford, UK, pp. 165–216.

Nilsson, O., Crozier, A., Schmülling, T., Sandberg, G. and Olsson, O. (1993) Indole-3-acetic acid homostasis in transgenic tobacco plants expressing the *Agrobacterium rhizogenes rolB* gene. *The Plant Journal* 3, 681–689.

Ostry, M., Hackett, W., Michler, C., Serres, R. and McCown, B. (1994) Influence of regeneration method and tissue source on the frequency of somatic variation in *Populus* to infection by *Septoria musiva*. *Plant Science* 97, 209–215.

Preece, J.E. (1992) Practical regulation of woody plant growth and development using biotechnology. In: Damiano, C. *et al.* (eds), *Acta Horticulturae 300. In Vitro Culture. Proceedings of the XXIIIrd International Horticultural Congress, International Society for Horticultural Science,* Florence, Italy, pp. 23–33.

Quail, P.H. (1991) Phytochrome: a light activated molecular switch that regulates plant gene expression. *Annual Review of Genetics* 25, 389–409.

Reid, J.B. (1993) Plant hormone mutants. *Journal Plant Growth Regulation* 12, 207–226.

Ritchie, G.A. and Long, A.J. (1986) Field performance of micropropagated Douglas fir. *New Zealand Journal of Forest Science* 16, 343–356.

Romano, C., Hein, M. and Kell, H. (1991) Inactivation of auxin in tobacco transformed with the indoleacetic acid–lysine synthetase gene of *Pseudomonas savastanoi*. *Genes and Development* 5, 438–446.

Romano, C.P., Cooper, M.L. and Klee, H.J. (1993) Uncoupling auxin and ethylene effects in transgenic tobacco and arabidopsis plants. *Plant Cell* 5, 181–189.

Saieed, N.Th., Douglas, G.C. and Fry, D.J. (1994) Somaclonal variation in growth, leaf phenotype and gas exchange characteristics of poplar: utilization of leaf morphotype analysis as a basis for selection. *Tree Physiology* 14, 17–26.

Schmülling, T., Schell, J. and Spena, A. (1988) Single genes from *Agrobacterium rhizogenes* influence plant development. *EMBO Journal* 7, 2621–2629.

Schwartzenberg, V.K., Doumas, P., Jouanin, L. and Pilate, G. (1994) Enhancement of the endogenous cytokinin concentration in poplar by transformation with *Agrobacterium* T-DNA gene *ipt*. *Tree Physiology* 14, 27–36.

Scorza, R., Zimmerman, T.W., Cordts, J.M., Footen, K.J. and Ravelonandro, M. (1994) Horticultural characteristics of transgenic tobacco expressing the *rolC* gene from *Agrobacterium rhizogenes*. *Journal of the American Society of Horticutural Science* 119, 1091–1098.

Serres, R., Stang, E., McCabe, D., Russell, D., Mahr, D. and McCown, B. (1992) Gene transfer using electric discharge particle bombardment and recovery of transformed cranberry plants. *Journal of the American Society of Horticultural Sci*ence 117, 174–180.

Shen, W.H., Davioud, E., David, C., Barbier-Brygoo, H., Tempe, J. and Guern, J. (1990) High sensitivity to auxin is a common feature of hairy root. *Plant Physiology* 94, 554–560.

Sitbon, F., Hennion, S., Sundberg, B., Little, C.H.A., Olsson, O. and Sandberg, G. (1992) Transgenic tobacco co-expressing the *A. tumefaciens iaaM* and *iaaH* genes display altered growth and IAA metabolism. *Plant Physiology* 99, 1062–1069.

Smart, C.M., Scofield, S.R., Bevan, M.W. and Dyer, T.A. (1991) Delayed leaf senescence in tobacco plants transformed with *tmr*, a gene for cytokinin production in *Agrobacterium*. *Plant Cell* 3, 647–665.

Smigocki, A.C. and Hammerschlag, F.A. (1991) Regeneration of plants from peach embryo cells infected with a shooty mutant strain of *Agrobacterium*. *Journal of the American Society of Horticultural Science* 116, 1092–1097.

Smigocki, A.C. and Owens, L.D. (1988) Cytokinin-to-auxin ratios and morphology of shoots and tissues transformed by a chimeric isopentenyl transferase gene. *Plant Physiology* 91, 808–811.

Spena, A., Aalen, R.B. and Schulze, S.C. (1989) Cell autonomous behaviour of the *rol C* gene of *Agrobacterium rhizogenes* during leaf development: a visual assay for transposon excision in transgenic plants. *Plant Cell* 1, 1157–1164.

Spena, A., Prinsen, E., Fladung, M., Schulze, S.C. and Van Onckelen, H. (1991) The indole acetic acid–lysine synthetase gene of *Pseudomonas syringae* subsp. *savastanoi* induces developmental alterations in transgenic tobacco and potato plants. *Molecular and General Genetics* 227, 205–212.

Thümmler, F., Dufner, M., Kreisl, P. and Dittrich, P. (1992) Molecular cloning of a novel phytochrome gene of the moss *Ceratodon purpureus* which encodes a putative light-regulated protein kinase. *Plant Molecular Biology* 20, 1003–1017.

Timmis, R. and Trotter, P.C. (1989) Biological and economic feasibility of genetically engineered trees for lignin properties and carbon allocation. In: Dhawan, V. (ed.), *Applications of Biotechnology in Forestry and Horticulture*. Plenum Press, New York, pp. 349–368.

Vansuyt, G., Vilaine, F., Tepfer, M. and Rossignol, M. (1992) *rolA* modulates the sensitivity to auxin of the proton translocation catalysed by the plasma membrane H^+-ATPase in transformed tobacco. *FEBS Letters* 298, 89–92.

Vierstra, R.D. (1993) Illuminating phytochrome functions. *Plant Physiology* 103, 679–684.

Wagner, D., Tepperman, J.M. and Quail, P.H. (1991) Overexpression of phytochrome B induces a short hypocotyl phenotype in transgenic *Arabidopsis*. *Plant Cell* 3, 1275–1288.

Yang, S.F. and Hoffman, N.E. (1984) Ethylene biosynthesis and its regulation in higher plants. *Annual Review of Plant Physiology* 35, 155–189.

Zimmerman, R.H. and Miller, S.S. (1991) Orchard growth and fruiting of micropropagated apple trees. *Journal of the American Society of Horticultural Sciences* 116, 780–785.

Molecular Events Associated with Floral Evocation

13

Brian R. Jordan[1] and Richard G. Anthony[2]
[1]New Zealand Institute for Crop & Food Research Limited, Levin Research Centre, Private Bag 4005, Levin, New Zealand; [2]Department of Biochemistry, Royal Holloway College, University of London, Egham, Surrey TW20 0EX, UK

INTRODUCTION

The transition from vegetative to reproductive meristem is a vital stage in plant development. It has been extensively studied and a wealth of descriptive information has been accumulated. The timing of flowering is controlled by an endogenous developmental regulation that determines the competence to flower. On top of this developmental programme there are environmental factors that can modify the flowering process (Thomas, 1993). These environmental cues can be very varied and include the most studied regulators of floral development, light and temperature. Unfortunately, although the factors involved in the physiological basis of this transition are relatively well characterized, the molecular mechanisms remain to be determined. Recently, through the use of molecular genetics, considerable progress has been made in our understanding of this process. To date, most of this research has been carried out on a limited number of species, notably *Arabidopsis* and *Antirrhinum* (Coen and Carpenter, 1993; Yanofsky 1995). However, the information obtained is likely to be widely relevant to a range of species, including ornamentals. This review will focus on the molecular mechanisms of one aspect of the transition to flowering; evocation. Evocation is the process that commits the apical meristem to floral morphology (Evans, 1969). This conversion is determined by a range of factors and involves a large number of cellular changes that lead to the gross alteration in morphology (Evans, 1971). These include changes at the cellular level such as an early reduction in the cell-cycle length, synchronization of cell division, increased synthesis of RNA and changes in growth regulator levels. In addition, a number of alterations in morphology take place; stem elongation and leaf production are arrested and the meristem becomes dome shaped. Two mechanisms that control evocation have been proposed (Bernier, 1988, and references therein). The

 Biotechnology of Ornamental Plants (eds R.L. Geneve, J.E. Preece and S.A. Merkle)

first mechanism is the clean-slate scenario in which evocation is a process that eliminates the pattern of vegetative morphogenesis. This process then allows the reproductive morphology to be established and involves changes in gene expression as a consequence. The alternative mechanism involves the primary action of changes in gene expression to establish the reproductive meristem. The subject of floral evocation has been discussed recently in a series of excellent reviews (Bernier, 1988; Bernier *et al.*, 1993; Kinet, 1993; O'Neill, 1993). The present review will focus primarily upon molecular aspects of this transition. Events before and after what can strictly be referred to as evocation will also be discussed as it is important to describe the continuum of the process that leads to flowering.

SIGNAL TRANSDUCTION AND EVOCATION

The environmental signal(s) to flower can be perceived in many different parts of the plant. For instance, photoperception is located predominantly in the leaves, water status is detected in the roots and low temperature (vernalization) in the apices themselves (Bernier *et al.*, 1993). It is therefore possible to distinguish between long-distance signals and those that function at the cellular level (Fig. 13.1). For example, phytochrome is involved in the perception of the photoperiodic signal in leaves (Vince-Prue, 1975; Lumsden, 1993). This photoreceptor may act by modifying gene expression through a signal transduction pathway(s) involving Ca^{2+}/calmodulin, G-proteins and cGTP (Bowler *et al.*, 1994). It is then necessary for a long-distance signal to transmit these changes in the leaf to the apical meristem. This signal must in turn stimulate the changes associated with evocation and is likely to involve another set of secondary messengers that establish the response. As many of the changes associated with evocation involve alteration in gene expression, the signals are likely to interact in some way with the regulation of transcription. This may involve changes in DNA such as alterations in the level of methylation. The role of DNA methylation and vernalization in the initiation of flowering has been studied in *Arabidopsis* (Burn *et al.*, 1993) by using the demethylating agent 5-azacytidine. These experiments suggest that methylation of a gene or genes involved in flowering prevents floral initiation. Vernalization, by its general demethylating effect, can release the block to flowering. Thus demethylation of a gene critical to flowering permits its transcription. Because vernalization is associated with the apex, the methylation regulation may be apex specific. These results also suggested that the transcription of a key enzyme involved in GA biosynthesis (kaurenoic acid hydroxylase) was regulated in this manner. Temperature has also been shown to have a major influence on floral development and meristem identity genes in *Brassica* species and *Arabidopsis* (Jordan *et al.*, 1994; Anthony *et al.*, 1996).

It is also apparent that each of these stages in signal transduction must be dependent on the ability of the tissue to respond. The putative role of different signal mechanisms in the transition to flowering has been well docu-

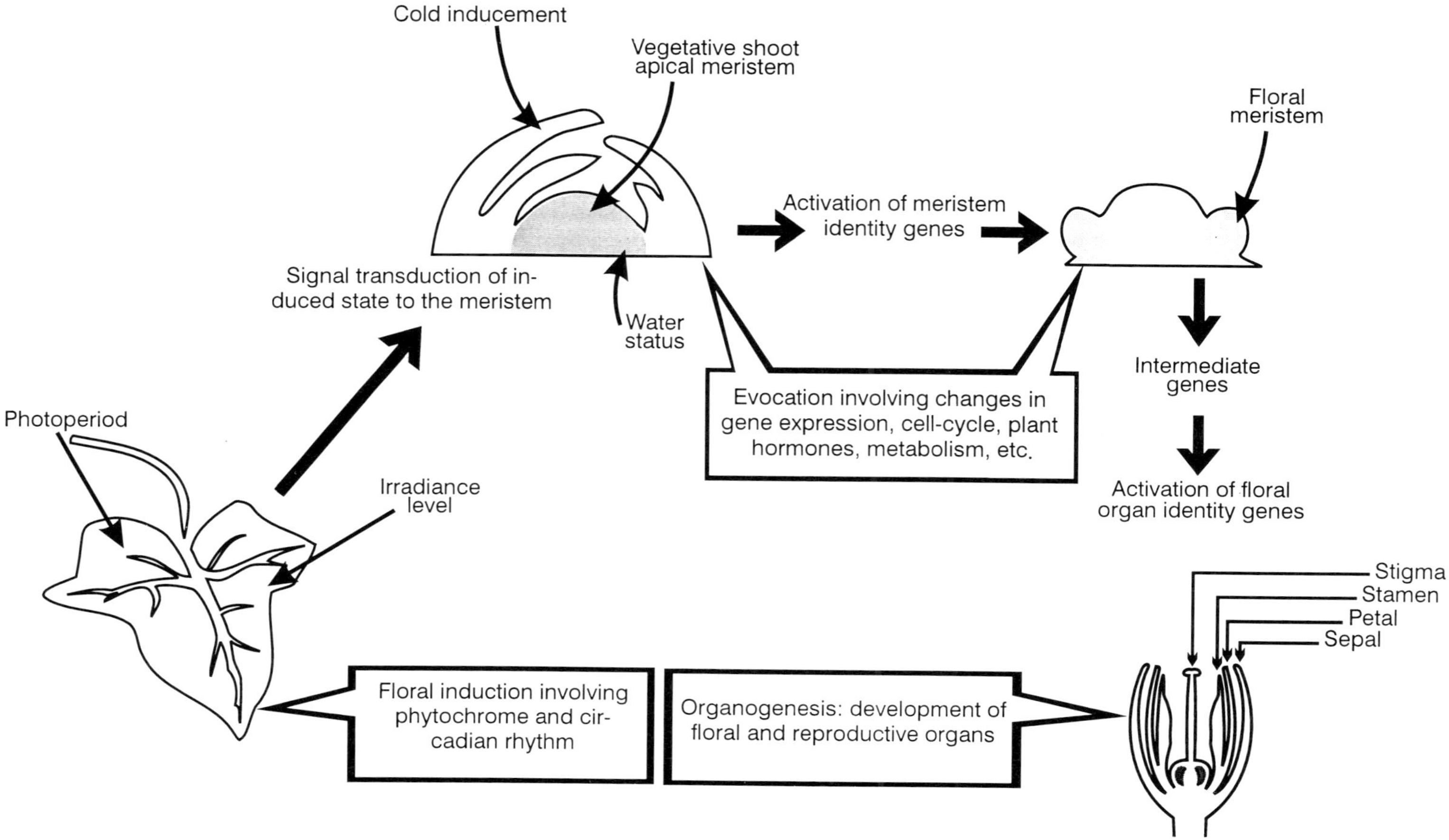

Fig. 13.1. Schematic representation of internal and external factors that influence floral development.

mented recently (see Bernier *et al.*, 1993 and Lumsden, 1993, for a discussion of signal transduction and flowering). These reviews describe a wide range of compounds that can play a role in this conversion, including carbohydrates, cytokinins, auxins, gibberellins and a range of secondary messengers. Changes in the levels of these compounds take place soon after the inductive stimulus and before cellular changes associated with evocation take place. For example, sucrose accumulates transiently in the apices of *Sinapis alba* at a very early stage after induction (Bodson and Outlaw, 1985). The role of these many and diverse compounds is, however, frequently inconsistent between species and may not replace the environmental stimulus precisely. Using transgenic *Arabidopsis* plants that overexpress cytokinin it was found that the time of flowering was not modified (Medford *et al.*, 1989). However, the levels of cytokinins were changed substantially throughout the tissue and at all stages of development. This contrasts with the increase in cytokinin levels associated with flowering (Bernier *et al.*, 1993) when there is both spatial and temporal regulation. The chemical structures of the molecules may also be important. For instance, hydroxylation of C12, C13 and C15 of GA will enhance flowering relative to stem growth, whereas C3 hydroxylation of GA will give the reverse response (Evans *et al.*, 1990, 1994). Furthermore, there may also be a prerequisite that a number of compounds change before evocation is initiated, such as GA, sucrose and cytokinins. This multifactorial system of control has been proposed by Bernier (1988) and explains how there can be a complex interaction of factors that can initiate evocation. In addition, the meristem may be required to be in a specific condition before these compounds can have activity. For instance, Besnard-Wibaut (1981) reported that GA was only effective on meristems that had undergone some cytological changes that could be considered as an intermediate phase in the transition to flowering.

Although the signal to flower usually results in floral development, there are examples of reversion, suggesting that once the signal is initiated the molecular processes controlling flowering are not necessarily fixed (Thomas, 1993, and references therein). For instance, in the SD plant *Impatiens balsamina*, floral reversion can take place at any stage of development and will give a range of reversion morphology. Plants that show reversion typically show no change in cellular organization between vegetative and floral apices. Thus it is possible that in these species there is a limitation to the cytological influences upon homeotic gene expression and/or hormone regulation as suggested by Besnard-Wibaut (1981). These studies also serve to emphasize the important role of physical changes at the apex during floral development, as described in the biophysical model of floral development proposed by Green (1988).

CELLULAR CHANGES DURING EVOCATION

Changes in cellular activity within different regions of the apex are essential for evocation (Bernier, 1988). One of the clearest indications of the transition to flowering is the change in the cell-cycle. This process is an orderly proliferation of cells which brings about the change in development. These changes have been extensively characterized in a number of species, notably in *Silene coeli-rosa* and *Sinapis alba* (Bernier, 1988; Francis, 1992), and involve shortening of the cell-cycle and its synchronization. For instance, in *Sinapis* the first change after an inductive LD is a shortening of the G2 phase of cycling cells and return to cycling of non-cycling cells (Gonthier *et al.*, 1987). This is followed by a shortening of G1 and S in cycling cells, leading to synchronization of the cell-cycle. The floral transition is the only developmental step in which synchrony is found in shoot meristems. In fact, in some species it is not possible to dissociate synchronization from flowering (e.g. *Sinapis*). However, this is not a consistent association and synchrony is likely to be a secondary effect stimulated by the floral promoter. It is, however, clear that changes in the cell-cycle are closely associated with the process of flowering. These changes are a prerequisite so that a critical proportion of cells are able to respond properly to floral induction and establish the process of evocation (see below for changes in gene expression for proteins involved in the cell-cycle). Moreover, changes in the cell-cycle are necessary for the conversion from vegetative to floral morphology and therefore are an integral part of evocation.

A wide range of changes in cellular composition also take place during early stages of the floral transition (Evans, 1971; Bernier, 1988; Kinet, 1993). These include increases in sucrose, invertase, ATP and mitochondria. These changes could be associated with a general alteration in cellular metabolism, as a consequence of the transition, not as a cause of it. However, data with some SD plants even contradict this assumption. Thus in *Pharbitis nil* there is a fall in ATP, and in *Xanthium* there is no increase in soluble sugars and related enzyme activities (Bernier, 1988). Therefore, although many of these changes clearly take place at early stages in evocation, they are unlikely to be causal.

Many studies have also been carried out on changes in protein biosynthesis during floral evocation (reviewed in O'Neill, 1993). These studies have largely involved radiolabelled precursor feeding, inhibitor studies and electrophoretic protein profiles. These techniques have well-documented problems of interpretation and frequently provide ambiguous conclusions. For example, inhibitors can cause a wide range of consequences not necessarily related to the process under investigation. One- or two-dimensional electrophoretic separation of polypeptides are relatively insensitive methods and are difficult to reproduce consistently. These studies however, have indicated that changes in protein biosynthesis are associated with floral evocation. For instance, Bonner *et al.* (1991) incorporated ^{3}H-uridine into RNA and carried out histochemical analysis of *Pharbitis nil* apices during the transition to flowering. They concluded that RNA synthesis was one of the earliest changes in evocation and that the temporal expression was centred on the rib meristem of the apex. These changes were thought to be closely associated with the initiation of evocation. In this context, it is particularly interesting that the earliest expression of meristem identity genes is not in the rib meristem (see Jordan and

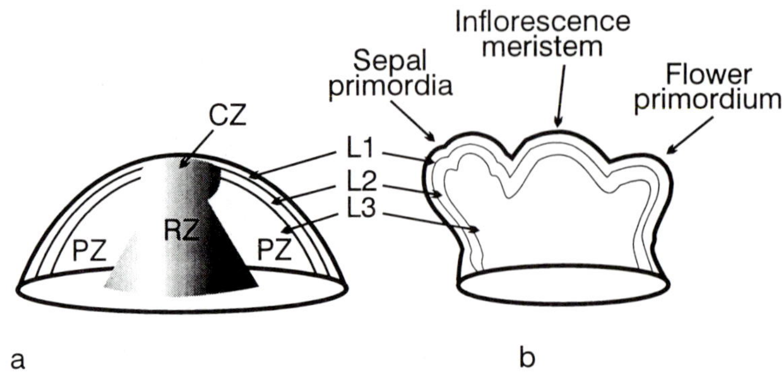

Fig. 13.2. Vegetative apical meristem (a) and inflorescence apical meristem (b) showing cytological zonation and tunica layers. CZ, central zone; PZ, peripheral zone; RZ, rib meristem; L, tunica layers.

Anthony, 1993 for expression of meristem identity genes). This suggests that other, as yet unidentified, upstream factors may be expressed in this region.

The development of more sophisticated techniques of molecular genetics have now confirmed that changes in gene expression are fundamental to the process of floral transition (see below) and have also started to unveil the true complexity of events in the apex as the meristem converts from vegetative to floral morphology.

THE ROLE OF CELL LAYERS IN FLORAL TRANSITION

There has been considerable research into the molecular basis of meristem development, in both vegetative and floral tissues. The shoot apical meristem is the most distal portion of the shoot apex. It typically comprises a group of 800 to 1200 cells and is approximately 100 μm in diameter. The size of the apical meristem varies, however, depending on both species and age (Medford *et al.*, 1991), with *Arabidopsis* having one of the smaller angiosperm meristems at 35 μm. One of the earliest changes in the meristem during floral transition is that it becomes dome shaped. This is partly caused by vacuolation and elongation of cells in the rib meristem (Nougarede, 1967). Changes in the shape and size are related to organogenesis and are usually not a determinant of flowering. However, in chrysanthemum, the receptacle must enlarge to a minimum diameter before it will flower (Horridge and Cockshull, 1979).

Cytological variation has led to the description of three regions in the apical meristem (Vaughan, 1952; Fig. 13.2.), a central zone (CZ), a peripheral zone (PZ) and the rib meristem (RZ). Organ primordium initiation occurs in the PZ, while the CZ consists of slowly dividing cells that replenish the PZ. Cells in the CZ are generally larger than in the PZ. The presence and extent of different cytological zones differs between species and the functional significance of zonation patterns is not known.

Changes in cellular activity during evocation are generally most apparent in the central zone and this typically leads to the loss of zonation during transition from the vegetative to the floral meristems (Fig. 13.2). The timing and extent of these changes within different meristematic zones also differs between species. For instance, the earliest activation is in the cells of the peripheral zone of *Pharbitis nil*, while for sunflower it is in the central zone (Bernier, 1988).

The apical meristem is further divided into three cell layers, called tunica layers, that in general remain clonally distinct throughout development (Miksche and Brown, 1965). The cells of L1 divide anticlinally throughout development and form the epidermis. The L2 cells divide predominantly anticlinally, although they will divide in any plane during differentiation. The L2 layer is also usually the source of the germ cells. The L3 layer divides in any plane and with L2 contributes to the body of the plant. The contribution of cells derived from the three layers to mature vegetative and floral organs is variable, and yet organs of regular morphology are produced (Stewart and Dermen, 1975). These cell layers persist between vegetative and reproductive growth, although the cytological zonation present in the vegetative tissue is lost in the inflorescence meristem (Huala and Sussex, 1993). The persistence of these layers and the ability of ectopic cells to differentiate in accordance with their new position suggests that the developmental fate of these cells is determined by their position rather than by cell lineage. However, substantial regulatory signalling between cell layers must also be involved (Huala and Sussex, 1993). This is well illustrated by grafting experiments on a *Camellia* chimera, generated from a *C. japonica* scion grafted onto a *C. sasangua* stock (Stewart *et al.*, 1972). The *C. sasangua* has a single whorl of petals and normal stamens and carpels. The *C. japonica* in contrast, has many whorls of petals, but no stamens and carpels. The chimera has cell layer 1 derived from *C. sasangua* and cell layers 2 and 3 from *C. japonica*. The phenotype of the flowers resemble the *C. sasangua* line in epidermal characteristics, such as petal colour. However, the chimera forms stamens and carpels despite the fact that its L2 and L3 are derived from the double-flowered *C. japonica*. Thus this study suggests that there is a signal transmitted from the L1 layer that can alter the differentiation normally associated with the *C. japonica* cell layers.

THE GENETIC REGULATION OF FLOWER DEVELOPMENT

Although it is clear that the environment plays a significant role in the determination of flowering, it is equally clear that the genetic control is also important (Law and Scarth, 1984; Law *et al.*, 1993). Although the genetics controlling flowering have not been comprehensively described, a number of crops, such as *Arabidopsis*, pea and wheat, have been analysed in some detail. Recently

Table 13.1. Genes affecting floral development in *Arabidopsis*.

Floral transition	
FCA, FD, FE, FHA, FPA, FRI, FT, FVE, FWA, FY, FG(CO), GI, KRY, TFL LD	
Meristem identity	
LFY(FLO*), AP1(SQUA*), CAL, AP2, FIM, TFL	
Cadastral genes	
SUP, LUG, AP2, AG(PLE*, FBP6**)	
Organ identity genes	
A function:	AP1(SQUA*), AP2
B function:	AP3(DEF*, GP**), PI(GLO*, FBP1**)
C function:	AG(PLE*, FBP6**)
Floral morphology	
CY*	
Other genes associated with the floral transition	
CLV1, EMF	

*Genes from *Antirrhinum*; **genes from *Petunia*.

there has also been substantial progress in identifying mutants and isolating genes that determine flowering in *Arabidopsis* (Table 13.1). Sung *et al.* (1992) have isolated a mutant, embryonic flower (*emf*), that will flower immediately upon seed germination without vegetative growth, suggesting that the EMF gene product is required for the inhibition of reproductive growth. These results suggest that reproductive development is a default pathway and that normal vegetative growth occurs when the reproductive programme is delayed as a result of inhibition. A number of other genes appear to regulate the photoperiodic or vernalization timing of floral development. Koornneef *et al.*, (1991) isolated 42 independently induced mutants which represented mutations at 11 different loci. Other late-flowering loci have also been mapped (Clarke and Dean, 1994) that are not allelic to those described by Koornneef *et al.* (1991). In wild-type *Arabidopsis*, vernalization has only a limited affect in promoting flowering. However, in late-flowering *fca, fve, fy* and *fpa* mutants, vernalization has large effects on flowering time in both SD and LD, whereas *co, gi, fwa* and *fd* are almost insensitive to the cold treatment. SD conditions extend the late flowering in most of the mutants, except *co* and *gi* which are insensitive to daylength. Thus, *co* and *gi* can be considered almost day neutral, flowering later than the wild type under LD and at the same time as the wild type under SD. The *co* gene has now been isolated (Putterill *et al.*, 1993, 1995) and has a putative zinc finger region, indicating that it may act as a transcription factor. Using RT-PCR, *co* RNA was detected in both leaf and shoot tissue. These results suggest a role in the early events of floral induction rather than acting specifically during evocation. In addition, RNA was found to be more abundant in tissue derived from LD than from SD seedlings. Thus the apparent lack of *co* function in SD could be related to the abundance of its RNA. Double mutants have been constructed (Coupland *et al.*, 1993; Putterill *et al.*,

1995) between *co* and mutations affecting gibberellin responsivity (*ga1-3* and *gai*), meristem identity (*lfy*) and phytochrome (*phyB* and *hy1*). Double mutants between *co* and mutants involved in gibberellin biosynthesis (*ga1-3*) or sensitivity (*gai*) generally enhanced the phenotype of either parental line. Supplying exogenous gibberellins, however, did not influence the phenotype of *co* mutants. These experiments indicated that *co* may mediate its action through an interaction with GA signal transduction. Phytochrome mutants cause early flowering, suggesting that phytochrome inhibits flowering. The double mutants flowered earlier than wild type and *co* mutants under SD, but considerably later than the phytochrome mutants. This suggests that phytochrome may be involved in reducing *co* activity under SD. Results in LD also indicate that *co* activity is required for early flowering of the phytochrome mutants. Double mutants of *co* and meristem identity genes show a greatly enhanced phenotype (increased inflorescence meristems) compared to *lfy* mutants, suggesting a strong influence of the *co* product on the meristem identity genes.

A similar gene, *luminidependens* (*ld*) has also been isolated from *Arabidopsis* (Lee *et al*., 1994). This gene is probably related to the mutants *fca*, *fve*, *fy* and *fpa* and acts through the pathway(s) proposed for these genes. The sequence of *ld* suggests that it is localized in the nucleus and has a glutamine-rich domain characteristic of some transcription factors. Recently, the factors influencing vernalization and flowering time in *Arabidopsis* mutants have been investigated (Chandler and Dean, 1994). The response to vernalization at different stages of development were qualitatively the same for wild-type, *fca* and *co*, although they differed in extent and timing. Vernalization was also more effective in darkness rather than in SD or continuous light. In addition, GA_3 and 5-bromo-deoxyuridine accelerated flowering for all mutants tested and were additive to a vernalization treatment for *fca*.

CHANGES IN GENE EXPRESSION DURING FLORAL TRANSITION

A number of genes involved in floral development have now been isolated and characterized to some extent (Table 13.1). These genes can, in general, be placed into two distinct classes. First there are 'early acting genes'; these can effect the transition of inflorescence meristems into floral meristems. Second, there are 'late acting genes'; these act alone or in combination to specify organ identity in a position-dependent fashion. There is considerable overlap between these classes. However, as evocation primarily involves early changes in gene expression, early acting genes will be discussed in most detail. It is possible to represent gene activity within shoot meristems in angiosperms as a progression through overlapping developmental stages: vegetative→inflorescence→floral (V→I→F). There is now evidence that there are also three interacting programmes influencing development (Meeks-Wagner, 1993). Some basic developmental programmes may remain fixed throughout the vegetative, inflorescence and floral stage, while other basic programmes may be modified to facilitate changes in meristem function during the V→I→F progression. Finally, some programmes are likely to be stage specific, and thus unique to a particular developmental stage.

FIXED PATTERNS OF MERISTEM GENE EXPRESSION

Several genes have been cloned that display fixed patterns of mRNA or protein expression in vegetative, inflorescence and floral meristems. The tobacco A3 clone (Kelly *et al.*, 1990), whose function is unknown, is specific to the shoot apex and remains at a nearly constant level throughout development. Studies suggest that although the expression pattern is fixed in the meristem dome throughout plant growth, the initiation of floral development permits A3 expression in positions outside the dome (Meeks-Wagner, 1993). Melzer *et al.* (1990) have isolated similar genes from *Sinapis alba. In situ* hybridization analysis with clone pSFD5.04, revealed hybridization to the peripheral zone of both inflorescence and early floral meristems. The gene was also located in vegetative shoots. Melzer *et al.* (1990) suggested that pSFD5.04 represents a gene with PZ meristem activity and that although involved in early floral meristem function, it is not specific to floral organogenesis.

Medford *et al.* (1991) have isolated a number of clones preferentially expressed in shoot apical meristems of cauliflower by differential screening. Transcriptional promoters from *Arabidopsis thaliana* were fused to the β-glucuronidase (GUS) reporter gene and introduced into plants. The *meri-5* promoter directed GUS expression in the meristematic dome but not in the surrounding leaf primordia. GUS expression was also directed at branching points in the shoot and root. Expression of *meri-5* was found to be continuous throughout the conversion to the floral state.

Another example of a gene exhibiting a fixed pattern of expression throughout the V→I→F is provided by Knotted-1 (KN1) isolated from the monocotyledonous maize (Smith *et al.*, 1992). KN1 protein has a similar pattern of expression in vegetative, inflorescence and floral meristems. The protein is found in all of the cell layers of the meristem, but is absent from the PZ cells that are specified to initiate organ primordia. KN1 appears to regulate development by specifying the meristematic state of cells. The KN1 gene product appears to be involved in the maintenance of the indeterminate state of plant meristems, and its down-regulation leads to the inhibition of determinate lateral organs such as leaves or floral organs. This is supported by the observation that KN1 protein is present in the cells of the floral meristem until the termination of meristem activity by the development of the ovule primordium (Smith *et al.*, 1992).

The recently identified *clavata1 (clv1)* gene has been shown to have a significant role in meristem development (Clark *et al.*, 1993). In *clavata1* plants, vegetative, inflorescence and floral meristems are all enlarged relative to wild type, suggesting that *clavata1* has a role throughout the vegetative to floral transitition. The apical meristem can fasciate (i.e. grow abnormally as a band or ring-like structure) in the more severe mutant alleles, and this fasciation can occur prior to the transition to flowering. Studies of double mutant combinations with mutants such as *agamous*, *apetala2* and *3* and *pistillata* indicate that *clavata1* controls the underlying floral meristem structure upon which these homeotic genes act. Double mutant combinations with *apetala1* and *leafy* also indicate that *clavata1* plays a role in determining meristem size.

MODIFIED PATTERNS OF MERISTEM GENE EXPRESSION

Some meristem-related developmental programmes are modified during the V→I→F progression. This alteration of basic developmental programmes allows for subtle changes in meristem activity during the different stages of plant development. An example of meristem gene expression pattern being modified is provided by a study of the *Arabidopsis cdc*2 gene (Martinez *et al.*, 1992). The *Arabidopsis cdc*2 gene was identified by complementation of *cdc*2 mutants of yeast. The yeast *cdc*2 gene encodes a protein kinase (p34) that functions in the control of the G1- to S-phase transition of the mitotic cell cycle (Nurse, 1990). The *Arabidopsis cdc*2 gene is likely to be an important regulator of cell division in shoot meristems, including the rib meristem region. *In situ* analysis revealed uniform distribution throughout all layers of the vegetative and inflorescence apical meristems. However, the pattern was significantly different in early floral meristems, where the transcript was absent from the surface layers, but was present at high levels in the cells of the rib meristem and the developing pedicel. These high levels of *cdc*2 transcript may reflect the increased mitotic and metabolic activity in this region during floral meristem initiation (Vaughan, 1955; Bonner *et al.*, 1991). Martinez *et al.* (1992) also noted that the expression pattern of *cdc*2 in the early floral meristem is the reciprocal of the pattern displayed by several of the floral homeotic genes. They suggested that antagonistic interactions between *cdc*2 and the homeotic genes may determine these expression patterns.

Another example of modified gene expression is provided by Pri-Hadash *et al.* (1992) and Shahar *et al.* (1992). Using arrested floral meristems from the cauliflower-like inflorescence of the tomato *anantha* mutant, antibodies were raised to two proteins which are abundant in *anantha* meristems, P2 and P18. The corresponding genes were isolated by screening a cDNA expression library. *In situ* hybridization revealed that P2 was localized in the L1 cell layer of the vegetative apical meristem, with low levels in L2, L3 and the rib meristem. By contrast, early floral meristems accumulate P2 protein in the L1 and L3 cell layers and in the rib meristem but not in the L2 cell layer. The transcript levels match the protein accumulation levels, suggesting P2 protein distribution is controlled at the transcriptional level (Shahar *et al.*, 1992). The P18 protein, was found to be a dUTPase. dUTPase catalyses the hydrolysis of dUTP to dUMP, which is a precursor of thymidine diphosphate: dUTPase is thus involved in the regulation of DNA metabolism. P18 was found to be abundant throughout the L2 and L3 cell layers of the meristem and the rib meristem but is absent from the L1 cell layer of the meristem. In early floral meristems, however, P18 accumulates in the L1 layer as well as in the L2 and L3 cell layers. It also accumulates in the rib meristem. This modification in the pattern of P18 expression is likely to be in anticipation of a specific, but unknown, role for the P18 dUTPase in the developing flower.

FLORAL STAGE-SPECIFIC GENE EXPRESSION

Genes that regulate floral meristem identity have now been well characterized in a number of species (for reviews see Coen and Carpenter, 1993; Jordan and Anthony, 1993; van der Krol *et al.*, 1993; Veit *et al.*, 1993; Ma, 1994; Yanofsky, 1995). Kelly *et al.* (1990) have isolated clones of several tobacco genes whose transcript levels in the shoot apex increase as the plant progresses through the V→I→F transitions. Clone FA2 displayed a dramatic increase in transcript accumulation in floral shoot apices. FA2 RNA was not detected in leaves or roots but high levels were detected in immature sepals, petals, stamens and pistils. The expression pattern was very similar to the pattern described for the floral meristem localization of the *Arabidopsis cdc*2 transcript. Sequence analysis, however, has shown that FA2 is not a *cdc*2 homologue.

Two cloned genes that function as important regulators of floral meristem identity in *Antirrhinum* are *floricaula (flo)* (Coen *et al.*, 1990) and *squamosa (squa)* (Huijser *et al.*, 1992). *Arabidopsis* homologues to these genes have also been cloned; *leafy (lfy)* is the homologue to *flo* (Weigel *et al.*, 1992) and *apetala1 (ap1)* is the homologue to *squa* (Mandel *et al.*, 1992). During inflorescence development *flo* and *squa* accumulate throughout the early floral meristem as well as in the young bracts that subtend maturing floral buds (Coen *et al.*, 1990; Huijser *et al.*, 1992). *Flo* and *squa* expression continues in all floral organs except the stamens. The *flo* transcript is present only during the early stages of floral organ initiation, whereas *squa* transcript persists through later stages of flower development. Neither *flo* nor *squa* have been detected in inflorescence or vegetative meristems. A very similar picture has emerged for *lfy* and *ap1* of *Arabidopsis*. Transcripts are localized to early floral meristems and to immature floral organs but are not detectable in inflorescence meristems (Mandel *et al.*, 1992; Weigel *et al.*, 1992). *lfy* is first detected in the cells of the emerging floral meristem and then continues in all regions of the early floral meristem. *lfy* transcript remains at significant levels in all floral organs through the early stages of flower development (Weigel *et al.*, 1992). *ap1* transcript shows a similar pattern of accumulation in the early floral meristem of *Arabidopsis* (Mandel *et al.*, 1992). Unlike *lfy* transcript, *ap1* transcripts are detected only in the sepal and petal primordia of the developing flower (Mandel *et al.*, 1992). Thus, *ap1* and *lfy* are now considered to be the primary genes involved in specification of floral meristem identity and other genes such as *cauliflower (cal)* and *apetala2 (ap2)* have secondary roles, possibly acting as modifiers (Bowman *et al.*, 1993).

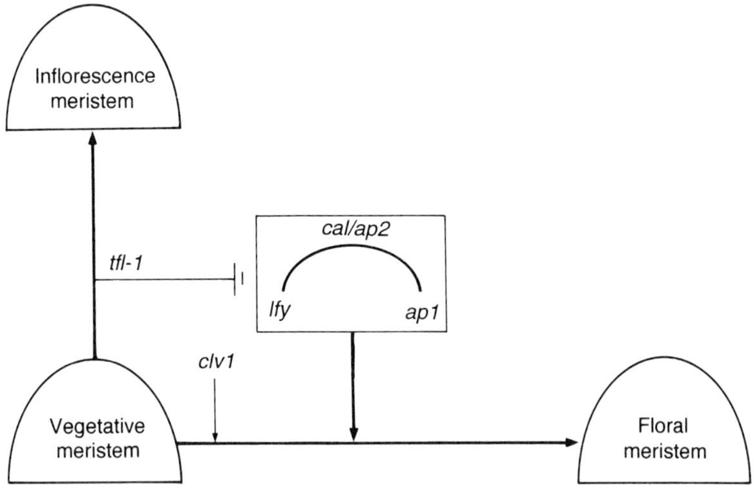

Fig. 13.3. The interaction of genes expressed during the vegetative to floral transition in *Arabidopsis*. The boxed genes positively interact with each other. *Terminal flower 1* (*tfl-1*) inhibits floral meristem development, antagonizing *lfy, ap1* and *ap2*. *Clavata1* (*clv1*) is expressed in vegetative and inflorescence meristems, but does enhance the phenotype of *lfy* and *lfy/ap1* mutants.

Figure 13.3 illustrates these interactions during the transition to flowering in *Arabidopsis*. Genes such as *terminal flower 1 (tfl-1)* influence phase transitions in *Arabidopsis*, and appear to regulate the timing of expression of *lfy*, *ap1* and *ap2* (Schultz and Haughn, 1993).

The evidence for strict floral stage-specific gene activity is not unequivocal. For example, although FA2 transcripts are not detected in tobacco vegetative meristems by *in situ* hybridization, their presence has been clearly demonstrated by quantitative polymerase chain reaction (PCR) analysis (A. Kelly and D.R. Meeks-Wagner, unpublished results: as cited in Meeks-Wagner, 1993). In addition, Kelly *et al.* (1995) have shown that *nfl*, the tobacco homologue of *floricaula* and *leafy*, is transcriptionally expressed in both vegetative and floral meristems (see also Rottmann *et al.*, 1993). This implies that the expression of *nfl* in the apical meristem of tobacco is not sufficient for the initiation of floral meristem development. Over-expression of *Arabidopsis lfy* in transgenic tobacco does result in earlier flowering, possibly because the transgene is expressed at higher effective levels than the endogenous gene (Weigel and Nilsson, 1995). Transgenic *Arabidopsis* plants that constitutively express either the *ap1* gene or the *lfy* gene also show transformation of apical and lateral shoots to flowers and establish earlier flowering. This indicates that both *ap1* and *lfy* alone can convert inflorescence shoot meristems into floral meristems (Mandel and Yanofsky, 1995; Weigel and Nilsson, 1995).

INTERACTION BETWEEN MERISTEM AND ORGAN IDENTITY GENES

The precise nature of how meristem and organ identity genes interact is poorly understood; however, some principles are beginning to emerge. A model based on the spatial distribution of control throughout the meristem has been proposed and can be used to explain many experimental observations (for discussion and further references see Jordan and Anthony, 1993; Weigel and Meyerowitz, 1994). The cells of the meristem are in three regions (A, B and C) capable, through the expression of homeotic genes, of determining organ identity (Fig. 13.4). Type *a* gene expression determines the formation of sepals in whorl 1. If *a* and *b* genes are active, petals are formed. Expression of genes *b* and *c* results in the formation of stamens and expression of *c* alone results in the formation of carpels. An important aspect of the model is that *a* inhibits *c* in whorls 1 and 2, and *c* inhibits *a* in whorls 3 and 4. The *Arabidopsis* and *Antirrhinum* mutant phenotypes indicate that *ap2* is essential for the *a* function, the *Arabidopsis ap3* and *pi* genes (*defa* and *glo* from *Antirrhinum*) are components of the *b* function, and the *Arabidopsis ag* gene (*plena* from *Antirrhinum*) is a prerequisite of the *c* function. Mutations in the *Arabidopsis a* function gene *ap2* can sometimes enhance the phenotype of mutations in the meristem identity genes *ap1* and *lfy*, suggesting that *ap2* may also play a role in establishing meristem identity (Irish and Sussex, 1990; Huala and Sussex, 1992). In addition, the strong expression of the meristem identity gene *flo* in the outer two whorls of the *Antirrhinum* flower has led to the suggestion that it may be involved in promoting the *a* function (Coen *et al.*, 1990). The requirement of the meristem identity gene *ap1* in perianth organ identity is now also well established (Mandel *et al.*, 1992). The flower-like secondary shoots formed in the upper inflorescences of all *ap1* alleles show transformation of perianth organs to reproductive organs. In stronger alleles third whorl organ number is reduced and gynoecial abnormalities often occur; such organ transforma-

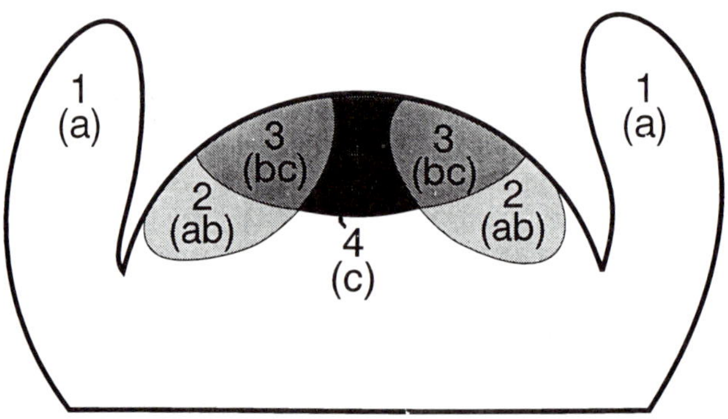

Fig. 13.4. Floral meristem showing the floral whorls, 1–4, and the regions expressing the *a, b* and *c* function genes.

tions are similar to *ap2*. This suggests that like *ap2, ap1* has a role in restricting class *c* gene expression to the inner whorls and that in its absence class *c* genes are ectopically expressed in the perianth. Mandel *et al.* (1992) have shown that *ap1* is strongly expressed in all whorls of *ag* mutant flowers, suggesting that *ag* restricts *ap1* expression to the perianth. Thus, *ap1* shares both aspects of *ap2–ag* regulation, being negatively regulated by *ag* and serving to negatively regulate *ag*. Schultz and Haughn (1993) have suggested that *ap1* and *ap2* gene products act together as class *a* genes specifying organ identity. Double mutant studies of *ap1*/*ap2* show that these genes are independently expressed and are not necessary for activation of each other.

It has been suggested that the meristem identity genes and organ identity genes such as *ap2* collaborate to establish the *a* function (Coen and Carpenter, 1993). However, meristem identity mutants can eventually produce an active *a* function. It may be that there is some genetic redundancy so that a mutation in a meristem identity gene may be overcome by an alternative route (Coen and Carpenter, 1993). Weigel *et al.* (1992) have suggested that *lfy* enhances transcription of *b* function genes such as *ap3*. Transformation of petals to sepals and stamens to carpels in the *lfy* mutant enhances transcription of *b* function genes such as *ap3*.

Transformation of petals to sepals and stamens to carpels in *lfy* single mutants indicate reduction in *pi*/*ap3* expression and this is supported by double mutant analysis (Huala and Sussex, 1992; Weigel *et al.*, 1992). As in *ap2* and *ap1* mutant phenotypes, the reduced class *b* gene expression may be the result of ectopic class *c* gene expression. The rare occurrence of stamen-carpels in the fourth whorl suggests that class *ab* organ identity genes are ectopically expressed in the fourth whorl. Thus, *lfy* may have some role in restricting *pi*/*ap3* expression, perhaps through activation of the *flo*10 gene (Schultz *et al.*, 1991; Bowman *et al.*, 1992). Transcripts of the *c* function gene *ple* are not detected in the meristems of *flo* mutants, although they are present in the rare carpelloid ‘flowers’ (D. Bradley, R. Carpenter and E.S. Coen, unpublished results, as cited in Coen and Carpenter, 1993). This indicates that *flo* is involved in activating *ple* transcription but that this requirement can eventually be overcome.

Complementary studies have also been carried out in *Arabidopsis*. The inflorescence shoots of *lfy* mutants always terminate in pistil-like structures consisting of fused carpels and carpel-leaves. Such structures form in *lfy*/*ap2*, *lfy*/*ap1*, *lfy*/*tfl1–14* and also to a lesser extent in *lfy*/*ag* double mutants. Their development suggests that *lfy* is required to repress class *c* organ identity gene activation late in inflorescence development (Schultz and Haughn, 1993).

A hypothesis has been proposed to explain the apparent paradoxical situation where *lfy* not only activates the floral progress early in floral-meristem development but also represses class *c* organ identity gene expression late in indeterminate inflorescence development. Schultz and Haughn (1993) suggest that certain factor(s) which control phase switching are highly expressed in apical meristems preventing floral initiation. The levels of these factors gradually decrease. At a certain level all of the meristem identity genes are initiated.

Initially, high levels of *lfy* suppresses class *c* gene function. Late in flower development the meristem identity gene levels decrease allowing expression of class *c* organ identity genes which, as well as activating reproductive organ development, continues maintenance of the floral programme by repressing the inflorescence programme.

Thus, it is possible that the meristem identity genes activate and/or act in parallel with the organ identity genes. The organ identity genes may in turn regulate the expression of the meristem identity genes. Evidence for this was provided by Mandel *et al.* (1992) who showed the absence of *ap1* expression in whorls 3 and 4 which has been proposed to be due to negative regulation by the *c* function gene *ag*. In addition, *flo* is not expressed in whorl 3, possibly because the *b* and *c* functions negatively regulate its expression (Coen *et al.*, 1990). Recently a gene *fimbriata* (*fim*) has been isolated that interacts between meristem and organ identity genes. Mutations in this gene result in reduced determinacy of the meristem and incomplete development of the floral organs. The expression and function of *fim* is determined by the meristem identity genes and it subsequently regulates the temporal and spatial expression of the organ identity genes. From the studies of gene expression during floral development, it is possible to envisage sequential gene expression ranging from genes such as *co* and *ld* through meristem identity genes, mediators such as *fim* and finally to the expression of organ identity genes. However, such a scenario cannot be the full explanation because genes such as *ap2* are expressed in vegetative tissue and in floral tissue. Furthermore, meristem identity genes such as *ap1* play significant roles in later stages of floral development.

FUTURE RESEARCH AND POTENTIAL FOR ORNAMENTALS

This short review illustrates the complexity of cellular events associated with floral evocation and particularly emphasizes the continuum of processes that finally lead to floral development. To date, this information has been gathered from a relatively small number of species. It is therefore important that the study of molecular events controlling evocation is extended to a wider range of species. Thus, although many of the genes controlling flower development appear to have been conserved, there are significant differences in the balance and interaction of gene activity. The control of these genes needs more study to elucidate mechanisms such as feedback-control, DNA binding and protein modifications. The proteins themselves need isolating and characterizing, so that their cellular function can be determined. In addition, more information must be gained on genes that are down-regulated or have mutations that render them ineffective. These studies will expand our knowledge and allow more detailed understanding and accurate explanations of the development of floral morphogenesis. Thus at present many wild-type flowers or putative homeotic mutants do not fit into the ABC model. For instance, in petunia a homeotic conversion is limited to one whorl only (van der Krol and Chua,

1993). Furthermore, there are homeotic mutations in ornamental species, such as lily, that could provide invaluable information. This range could also be extended by chemical or transposon mutagenesis.

Another important area of study is to determine how the environmental and developmental cues connect to the meristem identity genes and in turn how the information is transmitted to genes involved in organogenesis. This information will provide the basis for engineering the response of ornamentals, such as simultaneous and controlled flowering. Crops whose harvesting is presently controlled by seasonal factors such as photoperiod or temperature may be genetically reprogrammed for more efficient production. Thus an ornamental with a LD requirement may be genetically transformed to remove this regulation. Other regulatory genes may be engineered into plants to allow the flowering process to be conveniently switched on in a similar manner to the copper or tetracycline inducible promoters presently in use.

There is also a requirement to characterize the activation of genes that alter floral morphology. By understanding the regulation of these processes, additional whorls of petals could be introduced by expression of genes such as *agamous* or different shapes could be created by expression of genes such as *cycloidea* which controls dorsoventral symmetry in *Antirrhinum* (Lister *et al.*, 1993). Ultimately, it may be possible to change the whole morphology of the plant structure once all of the factors are understood. For instance, floral meristems could be induced in ectopic positions at different stages of development. Combination of these approaches with the genetic engineering of flower colour (see chapter 15, this volume) could lead to a whole range of new ornamental phenotypes. Clearly the potential for genetically engineering ornamentals is enormous. There is, however, still a requirement for further understanding of processes such as evocation before this potential can realistically be realized.

ACKNOWLEDGEMENTS

The authors would like to acknowledge the financial support of the Foundation for Scientific Research and Technology and the Biotechnology and Biological Sciences Research Council. We would also like to acknowledge the many helpful discussions that we have had with colleagues at the New Zealand Institute for Crop and Food Research Limited and Horticulture Research International.

REFERENCES

Anthony, R.G., James, P.E. and Jordan, B.R. (1996) Cauliflower (*Brassica oleracea* var. *botrytis* L.) curd development: the expression of meristem identity genes. *Journal of Experimental Botany* 47(295), 181–188.

Bernier, G. (1988) The control of floral evocation and morphogenesis. *Annual Review of Plant Physiology and Plant Molecular Biology* 39, 175–219.

Bernier, G., Havelange, A., Houssa, C., Petitjean, A. and Lejeune, P. (1993) Physiological signals that induce flowering. *Plant Cell* 5, 1147–1155.

Besnard-Wibaut, C. (1981) Effectiveness of gibberellins and 6-benzyladenine on flowering of *Arabidopsis thaliana. Physiologia Plantarum* 53, 205–212.

Bodson, M. and Outlaw, W.H. Jr (1985) Elevation in the sucrose content of the shoot apical meristem of *Sinapis alba* at floral evocation. *Plant Physiology* 79, 420–424.

Bonner, B.A., Gifford, E.M. and Reid, N.-M.R. (1991) Differential RNA synthesis in the shoot apex of *Pharbitis nil* during floral evocation. *American Journal of Botany* 78, 401–407.

Bowler, C., Neuhaus, G., Yamagata, H. and Chua, N.-H. (1994) Cyclic GMP and calcium mediate phytochrome phototransduction. *Cell* 77, 73–81.

Bowman, J.L., Sakai, H., Jack, T., Weigel, D., Mayer, U. and Meyerowitz, E.M. (1992) SUPERMAN, a regulator of floral homeotic genes in *Arabidopsis. Development* 114, 599–615.

Bowman, J.L., Alvarez, J., Weigel, D., Meyerowitz, E.M. and Smyth, D.R. (1993) Control of flower development in *Arabidopsis thaliana* by *APETALA1* and interacting genes. *Development* 119, 721–743.

Burn, J.E., Bagnall, D.J., Metzger, J.D., Dennis, E.S. and Peacock, W.J. (1993) DNA methylation, vernalization, and the initiation of flowering. *Proceedings of the National Academy of Sciences, USA* 90, 287–291.

Chandler, J. and Dean C. (1994) Factors influencing the vernalization response and flowering time of late flowering mutants of *Arabidopsis thaliana* (L.) Heynh. *Journal of Experimental Botany* 45, 1279–1288.

Clark, S.E., Running, M.P. and Meyerowitz, E.M. (1993) *CLAVATA1* a regulator of meristem and flower development in *Arabidopsis. Development* 119, 397–418.

Clarke, J.H. and Dean, C. (1994) Mapping FRI, a locus controlling flowering time and vernalization response in *Arabidopsis thaliana. Molecular and General Genetics* 242, 81–89.

Coen, E.S. and Carpenter, R. (1993) The metamorphosis of flowers. *Plant Cell* 5, 1175–1181.

Coen, E.S., Romero, J.M., Doyle, S., Elliott, R., Murphy, G. and Carpenter, R. (1990) *floricaula*: a homeotic gene required for flower development in *Antirrhinum majus. Cell* 63, 1311–22.

Coupland, G., Dash, S., Goodrich, J., Lee, K., Long, M.M., Puangsomlee, P., Putterill, J., Robson, F., Sunberg, E. and Wilson, K. (1993) Molecular and genetic analysis of the control of flowering time in *Arabidopsis thaliana. Flowering Newsletter* 16, 27–32.

Evans, L.T. (1969) *The Induction of Flowering.* Macmillan, Melbourne.

Evans, L.T. (1971) Flower induction and the florigen concept. *Annual Review of Plant Physiology* 80, 1025–1029.

Evans, L.T., King, R.W., Mander L.N., Chu, A. and Pharis, R.P. (1990) Gibberellin structure and florigenic activity in *Lolium temulentum*, a long-day plant. *Planta* 182, 97–106.

Evans, L.T., King, R.W., Mander, L.N. and Pharis, R.P. (1994) The relative significance for stem elongation and flowering in *Lolium temulentum* of 3β-hydroxylation of gibberellins. *Planta* 192, 130–136.

Francis, D. (1992) Tansley Review No. 38, The cell cycle in plant development. *New Phytologist* 122, 1–20.

Gonthier, R., Jacqmard, A. and Bernier, G. (1987) Changes in cell-cycle duration and growth fraction in the shoot meristem of *Sinapis* during floral transition. *Planta* 170, 55–59.

Green, P.B. (1988) A theory for inflorescence development and flower formation based on morphological and biophysical analysis in *Echeveria. Planta* 175, 153–169.

Horridge, J.S. and Cockshull, K.E. (1979) Size of the chrysanthemum shoot apex in relation to inflorescence initiation and development. *Annals of Botany* 44, 547–556.

Huala, E. and Sussex, I.M. (1992) *LEAFY* interacts with floral homeotic genes to regulate *Arabidopsis* floral development. *Plant Cell* 4, 901–913.

Huala, E. and Sussex, I.M. (1993) Determination and cell interactions in reproductive meristems. *Plant Cell* 5, 1157–1165.

Huijser, P., Klein, J., Lonnig, W.-E., Meijer, H., Saedler, H. and Sommer, H. (1992) Bracteomania, an inflorescence anomaly, is caused by the loss of function of the MADS-box gene *squamosa* in *Antirrhinum majus*. *EMBO Journal* 11, 1239–1249.

Irish, V.F. and Sussex, I.M. (1990) Function of the *apetala-1* gene during *Arabidopsis* floral development. *Plant Cell* 2, 741–753.

Jordan, B.R. and Anthony, R.G. (1993) Floral homeotic genes: isolation, characterization, and expression during floral development. In: Jordan, B.R. (ed.), *The Molecular Biology of Flowering*. CAB International, Wallingford, UK, pp. 93–116.

Jordan, B.R., Anthony, R.G. and James, P.E. (1994) Control of floral morphogenesis in cauliflower (*Brassica oleracea* L. var. *botrytis*): the role of homeotic genes. In: Scott R.J. and Stead A.D. (eds), *Molecular and Cellular Aspects of Plant Reproduction*. SEB Seminar Series 55, Cambridge University Press, Cambridge, pp. 17–29.

Kelly, A.J., Zagotta, M.T., White, R.A., Chang, C. and Meeks-Wagner, D.R. (1990) Identification of genes expressed in the tobacco shoot apex during the floral transition. *Plant Cell* 2, 963–972.

Kelly, A.J., Bonnlander, M.B. and Meeks-Wagner, D.R. (1995) NFL, the tobacco homolog of *floricaula* and *leafy*, is transcriptionally expressed in both vegetative and floral meristems. *Plant Cell* 7, 225–234.

Kinet, J.-M. (1993) Environmental, chemical and genetic control of flowering. *Horticultural Reviews* 15, 279–334.

Koornneef, M., Hanhart, C.J. and van der Veen J.H. (1991) A genetic and physiological analysis of late flowering mutants in *Arabidopsis thaliana*. *Molecular and General Genetics* 229, 57–66.

Law, C.N. and Scarth, R. (1984) Genetics and its potential for understanding the action of light in flowering. In: Vince-Prue, D., Thomas, B. and Cockshull, K.E. (eds), *Light and the Flowering Process*. Academic Press, London, pp. 193–209.

Law, C.N., Dean C. and Coupland G. (1993) Genes controlling flowering and strategies for their isolation and characterisation. In: Jordan, B.R. (ed.), *The Molecular Biology of Flowering*. CAB International, Wallingford, UK, pp. 47–68.

Lee, I., Auckerman, M.J., Gore, S.L., Lohmann, K.N., Michaels, S.D., Weaver, L.M., Manorama, J.C., Feldmann, K.A. and Amasino, R.M. (1994) Isolation of Luminidependens: a gene involved in the control of flowering time in *Arabidopsis*. *Plant Cell* 6, 75–83.

Lister, C., Jackson, D. and Martin, C. (1993) Transposon-induced inversion in *Antirrhinum* modifies *nivea* gene expression to give a novel flower colour pattern under the control of *cycloidea*^{radialis}. *Plant Cell* 5, 1541–1553.

Lumsden, P. (1993) Mechanisms of signal transduction. In: Jordan, B.R. (ed.), *The Molecular Biology of Flowering*. CAB International, Wallingford, UK, pp. 21–45.

Ma, H. (1994) The unfolding drama of flower development: recent results from genetic and molecular analyses. *Genes and Development* 8, 745–756.

Ma, H., Yanofsky, M.F. and Meyerowitz, E.M. (1991) AGL1–AGL6 an *Arabidopsis* gene family with similarity to floral homeotic and transcription factor genes. *Genes and Development* 5, 484–495.

Mandel, M.A., and Yanofsky, M.F. (1995) A gene triggering flower formation in *Arabidopsis*. *Nature* 377, 522–524.

Mandel, M.A., Gustafson-Brown, C., Savidge, B. and Yanofsky, M.F. (1992) Molecular characterization of the *Arabidopsis* floral homeotic gene *APETALA1*. *Nature* 360, 273–277.

Martinez, M.C., Jorgensen, J.-E., Lawton, M.A., Lamb, C.J. and Doerner, P. (1992) Spatial pattern of *cdc2* expression in relation to meristem activity and cell proliferation during plant development. *Proceedings of the National Academy of Sciences, USA* 89, 7360–7364.

Medford, J.I., Horgan, R., El-Sawi, Z. and Klee, H.J. (1989) Alterations of endogenous cytokinins in transgenic plants using a chimeric isopentenyl transferase gene. *Plant Cell* 1, 403–413.

Medford, J.I., Elmer, J.S. and Klee, H.J. (1991) Molecular cloning and characterization of genes expressed in shoot apical meristems. *Plant Cell* 3, 359–370.

Meeks-Wagner, D.R. (1993) Gene expression in the early floral meristem. *Plant Cell* 5, 1167–1174.

Melzer, S., Majewski, D.M. and Apel, K. (1990) Early changes in gene expression during the transition from vegetative to generative growth in the long-day plant *Sinapis alba*. *Plant Cell* 2, 953–961.

Miksche, J.P. and Brown, J.A.M. (1965) Development of vegetative and floral meristems of *Arabidopsis thaliana*. *American Journal of Botany* 52, 533–537.

Nougarede, A. (1967) Experimental cytology of the shoot apical cells during vegetative growth and flowering. *International Review of Cytology* 21, 203–351.

Nurse, P. (1990) Universal control mechanisms regulating onset of M-phase. *Nature* 344, 503–508.

O'Neill, S.D. (1993) Changes in gene expression associated with floral induction and evocation. In: Jordan, B.R. (ed.), *The Molecular Biology of Flowering*. CAB International, Wallingford, UK, pp. 69–92.

Pri-Hadash, A., Hareven, D. and Lifschitz, E. (1992) A meristem-related gene from tomato encodes a dUTPase: analysis of expression in vegetative and floral meristems. *Plant Cell* 4, 149–159.

Putterill, J., Robson, F., Lee, K. and Coupland, G. (1993) Chromosome walking with YAC clones in *Arabidopsis* – isolation of 17 kb continuous DNA on chromosome-5, including a 300 kb region containing the flowering-time gene CO. *Molecular and General Genetics* 239, 145–157.

Putterill, J., Robson, F., Lee, K., Simon, R. and Coupland G. (1995) The *CONSTANS* gene of *Arabidopsis* promotes flowering and encodes a protein showing similarities to zinc finger transcription factors. *Cell* 80, 847–857.

Rottmann, W.H., Boes,T.K. and Strauss, S.H. (1993) Structure and expression of a *leafy* homolog from *Populus*. *Journal of Cellular Biochemistry* S17B, 23.

Schultz, E.A. and Haughn, G.W. (1993) Genetic analysis of the floral initiation process (FLIP) in *Arabidopsis*. *Development* 119, 745–765.

Schultz, E.A., Pickett, F.B. and Haughn, G.W. (1991) The *FLO*10 gene product regulates the expression domain of homeotic genes *AP3* and *PI* in *Arabidopsis* flowers. *Plant Cell* 3, 1221–1227.

Shahar, T., Hennig, N., Gutfinger, T., Hareven, D. and Lifschitz, E. (1992) The tomato 66.3-kD polyphenoloxidase gene: molecular identification and developmental expression. *Plant Cell* 4, 135–147.

Smith, L., Greene, B., Veti, B. and Hake, S. (1992) A dominant mutation in the maize homeobox gene, *Knotted-1*, causes its ectopic expression in leaf cells with altered fates. *Development* 116, 21–30.

Stewart, R.N. and Dermen, H. (1975) Flexibility in ontogeny as shown by the contribution of the shoot apical layers to leaves of periclinal chimeras. *American Journal of Botany* 62, 935–947.

Stewart, R.N., Meyer, F.G. and Dermen, H. (1972) Camellia + 'Daisy Eagleson' a graft chimera of *Camellia sasangua* and *C. japonica*. *American Journal of Botany* 59, 515–524.

Sung, Z.R., Belachew, A., Shunong, B. and Betrand-Garcia, R. (1992) EMF, an *Arabidopsis* gene required for vegetative shoot development. *Science* 258, 1645–1647.

Thomas, B. (1993). Internal and external controls on flowering. In: Jordan, B.R. (ed.), *The Molecular Biology of Flowering*. CAB International, Wallingford, UK, pp. 1–19.

van der Krol, A.R. and Chua, N.-H. (1993) Flower development in *Petunia*. *Plant Cell* 5, 1195–1203.

van der Krol, A.R., Brunelle, A., Tsuchismolo, S. and Chua, N.H. (1993) Functional analysis of petunia floral homeotic MADS-box gene pMADS1. *Genes and Development* 7, 1214–1228.

Vaughan, J.G. (1952) Structure of the angiosperm apex. *Nature* 169, 458–459.

Vaughan, J.G. (1955) The morphology and growth of the vegetative and reproductive apices of *Arabidopsis thaliana* (L.) Heynh., *Capsella bursa-pastoris* (L.) Medic. and *Anagallis arvensis* L. *Journal of the Linnean Society, London. Botany*. 55, 279–300.

Veit, B., Schmidt. R.J., Hake S. and Yanofsky, M.F. (1993) Maize floral development: new genes and old mutants. *Plant Cell* 5, 1205–1215.

Vince-Prue, D. (1975) *Photoperiodism in Plants.* McGraw–Hill, London.

Weigel, D., Alvarex, J., Smyth, D.R., Yanofsky, M.F. and Meyerowitz, E.M. (1992) *LEAFY* controls floral meristem identity in *Arabidopsis*. *Cell* 69, 843–859.

Weigel, D. and Meyerowitz, E.M. (1994) The ABCs of floral homeotic genes. *Cell* 78, 203–209.

Weigel, D. and Nilsson, O. (1995) A developmental switch sufficient for flower initiation in diverse plants. *Nature* 337, 495–500.

Yanofsky, M.F. (1995) Floral meristems to floral organs: genes controlling early events in *Arabidopsis* flower development. *Annual Review of Plant Physiology and Molecular Biology* 46, 167–188.

Manipulation of Flower Shape

14

A.R. van der Krol and O. Vorst
Department of Plant Physiology, Agricultural University Wageningen, Arboretumlaan 4, 6703 BD Wageningen, The Netherlands

INTRODUCTION

For centuries conventional breeding programmes of ornamental plants have been aimed at the enhancement of the floral qualities and not without spectacular results! So why use molecular genetics in horticulture? In many chapters of this book, examples of alterations of a certain plant trait by manipulation of specific genes are given. It is this specificity that is the great power of genetic engineering; in conventional breeding practice the entire genome of two parents get mixed resulting in progeny plants with (more or less) unpredictable phenotypes. Out of this array of new phenotypes, the most desirable plants are selected and propagated. In this way it can take several generations of back crossing before a newly selected phenotype is stabilized. In genetic engineering, however, only a single gene or, at the most, a few genes are changed; the rest of the plant genome stays the same. This means that if one starts out with a plant that already has all the desired characteristics, changing a single gene or just a few genes will hardly disturb the other characteristics of the plant. This implies that there is no need for backcrossing to stabilize the new phenotype. However, since many traits in plants are determined by a combinatory effect of many genes, not all traits are easily manipulated by a genetic engineering approach.

The other great power of genetic engineering, which surpasses any possibility with conventional breeding, is the introduction of genetic material from non-related (plant) species. Introduction of foreign genetic material into a plant can be used to convey disease resistance, to open up new pathways of pigmentation, or to create new regulatory pathways that change development. In this chapter we will discuss what has been accomplished with the specific manipulation of flower development. The genetic engineering of floral form may eventually serve as a supplement to conventional breeding programmes that are aimed at the enhancement of the floral quality of ornamental plants.

MANIPULATION OF FLORAL FORM

What do we mean when we say that genetic engineering can change flower shape? Different plant species have different flowers, varying in size, shape, symmetry and compositional array. Not all these aspects of floral form are within reach of genetic manipulation yet. The floral manipulations that will be mainly discussed here are of a limited but none the less spectacular kind; changing the identity of organs within a flower or specifically changing the shape of a particular floral organ.

How can something as fundamental as the identity of a floral organ be changed? Surely a structure as complex as a flower is determined by the (inter)action of thousands of different genes. As stated above, genetic engineering only allows the manipulation of a handful of genes. Indeed this is true and it means that these manipulations involve very special types of genes. Floral identity genes themselves can control the expression of a great number of other genes which together define a particular floral organ. A gene that controls the identity of a floral organ is an example of a homeotic gene, as a mutation in the gene leads to a change in the identity of an organ. In past decades, many mutant plants have been isolated which were noted for their changed floral morphology (Meyerowitz *et al.*, 1989; Coen, 1991). Initially, they may have been collected as a curiosity, for most floral homeotic mutations are not an improvement to the attractiveness of the flower. However, some of the floral homeotic mutations have great value for the breeder of ornamental species. For instance, there are mutations that lead to a reiteration of the petal setting and can change a flower with a single whorl of petals into a spectacular floral structure filled with many whorls of petals. The flower of the *double* mutant in petunia, the cultivated rose and carnation are examples of this type of mutation. One could envision the great interest of horticulturists if such a mutation could be induced in a plant species at will through genetic engineering.

THE GENETIC BASIS OF FLOWER SHAPE

ANALYSIS OF FLORAL HOMEOTIC MUTANTS IN ARABIDOPSIS *AND* ANTHIRRHINUM

It has only been in the past decade that floral homeotic mutants have been systematically analysed and an inventory of their genetic interactions made. As might be expected, these studies were done on the model plant *Arabidopsis* (Bowman *et al.*, 1989). This plant may not be prized for its horticultural value, but the fast genetics of this small weed has become an invaluable tool to many plant geneticists. Most likely the genome of *Arabidopsis* (100,000 kilobases) will become the first plant genome to be fully sequenced! Genes that are identified in this genome project can then be used as a probe to search for homologues in commercially more interesting species.

The other plant that has contributed to the unravelling of floral morphogenesis is *Anthirrhinum* (snapdragon). Its attractive big flower makes this plant an ideal model to study the genetics of floral pigmentation (Stickland and Harrison, 1974), while the occurrence of a transposable element and actual cloning of this element (Sommer *et al.*, 1985) also facilitated the means of isolating tagged genes in *Anthirrhinum* (Carpenter and Coen, 1990). It was from the careful analysis of the *Anthirrhinum* and the *Arabidopsis* floral homeotic mutants, and the analysis of genetic interaction between homeotic mutants, that a model of floral morphogenesis could be postulated. The model describes the genetic interactions that determine the identity for each of the floral organs (Coen and Meyerowitz, 1991, Meyerowitz *et al.*, 1991).

THE ABC MODEL OF FLORAL ORGAN IDENTITY

In order to devise a comprehensive model for flower development, the intricate structure of a real flower, whether part of a compound flower or solitary, is schematized and reduced to its basic components. In this schematic view, the flower is an organized collection of organs that can have four different identities: sepal, petal, stamen and carpel. Usually, the organization of these organs is in concentric whorls that, by convention, are numbered from the outside to the inside. In the first floral whorl we usually find the sepals, in the second whorl the petals, in the third whorl the stamens and in the fourth floral whorl the carpels. Thus, the problem of flower development can be reformulated in the question of how a floral primordium develops into the right floral organ type. One of the first things that was noted from the analysis of the homeotic mutants of *Arabidopsis* and *Anthirrhinum* was that the homeotic mutations often affected two adjacent floral whorls. For instance, in the *Anthirrhinum* mutant *Deficiens* a conversion from petals to sepals has occurred in the second whorl, while at the same time in the third whorl a conversion of stamens to carpelloid structures is observed (Schwarz-Sommer *et al.*, 1990). The mutation *APETALA2* (*ap2*) in *Arabidopsis* leads to a conversion of the first whorl sepals to carpelloid structures, and of second whorl petals to stamens. So, this gene functions in both the first and second whorl (Kunst *et al.*, 1989). A third type of mutation (*AGAMOUS* in *Arabidopsis*) leads to a conversion of the third whorl stamens to petals and of fourth whorl carpel to sepals. This type of mutation is more pleiotropic, also affecting the determinate nature of the floral meristem. Instead of ending the floral meristem in the fourth whorl with a carpel, a reiteration of the floral pattern occurs, resulting in a 'flower within a flower' (Bowman *et al.*, 1989).

From all these observations, it was inferred that each type of floral homeotic gene functions in two consecutive whorls, and that there are basically

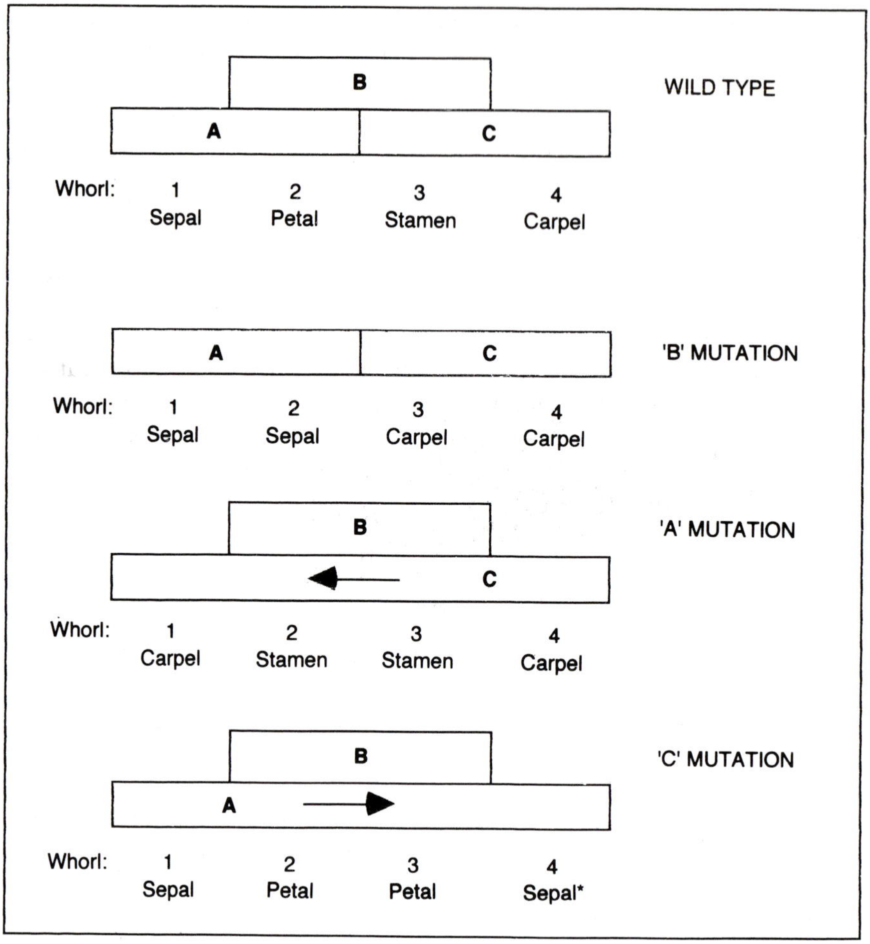

Fig. 14.1. The ABC model of floral organ identity determination (Coen and Meyerowitz, 1991). The flower is schematically divided into whorls, which are numbered from the outside to the centre of the flower. The identity of the organs in each of the floral whorls is determined by the action or interaction of an A, B and/or a C function. (*Reiteration of floral organs when no carpel is formed in fourth whorl.)

three types: type A functions in whorls 1 and 2, type B in whorls 2 and 3 and type C in whorls 3 and 4. The four different organ types in the flower can now be explained by the action of three groups of genes, named A, B and C, which act and interact as depicted in Fig. 14.1 (for review see Coen and Meyerowitz, 1991; Ma, 1994). Expression of group A genes in a floral primordium leads to sepal development, a combination of functions A plus B leads to a petal development, a combination of functions B and C to stamen development and function C alone to carpel development. A special feature of A and C is that they control each other's functional domain. If the C function is repressed, the action domain of A which is normally restricted to whorls 1 and 2, expands to the third and fourth whorl (see Fig. 14.1, C mutation). As a result, the A and B functions are now combined in both whorls 2 and 3. Thus the model predicts a flower consisting of sepals (A), petals (A+B), petals (A+B) and sepals (A), respectively. Conversely, when the A function is suppressed, the C functional domain expands towards the first and the second whorl, thus resulting in a flower with carpels (C), stamens (B+C), stamens (B+C) and a central carpel (C) (see Fig. 14.1, A mutation).

Thus, the identity of the four floral organs is determined by three functions that are expressed in different overlapping regions of the floral meristem. Each function or combination of functions controls the action of a whole group of genes which together define all the aspects of a particular floral organ. Although the ABC model is an oversimplification of a very complex developmental programme, it serves well as a basis for the interpretation of the process of floral organ identity determination. Obviously, not all flowers in nature fit this simplified scheme of the ABC model, and it will be a challenge indeed to adjust this floral genetic model to accommodate all possible scenarios of floral form.

COMPLICATIONS IN THE ABC MODEL

One of the crucial facts that led to the formulation of the ABC model was the observation that floral homeotic mutations in *Anthirrhinum* and *Arabidopsis* affect the floral organ identity in two adjacent whorls. However, in many plant species floral homeotic mutants exist in which only one of the floral whorls is affected. For instance, in the petunia mutant *green petals*, a homeotic conversion of petal to sepal has occurred, reminiscent of a B-type mutation. Unlike the B gene mutation in *Anthirrhinum* and *Arabidopsis* however, this mutation in petunia does not lead to a homeotic conversion of the stamens to carpels. This differential effect in the petunia mutant is not due to a weak allele of the gene, since it was shown that in this mutant the gene (*pMADS1*, see below) was completely deleted.

On the other hand, there are also homeotic genes that affect the organ identity of three instead of two floral whorls. The tomato MADS-box gene *TM5* and the petunia homologue *FBP2* are both expressed in the second, third and the fourth floral whorl. Inhibition of expression of these genes resulted in a partial homeotic transformation in the second whorl, changing petals to green petaloid structures. In the third whorl the stamens are replaced by green petaloid organs and in the fourth floral whorl the development of the gynoecium was affected. Most notably new flowers developed within the gynoecium (Angenent *et al.*, 1994; Pnueli *et al.*, 1994).

Thus, the genetic functions of the ABC model cannot always directly be translated into the action domain of individual genes: the petunia gene *pMADS1* is essential for the B function in the second, but not in the third whorl. On the other hand, the tomato *TM5* gene is part of the B function in the second and third whorl, but also part of the C function in the fourth floral whorl. This makes it difficult to group either of these genes under the A, B or C type homeotic genes and illustrates the limitations of the ABC model.

IDENTIFICATION AND ISOLATION OF HOMEOTIC A, B AND C GENES

In 1989, the first B and C type floral homeotic genes were isolated from *Anthirrhinum* (*DEFICIENS*; Sommer *et al.*, 1990) and *Arabidopsis* (*AGAMOUS*; Yanofsky *et al.*, 1990). Sequence analysis revealed that both genes encode a protein that can bind to DNA and thus probably are involved in the activation of gene transcription. The amino acid sequence encoding the DNA binding domain within these proteins are nearly identical, and very similar to the DNA binding domain of some mammalian and yeast transcription factors. Therefore, this DNA binding domain was subsequently named the MADS-box (*MCM1*, *AGAMOUS*, *DEFICIENS* and *SRF*) after the first four identified factors with this domain. With the MADS-box domain as a probe, many different MADS-box genes were isolated from a wide range of plant species (*Arabidopsis*, Ma *et al.*, 1991; *Anthirrhinum*, Tröbner *et al.*, 1992; tomato, Pnueli *et al.*, 1991; petunia, Angenent *et al.*, 1992, Kush *et al.*, 1993; *Brassica*, Mandel *et al.*, 1992; tobacco, Kempin *et al.*, 1993, Mandel *et al.*, 1994; maize, Schmidt *et al.*, 1993; potato, Garcia-Maroto *et al.*, 1993; rice, Chung *et al.*, 1994). It is estimated that 10–20 different MADS-box genes are present in plants per haploid genome. Outside the MADS-box the MADS-box proteins are quite different; also their expression pattern varies widely. Some MADS-box genes are expressed in floral organs, some in subdomains of floral organs, while still others are expressed in leaf and stem.

It took almost five more years to isolate the first A type gene. Cloning and sequencing of the *APETALA2* gene from *Arabidopsis* revealed that the gene is part of a small gene family and encodes a protein with a novel structure (Jofuku *et al.*, 1994). The putative AP2 protein contains a 37 amino acid serine-rich acidic domain that is analogous to regions that function as activation domains in a number of RNA polymerase II transcription factors. Furthermore, it has a 68 amino acid repeat, the AP2 domain, and the protein contains a putative nuclear localization signal (Jofuku *et al.*, 1994). The AP2 domain is conserved between highly divergent *AP2*-like gene family members in *Arabidopsis* and is also present in *AP2*-like genes isolated from petunia, *Anthirrhinum* and rice (Jofuku *et al.*, 1994)

Expression pattern of the floral homeotic genes

Characterization of the A type floral homeotic gene AP2 (a non MADS-box gene) shows that the AP2 is transcribed throughout the plant (Okamuro *et al.*, 1993; Jofuku *et al.*, 1994). It should be noted that the effect of a mutation in the AP2 gene is only scored in the first and second floral whorl. This discrepancy between transcription and functional domain of a gene can have several causes. It could be that in certain cell types the function of the gene is redundant. In this case, the effect of a genetic mutation in the gene would only be scored in those domains in which the gene function is limiting. Differences in expression domain and functional domain could also imply a translational control mechanism or control of protein stability in different parts of the flower. Similarly there could be a difference in protein stability or protein activation between different whorls.

The spatial expression of most B and C MADS-box genes coincides with the floral domains in which the B and C function are active; the B type genes are expressed in the second and third floral whorl, while the C type gene is expressed in the third and fourth floral whorl. There is an interesting feature to the expression profile of B MADS-box genes in a mutant background. Both *Anthirrhinum* and *Arabidopsis* have two homeotic B genes. When either one of these is mutated, the transcription of the other homeotic B gene is not affected at the very early stages of floral development. However, the expression of the non-mutated B gene fails upon maturation of the mutant flower. At later stages of flower development, no mRNA of this gene can be detected and the B function is completely lost. From this observation several conclusions can be drawn. First, it indicates that the signal that activates the B type MADS-box genes is only transiently present. Second, it means that in *Anthirrhinum* and *Arabidopsis* the maintenance of B type MADS-box gene expression is dependent on the functionality of both MADS-box proteins. Third, it indicates that the expression of B type MADS-box genes at the early developmental stages is not sufficient to impose B function in the organs of the second and third floral whorl. The B genes need to be expressed throughout maturation of the floral organs to complete the petal and stamen identity.

While in *Arabidopsis* and *Anthirrhinum* maintenance of expression of the two homeotic B genes is interdependent, in petunia this interdependency of homeotic B genes only exists in the second whorl. In the third whorl, the expression of two other petunia B genes, *FBP1* and *pMADS2* (see below), is maintained after the induction phase, even in the absence of a functional *pMADS1* gene. This is another illustration of a difference in domain of genetic function and domain of expression of a floral homeotic gene; *pMADS1* is expressed in the second and third floral whorl but when this gene gets mutated it is only the second whorl organ that changes organ identity (from petal to sepal).

DNA BINDING PROPERTIES OF B TYPE MADS-BOX PROTEINS

Floral MADS-box genes have a DNA binding domain similar to that of the human serum response factor (*SRF*), the regulator of yeast cell type specific genes, *MCM1* (Norman *et al.*, 1988; Passmore *et al.*, 1988). Extensive *in vivo* and *in vitro* studies on the SRF, MCM1 and AGAMOUS proteins have shown

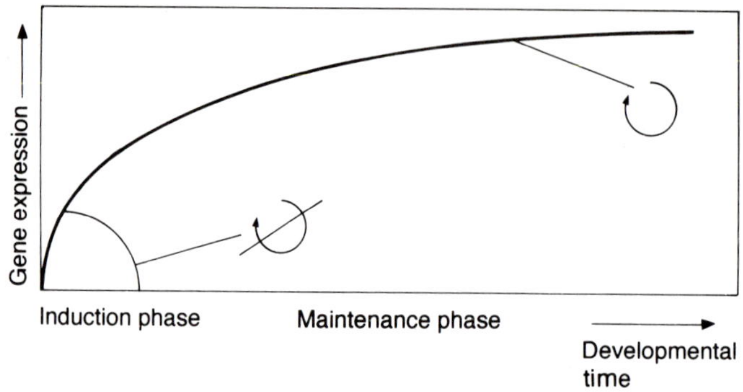

Fig. 14.2. Gene expression in the presence and absence of feedback regulation. During the induction phase the gene is switched on by factors that are transiently active. The gene expression ceases when the activity of the inducing factors stops, unless through a feedback mechanism the gene expression is maintained.

that they have overlapping DNA binding specificity and can bind to DNA which contains the sequence $CC(A/T)_6GG$ (CArG box; Mueller and Nordheim, 1991; Shiraishi *et al.*, 1993). Potential MADS-box binding sites have been found in the promoter of structural genes like chalcone isomerase, regulatory genes like *EPF2-5*, a gene encoding a zinc finger protein (Takatsuji *et al.*, 1994) and, most interestingly, in the promoter of MADS-box genes themselves (Tröbner *et al.*, 1992; Angenent *et al.*, 1994). The presence of MADS-box recognition sequences in the promoter of B type MADS-box genes suggests that these proteins can bind to their own promoter and regulate their own expression (feedback control).

DNA-binding studies with DEFICIENS and GLOBOSA protein have shown that neither the DEFICIENS nor the GLOBOSA protein can bind to DNA alone. Only the DEFICIENS/GLOBOSA heterodimer is capable of binding to DNA (Schwarz-Sommer *et al.*, 1992; Tröbner *et al.*, 1992). If the DEFICIENS and GLOBOSA protein are only functional as a heterodimer *in vivo*, this would explain the interdependency in the maintenance of gene expression. When either gene is mutated, feedback regulation of the expression of the two genes will fail and B gene expression will cease. Figure 14.2 shows the expression profile of such a transiently induced gene, in the absence and presence of feedback regulation. It should be noted that feedback control can act like a switch, sensitive to small differences in protein--protein or protein–DNA interactions. Thus, small differences between homeotic proteins from different plant species that influence feedback control may get amplified into an all or nothing effect. That B proteins may vary in specific properties is illustrated by the results obtained with the petunia pMADS1. Preliminary experiments have shown that this protein can bind to DNA as homodimer (A. Kush and van der Krol, unpublished), unlike the *Anthirrhinum* and *Arabidopsis* B type

MADS-box proteins (Tröbner *et al.*, 1992). Variations that exist in the effects of B gene mutations between different plant species may thus be explained.

CHANGING FLOWER SHAPE THROUGH GENETIC ENGINEERING

CHANGING THE EXPRESSION OF A GENE: ECTOPIC EXPRESSION AND GENE SILENCING

In genetic engineering only a limited number of genes can be manipulated. The manipulation can either involve the expression of a gene in a position where it normally is not expressed (ectopic expression) or the suppression of the expression of a gene. The promoter determines where and when a gene is expressed. Ectopic expression of a homeotic gene can therefore be accomplished by fusing the coding portion of a gene with a heterologous promoter that gives the desired expression pattern. A commonly used promoter is the viral cauliflower mosaic virus (CaMV) 35S RNA promoter; this promoter drives expression in virtually all plant tissues (Benfey *et al.*, 1990). Ectopic expression of a floral homeotic genes could also be accomplished by exchanging promoters of the different type of genes. For instance, if the promoter of a C type gene were to be fused to the coding region of a B type gene, the chimeric gene would be transcribed in the third and fourth whorl when introduced into a plant. In the third whorl, the B function is already present, so no changes are to be expected. The ectopic expression of the B gene in the fourth whorl could change carpels to stamens and cause a reiteration of the flower in the centre (see below).

Gene suppression can either be obtained by introduction into the plant of an antisense gene or through cosuppression (van der Krol *et al.*, 1988, 1990; Napoli *et al.*, 1990). An antisense gene encodes an RNA complementary to the RNA strand encoded by the target gene. The rationale of antisense genes is that the antisense RNA can form a duplex with the messenger RNA of the target gene. Formation of the duplex RNA would prevent translation of the messenger RNA into protein. Although it has not been proved that antisense genes do indeed function this way, introduction of an antisense gene into the plant genome can specifically block the expression of a target gene (van der Krol *et al.*, 1988). The second way to silence the expression of a gene is through cosuppression. For some unknown reason the introduction of a second copy of a gene can sometimes lead to a silencing of both the transgene and the endogenous gene, instead of overexpression. This combined gene silencing has been named cosuppression (Napoli *et al.*, 1990; van der Krol *et al.*, 1990). Both in antisense and cosuppression gene silencing, a severe reduction in mRNA steady-state level of the target gene is observed.

It should be noted that the introduction of a transgene into plants may therefore lead to two opposing effects: in some transformants the ectopic

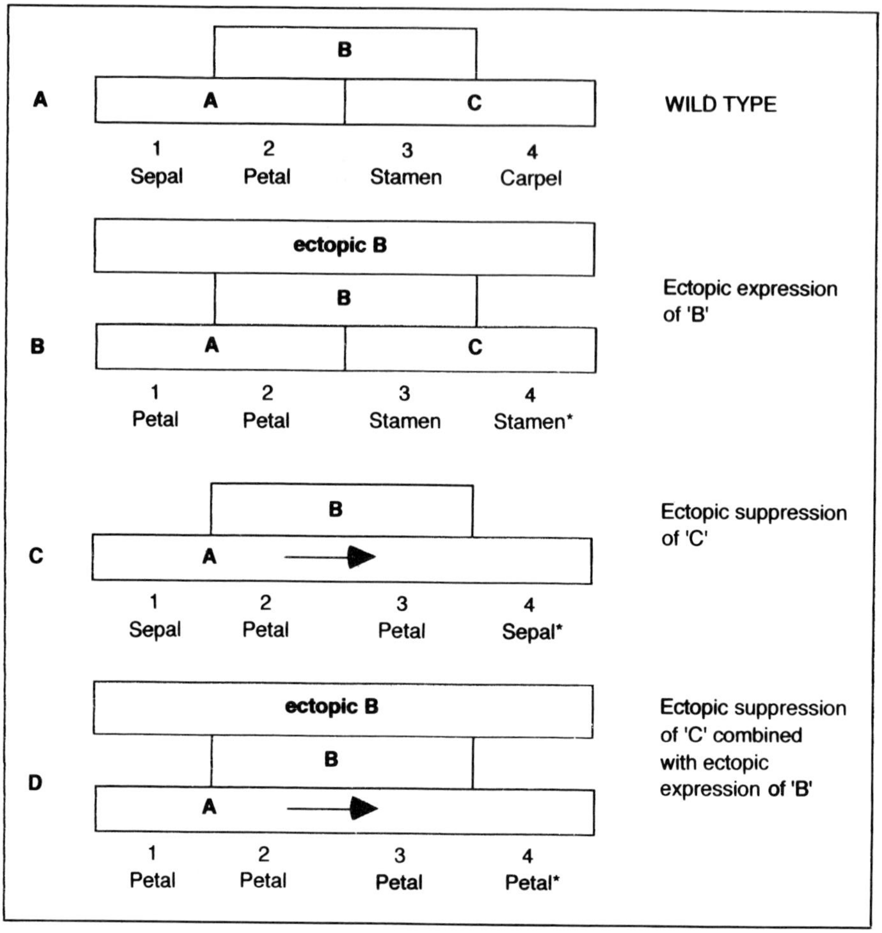

Fig. 14.3. Different possibilities to increase the number of petals in a flower through genetic engineering, according to the ABC model. (*Reiteration of floral organs when no carpel is formed in fourth whorl.)

expression of the transgene may lead to higher expression levels (endogenous gene plus the transgene), in other transformants the combination of the transgene and endogenous gene may lead to cosuppression, and thus to a specific loss in gene function.

TARGETED MANIPULATION OF FLORAL ORGAN IDENTITY

The ABC model can serve as basis for strategies devised at targeted manipulation of floral form (Fig. 14.3A). There are two strategies in genetic engineering that can be followed: either introduce an A, B or C function in a floral whorl where it normally is not expressed (ectopic expression of an A, B or C gene), or block an A, B or C function in a floral whorl where it is normally active. If the aim is the augmentation of the number of petals this could be obtained by introduction of a B function in the first whorl. Expression of a B gene in the first whorl could lead to flowers with the desired first whorl petals (A+B function). For this purpose, the B gene could be put under control of a constitutive promoter. Such a transgene is also expressed in the fourth whorl which could result in a conversion of carpel to stamen (B+C function). Because in this case the floral meristem is no longer determinate, a reiteration of the floral organs in the centre could occur (see Fig. 14.3B).

Another way to increase the number of petals is by blocking the C function and thus create an *AGAMOUS*-like mutation (see Fig. 14.3C). If we combine a genetically engineered suppression of the C function with a genetically engineered ectopic expression of the B function, the model predicts a flower with only petals that keeps reiterating itself in the centre (see Fig. 14.3D).

IS ECTOPIC EXPRESSION OF A SINGLE B GENE SUFFICIENT TO INDUCE B FUNCTION?

The above suggests that it should be quite easy to change the expression of a floral homeotic gene. The question is, however, whether ectopic expression of a single homeotic gene is sufficient to manipulate the identity of floral organs. For instance, Fig. 14.3 shows a scheme in which the presence a B function in the first whorl converts the sepals to petals. Ectopic expression of a single homeotic B gene will not necessarily be sufficient to impose B function in this whorl. Both *Arabidopsis* and *Anthirrhinum* have two B type genes and mutation of either one of these two is sufficient to completely abolish the B function (see section on DNA binding properties of B type MADS-box proteins). However, the fact that the loss of function of a single B gene is sufficient to lose B function, does not imply that the ectopic expression of a single B gene is sufficient to induce full B function.

Several experiments have been done to test the effect of ectopic expression of a single B gene. In *Arabidopsis*, the B gene *APETALA3* (*AP3*) was put under control of the CaMV 35S promoter and introduced into *Arabidopsis* plants (Jack *et al.*, 1994). First of all, ectopic expression of the *AP3* gene did not influence the vegetative stages of plant development. In the flower the sepals, petals and stamens remained virtually unaltered, while in the fourth floral whorl an incomplete conversion from carpel to stamen occurred (Jack *et al.*, 1994). This means that some, but not all of the homeotic changes that are predicted by the ABC model can be observed in these transformants (compare Fig. 14.3B). Intriguingly, the transcripts of the 35S-*AP3* gene could be detected throughout the transgenic flowers, while the AP3 protein was only detected in the second, third and fourth floral whorl where the B function was (partially) active (Jack *et al.*, 1994). This indicates either a translational control on the *AP3* mRNA in the first whorl or an extreme instability of the AP3

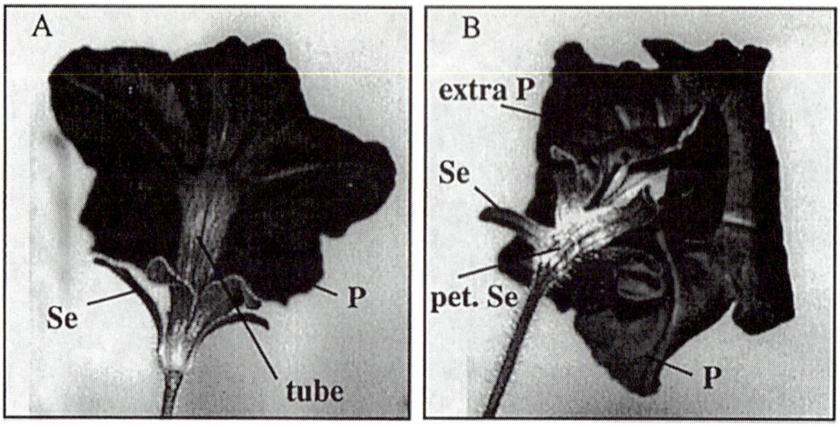

Fig. 14.4. Wild-type (A) and transgenic petunia flower (B). In the transgenic flower pMADS1 is overproduced in the first and second whorl, leading to a homeotic conversion at the base of the sepals (Se) to petaloid tissue (pet. Se). The resulting tube structure is similar to the tube that is formed by the fusion of the second whorl petals. In the second whorl extra petaloid structures (extra P) are formed on the otherwise normal petals (P).

protein in this whorl. Since AP3 functions only together with PISTILLATA (PI), presumably by forming a heterodimer (see section on 'DNA-binding properties of B type MADS-box proteins'), it could be that only the AP3/PI heterodimer is stable in a plant cell. Therefore it will be interesting to see whether a combination of ectopic expression of *AP3* and *PI* in the first whorl will be sufficient to induce a homeotic conversion of sepal to petal, accompanied by a stabilization of AP3 protein in this whorl.

A similar experiment was done in petunia; the *pMADS1* coding region was combined with the CaMV 35S promoter and introduced into the petunia lines W115 and V26 (Halfter *et al.*, 1994). In both lines some of the transformants showed a conversion of the base of the sepals to petaloid tissue and a delay in the fusion of the petals (see Fig. 14.4). The third and fourth whorl organs remained unchanged in these plants. The same transgene in tobacco had no phenotypic effect (van der Krol, unpublished). So again some, but not all, of the homeotic changes were predicted by the ABC model. Ectopic expression of the other petunia B type genes (*pMADS2* and *FBP1*) did not lead to any change in flower morphology that involved a gain of B function (Angenent *et al.*, 1993; Van der Krol *et al.*, 1993).

INDUCED SUPPRESSION OF FLORAL HOMEOTIC B GENES

The introduction of a transgene into plants can also lead to cosuppression. Introduction of a 35S-*pMADS1* gene into wild type petunia resulted in a reduced petal development in 7 out of 20 transformants (van der Krol *et al.*,

1993). In the most extreme case, the expression of the *pMADS1* gene was completely blocked resulting in a homeotic conversion of the second floral whorl organs into sepals. This transformant mimicked the phenotype of the petunia mutant *green petals* that had been obtained previously through mutagenesis. In the *green petals* mutant and the *pMADS1* cosuppression plants, the stamens in the third floral whorl still developed. This is in striking contrast with *Antirrhinum* and *Arabidopsis* where a B mutation affects both the second and third floral whorl (see Fig. 14.1). When the *FBP1* cDNA was reintroduced into wild type petunia under control of the CaMV 35S promoter, in 2 out of 20 plants a partial reduction in petal differentiation was observed. In the same flowers carpelloid structures appeared on top of the stamens (Angenent *et al.*, 1994). In this case, the effect of B gene suppression was as predicted by the ABC model: reduction of the B function in the second and third whorl led to a homeotic conversion of petal to sepal and stamen to carpel (see Fig. 14.1). These results suggest that the B function is, at least partially, under control of the *FBP1* gene. In turn, the *pMADS1* gene may control the expression of *FBP1* in the second whorl but not, or partially, in the third whorl.

ECTOPIC EXPRESSION OF A SINGLE HOMEOTIC C GENE CAN INDUCE C FUNCTION

Although the flowers that are obtained by ectopic expression of C type floral homeotic genes may not be of primary interest to ornamental breeders, some examples of these experiments will be discussed as illustration of the variation that may occur between different plant species. Fig. 14.5 shows the prediction, according to the ABC model, if ectopic expression of a single C type gene is sufficient to induce full C function. The ectopic expression will suppress A function in the first, and second whorl, resulting in C function in the first and

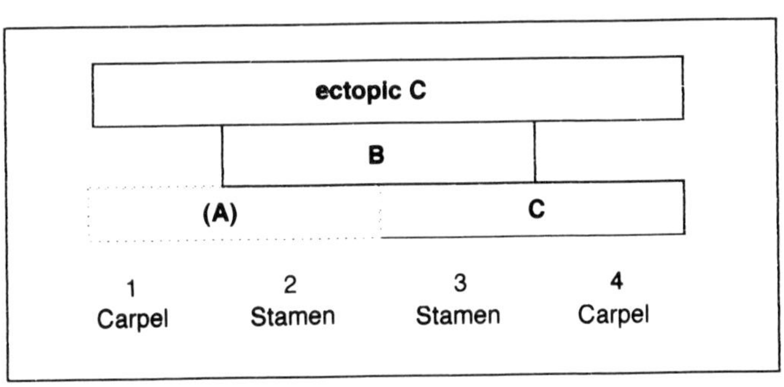

Fig. 14.5. The effect of ectopic expression of a C type floral homeotic gene on floral organ identity according the ABC model. If expression of the single C gene is sufficient to impose full C function, the A function in the first and second whorl will be suppressed, leaving only C function in the first whorl and the B+C function in the second whorl.

B+C function in the second whorl. Thus, a flower structure of carpels, stamens, stamens and carpels is to be expected.

Ectopic expression in petunia of *pMADS3*, a petunia gene homologous to *AGAMOUS*, leads to the formation of stigmatoid cells in the first floral whorl. In the second whorl, the limb of the petals is converted to antheroid structures (Tsuchimoto *et al.*, 1993). The phenotype of these plants thus mimics the petunia homeotic mutant *blind* (Vallade *et al.*, 1987). These homeotic changes are reminiscent of a partial gain of C function in the first and second whorl (cf. Fig. 14.5). Interestingly, the petunia *pMADS3* gene had a far more dramatic effect in transgenic tobacco. In these plants, the first whorl was completely transformed into a multilocular carpel which actually enclosed the inner-whorl organs. In the second whorl, as in transgenic petunia plants, the petal tube remained normal but the limb was transformed into antheroid structures (Tsuchimoto *et al.*, 1993).

A similar interspecific enhancement was seen by ectopic expression of a CaMV 35S–*AGAMOUS* gene construct in transgenic *Arabidopsis* plants (Bowman *et al.*, 1991) and ectopic expression in tobacco of *BAG1*, a *Brassica* gene homologous to *AGAMOUS* (Mandel *et al.*, 1992). These transgenic plants carried flowers with a carpelloid first and a stamenoid second whorl, as predicted by the ABC model (Fig. 14.5).

INDUCED SUPPRESSION OF HOMEOTIC C GENES

As mentioned above, introduction of a transgene can sometimes also lead to cosuppression instead of overexpression. In most cases, no cosuppression with C type transgenes was reported (35S–*AGAMOUS* in *Arabidopsis*, Mizukami and Ma, 1992; 35S–*NAG1* in tobacco, Kempin *et al.*, 1993; 35S–*TAG1* in tomato, Pnueli *et al.*, 1994; 35S–*pMADS3* in petunia and tobacco, Tsuchimoto *et al.*, 1993). However, expression of a 35S–*NAG1* antisense gene was able to suppress the C function in the third whorl resulting in an expansion of the functional A domain. This changes the B+C function in the third whorl into an A+B function and leads to a homeotic conversion of the stamens to petals (Kempin *et al.*, 1993). In the same plants, the antisense *NAG1* gene was not able to suppress C function in the fourth whorl, as no homeotic effect on carpel development was observed. This could be due to a differential activity of the CaMV 35S promoter that drives the expression of the antisense *NAG1* gene in the third and fourth whorls.

An antisense tomato C type gene, 35S-antisense *TAG1*, in tomato plants, resulted in the conversion of the third whorl stamens into petaloid structures and a pseudo, multilocular carpel in the fourth whorl. Apparently, in these plants the C function was suppressed in the third as well as in the fourth whorl, leading to an indeterminate floral meristem. Within the enclosed space of the central pseudo-carpel structure, new flowers developed (Pnueli *et al.*, 1994).

CHANGING THE SHAPE OF A PETAL

In the examples discussed above the form of a flower was changed by changing the identity of the floral organs. The appearance of a flower can also be changed by modifying the shape of a floral organ without changing the identity. The form of a floral organ is the result of the coordinated action of cell division, cell elongation and adhesion between cells. Altering this coordination changes the growth pattern and the organ will obtain a different shape. Potentially any gene that is involved in either cell division, cell elongation or cell adhesion that can be manipulated during floral organogenesis, can be used to change floral organ shape. For instance, if one could stimulate the cells of the second whorl to undergo extra divisions at an early stage of development, this could result in flowers with extra large petals.

Manipulating petal fusion

In petunia, the petals emerge as a congenitally fused organ; they are fused to form a tube. Overexpression of *pMADS1* delays this fusion, allowing for more lateral growth of the petals (Halfter *et al.*, 1994; see Fig. 14.4). The petal tissue that results from this extra lateral growth folds outwards. Eventually, however, the five petals fuse to form a tube. The outward folded petal tissue forms inverted petals that are fused with the tube (see Fig. 14.4). Do these results indicate that fusion is a signal to stop lateral growth of the petals? If they do, complete inhibition of petal fusion in petunia could lead to flowers with separate, but extra wide petals. Very little is known about the molecular processes that are involved in fusion in plants. Recently, the *Arabidopsis* gene *EMB30*, which is essential for normal cell division, cell expansion and cell adhesion in *Arabidopsis,* was cloned (Shevell *et al.*, 1994). A petal-specific manipulation of the expression of this gene could therefore potentially influence petal development. A better understanding of EMB30 function is needed to predict the effect of such manipulations.

Manipulating the dynamics of petal development

Introduction of 35S-*pMADS1* into the *green petals* mutant (which lacks the *pMADS1* gene) resulted in a restoration of petal development (van der Krol *et al.*, 1993). As well as plants that showed normal petals (full complementation of the mutation), some of the transgenic plants showed none or only a partial restoration of petal development. Some of the transgenic plants displayed aberrant, although fully differentiated, petals (see Fig. 14.6). On these plants the petal had a short tube and reduced corolla size. This difference in petal development may be attributed to a change in coordination between cell division and elongation during floral development. The altered promoter that is driving the *pMADS1* transgene expression, in combination with the specific integration site, is most likely responsible for this altered petal development.

Fig. 14.6. Manipulation of the petunia petal shape: the petunia mutant *green petal*, which lacks the wild-type *pMADS1* gene, was transformed with a 35S-*pMADS1* gene. Expression of the *pMADS1* transgene leads to a restoration in petal formation but the corolla tube is short compared to the wild-type petunia flower, while also the corolla limb remains small (compare to Fig. 14.4A).

These results indicate that different promoters, driving B type transgenes, can give rise to differently shaped petals.

Manipulating flower symmetry

The ABC model of flower development gives a good framework to study floral organ identity, but does not account for differences between actinomorphic flowers (radially symmetric) and zygomorphic flowers (mirror-image symmetric). Several mutations in *Anthirrhinum* result in flowers which are almost actinomorphic (Schwarz-Sommer *et al.*, 1990). The *cycloidea* mutant gives flowers with five (instead of four) stamens and all petals are like the lowest petal of the wild-type flower. Recently, the *CYCLOIDEA* gene has been identified and isolated, so experiments in the genetic manipulation of floral symmetry may soon be undertaken (Coen *et al.*, 1995).

MANIPULATION OF FLOWER SHAPE: PROCEED WITH CAUTION

Although genetic engineering may hold great promises for the future, one should keep in mind its limitations. At the moment genetic engineering is still very much restricted by the number of isolated genes linked to specific traits. However, the number of isolated and characterized plant genes is growing exponentially. Furthermore, for many of the horticulturally interesting plant species, an adequate transformation protocol is lacking.

Mutations in floral homeotic homologues from different plant species can have different effects (*Anthirrhinum Deficiens* vs. petunia *green petals*), while the effect of an identical homeotic transgene is not the same in different plant species (35S-*pMADS3* in petunia vs. in tobacco, 35S-*pMADS1* in petunia vs. in tobacco). These variations seriously hamper the reliable prediction of the effect of floral homeotic transgenes. Ectopic expression of floral homeotic genes has no dramatic effect on the vegetative parts of a plant. However, it is still uncertain whether all other aspects of plant growth and development are unaffected by these manipulations and whether in all cases the change in floral phenotype is stably inherited.

There are countless possibilities of crossing interspecific barriers with genetic engineering, but the effect of exchanging floral homeotic genes between non-related plant species is difficult to predict. The introduced homeotic protein may have altered stability, which in combination with the interactions with endogenous genes, may give rise to changes in regulatory pathways that influence development. Only when the molecular details of floral homeotic protein interactions are fully understood, may one be able to foresee all the effects of floral homeotic transgenes on floral shape. A more fundamental understanding of the functioning of homeotic proteins can only come from further experiments. Although these experiments are most easily done in model plants like *Arabidopsis*, tobacco and petunia, this will not give all the insights that are needed for an accurate prediction of the effect of homeotic transgenes in other plant species. This means that, even though genetic engineering of floral homeotic genes is a very targeted approach to change flower shape, for each plant species it still involves an element of trial and error. It therefore still remains to be seen whether the potential increase in commercial value of an ornamental species can outweigh the costs and efforts of genetic engineering that may be needed to obtain useful novel flower shapes.

REFERENCES

Angenent, G.C., Busscher, M., Franken, J., Mol, J.N.M. and van Tunen, A.J. (1992) Differential expression of two MADS-box gene in wild type and mutant petunia flowers. *Plant Cell* 4, 983–993.

Angenent, G.C., Franken, J., Busscher, M., Colombo, L. and van Tunen, A.J. (1993)

Petal and stamen formation in petunia is regulated by the homeotic gene fbp1. *Plant Journal* 4, 101–112.

Angenent, G.C., Franken J., Busscher, M., Weiss, D. and van Tunen, A.J. (1994) Cosuppression of the petunia homeotic gene fbp2 affects the identity of the generative meristem. *Plant Journal* 5, 33–44.

Benfey, P.N., Ren, L. and Chua, N.-H. (1990) Combinatorial and synergistic properties of CaMV 35S enhancer subdomains. *EMBO Journal* 9, 1685–1696.

Bowman, J.L., Smyth, D.R. and Meyerowitz, E.M. (1989) Genes directing flower development in *Arabidopsis*. *Plant Cell* 1, 37–52.

Bowman, J.L., Drews, G.N. and Meyerowitz, E.M. (1991) Expression of the *Arabidopsis* floral homeotic gene *AGAMOUS* is restricted to specific cell types late in flower development. *Plant Cell* 3, 749–758.

Bradley, E., Carpenter, R., Sommer, H., Hartley, N. and Coen, E.S. (1993) Complementary floral homeotic phenotypes result from opposite orientations of a transposon at the plena locus of *Anthirrhinum*. *Cell* 72, 85–95.

Carpenter, R. and Coen, E.S. (1990) Floral homeotic mutations produced by transposon-mutagenesis in *Anthirrhinum majus*. *Genes and Development* 4, 1483–1493.

Chung, Y.-Y., Kim, S.-R., Finkel, D., Yanofsky, M.F. and An, G. (1994) Early flowering and reduced apical dominance result from ectopic expression of a rice MADS box gene. *Plant Molecular Biology* 26, 657–665.

Coen, E.S. (1991) The role of homeotic genes in flower development and evolution. *Annual Review Plant Physiology Plant Molecular Biology* 42, 241–279.

Coen, E.S. and Meyerowitz, E.M. (1991) The war of the whorls: genetic interactions controlling flower development. *Nature* 353, 31–37.

Coen, E., Nugent, J.M. Luo, D., Cubas, P., Bradley, D., Chadwick, M., Copsey, L. and Carpenter, R. (1995) Evolution of floral symmetry. In: *Philosophical Transactions: Biological Sciences* 350(1331), 35–37.

Davies, R. and Schwarz-Sommer, Z. (1994) Control of floral organ identity by homeotic MADS-box transcription factors. In: Nover, L. (ed.), *Results and Problems in Cell Differentiation 20. Plant Promoters and Transcription Factors*. Springer-Verlag, Berlin.

Drews, G.N., Bowman, J.L. and Meyerowitz, E.M. (1991) Negative regulation of the *Arabidopsis* homeotic gene *AGAMOUS* by the *APETELA2* product. *Cell* 65, 991–1002.

Garcia-Maroto, F., Salamini, F. and Rohde, W. (1993) Molecular cloning and expression patterns of three alleles of the *Deficiens*-homologous *St-Deficiens* from *Solanum tuberosum*, *Plant Journal* 4, 771–780.

Halfter, U., Ali, N., Stockhaus, J., Ren, L. and Chua, N.-H. (1994) Ectopic expression of a single homeotic gene, the *Petunia* gene *green petal*, is sufficient to convert sepals to petaloid organs. *EMBO Journal* 13, 1443–1449.

Hansen, G., Estruch, J.J., Sommer, H. and Spena, A. (1993) NTGLO: tobacco homologue of the globosa floral homeotic gene of *Anthirrhinum majus*, cDNA sequence and expression pattern. *Molecular and General Genetics* 239, 310–312.

Huang, H., Mizukami, Y., Hu, Y. and Ma, H. (1993) Isolation and characterisation of the binding sequences for the product of the *Arabidopsis* floral homeotic gene *AGAMOUS*. *Nucleic Acids Research* 21, 4769–4776.

Jack, T., Brockman, L.L. and Meyerowitz, E.M. (1992) The homeotic gene *APETALA3* of *Arabidopsis thaliana* encodes a MADS box and is expressed in petals and stamens. *Cell* 68, 683–697.

Jack, T., Fox, G.L. and Meyerowitz, E.M. (1994) *Arabidopsis* homeotic gene *APETALA*-3 ectopic expression: transcriptional and post-transcriptional regulation determine floral organ identity. *Cell* 67, 703–716.

Jofuku, K.D., Boer, B.G.W. den, Montagu, M. van and Okamuro, J.K. (1994) Control of *Arabidopsis* flower and seed development by the homeotic gene *APETALA-2*. *Plant Cell* 6, 1211–1225.

Kempin, S.A., Mandel, M.A. and Yanofsky, M.F. (1993) Conversion of perianth into reproductive organs by ectopic expression of the tobacco floral homeotic gene NAG1. *Plant Physiology* 103, 1041–1046.

van der Krol, A.R. and Chua, N.-H. (1993) Flower development in petunia. *Plant Cell* 5, 1195–1203.

Kunst, L., Klenz, J.E., Martinez-Zapater, J. and Haughn, G.W. (1989) AP-2 gene determines the identity of perianth organs in flowers of *Arabidopsis thaliana*. *Plant Cell* 1, 1131–1135.

Kush, A., Brunelle, A., Shevelle, D. and Chua, N.-H. (1993) The cDNA sequence of two MADS-box proteins in petunia. *Plant Physiology* 102, 1051–1052.

Ma, H. (1994) The unfolding drama of flower development: recent results from genetic and molecular analyses. *Genes and Development* 8, 745–756.

Ma, H., Yanofsky, M.F. and Meyerowitz, E.M. (1991) *AGL1–AGL6*, an *Arabidopsis* gene family with similarity to floral homeotic and transcription factor genes. *Genes and Development* 5, 484-495.

Mandel, M.A., Bowman, J.L., Kempin, S.A., Ma, H., Meyerowitz, E.M. and Yanofsky, M.F. (1992) Manipulation of flower structure in transgenic tobacco. *Cell* 71, 133–143.

Mandel, T., Lutziger, I. and Kuhlemeier, C. (1994) A ubiquitously expressed MADS-box gene from *Nicotiana tacbacum*. *Plant Molecular Biology* 25, 319–321.

Meyerowitz, E.M., Smyth, D.R. and Bowman, J.L. (1989) Abnormal flowers and pattern formation in floral development. *Development* 106, 209–217.

Meyerowitz, E.M., Bowman, J.L., Brockman, L.L., Drews, G.N, Jack, T., Sieburth, L.E. and Weigel, D. (1991) A genetic and molecular model for flower development in *Arabidopsis thaliana*. *Development (Supplement)* 1, 157–167.

Mizukami, Y. and Ma, H. (1992) Ectopic expression of the floral homeotic gene *AGAMOUS* in transgenic *Arabidopsis* plants alters floral organ identity. *Cell* 71, 119–131.

Mueller, C.G. and Nordheim, A. (1991) A protein domain conserved between yeast MCM1 and human SRF directs ternary complex formation. *EMBO Journal* 10, 4219–4229.

Napoli, C., Lemieux, C. and Jorgensen, R. (1990) Introduction of a chimeric chalcone synthase gene into petunia results in reversible co-suppression of homologous genes in trans. *Plant Cell* 2, 279–289.

Norman, C., Runswick, M., Pollock, R. and Treisman, R. (1988) Isolation and properties of cDNA clones encoding SRF, a transcription factor that binds to the *c-fos* serum response element. *Cell* 55, 989–1003.

Okamuro, J.K., Boer, B.G.W. den and Jofuku, K.D. (1993) Regulation of *Arabidopsis* flower development. *Plant Cell* 5, 1183–1193.

Passmore, S., Maine, G.T., Elble, R., Christ, C. and Tye, B.K. (1988) A *Saccharomyces cerevisiae* protein involved in plasmid maintenance is necessary for mating of MATa cells. *Journal of Molecular Biology* 204, 593–606.

Pnueli, L., Abu-Abeid, M., Zamir, D., Nacken, W., Schwarz-Sommer, Z. and Lifschitz, E. (1991) The MADS box gene family in tomato: temporal expression during floral development, conserved secondary structures and homology with homeotic genes from *Antirrhinum* and *Arabidopsis*. *Plant Journal* 1, 255–266.

Pnueli, L., Hareven, D., Rounsleay, S.D. and Yanofsky, M.F. (1994) Isolation of the tomato *AGAMOUS* gene TAG1 and analysis of its homeotic role in transgenic plants. *Plant Cell* 6, 163–173.

Schmidt, R.J., Veit, B., Mandel, M.A., Mena, M., Hake, S. and Yanofsky, M.F. (1993) Identification and molecular characterization of *ZAG1*, the maize homolog of the *Arabidopsis* floral homeotic gene *AGAMOUS*. *Plant Cell* 5, 729–737.

Schwarz-Sommer, Z., Huijser, P., Nacken, W., Saedler, H. and Sommer, H. (1990) Genetic control of flower development: homeotic genes in *Antirrhinum majus*. *Science* 250, 931–936.

Schwarz-Sommer, Z., Hue, I., Huijser, P., Flor, P.J., Hansen, R., Tetens, F., Lönnig, W.-E., Saedler, H. and Sommer, H. (1992) Characterization of the *Antirrhinum* homeotic MADS-box gene *Deficiens*: evidence for DNA binding and autoregulation of its persistent expression throughout flower development. *EMBO Journal* 11, 251–263.

Shevell, D.E., Leu, W.-M., Gillmor, C.S., Xia, G., Feldmann, K.A. and Chua, N.-H. (1994) *EMB30* is essential for normal cell division, cell expansion, and cell adhesion in *Arabidopsis* and encodes a protein that has similarity to Sec7. *Cell* 77, 1051–1062.

Shiraishi, H., Okada, K. and Shimura, Y. (1993) Nucleotide sequences recognized by the *AGAMOUS* MADS domain of *Arabidopsis thaliana in vitro*. *Plant Journal* 4, 385–398.

Sommer, H., Carpenter, R., Harrison, B.J. and Saedler, H. (1985) The transposable element Tam3 of *Anthirrhinum majus* generates a novel type of sequence alteration upon excision. *Molecular and General Genetics* 199, 225–231.

Sommer, H., Beltrán, J.-P., Huiser, P., Pape, H., Lönnig, W.-E., Saedler, H. and Schwarz-Sommer, Z. (1990) *Deficiens*, a homeotic gene involved in the control of flower morphogenesis in *Antirrhinum majus:* the protein shows homology to transcription factors. *EMBO Journal* 9, 605–613.

Stickland, R.G. and Harrison, B.J. (1974) Precursors and genetic control of pigmentation. Induced biosynthesis of pelargonidin, cyanidin and delphinidin in *Antirrhinum majus*. *Heredity* 33, 108–114.

Takatsuji, H., Nakamura, N. and Katsumoto, Y. (1994) A new family of zinc finger proteins in petunia: structure, DNA sequence recognition, and floral organ-specific expression. *Plant Cell* 6, 947–958.

Tröbner, W., Ramirez, L., Motte, P., Hue, I., Huijser, P., Lönnig, W.-E., Saedler, H., Sommer, H. and Schwarz-Sommer, Z. (1992) *GLOBOSA*: a homeotic gene which interacts with *DEFICIENS* in the control of *Antirrhinum* floral organogenesis. *EMBO Journal* 11, 4693–4704.

Tsuchimoto, S., Krol, A.R. van der and Chua, N.-H. (1993) Ectopic expression of *pMADS3* in transgenic petunia phenocopies the petunia *blind* mutant. *Plant Cell* 5, 843–853.

Vallade, J., Maizonnier, D. and Cornu, A. (1987) La morphogenèse florale chez le pétunia. I Analyse d'un mutant à corolle staminée. *Canadian Journal of Botany* 65, 761–764.

van der Krol, A.R. and Chua, N.-H. (1993) Flower development in petunia. *Plant Cell* 5, 1195–1203.

van der Krol, A.R., Mol, J.N.M. and Stuitje, A.R. (1988) Antisense genes in plants: an overview. *Gene* 72, 45–50.

van der Krol, A.R., Mur, L.A., Beld, M., Mol, J.N.M. and Stuitje, A.R. (1990) Flavonoid genes in petunia: addition of a limited number of gene copies may lead to a suppression of gene expression. *Plant Cell* 2, 291–299.

van der Krol, A.R., Brunelle, A., Tsuchimoto, S. and Chua, N.-H. (1993) Functional analysis of petunia floral homeotic MADS-box gene pMADS1. *Genes and Development* 7, 1214–1228.

Yanofsky, M.F., Ma, H., Bowman, J.L., Drews, G.N., Feldmann, K.A. and Meyerowitz, E.M. (1990) The protein encoded by the *Arabidopsis* homeotic gene *AGAMOUS* resembles transcription factors. *Nature* 346, 35–39.

Flower Colour 15

Kevin M. Davies and Kathy E. Schwinn
New Zealand Institute for Crop & Food Research Ltd, Levin Research Centre, Private Bag 4005, Levin, New Zealand

INTRODUCTION

Flower colour is a key element in consumer selection between ornamental varieties available in the market-place. For some ornamental species, however, only a narrow colour spectrum is available, while in others specific colours, for instance blue, are lacking. Thus, the introduction of new colours and forms through genetic engineering is likely to have a large influence on the industry, and this research area has attracted the attention of several biotechnology companies. The extensive information available on the genetics and biochemistry of pigment biosynthesis gives a strong base to such research, increasing the probability of success.

The major pigments responsible for flower colour are the flavonoids, the carotenoids and the taxonomically restricted betalains. Although the chlorophylls may influence colour and are responsible for a few cases of green flowers (e.g. *Helleborus*), they are of lesser importance in this respect and will not be considered here.

Flavonoids often account for pink, red, orange, scarlet, purple, blue and blue–black flower colours, and also some yellow ones. In addition, they provide 'depth' to most white or cream flowers. Carotenoids alone provide yellow, orange and red colours. In many important ornamental species both carotenoids and flavonoids contribute to the range of available colours, in combination producing orange, yellow, scarlet and brown/black flowers. The betalains produce yellow, orange, red and purple colours in the floral tissues of only a few plant species, being restricted to the *Caryophyllales*. Many prokaryotes and lower eukaryotes also synthesize pigments, and the genes involved in these reactions could be used for the genetic modification of flower colour. The genes involved in the bacterial synthesis of carotenoids (e.g. Armstrong *et al.*, 1989; Misawa *et al.*, 1990) and indigo (e.g. Mermod *et al.*, 1986) have been particularly well defined.

(eds R.L. Geneve, J.E. Preece and S.A. Merkle)

In this chapter we will consider the role of the different plant pigments in producing flower colour, what is known about the biochemistry and molecular biology of their biosynthesis, and the prospects for genetic modification of the biosynthetic pathway. Of the pigments responsible for flower colour in ornamental crops, the flavonoids are the most common, commercially the most important, and from a biotechnology point of view the most amenable. Thus, they are the main focus of this chapter.

THE FLAVONOIDS

THE BIOSYNTHESIS OF FLAVONOIDS

The flavonoids are phenylpropanoid compounds that are generally water soluble and stored in the vacuole. Their biosynthesis, as part of the larger phenylpropanoid pathway, is one of the best characterized aspects of plant secondary metabolism. Many years of genetic and biochemical work on the enzymatic steps involved has been complemented more recently by the work of molecular biologists. Genes encoding most of the anthocyanin biosynthetic enzymes have now been cloned, and progress on the factors regulating the pathway is well advanced. Here we give only an outline of flavonoid biosynthesis. Several excellent reviews of the biochemistry and genetics of the pathway have been published recently (van Tunen and Mol, 1991; Forkmann, 1993; Heller and Forkmann, 1993; Martin and Gerats, 1993a,b), and the reader is referred to these for further information.

The basic flavonoid structure is a 15-carbon (C_{15}) nucleus composed of two aromatic rings (called the A- and B-rings) joined by a three-carbon unit (which usually forms a third ring called the C-ring; Fig. 15.1). The various

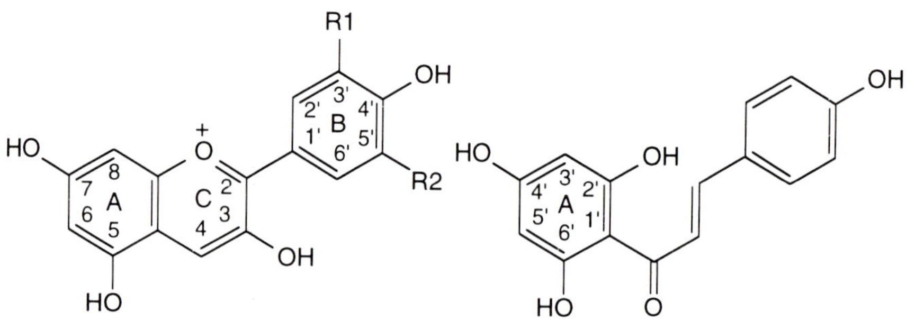

Fig.15.1. Structures of the main anthocyanidins (left) and a chalcone (right). The numbering of some of the carbons for the anthocyanidins (and the majority of flavonoids) and chalcones, and the lettering of the carbon rings are shown. Substitutions of R1 and R2 determine the various anthocyanidins as: R1 and R2 = H, pelargonidin; R1 = OH and R2 = H, cyanidin; R1 and R2 = OH, delphinidin; R1 = OCH_3 and R2 = H, peonidin; R1 = OCH_3 and R2 = OH, petunidin; R1 and R2 = OCH_3, malvidin.

classes of flavonoids are determined by the degree of oxidation of the C-ring, while the individual aglycone compounds within each class are determined by the extent of hydroxylation, or other substitution, of the A- and B-rings. Figure 15.2 presents a pictorial representation of a section of the phenylpropanoid pathway resulting in the formation of various flavonoid types, including the aurones, flavones, flavonols, and the simplest anthocyanins.

The first dedicated step to flavonoid biosynthesis is the formation of the C_{15} structure by the condensation of one hydroxycinnamic acid-CoA ester unit, usually *p*-coumaroyl-CoA, with three units of malonyl-CoA, a reaction catalysed by the enzyme chalcone synthase (CHS). The chalcones, yellow themselves, may also be converted to the brighter yellow aurones. The reactions involved in aurone biosynthesis have not been characterized, although enzyme preparations able to convert chalcones into the respective aurones have been studied (Wong, 1976). More commonly the chalcones are converted to the flavanones. This isomerization reaction may occur spontaneously, but is generally catalysed by the enzyme chalcone isomerase (CHI).

The flavanones are converted to the dihydroflavonols by the enzyme flavanone 3-hydroxylase (F3H). The flavanones and dihydroflavonols are also the substrates for flavone and flavonol formation, respectively, through the action of the flavone synthase (FNS) and the flavonol synthase (FLS). While occasionally these compounds are directly responsible for the perceived flower colour, they more commonly exert their influence through the formation of molecular complexes with the anthocyanins, a process called copigmentation (Brouillard and Dangles, 1993).

Reduction of the dihydroflavonols at the 4 position, catalysed by the dihydroflavonol 4-reductase (DFR), yields the corresponding leucoanthocyanidins. The leucoanthocyanidins are the direct precursors of anthocyanidins and generally do not accumulate in plant cells. The enzyme reactions involved in the conversion of the leucoanthocyanidins to the anthocyanidins have not been clearly defined. Genetic and biochemical evidence suggests that a hydroxylase and a dehydratase may be involved (Heller and Forkmann, 1988). The *A2* and *Candica* genes from *Zea mays* (maize) and *Antirrhinum majus* (snapdragon), respectively, may encode the hydroxylase, recently termed the anthocyanidin synthase (ANS). The product of another gene from *Antirrhinum, Incolorata I,* may also act at this stage (Martin and Gerats, 1993a).

Anthocyanidins are unstable under normal physiological conditions and do not accumulate in plant cells. They are normally stabilized to form the corresponding anthocyanins by the addition of a sugar residue(s), usually at the 3 and/or 5 position. Anthocyanidin 3-O-glycosyltransferases have been characterized from several plants (Heller and Forkmann, 1993).

Marrs *et al.* (1995) have recently proposed a mechanism by which anthocyanins may be targeted to the vacuole. A glutathione-S-transferase, encoded by the *Bronze2* gene in *Zea* (McLaughlin and Walbot, 1987), transfers a glutathione residue to the anthocyanin. The modified anthocyanin may then be transported into the vacuole by a vacuolar glutathione pump.

From the basic flavonoid structures an enormous range of derivatives can

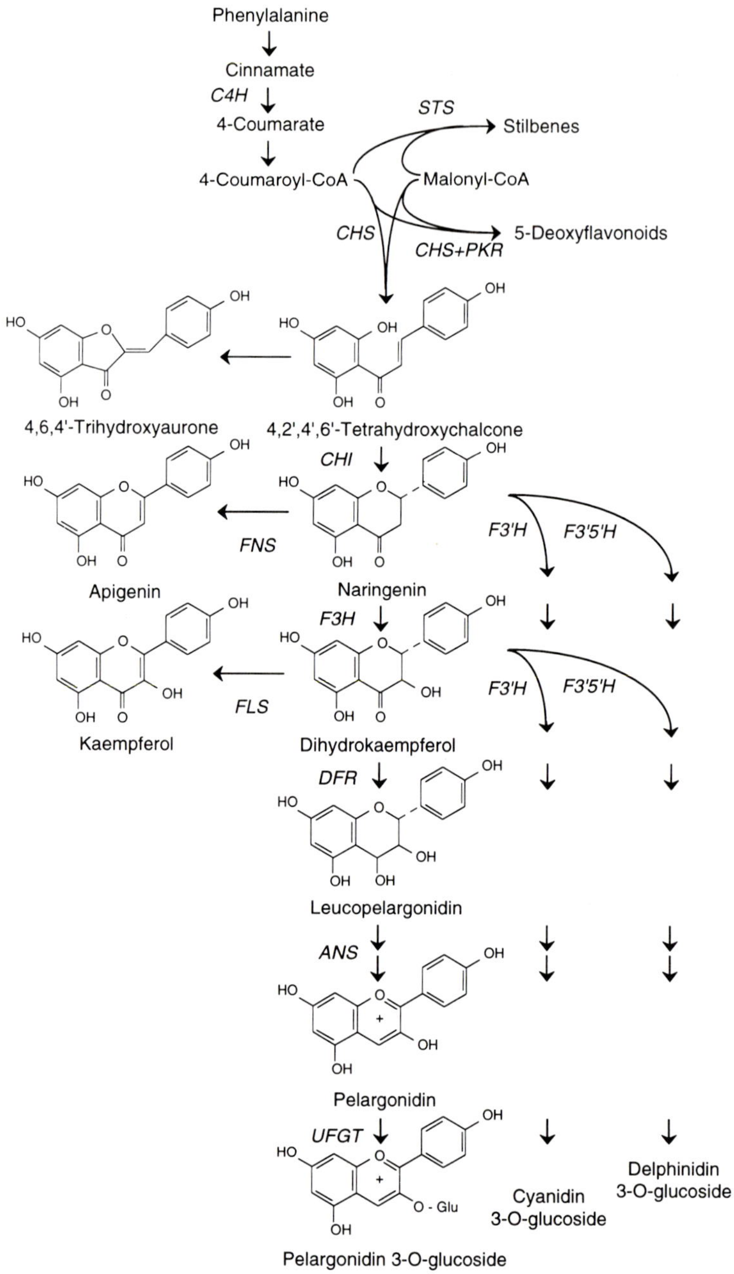

Fig. 15.2. Schematic representation of a section of the phenylpropanoid biosynthetic pathway leading to the production of various flavonoid types, including the anthocyanins, anthocyanidins, flavonols, flavones and aurones. Enzyme abbreviations are as given in the text except for STS = stilbene synthase and PKR = polyketide reductase.

be formed through glycosylation, acylation, methylation and even sulphation. For example, methylation of the B-ring hydroxyls of the anthocyanins yields the common derivatives, peonidin, petunidin and malvidin glycosides (Fig. 15.1) . Enzymes involved in some secondary modification reactions have been characterized from several species (Heller and Forkmann, 1993), and many related genetic loci are known (Forkmann, 1993). Besides the common flavonoids, a wide range of less common flavonoid derivatives are known that vary from the typical substitution pattern of the A- and C-rings.

The isolation of genomic and/or cDNA clones has been reported for most of the major biosynthetic enzymes, and some modifying enzymes (Forkmann, 1993). Table 15.1 presents a list of those reported from ornamental species. Generally, the clones have first been identified from one of the good genetic systems, *Petunia, Antirrhinum* or *Zea,* and then these subsequently used to isolate homologous clones from other species.

THE FLAVONOIDS AS FLOWER COLOUR PIGMENTS

Of the common flavonoids only the chalcones and the anthocyanins have significant colour, and of these the anthocyanins are by far the most common and important pigments. In petal tissue, they are usually located specifically in the epidermal cells (Kay *et al.*, 1981). Only three basic anthocyanin groups are commonly found, derivatives of pelargonidin, cyanidin or delphinidin. In general pelargonidin-derived pigments produce orange, pink or red colours, cyanidin-derived pigments produce red or mauve colours, and delphinidin-derived pigments produce purple, blue, or blue–black colours. Thus, the three basic anthocyanin types can produce a wide range of flower colours. Several factors are important in determining the final colour, the most important ones being the extent of oxygenation and the molecular environment of the anthocyanin and the pH of the vacuole. Surprisingly, when placed *in vitro* under conditions similar to those commonly found in the vacuoles of floral cells, the anthocyanins alone do not exist in a stable coloured form. To achieve stability hydration of the anthocyanin must be prevented. This is usually achieved by copigmentation (Brouillard and Dangles, 1993). The petal colour is also dependent on the amount of anthocyanin present. Thus, for instance in *Viola* and *Tulipa,* very high anthocyanin levels result in near-black flower colours (Harborne, 1976).

We will now consider some of these aspects of flavonoid biosynthesis and flower colour in more detail.

B-RING HYDROXYLATION

The three major anthocyanidin types, pelargonidin, cyanidin and delphinidin differ only in the extent of B-ring hydroxylation, being 4′, 3′4′ and 3′4′5′ hydroxylated, respectively (Fig. 15.1). The 4′-hydroxyl group is incorporated

Table 15.1. Flavonoid cDNA and genomic clones from ornamental species.

Plant species	*Gene function	Reference
Antirrhinum majus	CHS	Wienand *et al.* (1982), Sommer and Saedler (1986)
	CHI	Martin *et al.* (1991)
	F3H	Martin *et al.* (1991)
	DFR	Martin *et al.* (1985), Beld *et al.* (1989)
	ANS	Martin *et al.* (1991)
	UFGT	Martin *et al.* (1991)
Callistephus chinensis	CHS	Forkmann (1991)
	CHI	Forkmann (1993)
	F3H	Dedio and Forkmann [in Forkmann (1993)]
	DFR	Min and Forkmann [in Forkmann (1993)]
	ANS	Forkmann (1993)
Dendranthema × grandiflorum	CHS	Courtney-Gutterson *et al.* (1994)
	DFR	Barri-Rewell[a]
	FLS	Barri-Rewell[a]
	ANS	Barri-Rewell[a]
Dianthus caryophyllus	CHS	Forkmann (1993)
	CHI	Forkmann and Dedio [in Forkmann (1993)]
	F3H	Britsch *et al.* (1993), Dedio *et al.* (1995)
	DFR	Min and Forkmann [in Forkmann (1993)]
	FLS	Keam[b]
	ANS	Keam[b]
Eustoma grandiflorum	CHS	Davies *et al.* (1993)
	CHI	Davies *et al.* (1993)
	DFR	Davies *et al.* (1993)
	FLS	Nielsen[c]
	F3′5′H	Nielsen[c]
Gerbera hybrida	CHS	Elomaa *et al.* (1993)
	DFR	Helariutta *et al.* (1993)
Matthiola incana	CHS	Epping *et al.* (1990)
	F3H	Dedio and Forkmann [in Forkmann (1993)]
	DFR	Min and Forkmann [in Forkmann(1993)]
	ANS	Forkmann (1993)
Petunia hybrida	CHS	Reif *et al.* (1985), Koes *et al.* (1986, 1987, 1989) Niesbach-Klösgen *et al.* (1987)
	CHI	van Tunen *et al.* (1988, 1989)
	F3H	Britsch *et al.* (1992)
	DFR	Beld *et al.* (1989)
	FLS	Holton *et al.* (1993c)
	F3′ 5′H	Holton *et al.* (1993a), Simada *et al.* (1993), Toguri *et al.* (1993)
	F3′H	Holton *et al.* (1993b)
	ANS	Weiss *et al.* (1993)
	3RT	Brugliera *et al.* (1994), Kroon *et al.* (1994)
	A3′MT	Quattrocchio *et al.* (1993), Huits *et al.* (1994)
	3′5′MT	Quattrocchio *et al.* (1993), Huits *et al. (1994)*
Ranunculus acer	CHS	Niesbach-Klösgen *et al.* (1987)
Rosa hybrida	CHS	Tanaka[d], Courtney-Gutterson (1994)
	DFR	Tanaka *et al.* (1995)
	FLS	Tanaka[d]
	ANS	Tanaka[d]

*Abbreviations (except A3′MT and A3′5′MT) are defined in the text. A3′MT = anthocyanin 3′-O-methyltransferase; A3′5′MT = anthocyanin 3′5′-O-methyltransferase.
[a]G. Barri-Rewell, Florigene Pty Ltd, 1995, personal communication.
[b]L. Keam, Florigene Pty Ltd, 1995, personal communication.
[c]K. Nielsen, New Zealand Institute for Crop and Food Research Limited, 1994, personal communication.
[d]Y. Tanaka, Suntory Ltd, 1995, personal communication.

during formation of the flavonoid precursor 4-coumarate by the cinnamate 4-hydroxylase (C4H), and is common to most flavonoids. A free hydroxyl at the 4′ position of the B-ring or the 5 or 7 position of the A-ring is thought necessary for stable formation of the coloured forms of the anthocyanins (Brouillard, 1988). The addition of the 3′ and 5′ hydroxyl groups is usually catalysed by the flavonoid 3′-hydroxylase (F3′H) and flavonoid 3′,5′-hydroxylase (F3′5′H). In general, the presence or absence of the different anthocyanidin types can be accounted for by the activity of these two enzymes. Of the anthocyanin precursors, both the flavanones and dihydroflavonols have been shown to be substrates for the F3′H and F3′5′H (Heller and Forkmann, 1993). Recently it has been found that the F3′H of *Petunia* and F3′5′H of *Eustoma grandiflorum* (lisianthus) can also accept some leucoanthocyanidins as substrates (Schwinn, 1994).

Other B-ring hydroxylation activities may also be present in flowers. In *Antirrhinum* the hydroxylation of the B-ring of the aurones appears independent of that of the anthocyanins. Lines recessive for the *Eosinea* gene, which controls F3′H activity, still produce the 3′4′-hydroxylated aurone aureusin (Harborne, 1963). In addition, the 3′4′5′-hydroxylated aurone, bracteatin 6-glucoside, is also found in *Antirrhinum* (Harborne, 1963), although this species lacks the usual F3′5′H activity. Hydroxylation of 4-coumaroyl-CoA to caffeoyl-CoA has been demonstrated for a few species, and the gene *P* has been shown to control this activity in *Silene dioica* (red campion; Kamsteeg *et al.*, 1980). If caffeoyl-CoA is subsequently used as a substrate by CHS, cyanidin derivatives may be formed. While it has been demonstrated *in vitro* that CHS may indeed use the 3′4′-hydroxylated cinnamic acid-CoA ester, it is thought that for most plants this route is of minor or no importance *in vivo* (Heller and Forkmann, 1988). Notable exceptions include *Verbena hybrida* and *Silene* (Kamsteeg *et al.*, 1980).

As the hydroxylation of the B-ring determines which of the anthocyanidin types is formed, the availability of the genes encoding the F3′H and F3′5′H is very important for the genetic modification of flower colour. The F3′H, F3′5′H and C4H are cytochrome P450 mono-oxygenases (P450's), a group of membrane-bound enzymes requiring the NADPH:cytochrome P450 reductase as a co-factor. Although they have been extensively studied in animals and lower eukaryotes, data on P450 enzymes in plants are limited. Plant P450 enzymes have only been purified in a few cases, in particular the C4H (Mizutani *et al.*, 1993; Teutsch *et al.*, 1993). Thus, although F3′H and F3′5′H activities have been demonstrated for several species (Heller and Forkmann, 1993), the corresponding proteins have not been isolated. Holton *et al.* (1993a), however, circumvented the need for protein purification by using a polymerase chain reaction (PCR) based strategy to obtain cDNA clones for the F3′5′H from *Petunia*. Degenerate oligonucleotides were designed to regions conserved in the haem-binding domain of P450 enzymes, and used to PCR-amplify sequences from *Petunia* flower petal cDNA. The putative P450 sequences were then used to isolate clones from a *Petunia* petal cDNA library. These cDNAs were RFLP mapped to the *Hf1* and *Hf2* loci, thought to encode the F3′5′H

(Wiering and de Vlaming, 1984), and cDNA clones isolated that corresponded to both loci. The identities of the *Hf1* and *Hf2* cDNA clones were confirmed by production of the encoded proteins in yeast and demonstration of F3′5′H activity, and transgenic complementation of a *hf1hf1hf2hf2 Petunia* line. The isolation of cDNAs for the F3′5′H from *Petunia* has also been reported by Simada *et al.* (1993) and Toguri *et al.* (1993).

A cDNA clone encoding a protein with F3′H activity has also been isolated from *Petunia* (Holton *et al.*, 1993b). By screening a *Petunia* petal cDNA library with a mixed probe of P450 cDNA clones (isolated during the F3′5′H screen), further P450 cDNA clones were isolated. One of these has been demonstrated, from yeast expression studies, to encode a protein with F3′H activity.

FORMATION OF LEUCOANTHOCYANIDINS

The leucoanthocyanidins are formed from the respective dihydroflavonols by the enzyme DFR. For most plants investigated DFR will reduce the three major dihydroflavonols; however, in some species strong substrate preference is shown. For example, DFR preparations from plants of the solanaceous genera *Petunia, Lycopersicon* and *Nicotiana* will not use dihydrokaempferol (Heller and Forkmann, 1993), thus preventing pelargonidin production. In the first case of successful engineering of flower colour, Meyer *et al.* (1987) overcame this biosynthetic block in *Petunia* by introducing the *Zea A1* cDNA encoding DFR. The *A1* cDNA was placed under the control of the cauliflower mosaic virus (CaMV) 35S promoter and introduced into a *Petunia* line that accumulated dihydrokaempferol. The flowers of the resultant transgenics produced pelargonidin-derived pigments. Subsequent crosses of the transgenic lines to the 'Red Titan' cultivar resulted in a range of plant lines in the F_2 generation that had flowers containing either pelargonidin-derived pigments, cyanidin-derived pigments, or a mixture of both (Meyer, 1991; Griesbach, 1993). Some lines had orange flower colours novel to *Petunia* (see Meyer (1991) for colour photographs of the F_2). Phenotypic instability was observed in some transgenic lines during glasshouse, and particularly, field trials (Linn *et al.*, 1990; Meyer *et al.*, 1992). These lines have since proved valuable for studies on factors influencing the stability of transgene expression (Meyer *et al.*, 1993; Meyer and Heidmann, 1994).

In non-solanaceous species the substrate specificity of the DFR could also affect the relative amounts of the different anthocyanidin types formed, particularly as in some cases the DFR may compete with the flavonol synthase (FLS) for substrate.

FLAVONE AND FLAVONOL SYNTHESIS

Although the commonly found flavone and flavonol glycosides very rarely account for strong flower colours, they play an important role in copigmentation and provide 'depth' to acyanic varieties. The enzyme responsible for flavonol synthesis, the FLS, has been characterized in several species and determined to be a 2-oxoglutarate-dependent dioxygenase (Heller and Forkmann, 1993). In *Petunia,* it is encoded by the *Fl* locus. Holton *et al.* (1993c) used a PCR approach to isolate a cDNA clone corresponding to *Fl.* Dioxygenase-related gene fragments were amplified from cDNA prepared from flowers of either *FlFl* or *flfl* plants and used to differentially screen a *Petunia* petal cDNA library. cDNA clones that had strong hybridization to the *Fl* PCR products but not the *fl* PCR products were isolated and analysed further. From this work a full-length cDNA clone was identified, and confirmed to encode FLS activity by expression in yeast.

Antisense expression of the FLS cDNA in *Petunia* lead to a marked reduction in flavonols and a shift in flower colour towards red, thought to be a result of altering the copigmentation (Holton *et al.*, 1993c). Introduction of the same antisense FLS construct into *Nicotiana* increased the level of anthocyanins in the flowers, indicating that in this species there is a competition for substrate supply between anthocyanin and flavonol production (Holton *et al.*, 1993c). Thus, reduction of the synthesis of side-branches of the flavonoid biosynthetic pathway can affect copigmentation and may also provide a means by which anthocyanin levels in weakly pigmented flowers can be increased.

Genes encoding the FNS have yet to be cloned. In *Antirrhinum* the FNS is thought to be encoded by the *daphne* locus. Recessive loci for *daphne* are known (Luo *et al.*, 1991), which may aid in the cloning of a FNS cDNA by differential approaches.

GLYCOSYLATION

The glycosylation of the anthocyanidins is important in stabilizing the molecule and increasing its water solubility. Following the identification of the gene in *Zea* (Fedoroff *et al.*, 1984), cDNA clones for the UDP glucose: flavonoid 3-O-glucosyltransferase (UFGT) have been isolated from several species (Forkmann, 1993). More recently cDNA and genomic clones encoding the UDP rhamnose:anthocyanidin 3-glucoside rhamnosyltransferase (3RT) from *Petunia* have been identified (Brugliera *et al.*, 1994; Kroon *et al.*, 1994).

These genes might be used to modify anthocyanin and copigment biosynthesis in two ways. First, inhibition of the gene activity might enable novel anthocyanin types to be produced. Brugliera *et al.* (1994) inhibited 3RT gene activity in *Petunia* by introducing an antisense RNA construct. Flowers of the transgenic plants had reduced levels of the malvidin and increased levels of delphinidin and petunidin pigments. Presumably, the anthocyanin 3′5′-O-methyltransferase, which acts after the addition of the sugar and acyl residues, is unable to act on the simpler anthocyanin structure. A second approach is to use the cDNA clones to increase, or introduce *de novo,* the corresponding glycosylation activities in other species. This has been demonstrated for *Eustoma* (Schwinn *et al.*, 1993). Flowers of *Eustoma* predominantly accumulate anthocyanins and flavonols that are 3-galactosylated, although a small amount of 3-glucosylated flavonols also occur. Further, 50% of the galactosylated flavonols are acylated at the 4-hydroxy position of the galactose. Novel anthocyanins glucosylated at the C-3 position were produced in petals of transgenic plants expressing the *Antirrhinum* UFGT cDNA under the control of the CaMV 35S promoter. In addition, a decrease in the amount of acylated flavonols occurred, which correlated with an increase in the amount of glucosylated flavonols. Although copigmentation occurs in *Eustoma* between the anthocyanins and the flavonols (Markham and Ofman, 1993), no change in the flower colour resulted. In general, it is the position the sugar residue occupies on the flavonoid, rather than the nature of the sugar, that has a more direct effect on the perceived flower colour. However, the nature of the sugar can indirectly impact on the colour by affecting subsequent modifications such as methylation, acylation and further glycosylation.

REDUCTION OF ANTHOCYANIN PRODUCTION

By reducing the activity of the flavonoid biosynthetic enzymes it is possible to reduce or even abolish anthocyanin production. This approach may be used to produce either white-flowered varieties in species that lack them, or novel shades of flower colour in a particular cultivar. In addition, experiments carried out to date show that lines with novel patterning of floral pigmentation can also be generated.

Two strategies have been successfully demonstrated to inhibit expression of the anthocyanin biosynthetic genes: production of antisense RNA (reviewed in van Blokland *et al.*, 1993; Elomaa and Holton, 1994), or introduction of additional copies of the target gene, termed sense- or co-suppression (see Jorgensen, 1991, 1994, 1995; van Blokland *et al.*, 1993; Elomaa and Holton, 1994). The first successful transgenic antisense experiment to be reported in plants was for CHS in *Petunia* and *Nicotiana* (van der Krol *et al.*, 1988). Introduction of an antisense gene construct reduced CHS mRNA levels in *Petunia* floral tissue to approximately 1% of normal, resulting in an inhibition of pigmentation. Antisense constructs have subsequently been used to down-regulate CHS and DFR in *Gerbera* (Elomaa *et al.*, 1993) and CHS in *Eustoma* (Deroles *et al.*, 1995; Fig. 15.3a-b). Co-suppression was first demonstrated for the inhibition of either CHS or DFR expression in *Petunia* (Napoli *et al.*, 1990; van der Krol *et al.*, 1990). In the first application of the technology for producing novel commercial plant lines, both sense and antisense gene constructs were used to inhibit the activity of CHS in *Dendranthema* × *grandiflorum* (florist's chrysanthemum), generating novel white-flowered lines from the usually pink-flowered 'Moneymaker' cultivar (Courtney-Gutterson, 1994; Courtney-Gutterson *et al.*, 1994). Co-suppression from a sense construct of CHS has also been used to obtain a range of lines with reduced flower colour intensity from a deep-red coloured *Rosa hybrida* cv. 'Royalty' (Courtney-Gutterson, 1994).

Unexpectedly, from both sense and antisense experiments, plant lines

Fig. 15.3. (a) and (b) Flowers from a control plant (a) or a CHS-antisense transgenic plant (b) of *Eustoma*. (c–f) Foliage and a flower from a control plant (c and e) and from a *Lc*-sense transgenic plant (d and f) of *Petunia*. (g) Florets from a *Dendranthema* flower fed a control (left) or dihydromyricetin-containing (right) solution.
See also colour frontispiece.

with patterning of the floral pigmentation have arisen. Both ordered and erratic pigmentation patterns have been obtained for CHS and DFR experiments in *Petunia* (van der Krol *et al.*, 1988, 1990; Napoli *et al.*, 1990) and CHS antisense in *Eustoma* (Deroles *et al.*, 1995, Fig. 15.3). Several distinct classes of pigmentation phenotype have been defined for CHS cosuppression experiments in *Petunia,* and a relationship established between the occurrence of a particular type and the structure of the transgene locus or loci in that plant (Jorgensen *et al.*, 1994). The pigmentation phenotype may also vary among flowers on a single plant. This somatic instability is thought to be dependent on the influence of developmental and morphological parameters, and has been studied extensively by Jorgensen (1994, 1995). These morphological influences may act both during formation of the floral cells in the meristem and also in mature structures, such as the corolla. No floral patterning was observed from antisense CHS experiments with *Dendranthema* (Courtney-Gutterson *et al.*, 1994), *Rosa* (Courtney-Gutterson, 1994) or *Gerbera* (Elomaa *et al.*, 1993), species that traditionally lack patterned varieties.

INCREASING ANTHOCYANIN LEVELS

The production of anthocyanins in a particular tissue is determined by the activity of the biosynthetic genes, which in turn is controlled by the expression of specific regulatory genes. Progress on characterization of such genes has been covered in some excellent recent reviews (Dooner *et al.*, 1991; Martin and Gerats, 1993a,b). In this section, we will consider only the potential use of the isolated regulatory genes in the modification of plant colour.

The cloning of two types of genes regulating flavonoid biosynthesis has been reported. The *C1/Pl* gene family from *Zea* encodes proteins with homology to *myb*-type transcription factors. The *R/Lc* gene family from *Zea* and the *delila* gene from *Antirrhinum* encode proteins with homology to the *myc*-type transcription factors (see Martin and Gerats, 1993a). In *Zea,* the simultaneous expression of one member from each of the gene families is both necessary and sufficient to induce anthocyanin biosynthesis. The *delila* product has high homology over the entire length of the protein to that of *Lc,* and seems to function analogously (Goodrich *et al.*, 1992). However, different groupings of the anthocyanin biosynthetic genes are regulated by *delila* and *Lc* in their respective species, perhaps reflecting the presence or absence of other *trans*-acting factors and/or differences in the promoter regions of the target genes. It is expected that factors equivalent to *C1* are also present in other species. Recently, *myb*-type factors from a dicotyledonous plant *(Antirrhinum;* Sablowski *et al.*, 1994) and a gymnosperm *(Picea mariana;* Xue *et al.*, 1994) have been demonstrated to activate transcription of phenylpropanoid genes.

As the activity of the target anthocyanin biosynthetic genes appears primarily determined by the expression patterns of the regulatory genes, altering these patterns may allow the temporal and spatial modification of anthocyanin biosynthesis. This prospect has been tested by placing the *C1* and *Lc* cDNAs from *Zea* under the control of the CaMV 35S promoter and introducing the constructs into *Nicotiana* and *Arabidopsis* (Lloyd *et al.*, 1992). The expression of *Lc* alone increased anthocyanin production in both species, particularly in the flowers of *Nicotiana,* but did not alter where in the plant it was made. Alone, *C1* had no effect. *Arabidopsis* plants containing both *Lc* and *C1* had both increased anthocyanin levels and novel tissue distribution patterns, with anthocyanins appearing in the petal, root and stamens. We have obtained similar results in our laboratory by introducing a construct containing the *Lc* cDNA driven by the CaMV 35S promoter into the white-flowered Mitchell line of *Petunia.* The introduction of *Lc* both greatly increased anthocyanin biosynthesis in the vegetative tissues and partially overcame the genetic blocks affecting anthocyanin production in the flower (Bradley *et al.*, 1995; Fig. 15.3c–f).

Constitutive expression of the *delila* cDNA under the CaMV 35S promoter produced altered pigmentation in both *Lycopersicon* and *Nicotiana* (Mooney *et al.*, 1995). In *Lycopersicon* pigmentation in the vegetative tissues was increased, but in *Nicotiana* an increase in anthocyanins was seen only in the flowers. In *Arabidopsis* no consistent or strong phenotypic effect of *delila* was observed. Robbins and Harbord (1994) introduced a similar *delila* construct into *Petunia.* Although pigmentation of the vegetative tissues increased, anthocyanin production in the corolla of the cyanic line V26 was reduced. Further, the acyanic phenotype of the Mitchell line, and other *Petunia anthocyanin* mutant lines, was unaffected by the *delila* transgene. The underlying causes for these observations have not been established.

Thus, evidently the availability of defined regulatory genes for the flavonoid biosynthetic pathway may provide tools for modulating both the amount and distribution of anthocyanins in floral and vegetative tissues.

VACUOLAR *pH*

The pH of the vacuole has a major effect on the secondary structure of the anthocyanins and thus on the resultant flower colour. A decrease in pH causes a shift towards the red cation form of the anthocyanin, while an increase in the pH causes a shift towards the bluer quinonoidal types (Brouillard and Dangles, 1993). The pH of vacuolar extracts from petal cells is most commonly mildly acidic. However, a range from pH 2.5 to pH 7.5 has been reported from a study of petal tissue of 20 species (Stewart *et al.*, 1975). In addition, the pH may also vary between cultivars within a species and even between developmental stages of an individual flower. Asen *et al.* (1977) found that the pH of corolla extract of the morning glory *Ipomea tricolor* cv. 'Heavenly Blue' changed from approximately pH 6.5 to pH 7.5 during flower opening, causing a shift in colour from reddish buds to light-blue open flowers. Yoshida *et al.* (1995) have been able to confirm and extend this observation using a proton-sensitive microelectrode to measure the pH of living petal cells.

Six genes are known for *Petunia* that affect the vacuolar pH of petal tissue *(Ph1* to *Ph6,* de Vlaming *et al.*, 1983; Chuck *et al.*, 1993). If one of the six genes is present as the recessive allele then the pH of the petal extract is increased by approximately 0.5 of a pH unit. The *Ph6* gene has been isolated by heterologous transposon tagging (Chuck *et al.*, 1993). As yet, the manner in which the *Ph6* gene might regulate the vacuolar pH has not been published. The availability of the *Ph6* gene raises the possibility of modifying pH by genetic engineering. However, in *Petunia*, a link between the action of the *Ph* genes and anthocyanin production has been observed (Gerats and Martin, 1992), which may complicate their use.

THE CONTRIBUTION OF FLAVONOIDS TO SPECIFIC COLOURS

Blue

There is an excellent correlation between blue flower colour and the presence of delphinidin-derived anthocyanins. Several of the most popular ornamental species, particularly within the *Asteraceae*, do not produce delphinidin and lack blue colours. Thus, an obvious approach for biotechnology is to introduce the F3′5′H into these species. Such experiments are in progress for *Rosa, Dianthus* and *Dendranthema* (Elomaa and Holton, 1994). However, the presence of delphinidin-derived anthocyanins does not necessarily result in a blue flower colour. *Tulipa, Impatiens, Cyclamen, Eustoma* and *Pelargonium* all produce delphinidin derivatives yet lack (true) blue colours. Further, blue flower colour may be obtained without delphinidin types, although this is uncommon. For example, the major pigment in petals of the blue cornflower *Centaurea cyanus* is an acylated cyanidin 3,5-diglycoside, and the 'Heavenly Blue Anthocyanin' from blue petals of *Ipomoea* consists of peonidin with six molecules of glucose and three molecules of caffeic acid attached (Goto and Kondo, 1991).

Chemical studies on several blue-flowering species have suggested that other factors besides the presence of delphinidin-derived anthocyanins are usually involved in producing blue. Generally these may include the vacuolar pH, self-association, intra- or intermolecular copigmentation and association with metal ions. Excellent reviews of the role of these factors have been published by Goto and Kondo (1991) and Brouillard and Dangles (1993).

Stabilization through self-association and intramolecular copigmentation has been shown for anthocyanins, the latter particularly for those substituted with aromatic acyl groups (Goto and Kondo, 1991; Brouillard and Dangles, 1993). In species such as *Ipomea, Platycodon* (Chinese bellflower) and *Senecio* (garden cineraria), the anthocyanins are highly acylated with cinnamic acid derivatives linked via the glycosyl groups. It is thought that these anthocyanins may be stabilized by sandwich-type stacking of the anthocyanin nucleus and the acyl groups, preventing hydration and thus loss of colour (see Goto and Kondo, 1991).

Intermolecular copigmentation is known to occur between anthocyanins and other phenolics, in particular flavonols and flavones. The flavones, especially the C-glycosides, seem particularly effective copigments for generating blue colours. Indeed, Asen *et al.* (1986) demonstrated that substituting the usual flavonol copigments found in *Eustoma* with flavone C-glycosides changed the colour of a solution containing the *Eustoma* delphinidin-based pigments from purple to blue. Flavone C-glycosides are also involved as copigments in blue flowers of several other species, including *Limonium* (Asen *et al.*, 1973), *Iris* (Asen *et al.*, 1970), *Centaurea* and *Commelina communis* (Goto and Kondo, 1991).

A few cases are known of metallic ions, either iron, magnesium or aluminium, interacting with anthocyanin–copigment complexes to form blue molecules. It is thought that metal chelation may strengthen the association of the pigment–copigment, or assist in the initial formation of the complex (Brouillard and Dangles, 1993). The best studied example is for the blue pigment commelinin from *Commelina* flowers. Commelinin is a supermolecule consisting of six molecules of a malonated delphinidin 3,5-diglucoside copigmented with six molecules of a methylated derivative of an apigenin 4-′glucoside-6-C-glucoside, and complexed with two magnesium atoms (Goto and Kondo, 1991; Kondo *et al.*, 1992). This pigment complex is very stable and is even blue *in vitro* at pH 4.0.

It may be concluded that the generation of a blue flower requires not only the production of a suitable anthocyanin (usually delphinidin-derived) but also the presence of an appropriate copigment and/or extensive modifications to the basic anthocyanin, in particular aromatic acylation. Successful generation of novel blue colours by introduction of the F3′5′H will depend on either the selection of a recipient cultivar with a suitable chemical background, or subsequent further genetic modification of the secondary factors. Unfortunately, from the view of biotechnology, these aspects are difficult to manipulate. Thus, while the flavone C-glycosides are excellent copigments for the generation of blue colours, no cDNA clones have been isolated for either the C-glycosylating enzymes (such as those studied by Kerscher and Franz, 1987) or the FNS. Further, no cDNA clones for the anthocyanin acyltransferases have been reported.

The vacuolar pH is obviously also an important consideration when attempting to generate novel blue-flowered lines. In *Pelargonium,* for example, although delphinidin-derived anthocyanins are present, the low pH of the vacuoles will make the generation of blue-flowered varieties difficult.

Red and orange

Most of the red colours seen in flowers are based on pelargonidin or cyanidin-derived anthocyanins, with the variation in the shade of red being due either to secondary modifications of the anthocyanin (e.g. methylation), pH effects or copigmentation. Pelargonidin and cyanidin-derived pigments, either alone or in combination with carotenoids, may also produce orange and scarlet colours. There are also less common types of flavonoid, differing in the oxygenation of the A- or C-rings, that can result in strong red or orange flower colours.

The 3-deoxyanthocyanins provide orange-red or orange-yellow colours in flowers of some species of the *Gesneriaceae,* including *Sinningia cardinalis, Gesneria cuneifolia, Kohleria eriantha* and *Columena hybrida* (Harborne, 1966a). They are also produced in maize pericarp, in which the pathway is regulated by the *P* locus (Grotewold *et al.*, 1991). Their precursors are formed by the flavanone 4-reductase through a reaction closely related to that of DFR (Stich and Forkmann, 1988a, b). The introduction of this enzyme activity into other species might enable 3-deoxyanthocyanins to be produced, though the substrate specificities of the ANS and glycosyltransferases of the host species would need to be considered. The presence of 6-hydroxypelargonidin produces striking tangerine colours in *Impatiens aurantiaca* (Clevenger, 1964). The analogous 6-hydroxycyanidin and 6-hydroxydelphinidin have been found in red and pink-purple flowers, respectively, of *Alstromeria* (Saito *et al.*, 1988). However, the enzyme(s) involved in the 6-hydroxylation reaction have yet to be characterized.

Copigmentation is also an important influence on the expression of the red anthocyanin colours. Among the reported cases there are two interesting examples in *Rhododendron*. The intense red colour of *R. griersonianum* petals was shown to be due to cyanidin 3-glycosides occurring in the absence of the usual flavonol copigments (Arisumi *et al.*, 1990). Similarly, an orange-flowered sport of Red Wing azalea (a *Rhododendron*) was also found to have greatly reduced levels of flavonol copigments in comparison to the parent (Asen *et al.*, 1971).

Several important ornamental species lack pelargonidin-derived anthocyanins. Except in the solanaceous species, this is most probably due to the dominant nature of the F3′H and F3′5′H activities. A high activity of these enzymes on the naringenin and/or dihydrokaempferol substrates could completely convert the available substrate to a 3′4′-hydroxylated form before it becomes committed to the formation of pelargonidin. Thus, in varieties in which the F3′H and/or F3′5′H activities are lacking, pelargonidin-derived pigments may accumulate rather than the usual cyanidin or delphinidin-derived anthocyanins (Forkmann, 1993). In some species, however, such lines are either absent or rare. The availability of the F3′H and F3′5′H cDNA clones allows the possibility of developing mutant lines in species that lack them by sense or antisense inhibition of the corresponding gene activities. *In vivo* chemical inhibition experiments have demonstrated the applicability of this approach to *Dendranthema* (Schwinn *et al.*, 1994).

Yellow

Bright yellow colours are most commonly associated with the presence of carotenoids. However, a range of yellow flavonoids, in particular the aurones, chalcones, and various flavonol derivatives, are also important flower pigments. In general the chalcones provide paler yellow colours, while the aurones and carotenoids provide stronger yellows. In most flowers, the flavonols produce cream or very pale yellow colours; however, a variety of flavonol derivatives can provide more intense yellow colours.

In yellow-flowered varieties of *Dianthus caryophyllus* (carnation) and *Callistephus chinensis* (China aster) the accumulation of chalcones is associated with a loss of CHI activity due to recessive mutations at the *I* (Forkmann and Dangelmayr, 1980) and *Ch* (Kuhn *et al.*, 1978) loci respectively. A similar mutation may be responsible for a yellow-flowered variant of *Cyclamen persicum* (Miyajima *et al.*, 1991). True yellow-flowered varieties are lacking for many commercial ornamentals, such as *Petunia, Pelargonium, Eustoma* and *Impatiens.* One might expect that inhibition of CHI gene activity via genetic approaches would lead to an accumulation of chalcones and thus yellow flower colours. However, the biochemistry of the target species needs to be taken into account when considering this approach. Two types of chalcones are of interest with regard to this, those with 2′4′6′-trihydroxylation of the A-ring (see Fig. 15.1 for the numbering of the chalcone positions), which are the chalcone substrates for anthocyanin biosynthesis, and the 6′-deoxychalcones, which are substrates for the formation of 5-deoxyflavonoids (Barz and Welle, 1992). Chalcones with unsubstituted 2′- and 6′-hydroxyl groups tend to spontaneously isomerize (involving closure of the carbon link to form the C-ring) to the corresponding flavanone, and may do so at significant rates under physiological conditions (Miles and Main, 1985). Chalcone stability is increased if the 2′-hydroxyl is substituted with a methyl or glycosyl residue, preventing isomerization, or the 6′-hydroxyl is absent, leading to the stabilizing association of the 2′-hydroxyl and the carbonyl group of the carbon link. Thus, chalcones glycosylated at the 2′ position generate yellow flower colours in some cultivars of *Dianthus, Helichrysum* (everlasting flower), *Paeonia* spp. and *Callistephus* (Harborne, 1966b; Bohm, 1988, 1993), and aglycones and glycosides of 6′-deoxychalcones occur, either alone or in some cases with aurones, as the main pigments in some yellow-flowered varieties of *Cosmos, Dahlia, Coreopsis* and *Bidens* (all *Asteraceae*). In a few instances, naringenin chalcone aglycone does accumulate. For instance, Hunt and Baker (1980) reported that naringenin chalcone accounts for the yellow colour of carotenoid-lacking tomato fruit. Mol *et al.* (1985) found that spontaneous isomerization of naringenin chalcone is reduced at pH $\leq$6.5 or by high protein concentrations, and it is possible that these factors may be important in such instances.

In many ornamental species, the 2′4′6′-hydroxy chalcones predominate. If they are to be useful for generating yellow flower colours, the presence of the 2′-glycosylation activity may be required. Unfortunately, if this is not present in the target species, it cannot currently be introduced, as no cDNA clones for the corresponding enzyme have been reported. However, Welle *et al.* (1991) have reported the isolation of cDNA clones for the enzyme responsible for the formation of the 6′-deoxychalcones. This activity may be useful for increasing chalcone stability in genetically modified plants. A cDNA clone for a 2′-methyltransferase, acting on 6′-deoxychalcones, has also been reported (Maxwell *et al.*, 1993). Surprisingly, attempts so far to reduce the expression of CHI genes in *Petunia* with either cosuppression or antisense RNA approaches have been unsuccessful (van Blokland *et al.*, 1993).

The aurones, in some cases in combination with the related chalcone, are responsible for attractive yellow flower colours in a few ornamental species. They occur primarily in flowers of the *Asteraceae* (e.g. *Cosmos, Coreopsis, Zinnia),* the *Scophulariaceae* (*Linaria* and *Antirrhinum*) and the *Plumbaginaceae* (*Limonium*). The ability to confer aurone biosynthesis in novel species would be of considerable commercial interest. At present, however, little is known about the biochemistry and genetics of aurone biosynthesis, and it is a research area that warrants further attention.

Over 1500 different flavonol and flavone derivatives have been reported (Williams and Harborne, 1993), some of which are highly coloured. In particular quercetin 7-glycosides, quercetin 4′-glucosides, 3′-methoxy-quercetin-glycosides, 3′,5′-dimethoxy-myricetin-glycosides, and 6- or 8-hydroxy derivatives of kaempferol or quercetin are known to provide yellow pigments in a range of ornamentals (Harborne, 1976). Unfortunately, although the isolation of cDNA clones has been reported for the FLS and UFGT, none are available for the enzymes involved in the biosynthesis of the more complex derivatives. Indeed, no successful modification of the biosynthesis of any of the yellow flavonoid types via genetic engineering has been reported so far.

THE CAROTENOIDS

Carotenoids are an ubiquitous group of plant pigments differing greatly from flavonoids in structure and compartmentation. They are hydrophobic, lipid-soluble pigments with a structure normally based on a 40-carbon chain derived from the general isoprenoid pathway (Fig. 15.4). Their major role is to protect photosynthetic tissues from photooxidation. However, they may also provide brightly coloured pigments to organs such as fruit and flowers. Two common groups of carotenoids associated with flower colour are the hydrocarbon carotenes and their oxygenated derivatives, the xanthophylls. These are often associated with orange and yellow colours, respectively (Goodwin, 1980). The oxygenation, commonly being the introduction of hydroxy, carbonyl and epoxy groups to the rings at the end of the structures (Fig. 15.4), occurs late in the biosynthetic pathway. Petal xanthophylls are usually partly or fully esterified (Goodwin, 1976).

Approximately 600 naturally occurring carotenoids have been isolated and identified to date (Pfander, 1987). Some ornamentals are very rich in the abundance or diversity of their carotenoids. For example, in daffodils (*Narcissus* spp.) β-carotene levels can reach 16.5% dry weight in the flower corona (Booth, 1957), and in approximately 40 yellow *Rosa* varieties and species, about 75 different carotenoids have been identified (Eugster and Märki-Fischer, 1991). Generally, carotenoid-producing flowers will synthesize predominantly either the hydrocarbon type, the highly oxidized xanthophylls or highly species-specific pigments, rather than a mix of these (Goodwin, 1976).

In flowers, carotenoids are synthesized and stored in chromoplasts, specialized plastids differentiated from chloroplasts or non-photosynthetic plastids. The first committed step in carotenoid biosynthesis is the formation

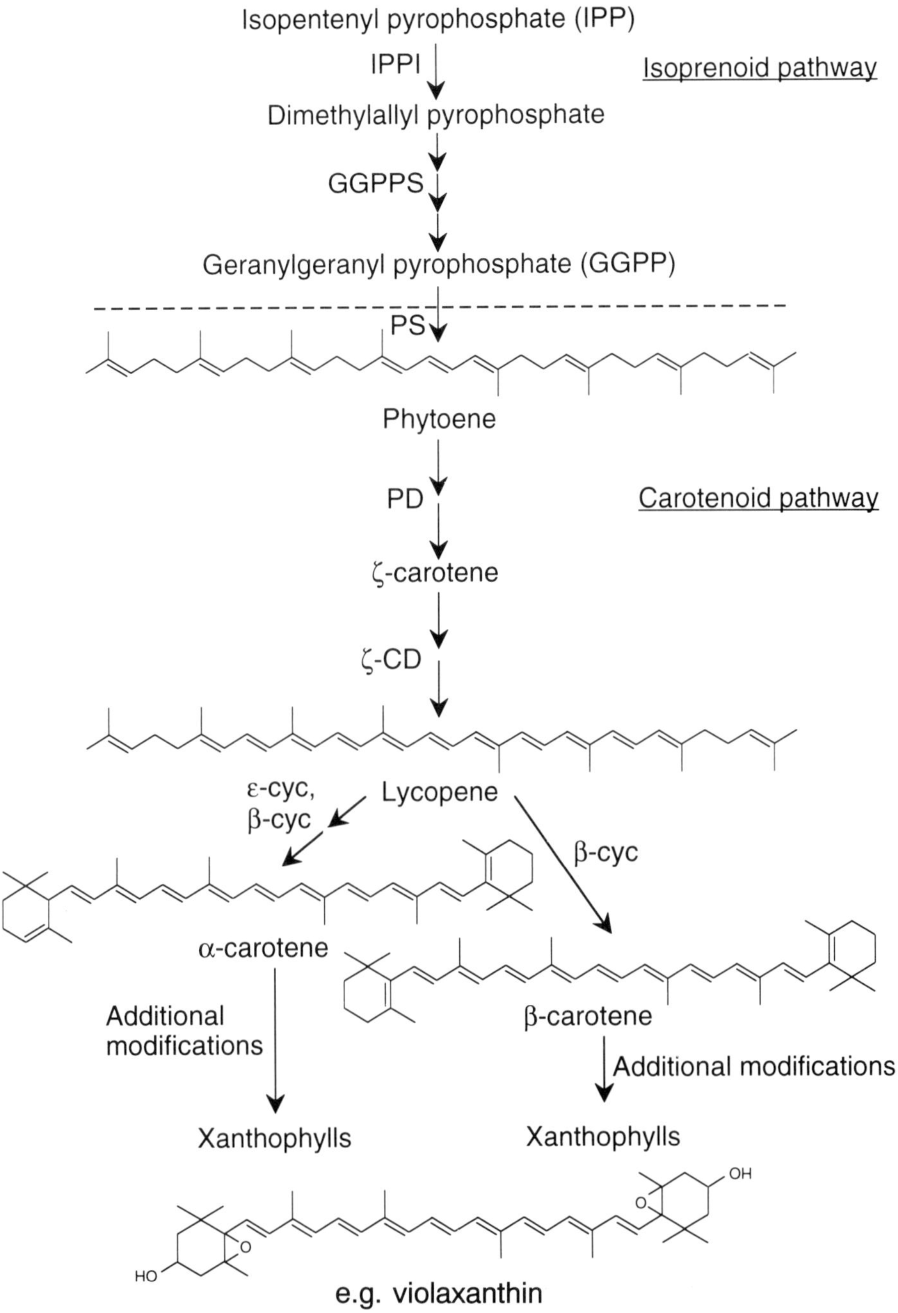

Fig. 15.4. Schematic representation of a section of the isoprenoid and carotenoid biosynthetic pathways of plants. Enzymes are shown are IPP isomerase (IPPI), GGPP synthase (GGPPS), phytoene synthase (PS), phytoene desaturase (PD), ζ-carotene desaturase (ζ-CD), lycopene ε-cyclase (ε-cyc) and lycopene β-cyclase (β-cyc).

of phytoene from two molecules of geranylgeranyl pyrophosphate (Fig. 15.4). This two-step reaction occurs via the intermediate prephytoene pyrophosphate and is catalysed by the enzyme phytoene synthase (PS, Fig. 15.4), which is associated with the stroma of the plastid (Camara *et al.*, 1982; Kreuz *et al.*, 1982; Fraser *et al.*, 1994). The colourless phytoene is converted into the coloured carotenoids by a series of membrane-bound enzymes (Fig. 15.4), the first of which is phytoene desaturase (PD). The possible association of the biosynthetic enzymes as a complex and their membrane localization make *in vitro* work difficult, as the enzymes are easily inactivated during isolation. However, much knowledge has been gained through work on isolated chromoplasts from *Narcissus, Lycopersicon* and *Capsicum.* A detailed discussion of carotenoid biosynthesis is beyond the scope of this chapter; therefore, we suggest the reviews of Britton (1988), Sandmann (1991, 1994) and Bartley and Scolnik (1995).

The molecular biology of plant carotenoid biosynthesis is in its infancy relative to that of flavonoid biosynthesis as, until recently, few cloned genes were available for study. This, however, is changing rapidly, largely as a result of the breakthroughs made in research on carotenogenic bacteria. The model fruit systems, *Lycopersicon* and *Capsicum,* have been the source of many of the plant cDNAs and genes cloned to date. Also, the availability of a variety of carotenogenic mutants of these plants will help in the isolation of further genes. The biosynthetic enzymes for which plant clones have been published are geranylgeranyl pyrophosphate synthase (Kuntz *et al.*, 1992), PS (Bartley *et al.*, 1992; Ray *et al.*, 1992; Schledz *et al.*, 1994), PD (Bartley *et al.*, 1991; Albabili *et al.*, 1994) and lycopene β-cyclase (Hugueney *et al.*, 1995; Scolnik and Bartley 1995). In addition, a cDNA encoding the *Capsicum* xanthophyll biosynthetic enzyme capsanthin-capsorubin synthase has been isolated (Bouvier *et al.*, 1994). This enzyme catalyses the conversion of antheraxanthin and violaxanthin into capsanthin and capsorubin, respectively. We have only included the earliest reports on cloned plant genes as well as reports pertaining to ornamental species. A more comprehensive list of the plant cDNAs and genes can be found in the review of Bartley and Scolnik (1995). Another useful review of the molecular biology of plant carotenoid biosynthesis can be found in Bartley *et al.* (1994).

Although molecular studies of carotenoid biosynthesis in plants are still in the early stages, both plant and bacterial derived genes have already been used to modify the biosynthetic pathway in transgenic plants. In the first successful example reported, introduction of an antisense RNA construct of the PS cDNA (pTOM5) from *Lycopersicon* into plants of the same species resulted in transgenic plants with yellow fruit and reduced pigmentation in the flowers (Bird *et al.*, 1991). The phenotype was correlated with a reduction in PS mRNA abundance and a dramatic fall in carotenoid levels (Bird *et al.*, 1991; Bramley *et al.*, 1992). Fray and Grierson (1993) used the same cDNA clone to complement two yellow-flesh mutant lines of *Lycopersicon,* and demonstrated that the mutant phenotype was due to aberrant PS mRNA. In some transgenic lines carotenoid accumulation was further suppressed in the fruit, foliage and flowers, leading to photobleaching of the leaves and fruit. The phenotype was thought to be due to co-suppression of other PS genes normally expressed in the green tissues.

Misawa *et al.* (1993, 1994) introduced the gene encoding PD from *Erwinia uredovora*, linked to the 35S CaMV promoter and the chloroplast transit peptide of ribulose-1,5-bisphosphate carboxylase small subunit, into *Nicotiana* plants. The PD enzyme from *Erwinia* differs from those of plants and cyanobacteria in carrying out four desaturation reactions, converting phytoene directly to lycopene. Foliage of the transgenic plants had both qualitative and quantitative changes in the leaf carotenoids, in particular increases in β-carotene and its xanthophyll derivatives. This result is encouraging with regard to using genes from carotogenic bacteria, for example that encoding ζ-carotene desaturase (Linden *et al.*, 1993, 1994), for which no plant homologue has been reported, to modify flower colour.

A number of points, however, need addressing when attempting to modify carotenoid biosynthesis in the floral tissue of transgenic plants. For example, the gene product must be located to the plastids, requiring the addition of the appropriate target sequences to potentially useful bacterial genes. In addition, the carotenoid precursors, and the chromoplasts themselves (or potential chromoplasts), must be present in the target tissue. These precursors are also involved in the synthesis of a wide range of other plant products, for example the gibberellins and monoterpenes, and altering carotenoid biosynthesis could affect the plant phenotype in other ways. Furthermore, as carotenoids serve an essential protective function in green tissues, preventing photooxidation, the perturbation of their biosynthesis in non-target organs could have undesirable effects on the plant.

THE BETALAINS

Betalains are nitrogenous vacuolar pigments providing yellow, orange, red and violet colours. They not only occur in plants but also in some fungi such as *Amanita muscaria*. Their importance as colour pigments in plants, however, is limited as they are restricted to the order *Caryophyllales* (*Centropermae*) (for distributions see Piattelli, 1976, and Clement *et al.*, 1994). Interestingly, their presence is mutually exclusive with anthocyanins though not other flavonoid types. In this section, we will mention only briefly various aspects of betalains. For further information on these pigments we refer the reader to reviews by Piattelli (1976, 1981), Mabry (1980) and Clement *et al.* (1994), and for an update on the analytical techniques we suggest Strack and Wray (1994).

Betalains contain betalamic acid (Fig. 15.5) as the chromophore. Betalamic acid itself is a yellow betalain pigment and accumulates in some species (Chang *et al.*, 1974). However, it is commonly conjugated with cyclo-3,4-dihydroxyphenylalanine (cyclo-DOPA) to form the red-violet betacyanins or with amino acids or amine side-chains other than cyclo-DOPA to form the

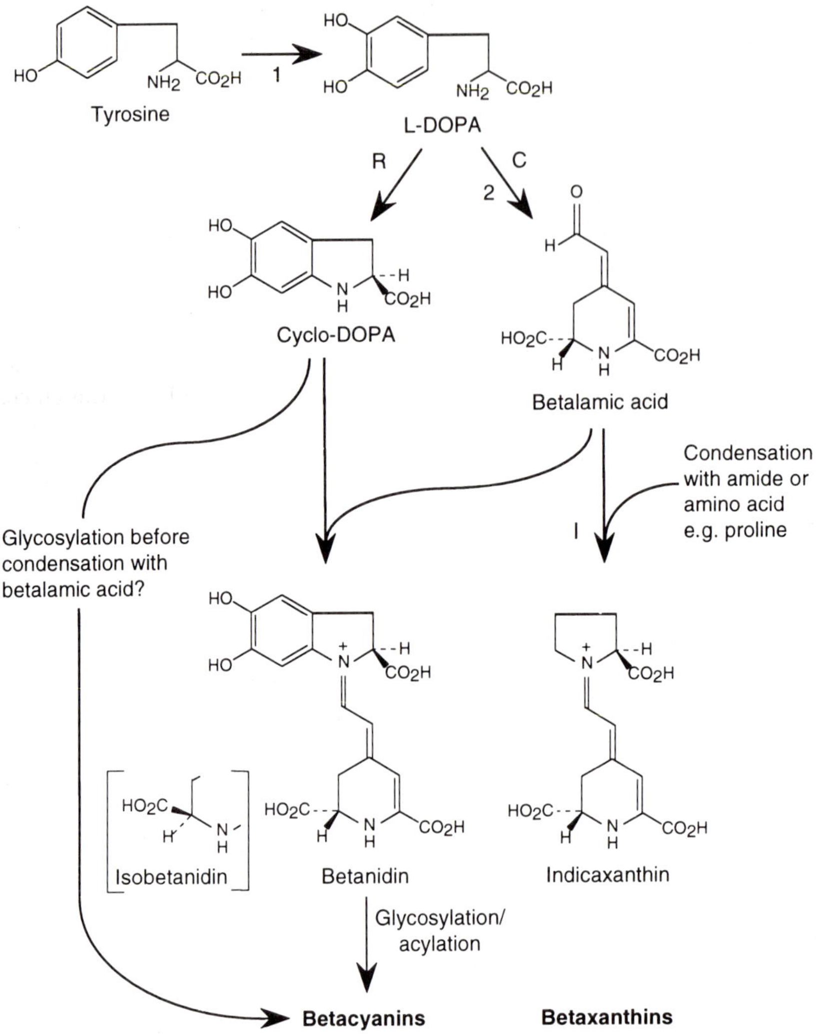

Fig. 15.5. Schematic representation of a section of the betalain biosynthetic pathway. Numbering represents the enzymes: tyrosine hydroxylase (1) and L-DOPA 4,5-dioxygenase (2). For betacyanins, the step at which glycosylation occurs has not been conclusively determined. The positions at which the genetic loci of *Portulacca* are thought to act are indicated: C is responsible for conversion of L-DOPA to betalamic acid, R is responsible for conversion of L-DOPA to cyclo-DOPA and may be involved in its glycosylation, and I prevents conjugation of amino acid residues with betalamic acid.

yellow betaxanthins. Betacyanins are further characterized by their glycosylation and acylation, features not shared with the betaxanthins.

Betacyanins and betaxanthins are derived from the amino acid tyrosine via the intermediates L-DOPA and betalamic acid (Fig. 15.5). While molecular and biochemical studies on betalain biosynthesis are not as advanced as for flavonoids and carotenoids, some progress is being made in defining the biosynthetic enzymes involved. In particular, the tyrosine hydroxylase and DOPA 4,5-dioxygenase have recently been purified from the fungus *Amanita muscaria* (Girod and Zrÿd, 1991; Zrÿd *et al.*, 1993); a betacyanin acyltransferase has been purified from *Chenopodium rubrum* (Brokern *et al.*, 1992), and betanidin glucosyltransferase activity has been found in protein extracts from *Dorotheanthus bellidiformis* cell cultures (Heuer and Strack, 1992). Despite this progress, there are still a few key biosynthetic steps that have not been characterized. The first concerns the point at which glycosylation of the betacyanins occurs. Evidence supporting glycosylation of cyclo-DOPA before its condensation with betalamic acid is seen in *Beta vulgaris,* in which cyclo-DOPA glucoside accumulates (Clement *et al.*, 1994). However, the presence of betanidin glucosyltransferase activity in *Dorotheanthus* (Heuer and Strack, 1992) supports glycosylation of the condensation product, betanidin. It is possible that both routes exist and that one or the other (or both) are used depending on the plant. Another key point to be resolved is whether the condensation reaction with betalamic acid is spontaneous or enzymatically controlled (Clement *et al.*, 1994).

With the exception of *Portulacca grandiflora,* little is known about the genetics controlling the pathway. In *Portulacca,* three main genetic loci (C, R, I) have been proposed (Fig. 15.5) (Trezzini and Zrÿd, 1990). Furthermore, it has been found that betacyanin production is dominant to betaxanthin production and that coloured forms are dominant to the white forms (Piattelli, 1981).

Although little information is currently available for the betalains compared to the flavonoids and carotenoids, researchers are poised to make great progress in the next few years. For instance, the field of betalain molecular genetics is now opening up with the isolation of a cDNA encoding DOPA 4,5-dioxygenase from *Amanita* (Fivaz *et al.*, 1994), and its use in plant transformation experiments (Müller *et al.*, 1995). It is likely that the cloning of a plant homologue will not be far behind.

Despite the limited role betalains play in flower pigmentation, their biosynthetic pathway is still an attractive target for biotechnology, for several reasons. The pathway produces attractive colours, the number of enzymic steps to the first pigments is small and the necessary precursor, tyrosine, occurs in all plants. Genetic modification of betalain biosynthesis could also be used to address some the questions about the pathway, for example why the presence of anthocyanins and betalains seem to be mutually exclusive.

EPIDERMAL CELL SHAPE

The perceived colour of flowers is also dependent on the physical nature of the epidermis. For example, in many species the epidermal cells of the petal are conical in shape (Kay *et al.*, 1981). This is thought to enhance the effect of the cellular pigments by increasing the proportion of light entering the cell

Fig. 15.6. Scanning electron micrograph of the inner epidermis of a flower from an *Antirrhinum* line somatically unstable for the *mixta* mutation, showing the effect on epidermal cell differentiation. Flattened mutant cells are shown next to conical-papillate revertant cells.

relative to the amount of reflected light. In *Antirrhinum,* this phenotype is under the control of the *mixta* gene (Noda *et al.*, 1994). In recessive *mixta* lines, the epidermal cells of the corolla fail to differentiate into their normal conical form, with the apparent effect of reducing pigment intensity. The *mixta* mutation was generated by the insertion of the transposon Tam4. In lines in which the transposon is somatically excising, patches of revertant, wild-type conical cells are seen against the background of flat mutant cells (Fig. 15.6). The *mixta* gene has been cloned by transposon tagging and shown to encode a putative *myb*-type protein (Noda *et al.*, 1994). Thus it could act as a regulatory factor controlling the expression of genes affecting epidermal cell shape. The usefulness of the *mixta* gene in beneficially altering flower colour in transgenic plants has yet to be assessed. However, recent experiments have shown that when introduced into tobacco under the CaMV 35S promoter, *mixta* causes dramatic changes in epidermal cell shape, including the production of trichomes in petal tissue from the conical-papillate cells that are normally present

(C. Martin, John Innes Centre, 1995, personal communication). These changes resulted in altered leaf and flower morphology.

A remarkable example of the effect of epidermal cell form has been reported for the fruits of *Elaeocarpus*. In this example, a modified epidermal structure, termed an iridosome, can generate an iridescent blue colour without any accompanying pigmentation. The iridosome consists of a three-dimensional lattice structure of electron-translucent strands, which may give blue light reflectance through a thin-film interference process (Lee, 1991).

BIOCHEMICAL ASPECTS OF GENETIC MODIFICATION PROGRAMMES

A thorough understanding of the biochemistry of a target species can be of great benefit in developing a genetic engineering strategy for the modification of flower colour. For instance, the successful experiments of Meyer *et al.* (1987), enabling the production of pelargonidin-based pigments in transgenic *Petunia*, were based on previous work characterizing the substrate specificities of the flavonoid biosynthetic enzymes. The type and amount of the different flavonoids present in a particular flower can be established using a variety of well-developed techniques (Markham, 1982; Harborne, 1988; Markham and Geiger, 1993). By carrying out analyses of a wide range of cultivars within a species, a picture can be formed of the flavonoid biosynthetic pathway present and the possible avenues for its genetic modification. This information may also assist in the selection of target cultivars for introduction of novel genes, such as the F3′5′H. Extensive flavonoid surveys have been carried out for some important ornamentals, for example *Dendranthema* (Schwinn *et al.*, 1994); however, similar research for flowers coloured by carotenoids has only been reported for a very few species, for instance *Rosa* (Eugester and Märki-Fischer, 1991).

Strategies developed for the modification of flavonoid biosynthesis can, to some extent, be tested using flavonoid precursor feeding, enzyme assays, inhibitor studies, or *in vitro* reconstitution experiments with isolated pigments. One strategy for producing novel flower colours is the introduction of the F3′5′H gene into species lacking delphinidin-derived anthocyanins. Questions regarding this approach include whether the biosynthetic enzymes of the target species can use the 3′4′5′-hydroxylated anthocyanin precursors, and what colours might result from delphinidin synthesis. These questions can be partially addressed by introducing the 3′4′5′-hydroxylated precursors into floret tissue, as demonstrated for *Dianthus* (Stich *et al.*, 1992) and *Rosa hybrida* (Holton and Tanaka, 1994). Figure 15.3 shows an example of such a feeding experiment for *Dendranthema*. Dihydromyricetin was introduced to the *Dendranthema* cultivar ‘Peach Margaret’, resulting in the formation of delphinidin-derived anthocyanins (Schwinn, unpublished data). Thus, the feeding results support the current experiments introducing the F3′5′H into *Dianthus, Rosa* and *Dendranthema* (Elomaa and Holton, 1994).

Strategies for producing pelargonidin-derived anthocyanins in species lacking them can similarly be tested before embarking on the genetic modification experiments. Lack of pelargonidin pigments might be due to either the substrate specificity of the DFR, as is the case in *Petunia,* or a high activity of the F3′H and/or F3′5′H. The results of Schwinn *et al.* (1994) suggest that in *Dendranthema,* the second explanation applies. Thus, inhibition of the F3′H might result in pelargonidin biosynthesis. This can be tested by feeding with a F3′H inhibitor, tetcyclacis. When tetcyclacis was introduced into intact *Dendranthema* flower buds using wick feeding, pelargonidin pigments were formed (Schwinn *et al.*, 1994). This indicates that *Dendranthema* has the capacity to produce pelargonidin pigments and supports genetic modification approaches.

CONCLUDING REMARKS

The prospects for the genetic modification of flavonoid-based flower colours in commercially important ornamentals are very good. Many years of work on the chemistry, biochemistry and genetics of flavonoid biosynthesis have formed a substantial knowledge base from which molecular strategies can be developed. Genes have been isolated corresponding to most of the important biosynthetic enzymes, including the F3′H and F3′5′H. Flower colour has been successfully modified in transgenic plants of *Petunia* (Meyer *et al.*, 1987; van der Krol *et al.*, 1988; Holton *et al.*, 1993c; Brugliera *et al.*, 1994), *Gerbera* (Elomaa *et al.*, 1993), *Eustoma* (Deroles *et al.*, 1995), *Nicotiana* (van der Krol *et al.*, 1988), *Rosa* (Courtney-Gutterson *et al.*, 1994) and *Dendranthema* (Courtney-Gutterson, 1994). In the case of the DFR *Petunia* experiment of Meyer *et al.* (1987), this resulted in a range of colours novel to this species. In addition, the isolation of genes regulating flavonoid biosynthesis has allowed the spatial and temporal modification of anthocyanin production in floral tissues of *Nicotiana* and *Petunia* (Lloyd *et al.*, 1992; Bradley *et al.*, 1995). However, while the amount or basic type of anthocyanin formed may be modified, the final colour of the flower is also dependent on many other factors, such as copigmentation and vacuolar pH. The technology for modifying these is still at an early stage.

Molecular knowledge on the biosynthesis of carotenoids and betalains is increasing rapidly, and carotenoid biosynthesis has been successfully modified in fruit and flowers of *Lycopersicon* (Fray and Grierson, 1993). However, additional knowledge is still required on the pigment biochemistry of several important ornamental species, particularly regarding the carotenoids.

ACKNOWLEDGEMENTS

We thank Tim Holton, Paula Elomaa, Richard Jorgensen, Yoshi Tanaka, Lisa Keam, Gina Barri-Rewell, Tim Robbins, Tom Mabry and John Clement for supplying us with reprints, preprints and unpublished information, and Cathie Martin, Murray Boase, Karen Nielsen, Ken Markham and Brian Jordan for their helpful comments on the drafts of this article. We are also grateful to Paul Linstead and Cathie Martin for supplying Figure 15.6.

REFERENCES

Albabili, S., Lintig, J., Kleinig, H. and Beyer, P. (1994) EMBL/GenBank Accession X78815.

Arisumi, K., Sakata, Y. and Takeshita, S. (1990) The pigment constitution of *R. griersonianum*. *American Rhododendron Society Journal* 44, 15–17.

Armstrong, G.A., Alberti, M., Leach, F. and Hearst, J.E. (1989) Nucleotide sequence, organisation, and nature of the protein products of the carotenoid biosynthesis gene cluster of *Rhodobacter capsulatus*. *Molecular and General Genetics* 216, 254–268.

Asen, S., Stewart, R.N., Norris, K.H. and Massie, D.R. (1970) A stable blue non-metallic copigment complex of delphanin and C-glycosylflavones in Prof. Blaauw iris. *Phytochemistry* 9, 619–627.

Asen, S., Stewart, R.N. and Norris, K.H. (1971) Co-pigmentation effect of quercetin glycosides on absorption characteristics of cyanidin glycosides and color of Red Wing azalea. *Phytochemistry* 10, 171–175.

Asen, S., Norris, K.H., Stewart, R.N. and Semeniuk, P. (1973) Effect of pH, anthocyanin, and flavonoid co-pigments on the color of statice flowers. *Journal of the American Society for Horticultural Science* 98, 174–176.

Asen, S., Stewart, R.N. and Norris, K.H. (1977) Anthocyanin and pH involved in the colour of 'Heavenly Blue' morning glory. *Phytochemistry* 16, 1118–1119.

Asen, S., Griesbach, R.J., Norris, K.H. and Leonhard, B.A. (1986) Flavonoids from *Eustoma grandiflorum* flower petals. *Phytochemistry* 25, 2509–2513.

Bartley, G.E. and Scolnik, P.A. (1995) Plant carotenoids: pigments for photoprotection, visual attraction, and human health. *Plant Cell* 7, 1027–1038.

Bartley, G.E., Viitanen, P.V., Pecker, I., Chamovitz, D., Hirschberg, J. and Scolnik, P.A. (1991) Molecular cloning and expression in photosynthetic bacteria of a soybean cDNA coding for phytoene desaturase, an enzyme of the carotenoid biosynthesis pathway. *Proceedings of the National Academy of Sciences, USA* 88, 6532–6536.

Bartley, G.E., Viitanen, P.V., Bacot, K.O. and Scolnik, P.A. (1992) A tomato gene expressed during fruit ripening encodes an enzyme of the carotenoid biosynthesis pathway. *Journal of Biological Chemistry* 267, 5036–5039.

Bartley, G.E., Scolnik, P.A. and Giuliano, G. (1994) Molecular biology of carotenoid biosynthesis in plants. *Annual Review of Plant Physiology and Molecular Biology* 45, 287–301.

Barz, W. and Welle, R. (1992) Biosynthesis and metabolism of isoflavones and pterocarpan phytoalexins in chickpea, soyabean and phytopathogenic fungi. In: Stafford, H.A. and Ibrahim, R.K. (eds), *Phenolic Metabolism in Plants*. Plenum Press, New York, pp. 139–164.

Beld, M., Martin, C., Huits, H., Stuitje, A.R. and Gerats, A.G.M. (1989) Flavonoid synthesis in *Petunia hybrida*: partial characterization of dihydroflavonol-4-reductase genes. *Plant Molecular Biology* 13, 491–502.

Bird, C.R., Ray, J.A., Fletcher, J.D., Boniwell, J.M., Bird, A.S., Teulieres, C., Blain, I., Bramley, P.M. and Schuch, W. (1991) Using antisense RNA to study gene function:

inhibition of carotenoid biosynthesis in transgenic tomatoes. *Bio/Technology* 9, 635–639.

Bohm, B.A. (1988) The minor flavonoids. In: Harborne, J.B. (ed.), *The Flavonoids: Advances in Research Since 1980*. Chapman & Hall, London, pp. 329–388.

Bohm, B.A. (1993) The minor flavonoids. In: Harborne, J.B. (ed.), *The Flavonoids: Advances in Research Since 1988*. Chapman & Hall, London, pp. 387–440.

Booth, V.H. (1957) β-carotene in the flowers of *Narcissus*. *Biochemical Journal* 65, 660–663.

Bouvier, F., Hugueney, P., d'Harlingue, A., Kuntz, M. and Camara, B. (1994) Xanthophyll biosynthesis in chromoplasts: isolation and molecular cloning of an enzyme catalyzing the conversion of 5,6-epoxycarotenoid into ketocarotenoid. *The Plant Journal* 6, 45–54.

Bradley, J.M., Davies, K.M., Deroles S.C., Schwinn K.E. and Manson D.G. (1995) Colour modification in petunia using the *Lc* regulatory gene from maize. In: *Proceedings of the XVIIIth Eucarpia Symposium, Ornamental Section*, Tel Aviv, Israel (in press).

Bramley, P., Teulieres, C., Blain, I., Bird, C. and Schuch, W. (1992) Biochemical characterization of transgenic tomato plants in which carotenoid synthesis has been inhibited through the expression of antisense RNA to pTOM5. *The Plant Journal* 2, 343–350.

Britsch, L., Ruhnau-Brich, B. and Forkmann, G. (1992) Molecular cloning, sequence analysis and *in vitro* expression of flavanone 3-β-hydroxylase from *Petunia hybrida*. *Journal of Biological Chemistry* 267, 5380–5387.

Britsch, L., Dedio, J., Saedler, H. and Forkmann, G. (1993) Molecular characterization of flavanone 3-β-hydroxylases, consensus sequence, comparison with related enzymes and the role of conserved histidines. *European Journal of Biochemistry* 217, 745–754.

Britton, G. (1988) Biosynthesis of carotenoids. In: Goodwin, T.W. (ed.), *Plant Pigments*. Academic Press, London, pp. 133–182.

Brokern, M., Heuer, S. and Strack, D. (1992) Hydroxycinnamic acid transferases in the biosynthesis of acylated betacyanins: purification and characterization from cell cultures of *Chenopodium rubrum* and occurrence in some other members of the Caryophyllales. *Botanica Acta* 105, 146–151.

Brouillard, R. (1988) Flavonoids and flower colour. In: Harborne, J.B. (ed.), *The Flavonoids: Advances in Research Since 1980*. Chapman & Hall, London, pp. 525–538.

Brouillard, R. and Dangles, O. (1993) Flavonoids and flower colour. In: Harborne, J.B. (ed.), *The Flavonoids: Advances in Research Since 1986*. Chapman & Hall, London, pp. 565–587.

Brugliera, F., Holton, T.A., Stevenson, T.W., Farcy, E., Lu, C.-Y. and Cornish, E.C. (1994) Isolation and characterization of a cDNA clone corresponding to the *Rt* locus of *Petunia hybrida*. *The Plant Journal* 5, 81–92.

Camara, B., Bardat, F. and Monéger, R. (1982) Sites of biosynthesis of carotenoids in *Capsicum* chromoplasts. *European Journal of Biochemistry* 127, 255–258.

Chang, C., Kimler, L. and Mabry, T.J. (1974) Biogenesis of betalamic acid. *Phytochemistry* 13, 2771–2775.

Chuck, G., Robbins, T., Nijjar, C., Ralston, E., Courtney-Gutterson, N. and Dooner, H.K. (1993) Tagging and cloning of a petunia flower colour gene with the maize transposable element *Activator*. *Plant Cell* 5, 371–378.

Clement J.S., Mabry, T.J., Wyler, H. and Dreiding, A.S. (1994) Chemical review and evolutionary significance of the betalains. In: Behnke, H.-D. and Mabry, T.J. (eds), *Caryophyllales; Evolution and Systematics*. Springer-Verlag, Berlin, pp. 247–261.

Clevenger, S. (1964) A new anthocyanidin in *Impatiens*. *Canadian Journal of Biochemistry* 42, 154–155.

Courtney-Gutterson, N. (1994) The biologist's palette: genetic engineering of anthocyanin biosynthesis and flower colour. In: Ellis, B.E., Kuroki, G.W. and Stafford, H. (eds), *Genetic Engineering of Plant Secondary Metabolism; Recent Advances in Phytochemistry, Vol. 28*, Plenum Press, New York, pp. 93–124.

Courtney-Gutterson, N., Napoli, C., Lemieux, C., Morgan, A., Firoozabady, E. and Robinson, K.E.P. (1994) Modification of flower color in florist's chrysanthemum: production of a white-flowering variety through molecular genetics. *Bio/Technology* 12, 268–271.

Davies, K.M., Bradley, J.M., Schwinn, K.E., Markham, K.R. and Podivinsky, E. (1993) Flavonoid biosynthesis in flower petals of five lines of lisianthus (*Eustoma grandiflorum* Grise.). *Plant Science* 95, 67–77.

Dedio, J., Saedler, H. and Forkmann, G. (1995) Molecular cloning of the flavanone 3β-hydroxylase gene (FHT) from carnation (*Dianthus caryophyllus*) and analysis of stable and unstable FHT mutants. *Theoretical and Applied Genetics* 90, 611–617.

Deroles, S.C., Bradley, J.M., Davies, K.M., Schwinn, K.E. and Manson, D.G. (1995) Generation of novel patterns in lisianthus flowers using an antisense chalcone synthase gene. In: *Proceedings of the XVIIIth Eucarpia Symposium, Ornamental Section*, Tel Aviv, Israel.

de Vlaming, P., Schram, A.W. and Wiering, H. (1983) Genes affecting flower colour and pH of flower limb homogenates in *Petunia hybrida*. *Theoretical and Applied Genetics* 66, 271–278.

Dooner, H.K., Robbins, T.P. and Jorgensen, R.A. (1991) Genetic and developmental control of anthocyanin biosynthesis. *Annual Review of Genetics* 25, 173–199.

Elomaa, P. and Holton, T. (1994) Modification of flower colour using genetic engineering. *Biotechnology and Genetic Engineering Reviews* 12, 63–88.

Elomaa, P., Honkanen, J., Puska, R., Seppänen, P., Helariutta, Y., Mehto, M., Kotilainen, M., Nevalainen, L. and Teeri, T.H. (1993) *Agrobacterium*-mediated transfer of antisense chalcone synthase cDNA to *Gerbera hybrida* inhibits flower pigmentation. *Bio/Technology* 11, 508–511.

Epping, B., Kittel, M., Ruhnau, B. and Hemleben, V. (1990) Isolation and sequence analysis of a chalcone synthase cDNA of *Matthiola incana* R. Br. (Brassicaceae). *Plant Molecular Biology* 14, 1061–1063.

Eugester, C.H. and Märki-Fischer, E. (1991) The chemistry of rose pigments. *Angewandte Chemie – International Edition in English* 30, 654–672.

Fedoroff, N., Furtek, D. and Nelson, O.E. (1984) Cloning of the *bronze* locus in maize by a simple and generalizable procedure using the transposable element *Activator (Ac)*. *Proceedings of the National Academy of Sciences, USA* 81, 3825–3829.

Fivaz, J., Hinz, U. and Zrÿd J.-P. (1994) DOPA-4,5-dioxygenase, an enzyme for *in vitro* production of betalain pigments. In: *Book of Abstracts, 4th International Congress of Plant Molecular Biology*, Amsterdam, Abstract 1304.

Forkmann, G. (1991) Flavonoids as flower pigments: the formation of the natural spectrum and its extension by genetic engineering. *Plant Breeding* 106, 1–26.

Forkmann, G. (1993) Genetics of flavonoids. In: Harborne, J.B. (ed.), *The Flavonoids: Advances in Research Since 1986*. Chapman & Hall, London, pp. 537–564.

Forkmann, G. and Dangelmayr, B. (1980) Genetic control of chalcone isomerase activity in flowers of *Dianthus caryophyllus*. *Biochemical Genetics* 18, 519–527.

Fraser, P.D., Truesdale, M.R., Bird, C.R., Schuch, W. and Bramley, P.M. (1994) Carotenoid biosynthesis during tomato fruit development. *Plant Physiology* 105, 405–413.

Fray, R.G. and Grierson, D. (1993) Identification and genetic analysis of normal and mutant phytoene synthase genes of tomato by sequencing, complementation and co-suppression. *Plant Molecular Biology* 22, 589–602.

Gerats, A.G.M. and Martin, C. (1992) Flavonoid synthesis in *Petunia hybrida*; genetics and molecular biology of flower colour. In: Stafford, H.A. and Ibrahim, R.K. (eds), *Recent Advances in Phytochemistry, Vol. 26: Phenolic Metabolism in Plants*. Plenum Press, New York, pp. 165–200.

Girod, P.-A. and Zrÿd, J.-P. (1991) Biogenesis of betalains: purification and partial characterization of DOPA 4,5-dioxygenase from *Amanita muscaria*. *Phytochemistry* 30, 169–174.

Goodrich, J., Carpenter, R. and Coen, E.S. (1992) A common gene regulates pigmentation pattern in diverse plant species. *Cell* 68, 955–964.

Goodwin, T.W. (1976) Distribution of carotenoids. In: Goodwin, T.W. (ed.), *Chemistry and Biochemistry of Plant Pigments,* 2nd edition, Vol. 1. Academic Press, London, pp. 225–261.

Goodwin, T.W. (1980) Carotenoids. In: Bell, E.A. and Charlwood, B.V. (eds), *Secondary Plant Products: Encyclopedia of Plant Physiology, New Series Vol. 8.* Springer-Verlag, Berlin, pp. 257–287.

Goto, T. and Kondo, T. (1991) Structure and molecular stacking of anthocyanins – flower color variation. *Angewandte Chemie – International Edition in English* 30, 17–33.

Griesbach, R.J. (1993) Characterization of the flavonoids from *Petunia* × *hybrida* flowers expressing the *A1* gene of *Zea mays*. *HortScience* 28, 659–660.

Grotewold, E., Athna, P. and Peterson, T. (1991) Alternatively spliced products of the maize P gene encode proteins with homology to the DNA binding domain of *Myb*-like transcription factors. *Proceedings of the National Academy of Sciences, USA* 88, 4587–4591.

Harborne, J.B. (1963) Plant polyphenols. X. Flavone and aurone glycosides of *Antirrhinum. Phytochemistry* 2, 327–334.

Harborne, J.B. (1966a) Comparative biochemistry of flavonoids – II. 3-desoxyanthocyanins and their systematic distribution in ferns and gesnerads. *Phytochemistry* 5, 589–600.

Harborne, J.B. (1966b) Comparative biochemistry of flavonoids – I. Distribution of chalcone and aurone pigments in plants. *Phytochemistry* 5, 111–115.

Harborne, J.B. (1976) Functions of flavonoids in plants. In: Goodwin, T.W. (ed.), *Chemistry and Biochemistry of Plant Pigments, Vol. 1.* Academic Press, New York, pp. 736–778.

Harborne, J.B. (1988) The flavonoids: recent advances. In: Goodwin, T.W. (ed.,) *Plant Pigments.* Academic Press, London, pp. 299–343.

Helariutta, Y., Elomaa, P., Kotilainen, M., Seppänen, P. and Teeri, T.H. (1993) Cloning of cDNA coding for dihydroflavonol-4-reductase (DFR) and characterization of *dfr* expression in the corollas of *Gerbera hybrida* var. Regina (Compositae). *Plant Molecular Biology* 22, 183–193.

Heller, W. and Forkmann, G. (1988) Biosynthesis of flavonoids. In: Harborne, J.B. (ed.), *The Flavonoids: Advances in Research Since 1980*. Chapman & Hall, London, pp. 399–425.

Heller, W. and Forkmann, G. (1993) Biosynthesis of flavonoids. In: Harborne, J.B. (ed.), *The Flavonoids: Advances in Research Since 1986*. Chapman & Hall, London, pp. 499–536.

Heuer, S. and Strack, D. (1992) Synthesis of betanin from betanidin and UDP-glucose by a protein preparation from cell suspension cultures of *Dorotheanthus bellidiformis* (Burm. f.) N. E. Br. *Planta* 186, 626–628.

Holton, T.A. and Tanaka, Y. (1994) Blue roses – a pigment of our imagination? *TIBTECH* 12, 40–42.

Holton, T.A., Brugliera, F., Lester, D.R., Tanaka, Y., Hyland, C.D., Menting, J.G.T., Lu, C.-Y., Farcy, E., Stevenson, T.W. and Cornish, E.C. (1993a) Cloning and expression of cytochrome P450 genes controlling flower colour. *Nature* 366, 276–279.

Holton, T.A., Cornish, E.C. and Tanaka, Y. (1993b) Genetic sequences encoding flavonoid pathway enzymes and uses therefor. International patent publication number WO 93/20206: International application number PCT/AU93/00127.

Holton, T.A., Brugliera, F. and Tanaka, Y. (1993c) Cloning and expression of flavonol synthase from *Petunia hybrida. The Plant Journal* 4, 1003–1010.

Hugueney, P., Badillo, A., Chen, H.-C., Klein, A., Hirschberg, J., Camara, B. and Kuntz, M. (1995) Metabolism of cyclic carotenoids: a model for the alteration of this biosynthetic pathway in *Capsicum annuum* chromoplasts. *The Plant Journal* 8, 417–424.

Huits, S.M., Gerats, A.G.M., Kreike, M.M., Mol, J.M.N. and Koes, R.E. (1994) Genetic control of dihydroflavonol 4-reductase gene expression in *Petunia hybrida. The Plant Journal* 6, 295–310.

Hunt, G.M. and Baker, E.A. (1980) Phenolic constituents of tomato fruit cuticles. *Phytochemistry* 19, 1415–1419.

Jorgensen, R. (1991) Silencing of plant genes by homologous transgenes. *AgBiotech News and Information* 4, 265N–273N.

Jorgensen, R. (1994) Developmental significance of epigenetic impositions on the plant genome: a paragenetic function for chromosomes. *Developmental Genetics* 15, 523–532.

Jorgensen, R.A. (1995) Cosuppression, flower color patterns, and metastable gene expression states. *Science* 268, 686–691.

Jorgensen, R., Cluster, P., Que, Q., Tang, F.-C. and Napoli, C. (1994) Why do chalcone synthase transgenes elicit complex, morphology-based patterns of coloration? In: *Abstracts, 4th International Congress of Plant Molecular Biology*, Amsterdam, Abstract 501.

Kamsteeg, J., van Brederode, J. and van Nigtevecht, G. (1980) Genetic and biochemical evidence that the hydroxylation pattern of the anthocyanin B ring in *Silene dioica* is determined at the p-coumaroyl-coenzyme-A stage. *Phytochemistry* 19, 1459–1462.

Kay, Q.O.N., Daoud, H.S. and Stirton, C.H. (1981) Pigment distribution, light reflection and cell structure in petals. *Botanical Journal of the Linnaean Society* 83, 57–84.

Kerscher, F. and Franz, G. (1987) Biosynthesis of vitexin and isovitexin: enzymatic synthesis of the C-glucosylflavones vitexin and isovitexin with an enzyme preparation from *Fagopyrum esculentum* M. seedlings. *Zeitschrift für Naturforschung* 42c, 519–524.

Koes, R.E., Spelt, C.E., Reif, H.J., van den Elzen, P.J.M., Veltkamp, E. and Mol, J.N.M. (1986). Floral tissue of *Petunia hybrida* (V30) expresses only one member of the chalcone synthase multigene family. *Nucleic Acids Research* 14, 5229–5239.

Koes, R.E., Spelt, C.E., Mol, J.N.M. and Gerats, A.G.M. (1987). The chalcone synthase multigene family of petunia hybrida (V30): sequence homology, chromosomal location and evolutionary aspects. *Plant Molecular Biology* 10, 159–169.

Koes, R.E., Spelt, C.E., van den Elzen, P.J.M. and Mol, J.N.M. (1989). The cloning and molecular characterization of the chalcone synthase multigene family of *Petunia hybrida. Gene* 81, 245–257.

Kondo, T., Yoshida, K., Nakagawa, A., Kawai, T., Tamura, H. and Goto, T. (1992) Structural basis of blue-colour development in flower petals from *Commelina communis. Nature* 358, 515–518.

Kreuz, K., Beyer, J. and Kleinig, H. (1982) The site of carotenogenic enzymes in chromoplasts from *Narcissus pseudonarcissus* L. *Planta* 154, 66–69.

Kroon, J., Souer, E., de Graaff, A., Xue, Y., Mol, J. and Koes, R. (1994) Cloning and structural analysis of the anthocyanin pigmentation locus *Rt* of *Petunia hybrida:* characterization of insertion sequences in two mutant alleles. *The Plant Journal* 5, 69–80.

Kuhn, B., Forkmann, G. and Seyffert, W. (1978) Genetic control of chalcone–flavanone isomerase activity in *Callistephus chinensis. Planta* 138, 199–203.

Kuntz, M., Römer, S., Suire, C., Hugueney, P., Weil, J.H., Schantz, R. and Camara, B. (1992) Identification of a cDNA for the plastid-located geranylgeranyl pyrophosphate synthase from *Capsicum annuum:* correlative increase in enzyme activity and transcript level during fruit ripening. *The Plant Journal* 2, 25–34.

Lee, D.W. (1991) Ultrastructural basis and function of iridescent blue colour of fruits in *Elaeocarpus. Nature* 349, 260–262.

Linden, H., Vioque, A. and Sandmann, G. (1993) Isolation of a carotenoid biosynthesis gene coding for ζ-carotene desaturase from *Anabaena* PCC7120 by heterologous complementation. *FEMS Microbiology Letters* 106, 99–104.

Linden, H., Misawa, N., Saito, T. and Sandmann, G. (1994) A novel carotenoid biosynthesis gene coding for ζ-carotene desaturase: functional expression, sequence and phylogenetic origin. *Plant Molecular Biology* 24, 369–379.

Linn, F., Heidmann, I., Saedler, H. and Meyer, P. (1990) Epigenetic changes in the expression of the maize *A1* gene in *Petunia hybrida:* role of numbers of integrated gene copies and state of methylation. *Molecular and General Genetics* 222, 329–336.

Lloyd, A.M., Walbot, V. and Davis, R.W. (1992) *Arabidopsis* and *Nicotiana* anthocyanin production activated by maize regulators *R* and *C1*. *Science* 258, 1773–1775.

Luo, D., Coen, E.S., Doyle, S. and Carpenter, R. (1991) Pigmentation mutants produced by transposon mutagenesis in *Antirrhinum majus. The Plant Journal* 1, 59–69.

Mabry, T.J. (1980) Betalains. In: Bell, E.A. and Charlwood, B.V. (eds), *Secondary Plant Products, Encyclopedia of Plant Physiology, New Series Vol. 8.* Springer-Verlag, Berlin, pp. 513–533.

Markham, K.R. (1982) *Techniques of Flavonoid Identification*. Academic Press, London.

Markham, K.R. and Geiger, H. (1993) ^{1}H nuclear magnetic resonance spectroscopy of flavonoids and their glycosides in hexadeuterodimethylsulfoxide. In: Harborne, J.B. (ed.), *The Flavonoids: Advances in Research Since 1986.* Chapman & Hall, London, pp. 441–498.

Markham, K.R. and Ofman, D.J. (1993) Lisianthus flavonoid pigments and factors influencing their expression in flower colour. *Phytochemistry* 34, 679–685.

Marrs, K.A., Alfenito, M.R., Lloyd, A.M. and Walbot, V. (1995) A glutathione S-transferase involved in vacuolar transfer encoded by the maize gene *Bronze-2. Nature* 375, 397–400.

Martin, C. and Gerats, T. (1993a) The control of flower colouration. In: Jordan, B.R. (ed.), *The Molecular Biology of Flowering.* CAB International, Wallingford, UK, pp. 219–255.

Martin, C. and Gerats, T. (1993b) Control of pigment biosynthesis genes during petal development. *Plant Cell* 5, 1253–1264.

Martin, C., Carpenter, R., Sommer, H., Saedler, H. and Coen, E.S. (1985) Molecular analysis of instability in flower pigmentation of *Antirrhinum majus*, following isolation of the *pallida* locus by transposon tagging. *EMBO Journal* 4, 1625–1630.

Martin, C., Prescott, A., Mackay, S., Bartlett, J. and Vrijlandt, E. (1991) Control of anthocyanin biosynthesis in flowers of *Antirrhinum majus*. *The Plant Journal* 1, 37–49.

Maxwell, C.A., Harrison, M.J. and Dixon, R.A. (1993) Molecular characterization and expression of isoliquiritigenin 2′-*O*-methyltransferase, an enzyme specifically involved in the biosynthesis of an inducer of *Rhizobium meliloti* nodulation genes. *The Plant Journal* 4, 971–981.

McLaughlin, M. and Walbot, V. (1987) Cloning of a mutable bz2 allele of maize by transposon tagging and differential hybridization. *Genetics* 117, 771–776.

Mermod, N., Harayama, S. and Timmis, K.N. (1986) New route to bacterial production of indigo. *Bio/Technology* 4, 321–324.

Meyer, P. (1991) Engineering of novel flower colours. In: Harding, J., Singh, F. and Mol, J.N.M. (eds), *Genetics and Breeding of Ornamental Species*. Kluwer Academic Publishers, Dordrecht, pp. 285–307.

Meyer, P. and Heidmann, I. (1994) Epigenetic variants of a transgenic petunia line show hypermethylation in the transgene DNA: an indication for specific recognition of foreign DNA in transgenic plants. *Molecular and General Genetics* 243, 390–399.

Meyer, P., Heidmann, I., Forkmann, G. and Saedler, H. (1987) A new *Petunia* flower colour generated by transformation of a mutant with a maize gene. *Nature* 330, 677–678.

Meyer, P., Linn, F., Heidmann, I., Meyer, A.H., Niedenhof, I. and Saedler, H. (1992) Endogenous and environmental factors influence 35S promoter methylation of a maize *A1* gene construct in transgenic petunia and its colour phenotype. *Molecular and General Genetics* 231, 345–352.

Meyer, P., Heidmann, I. and Niedenhof I. (1993) Differences in DNA methylation are associated with a paramutation phenomenon in transgenic petunia. *The Plant Journal* 4, 89–100.

Miles, C.O. and Main, L. (1985) Kinetics and mechanism of the cyclisation of 2′, 6′-dihydroxy-4,4′-dimethoxychalcone; influence of the 6′-hydroxyl group on the rate of cyclisation under neutral conditions. *Journal of the Chemistry Society Perkin Transactions II*, 1639–1642.

Misawa, N., Nakagawa, M., Kobayashi, K., Yamano, S., Izawa, Y., Nakamura, K. and Harashima, K. (1990) Elucidation of the *Erwinia uredovora* carotenoid biosynthetic pathway by functional analysis of gene products expressed in *Escherichia coli*. *Journal of Bacteriology* 172, 6704–6712.

Misawa, N., Yamano, S., Linden, H., de Felipe, M.R., Lucas, M., Ikenaga, H. and Sandmann, G. (1993) Functional expression of the *Erwinia uredovora* carotenoid biosynthesis gene *crtI* in transgenic plants showing an increase of β-carotene biosynthesis activity and resistance to the bleaching herbicide norflurazon. *The Plant Journal* 4, 833–840.

Misawa, N., Masamoto, K., Hori, T., Ohtani, T., Böger, P. and Sandmann, G. (1994) Expression of an *Erwinia* phytoene desaturase gene not only confers multiple resistance to herbicides interfering with carotenoid biosynthesis but also alters xanthophyll metabolism in transgenic plants. *The Plant Journal* 6, 481–489.

Miyajima, I., Maehara, T., Kage, T. and Fujieda, K. (1991) Identification of the main agent causing yellow color of yellow-flowered cyclamen mutant. *Journal of the Japanese Society for Horticultural Science* 60, 409–414.

Mizutani, M., Ward, E., DiMaio, J., Ohta, D., Ryals, J. and Sato, R. (1993) Molecular cloning and sequencing of a cDNA encoding mung bean cytochrome P450 (P450C4H) possessing cinnamate 4-hydroxylase activity. *Biochemical and Biophysical Research Communications* 190, 875–880.

Mol, J.N.M., Robbins, M.P., Dixon, R.A. and Veltkamp, E. (1985) Spontaneous and enzymatic rearrangement of naringenin chalcone to flavanone. *Phytochemistry* 24, 2267–2269.

Mooney, M., Desnos, T., Harrison, K., Jones, J., Carpenter, R. and Coen, E. (1995) Altered regulation of tomato and tobacco pigmentation genes caused by the *delila* gene of *Antirrhinum. The Plant Journal* 7, 333–339.

Müller, L., Hinz, U. and Zrÿd, J.-P. (1995) Transformation of plants with a dopa-4, 5-dioxygenase cDNA from *Amanita muscaria. Experientia* 51, A16.

Napoli, C., Lemieux, C. and Jorgensen, R. (1990) Introduction of a chimeric chalcone synthase gene into *Petunia* results in reversible co-suppression of homologous genes *in trans. Plant Cell* 2, 279–289.

Noda, K.-I., Glover, B.J., Linstead, P. and Martin, C. (1994) Flower colour intensity depends on specialized cell shape controlled by a Myb-related transcription factor. *Nature* 369, 661–664.

Niesbach-Klösgen, V., Barzen, E., Bernhardt, J., Rohde, W., Schwarz-Sommer, Zs., Rief, H.J., Wienand, V. and Saedler, H. (1987) Chalcone synthase genes in plants: a tool to study evolutionary relationships. *Journal of Molecular Evolution* 26, 213–225.

Pfander, H. (1987) In: Straub, O. (ed.), *Key to Carotenoids*, 2nd edition. Birkhaeuser Verlag, Basel.

Piattelli, M. (1976) Betalains. In: Goodwin, T.W. (ed.), *Chemistry and Biochemistry of Plant Pigments,* 2nd edition Vol. 1. Academic Press, London, pp. 560–596.

Piattelli, M. (1981) The betalains: structure, biosynthesis and chemical taxonomy. In: Conn, E.E. (ed.), *The Biochemistry of Plants. Vol. 7. Secondary Plant Products.* Academic Press, London, pp. 557–575.

Quattrocchio, F., Wing, J.F., Leppen, H.T.C., Mol, J.N.M. and Koes, R.E. (1993) Regulatory genes controlling anthocyanin pigmentation are functionally conserved among plant species and have distinct sets of target genes. *Plant Cell* 5, 1497–1512.

Ray, J., Moureau, P., Bird, C., Bird, A., Grierson, D., Maunders, M., Truesdale, M., Bramley, P. and Schuch, W. (1992) Cloning and characterization of a gene involved in phytoene synthesis from tomato. *Plant Molecular Biology* 19, 401–404.

Reif, H.J., Niesbach, U., Deumling, B. and Saedler, H. (1985) Cloning and analysis of two genes for chalcone synthase from *Petunia hybrida. Molecular and General Genetics* 199, 208–215.

Robbins, T. and Harbord, R. (1994) A regulatory gene from *Antirrhinum* disrupts anthocyanin biosynthesis in *Petunia hybrida.* In: *Abstracts, 4th International Congress of Plant Molecular Biology*, Amsterdam, Abstract 263.

Sablowski, R.W.M., Moyano, E., Culianez-Macia, F.A., Schuch, W., Martin, C. and Bevan, M. (1994) A flower-specific Myb protein activates transcription of phenylpropanoid biosynthetic genes. *EMBO Journal* 13, 128–137.

Saito, N., Yokoi, M., Ogawa, M., Kamijo, M. and Honda, T. (1988) 6-hydroxyanthocyanidin glycosides in the flowers of *Alstromeria. Phytochemistry* 27, 1399–1401.

Sandmann, G. (1991) Biosynthesis of cyclic carotenoids: biochemistry and molecular genetics of the reaction sequence. *Physiologia Plantarum* 83, 186–193.

Sandmann, G. (1994) Phytoene desaturase: genes, enzymes and phylogenetic aspects. *Journal of Plant Physiology* 143, 444–447.

Schledz, M., Lintig, J., Albabili, S., Kleinig, H. and Beyer, P. (1994) EMBL/GenBank Accession X78814.

Schwinn, K.E. (1994) Novel flavonoid 3′-hydroxylase activity in petunia. *Polyphenols Actualities* 11, 58.

Schwinn, K.E., Markham, K.R., Miller, R.M., Davies, K.M., Ofman, D.J., Deroles, S.C., Bradley, J.M., Manson, D.G. and Given, N.K. (1993) Floral flavonoids and their modification in lisianthus. *Proceedings of the XVIIth Eucarpia Symposium, Ornamental Section,* San Remo, Italy, pp. 337–341.

Schwinn, K.E., Markham, K.R. and Given, N.K. (1994) Floral flavonoids and the potential for pelargonidin biosynthesis in commercial chrysanthemum cultivars. *Phytochemistry* 35, 145–150.

Scolnik, P.A. and Bartley, G.E. (1995) Nucleotide sequence of lycopene cyclase (Genbank L40176) from *Arabidopsis*. Plant Gene Register PGR95-019. *Plant Physiology* 108, 1343.

Simada, Y., Ohbayashi, M., Nakano, R., Okinaka, Y., Kiyokawa, S. and Kikuchi, Y. (1993) Molecular cloning of a cDNA encoding flavonoid-3′5′-hydroxylase from *Petunia hybrida* II. *Plant Cell Physiology* 34, s17.

Sommer, H. and Saedler, H. (1986) Structure of the chalcone synthase gene of *Antirrhinum majus*. *Molecular and General Genetics* 202, 429–434.

Stewart, R., Norris, K. and Asen, S. (1975) Microspectrophotometric measurement of pH and pH effect on color of petal epidermal cells. *Phytochemistry* 14, 937–942.

Stich, K. and Forkmann, G. (1988a) Biosynthesis of 3-deoxyanthocyanidins with flower extracts from *Sinningia cardinalis*. *Phytochemistry* 27, 785–789.

Stich, K. and Forkmann, G. (1988b) Studies on columnidin biosynthesis with flower extracts from *Columena hybrida*. *Zeitschrift für Naturforschung* 43c, 311–314.

Stich, K., Eidenberger, T., Wurst, F. and Forkmann, G. (1992) Enzymatic conversion of dihydroflavonols to flavan-3-4-diols using flower extracts of *Dianthus caryophyllus* L. (carnation). *Planta* 187, 103–108.

Strack, D. and Wray, V. (1994) Recent advances in betalain analysis. In: Behnke, H.-D. and Mabry, T.J. (eds), *Caryophyllales; Evolution and Systematics*. Springer-Verlag, Berlin, pp. 263–277.

Tanaka, Y., Fukui, Y., Fukuchimizutani, M., Holton, T.A., Higgins, E. and Kusumi, T. (1995) Molecular cloning and characterisation of *Rosa hybrida* dihydroflavonol 4-reductase gene. *Plant and Cell Physiology* 36, 1023–1031.

Teutsch, H.G., Hasenfratz, M.P., Lesot, A., Stoltz, C., Garnier, J.-M., Jeltsch, J.-M., Durst, F. and Werck-Reichhart, D. (1993) Isolation and sequence of a cDNA encoding the Jerusalem artichoke cinnamate 4-hydroxylase, a major plant cytochrome P450 involved in the general phenylpropanoid pathway. *Proceedings of the National Academy of Sciences, USA* 90, 4102–4106.

Toguri, T., Azuma, M. and Ohtani, T. (1993) The cloning and characterization of a cDNA encoding a cytochrome P450 from the flowers of *Petunia hybrida*. *Plant Science* 94, 119–126.

Trezzini, G.F. and Zrÿd, J.-P. (1990) *Portulaca grandiflora:* a model system for the study of the biochemistry and genetics of betalain synthesis. *Acta Horticulturae* 280, 581–585.

van Blokland, R., de Lange, P., Mol, J.N.M. and Kooter, J.M. (1993) Modulation of gene expression in plants by antisense genes. In: Crooke, S.T. and Lebleu, B. (eds), *Antisense Research and Applications*. CRC Press, London, pp. 125–148.

van der Krol, A.R., Lenting, P.E., Veenstra, J.G., van der Meer, I.M., Koes, R.E., Gerats, A.G.M., Mol, J.N.M. and Stuitje, A.R. (1988) An antisense chalcone synthase gene in transgenic plants inhibits flower pigmentation. *Nature* 333, 866–869.

van der Krol, A.R., Mur, L.A., Beld, M., Mol, J.N.M. and Stuitje, A.R. (1990) Flavonoid genes in *Petunia*: addition of a limited number of additional copies may lead to a suppression of gene activity. *Plant Cell* 2, 291–299.

van Tunen, A.J. and Mol, J.N.M. (1991) Control of flavonoid synthesis and manipulation of flower colour. In: Grierson, D. (ed.), *Developmental Regulation of Plant Gene Expression.* Blackie, Glasgow, pp. 94–130.

van Tunen, A.J., Koes, R.E., Spelt, C.E., van der Krol, A.R., Stuitje, A.R. and Mol, J.N.M. (1988) Cloning of the two chalcone flavanone isomerase genes from *Petunia hybrida:* coordinate, light-regulated and differential expression of flavonoid genes. *EMBO Journal* 7, 1257–1263.

van Tunen, A.J., Hartman, S.A., Mur, L.A. and Mol, J.N.M. (1989) Regulation of chalcone flavanone isomerase (CHI) gene expression in *Petunia hybrida:* the use of alternative promoters in corolla, anthers and pollen. *Plant Molecular Biology* 12, 539–551.

Weiss, D., van der Luit, A.H., Kroon, J.T.M., Mol, J.N.M. and Kooter, J.M. (1993) The petunia homologue of the *Antirrhinum majus candi* and *Zea mays A2* flavonoid genes; homology to flavanone 3-hydroxylase and ethylene-forming enzyme. *Plant Molecular Biology* 22, 893–897.

Welle, R., Schröder, G., Schiltz, E., Grisebach, H. and Schröder, J. (1991) Induced plant responses to pathogen attack, analysis and heterologous expression of the key enzyme in the biosynthesis of phytoallexin in soyabean (*Glycine max* Merr. cv Harosoy 63). *European Journal of Biochemistry* 196, 423–430.

Wienand, U., Sommer, H., Schwarz, Zs., Shepherd, N., Saedler, H., Kreuzaler, F., Ragg, H., Fautz, E., Hahlbrock, K., Harrison, B. and Peterson, P.A. (1982). A general method to identify plant structural genes among genomic DNA clones using transposable element induced mutations. *Molecular and General Genetics* 187, 195–201.

Wiering, H. and de Vlaming, P. (1984) Inheritance and biochemistry of pigments. In: Sink, K.C. (ed.), *Petunia*. Springer-Verlag, Berlin, pp. 49–76.

Williams, C.A. and Harborne, J.B. (1993) Flavone and flavonol glycosides. In: Harborne, J.B. (ed.), *The Flavonoids: Advances in Research Since 1986.* Chapman & Hall, London, pp. 337–386.

Wong, E. (1976) Biosynthesis of flavonoids. In: Goodwin, T.W. (ed.), *Chemistry and Biochemistry of Plant Pigments, Vol. 1.* Academic Press, New York, pp. 464–526.

Xue, B., Charest, P. and Rutledge, B. (1994) Isolation and characterization of a C1-related myb cDNA clone from black spruce and its functional assay via transient gene expression. In: *Abstracts, 4th International Congress of Plant Molecular Biology*, Amsterdam, Abstract 765.

Yoshida, K., Kondo, T., Okazaki, Y. and Katou, K. (1995) Cause of blue petal colour. *Nature* 373, 291.

Zrÿd, J.-P., Hinz, U., Bindschaedler, L. and Müller, L. (1993) Molecular genetics of the betalain pathway in higher plants and fungi. *Phytochemical Society of North America Newsletter* 33, 15.

Prospects for the Improvement of Tolerance to Abiotic Stresses Using Recombinant DNA Approaches

16

Charles L. Guy

Plant Molecular and Cellular Biology Program, Department of Environmental Horticulture, Institute of Food and Agricultural Sciences, University of Florida, Gainesville, Florida 32611, USA

INTRODUCTION

All of the various physical components of the natural environment are rarely optimal with respect to growth and development during any plant's entire life cycle. Because terrestrial environments are dynamic, most plants have evolved mechanisms to cope with marginally unfavourable conditions that may otherwise be deleterious.

Unlike animals, plants are sessile and unable to move from unfavourable to more optimal environmental conditions. As a consequence, for plants to survive and continue from one generation to the next under the duress of deleterious environmental conditions, their fate has to be largely determined from within. Because natural environments historically are inherently dynamic within reasonably defined limits over the time-scale of the life cycle of most higher plants, the homeostatic processes that support growth and development must possess a level of flexibility. Evidence of this flexibility can best be seen in the wide geographical distribution of many species, including a number of perennials, e.g. red maple, smooth sumac and black walnut, just to name a few. How widely spread a given species is distributed may largely depend on the magnitude of flexibility in its homeostatic processes and how amenable these basic processes are to mutagenic optimization to permit ecotypic variation necessary for successful growth and development over widely divergent environments.

TYPES OF ENVIRONMENTAL STRESSES

There are many components that make up the physical environment from which all of earth's biota has arisen. For terrestrial plants, the major abiotic elements of the environment that influence their continued existence in nature are water, light, temperature and essential nutrients. At any given time, one of these major factors, or one of a number of other minor factors, may be less than optimal for growth and development. The environmental factors most likely to impose stress in plants are listed in Table 16.1. Of these stresses, those associated with water and temperature are the most pervasive. In addition to being present in excess, or inadequate amounts, water can exist in two physical forms, a liquid or solid, depending on the temperature.

Table 16.1. Typically encountered abiotic environmental stresses of plants growing in nature.

1. Temperature
   - High
     - Heat shock
     - Unremitting
   - Low
     - Chilling
     - Freezing
2. Water
   - Drought
   - Flooding
3. Salt
   - Ionic
   - Osmotic
4. Nutrient
   - Excess
   - Deficiency
5. Heavy metal
6. Atmospheric
   - Acid rain
   - Sulphur dioxide
   - Ozone
   - Wind
   - Hail
7. Radiation
   - Photosynthetically active
   - Ultraviolet (UV)
8. Biological
   - Anthropogenic
   - Insect
   - Microbial

Together, temperature and water exert the greatest influence on where, when and how the vast majority of the earth's flora exists.

ORNAMENTAL PLANT PRODUCTION AND ENVIRONMENTAL STRESS

The production of ornamental plants is perhaps more controlled with respect to the cultural requirements of plants than any other form of crop production in agriculture. Production conditions can range from the most highly controlled state, that of *in vitro* culture, to the more variable and less controlled conditions of field-grown materials. In greenhouse situations, most, if not all environmental stresses can be completely eliminated with appropriate care, even in unusual geographic settings. If ambient temperature is low, the growing area can be heated; if natural light or daylength is suboptimal, supplemental artificial lighting can be provided or withheld, and so on with a number of other potential stress factors. Operations that emphasize field grown ornamental materials are usually reasonably equipped to prevent or significantly mitigate many of the common environmental stresses that may occur in nature. Therefore, environmental stress is seldom the serious obstacle to production nor is it the annual spectre that producers of field crops face. The one exception to this may be the impact of extreme low temperatures on field grown materials, especially if the operation is large and makes extensive use of field settings of non-native species (Richer-Leclerc *et al.*, 1993). Operations that fit a given situation are most often involved in production of sizeable landscape materials like shrubs and trees (Lindstrom and Dirr, 1989). If plant materials are small, they can be temporarily moved to a protective environment, grouped together, and covered or protected in another way. Plants native to the locale are usually not at risk from low temperature injury unless the temperature is unseasonably extreme, with perhaps one exception. Roots can be one of the most sensitive parts of a plant to freeze injury. The roots of containerized plants may be damaged because of the inability of the limited soil volume to adequately buffer the reduction in soil temperature that would otherwise occur if the plant were in the ground. However, when plants are containerized, producers can move the materials into shelters or they can irrigate or temporarily flood to prevent low temperature damage to roots.

While exposure of ornamental plants to environmental stresses during production can be minimized, their greatest vulnerability most often occurs post-production during shipping and marketing or after installation. For certain types of ornamentals, shipping and the period for retail marketing is when the plants are subjected to the greatest variety of stressful conditions. During shipping, plants may be chilled too much or heat stressed; they may be exposed to high concentrations of gases which can be deleterious, like CO_2 or endogenously produced ethylene, light levels invariably are markedly altered (most often not provided), and mechanical damage from handling is common, as is inadequate or excessive irrigation. Thus, it is during the post-production stage where introducing stress tolerance is likely to have the greatest benefit for ornamentals. Beyond this post-production stage, plants that are intrinsically more impervious to the vagaries of environmental stresses will perform better once in consumer hands, be it a potted flowering plant, an indoor foliage plant, a landscape plant or any other ornamental.

MECHANISMS OF STRESS INJURY

How plants respond to and cope with the variety of environmental stresses is of critical importance to agricultural interests and to long-term human welfare. To understand how plants withstand stressful conditions, we must understand the nature of each type of stress. Unfavourable abiotic conditions impose stress in plants in two major ways, via mechanical and/or chemical mechanisms (Table 16.2). By far the easiest forms of stress to understand are those that cause damage by mechanical means on a macroscopic scale. Wind, hail and snow represent three examples of this kind of mechanical stress. More difficult to understand are those stresses that are mediated strictly by chemical mechanisms. In this grouping are ultraviolet (UV) irradiation, atmospheric pollutants, nutrient imbalances, heavy metals, and high temperature. The most difficult to understand are stresses that are mediated by both mechanical and chemical mechanisms. Included in this grouping are drought, freezing and salinity. It is not surprising that, in spite of extensive research efforts, these stresses still remain the least understood. Table 16.3 outlines some of the biochemical and molecular consequences of the various stresses. Some stress mechanisms are quite well understood, while others remain poorly defined. While each type of abiotic stress may generally act by a unique physical and/or chemical mechanism, it is clear that different stresses can result

Table 16.2. Modes of action for the major abiotic environmental stresses.

	Mechanical	Chemical
Wind	✔	
Snow	✔	
Hail	✔	
Anthropogenic	✔	
Drought	✔	✔
Osmotic	✔	✔
Salt	✔	✔
Freezing	✔	✔
Chilling		✔
Flooding		✔
Air pollutants		✔
Ultraviolet light		✔
Excess photosynthetically active radiation		✔

Table 16.3. An abbreviated summary of the metabolic and biochemical alterations during abiotic stress.

Type	Macroscopic	Biochemical/molecular
Ultra violet light	Pigmentation → no symptoms	Thymine dimerization, Lipid peroxidation
Flooding	Chlorosis → wilting → death	Anaerobic respiration Photoinhibition Toxic metabolites pH imbalance
High temperature	Growth inhibition → death	Protein denaturation Membrane liquefaction Photoinhibition ?
Low temperature	Growth cessation → photobleaching → death	Lipid peroxidation Membrane dysfunction Photoinhibition Protein denaturation ?
Drought	Growth inhibition → wilting → death	Ionic imbalance Photoinhibition Enzyme inhibition ?
Salt	Growth inhibition → leaf drop → death	Lipid peroxidation Ionic imbalance Photoinhibition Enzyme inactivation ?
Heavy metal	Growth inhibition → necrosis → death	Enzyme inactivation ?

in similar pathological symptoms. Perhaps the best examples of this involve photoinhibition, photooxidation and lipid peroxidation caused by UV irradiation, low temperature/high light, and salt stress. Other common mechanisms of injury may be a general membrane dysfunction, enzyme inactivation, or protein denaturation. The last three may be secondary injuries that occur in dying cells, or they may represent a primary lesion resulting from stress. Regardless of whether they represent a primary or secondary response, a number of metabolic or physiologic responses appear to be influenced by a variety of stresses.

STRESS TOLERANCE AND A DEVELOPMENTAL WINDOW

The life cycle of a plant has many stages: dormant seed, germination, emergence, vegetative growth, reproductive phase including the complex development of inflorescences, gametogenesis, sporogenesis, fertilization, fruit and seed formation. For most plants, each developmental stage will have its own characteristic sensitivity, with some having more and others having less tolerance to environmental stresses. Often the switch to the reproductive phase is the most vulnerable time for plants (Hague, 1988). Similarly, certain stages in the life cycle might be especially adaptable to the imposition of stress in such a way as to alter morphology and permanently express a level of tolerance that is not possible if the stress is imposed at any other time in the life cycle (Amzallag *et al.*, 1993). A good example of this is seen in the response of leaf development to incident light. During emergence and expansion of the leaf lamina, irradiance level exerts a profound influence on morphology and physiology that endures until the leaf undergoes senescence (Boardman, 1977). Under high light conditions, 'sun leaves' are produced with lower chlorophyll levels, higher chlorophyll *a*/chlorophyll *b* ratio, greater photosynthetic capacity, high stomata density per unit area, and thick lamina. Development under low light conditions results in the formation of 'shade leaves' with higher chlorophyll content, lower chlorophyll *a*/chlorophyll *b* ratio, lower photosynthetic capacity, decreased stomata density per unit area and thin lamina. While pigment content and photosynthetic capacity can be adjusted in response to changing light levels, stomatal density and lamina thickness are fixed for the particular light conditions that prevailed during development. This example constitutes what is termed a developmental window. In some situations, the window is a stage of maximum sensitivity, in others it may be the stage of maximum adaptability. This phenomenon highlights the fact that the interaction between the plant and its environment is indeed a complex interplay of factors.

PHYSIOLOGICAL AND METABOLIC CONSEQUENCES OF ENVIRONMENTAL STRESS

PHOTOSYNTHETIC CARBON ASSIMILATION

Perhaps the single most important process occurring in plants, to life as we know it, is photosynthesis. Here light energy is harvested and utilized in a host of ways, not the least of which is the assimilation of CO_2 into triose phosphates. Production of triose phosphates in conjunction with the synthesis of ATP and NADPH from photosynthetic electron transport processes provides the basic building blocks for all aspects of metabolism needed to support growth and development. Thus, it is not surprising that factors that restrict a plant's ability to assimilate carbon into usable form will ultimately affect growth processes. Usually one of the earliest discernible consequences of any acute environmental stress is a reduction in the growth rate (Levitt, 1972). Growth inhibition may result from the direct impairment of a fundamental growth-related cellular process or function common to all cells, or result indirectly from an alteration in processes ancillary for growth such as photosynthesis. Stresses that alter plant water relations (osmotic, salinity, drought, flooding, low and high temperature) all are known to impair growth and photosynthetic carbon assimilation. Reductions in growth caused by these stresses could be a direct effect from loss of turgor and turgor-driven cell expansion, or an indirect effect from the impairment of CO_2 assimilation.

Carbon assimilation (A) is impaired by many stresses. Water deficits during drought conditions limit assimilation. One of the best known responses of plants to water stress is the reduction of stomatal aperture and conductance to limit the loss of water (Levitt, 1972). The closing of the stomata reduces water loss, but at the same time restricts the availability of CO_2 for carbon assimilation (Bassman and Zwier, 1991) by lowering intercellular CO_2 (C_i) concentration (Farquhar and Sharkey, 1982). While droughted plants may exhibit greater water use efficiency and less stomatal limitation of A than non-stressed plants (Heitholt *et al.*, 1991), CO_2 assimilation invariably is less than in non-stressed plants (Hsiao, 1973) and this will ultimately limit growth and decrease yield. Drought stress also may limit A by a yet undefined non-stomatal limitation component related to optimal organellar function (Hsiao, 1973). Contrasting drought stress with flooding stress, instead of water being limited, it is in excess. Yet stomatal aperture and conductance are reduced in drought and during flooding (Pezeshki and Chambers, 1986), and with flooding the reduction in stomatal conductance is not caused by a water deficit (Pezeshki and Chambers, 1986). However, as with drought, A is reduced and its reduction is strongly correlated with decreased stomatal conductance (Davies and Flore, 1986). Salt stress also causes a reduction in stomatal conductance and lowers C_i, which in turn reduces A (Downton *et al.*, 1985; Brugnoli and Lauteri, 1991). Since high salinity can lead to the accumulation of elevated levels of cations and anions, there is the potential for biochemical inhibition of photosynthesis. High levels of Cl^- appear to reduce A without altering stomatal conductance by causing a decline in RuBisCO activity, largely by reducing the amount of RuBisCO protein (Ziska *et al.*, 1990). High chloride levels also can impair regeneration of RuBP (Brugnoli and Lauteri, 1991), and lower pool size of RuBP, triose phosphates and phosphoglycerate (Ziska *et al.*, 1990). Low temperature also can reduce photosynthesis. In chilling sensitive soybean, a single night of 5°C can impair assimilation by nearly 100% on the following day, even if all conditions are favourable for photosynthesis (Purcell *et al.*, 1987). Interestingly, this photosynthetic response of soybean to low temperature is largely attributed to a chill-induced water stress which would have effects analogous to a typical warm temperature drought stress. When tomato is chilled in the dark for 16 h at 1°C, photosynthesis is impaired about equally by decreased stomatal conductance and a non-stomatal impairment of chloroplast function (Martin *et al.*, 1981). On the other hand, as temperature increases, A increases to some optimum, then decreases as superoptimal temperatures are experienced (Kobza and Edwards, 1987). The limitations imposed by high temperatures on photosynthesis have been attributed to enhanced photorespiratory activity, feedback limitation caused by reduced photosynthate utilization, enzyme inactivation, or diminished photochemistry. In wheat, high temperature reduced A by causing a decline in RuBisCO and

FBPase activities, and a loss of ability to utilize triose phosphates (Kobza and Edwards, 1987). With potato, heat stress appeared to influence A first by reducing stomatal conductance, which in turn lowered C_i, and at longer durations of stress, by a loss of chlorophyll (Reynolds *et al.*, 1990).

PHOTOINHIBITION

Photochemistry and the ability to carry on electron transport are also sensitive to environmental stresses. Most frequently impairment of photosynthetic electron transport leads to a condition termed photoinhibition. Photoinhibition, as defined by Powles (1984) is 'the reduction of photosynthetic capacity, independent of gross changes in pigment concentration, induced by visible light (400–700 nm)'. Another form of injury to the photosynthetic machinery is the photodestruction of the light-harvesting pigments, known as photooxidation. Photooxidation is the light-dependent oxidative bleaching of photosynthetic pigments (Powles, 1984). Photooxidation is an irreversible secondary event occurring well after the onset of photoinhibition. Thus, photoinhibition can be viewed as a primary response or injury caused by stress. Photoinhibition can occur whenever light levels are in excess of the electron transport system's ability to handle the electron flux, even in the absence of other stress conditions. However, environmental stresses tend to exacerbate the propensity for photoinhibition. Those stresses known to trigger photoinhibition include heat, cold, drought, flooding and salinity. When oleander was drought stressed under high and low light, photoinhibition occurred, but it occurred earlier and was more severe with high light (Powles and Björkman, 1983). It appears that photoinhibition during drought affects either photosystem II photochemistry or plastoquinone reoxidation in photosystem I (Genty *et al.*, 1987). The degree of drought stress is also paralleled by the severity of photoinhibition in sorghum and pearl millet (Masojidek *et al.*, 1991). Similarly, low or high temperature increases the potential for photoinhibition. Plants that are most sensitive to chilling temperatures are most susceptible to photoinhibition when photon flux density is high, and even very cold tolerant plants like wheat can be photoinhibited if the temperature is low enough (Hetherington *et al.*, 1989). Photoinhibition during chilling appears to arise from inhibition of PS II activity and perhaps a disruption of electron transfer from PS II to the plastoquinone pool (Tyystjärvi *et al.*, 1989). High temperature likewise accelerates the onset of photoinhibition during high (photon flux density) PFD. As with cold, PS II appears to be most sensitive to heat while PS I is relatively stable (Berry and Björkman, 1980). Heat stress in wheat lowers whole chain electron transport and quantum yield, but its greatest impact appears to reside near the water splitting complex (Al-Khatib and Paulsen, 1989).

MEMBRANES

Membranes are often linked with deleterious effects caused by environmental stresses. To a large extent, the mechanisms leading to photoinhibition are connected with some type of dysfunction in the thylakoid membrane. Not surprisingly, those stresses that exacerbate photoinhibition (heat, cold, drought, flooding and salt) all alter the function of membranes. These same stresses induce changes in membrane composition. Low temperature exposure increases the unsaturation level of many of the lipid constituents of membranes (Alberdi *et al.*, 1991) although this is but one aspect of a very complex biochemical restructuring (Lynch and Steponkus, 1987). While no other studies have been conducted in the same degree of detail for other stresses as those conducted by Lynch and Steponkus, it is clear that some of the more gross changes in composition also occur (Pearcy, 1978; Misra *et al.*, 1986; Vigh *et al.*, 1986; Blits and Gallagher, 1990). The alterations in membrane composition are considered to be part of an acclimation process aimed at maintaining homeostatic function.

PROTEIN BIOGENESIS

Along with membrane dysfunction, proteins have long been implicated as possible targets for inactivation by environmental stresses (Levitt, 1972). The best biochemically defined response of proteins to any of the environmental stresses considered here is that to heat. It is well established that the native structure of proteins is lost at some point when the temperature is continuously increased. Denaturation and loss of the native structure of globular proteins often result in the irreversible loss of biological function. For most of the major environmental stresses, there is little or no evidence that protein inactivation is a primary injury. More likely, loss of enzymatic activity or protein function is the result of a cell or tissue that is near to death and that can either no longer maintain a normal intracellular environment or is incapable of reversing the mounting metabolic dysfunctions. However, there is a growing linkage between protein denaturation and heat shock (Pelham, 1986). In recent years, some of the proteins induced in response to heat shock conditions have been classified as molecular chaperones (Gething and Sambrook, 1992). These proteins were originally studied as a result of their strong induction during heat stress, but now it is known that they are always present in all cells, and they are essential for cell survival in the absence of stress. It appears that their role is to participate in a variety of functions linked to protein biogenesis and turnover. Some of these processes include protein folding, maintaining proteins in a translocation competent state (mostly unfolded), prevention of aggregation of unfolded proteins, facilitating assembly of oligomeric complexes, and disassembly of oligomer complexes (Gething and Sambrook, 1992). So far, these functions of the heat shock proteins have not been intensively studied in plants (Vierling, 1991). While most proteins can spontaneously fold into a functional or native state *in vitro*, it appears that within the context of a cell the process may be more fragile to environmental stress than previously considered. It has recently been demonstrated that a number of plants show increased levels of 70 kDa heat shock proteins after a period of chilling at 5°C (Neven *et al*., 1992). Perhaps when this area is more intensely investigated, some of the membrane dysfunctions will be found to arise from a disruption of the normal biogenesis of membrane proteins.

GENETICS OF STRESS TOLERANCE

Tolerance to many environmental stresses is considered to be the product of not one or a few genes, but of many genes (Tal, 1965; Sutka, 1981; Stone *et al*., 1993). Traits influenced by numerous genes are termed quantitative traits. Tolerance to drought, heat, chilling or freezing, and salinity may well be quantitative in nature (Levitt, 1980). For tolerance to these stresses, two possibilities exist. For a trait that is the product of many genes, each gene that contributes to the trait, alone, may contribute only a small degree of influence that may be all but imperceptible against a background lacking all of the other additive genes. By itself, it may be totally ineffectual. Alternatively, while tolerance may be the product of numerous genes, a few of these may furnish the majority of the tolerance capacity, and they may or may not contribute to tolerance independently and efficiently in the absence of the other additive genes. It is this case that is most often considered in recombinant approaches to stress tolerance modification. If the genes that contribute the majority of the tolerance act independently of one another and can be identified, isolated and characterized, these are the genes that have the potential to be most useful in recombinant approaches to stress tolerance enhancement. There has yet to be a clear demonstration that any of the genes that contribute to tolerance mechanisms for any major environmental stress will act in such a favourable manner. While no breakthrough using this strategy has occurred, this does not constitute a sufficient basis to abandon this avenue of research. For the most part, the characterization of environmental stress related genes has been under-way for less than ten years, and the majority of the advances have come about in the last five years.

MOLECULAR APPROACHES TO ENHANCE STRESS TOLERANCE

The complexity of how environmental stresses injure plants and the mechanisms that plants use to tolerate or minimize injury has hampered efforts to understand and improve plant stress tolerance. One tool that holds promise to help researchers understand stress responses and to possibly enhance tolerance is recombinant DNA. Several attempts to alter stress tolerance in transgenic plants have been reported (Table 16.4). At this early stage, the results are mixed but most encouraging. In transgenic tobacco, expression of several 'dehydrin' genes from the desiccation-tolerant resurrection plant failed to alter drought tolerance (Iturriaga *et al.*, 1992). Overexpression of a mitochondrial Mn superoxide dismutase in chloroplasts or mitochondria has been shown to improve resistance to oxygen radical stress (Bowler *et al.*, 1991) in transgenic tobacco, while overexpression of a chloroplastic Cu/Zn superoxide dismutase failed to provide any measurable benefit against ozone stress (Pitcher *et al.*, 1991). Transgenic tobacco expressing high chloroplastic glutathione reductase activity, another component of antioxidant metabolism, exhibited markedly lower susceptibility to paraquat or sulphur dioxide stress (Aono *et al.*, 1993). More promising results have come from efforts to alter chilling and salt tolerance. When chilling sensitive *Anacystis nidulans* was transformed with *desA*, a gene encoding a Δ^{12} desaturase from chilling-tolerant *Synechocystis*, it became more resistant to chilling (Wada *et al.*, 1990). Even more striking was the benefit to tobacco expressing a bacterial gene for mannitol-1-phosphate dehydrogenase. Plants ostensibly able to produce mannitol were able to grow in 250 mM NaCl almost as well as plants in the absence of salt stress (Tarczynski *et al.*, 1993). These initial results show that together with biochemical and physiological studies, recombinant DNA approaches hold much promise to provide a better understanding of stress injury mechanisms and possibly enhance tolerance in ways not otherwise possible.

GENETIC RESOURCES FOR ALTERING STRESS TOLERANCE

There are several categories in the gene pool concept (Hodgkin and Debouck, 1992). Genes from within the species constitute the primary gene pool; those found in related species with perhaps some degree of sexual compatibility belong to the secondary gene pool; and the tertiary gene pool includes sources that are evolutionarily related but sexually incompatible. With the advent of modern molecular genetics, plant scientists can now access additional biological sources for novel genes. Taking the gene pool concept further through the use of molecular genetics and transformation, we can envision two new levels; a quaternary pool to include all plant sources and a quinternary pool of genes from all other organisms, including microorganisms and animals.

Generally, sources for stress-tolerant germplasm have been sought from environments where a given stressful condition has existed for a long period of time. Two examples of traditional approaches to stress tolerance improvement in crop plants are especially illustrative. Charles Rick's group at the University of California at Davis has used a wild tomato relative, *Lycoperscion cheesmanii*, found on the seashore of the Galapagos Islands, for the genetic improvement of salt tolerance in tomatoes. Not surprisingly, plants growing at the seashore must have developed over time a level of salt tolerance that is generally lacking in plants whose normal habitat lacks saline conditions. Similarly, a species of the potato family, *Solanum acaule*, found at 3500–4000 m

Table 16.4. Recombinant approaches to study the mechanism or enhance plant stress tolerance.

Stress	Gene	Function	Source	Target	Outcome	Reference
Heat		Heat shock transcription factor	*Arabidopsis*	*Arabidopsis*	Enhanced tolerance	Lee *et al.* (1995)
Chilling	*desA*	Desaturase	*Synechocystis*	*Anacystis*	Enhanced tolerance	Wada *et al.* (1990)
Chilling		Glycerol-3-P acyltransferase	*Arabidopsis* Squash	Tobacco	Enhanced tolerance	Murata *et al.* (1992)
Chilling	*plsB*	Glycerol-3-P acyltransferase	*E. coli*	*Arabidopsis*	Chilling sensitivity	Wolter *et al.* (1992)
Freezing	*kin1*	'Antifreeze'	*Arabidopsis*	Tobacco, petunia	No change	M. Borg (pers. comm.)
Freezing		'Antifreeze'	Fish	Tobacco	No change	P.L. Davies (pers. comm.)
Freezing	*inaZ*	Ice nucleation protein	*P. syringae*	Tobacco	No change	Baertlein *et al.* (1992)
Freezing		Mn SOD	Tobacco	Alfalfa	Enhanced tolerance	McKersie *et al.* (1993)
Drought	pcC6-19 pcC3-06 pcC27-45	'Dehydrin' 'LEA' Unknown	Resurrection plant	Tobacco	No change	Iturriaga *et al.* (1992)
Drought		Carbonic anhydrase	Tobacco	Tobacco	Enhanced tolerance	Majeau *et al.* (1995)
Salt	*mtlD*	Mannitol-1-P dehydrogenase	*E. coli*	Tobacco	Enhanced tolerance	Tarczynski *et al.* 1993)
Osmotic	*p5cs*	Δ^1-Pyrroline-5-carboxylate synthatase	*Vigna*	Tobacco	Enhanced tolerance	Kishor *et al.* (1995)
Ozone SO_2	*gor*	Glutathione reductase	*E. coli*	Tobacco	Enhanced tolerance	Aono *et al.* (1993)
Ozone		Cu/Zn SOD	Petunia	Tobacco	No change	Pitcher *et al.* (1991)
Oxidative	*gor*	Glutathione reductase	*E. coli*	Poplar	Enhanced tolerance	Foyer *et al.* (1995)
Oxidative	pMitSOD	Mn SOD	Tobacco	Tobacco	Enhanced tolerance	Bowler *et al.* (1991)
Oxidative		MN SOD	Tobacco	Tobacco	Enhanced tolerance	Slooten *et al.* (1995)
Oxidative	Cu/Zn SOD	Cu/Zn SOD	Pea	Tobacco	Enhanced tolerance	Gupta *et al.* (1993)
Oxidative		Coproporphyrinogen oxidase	Tobacco	Tobacco	Enhanced tolerance	Kruse *et al.* (1995)

elevations in the Andes possesses a greater degree of frost tolerance than the commercially cultivated potato and could provide a source of germplasm or genes with which to improve the cold tolerance of potato (Palta and Li, 1979).

SOURCES FOR STRESS TOLERANCE GENES

As has been the case for many of the transgenic experiments performed to date, many genes useful for stress tolerance are already present and available in most commercially important crop plants. However, it is clear that some exotic sources for stress tolerance genes have already been utilized, such as the resurrection plant (Iturriaga *et al.*, 1992). It should come as no surprise that genetic resources for abiotic stress tolerance may be found anywhere including in Third World countries. Thus, accessing genetic resources from natural habitats in areas of genetic diversity would constitute an additional form of material that could be transported out of developing countries to developed or technologically advanced countries (Abbiw, 1992). Frequently, developed countries rely upon plant materials obtained initially from developing or Third World countries for industry and commerce. There are at least 12 main groups of materials that are transported from developing to developed countries (Abbiw, 1992). These include timber, charcoal, cash crops (e.g. coffee, tobacco, sugar), bamboo and palms, medicinal plants, dye-making plants, ornamentals, sweeteners (fruits), food plants, spices and flavourings, and other (gums, resins). Since many economically important ornamental plants have their origins in developing countries, it is not unexpected that these areas of plant diversity also possess a potential treasure trove of genetic resources. This ultimately raises issues of ownership of the these resources, not to mention the potential for extinction of rare and endangered plants.

Wild plant extinction as a result of human activity is now commonplace and we are losing potential sources of economically valuable genes. While gene transfer is becoming common and offers the potential to improve plants, many wild sources of useful genes are becoming extinct and with them perhaps novel genes of value. Therefore, we may miss some opportunities (Lin *et al.*, 1992) with respect to altering crop or ornamental plant stress tolerance.

ENHANCING STRESS TOLERANCE IN ORNAMENTALS

Is there a benefit in enhancing environmental stress tolerance of ornamental plants from a production standpoint? The answer is yes, but it should be a qualified yes. In special cases, like cold stress on containerized plants, there would be clear and important advantages that would accrue from enhancing stress resistance. For at-risk field grown materials that are produced on a large scale, reducing the perils of stress injury by increasing resistance could lead to a change in cultural and/or protective strategies that could conceivably reduce production costs.

Engineering enhanced stress tolerance in ornamentals may have its greatest benefit as a value-added attribute. If the durability of ornamentals against post-production stresses can be improved, the product delivered to the consumer will be much healthier and of much higher quality. After purchase, the durability and versatility of more-stress-tolerant ornamentals will be increased, through improved quality traits like longevity, or floriferousness, or lowered maintenance requirements during periods of mild stress. Further, overall survival under less than optimal post-consumer conditions may be improved, and, as such, enhanced performance capabilities would represent real value-added input to the commodity. Therefore, the greatest benefit of biotechnological based improvements in tolerance to abiotic stresses of ornamental plants may be through the value-added factor to the consumers, and not directly on the production side as one might initially conclude. Nevertheless, such value-added features will translate into higher marketability, improved consumer satisfaction, and greater overall economic returns.

PRACTICALITY OF BIOTECHNOLOGICAL APPROACHES TO ENHANCED ENVIRONMENTAL STRESS TOLERANCE

At present, much effort is devoted to understanding the basic mechanisms by which plants are injured by environmental stress. Coupled with understanding of how plants are injured is the scientific pursuit of the mechanisms by which plants tolerate the stresses and strains imposed, from the level of the whole plant all the way to the molecular interactions and consequences. It is presumed that once an understanding of how injury occurs and how tolerant plants resist injury is achieved, strategies for genetically altering tolerance in plants of economical importance can be developed. While transformation methodologies have been developed for a wide variety of plants, much of the attention has focused on food crops, and mostly on a subset of food crops of the greatest economic importance. In direct contrast to this, ornamental plants encompass a much wider and larger array. Although most of the transformation and recombinant DNA technologies can, and will, be utilized with ornamentals, they will require some level of adaptation and modification. Since these technologies have not yet been simplified to a level that makes their use on any plant completely feasible, an economic barrier will exist to the genetic improvement for many ornamentals for at least some time into the future. While efforts to use recombinant approaches in stress tolerance enhancement will surely proceed for high-value and important ornamentals, I do not foresee a general or universal application of this technology in the very near future. However, if the rate of technological advancement and simplification of recombinant methodologies continues at its current pace, in the years to come, little will stand in the way of molecular genetic approaches to improved stress tolerance of virtually all ornamentals, other than establishing a clear economic need and benefit.

REFERENCES

Abbiw, D. (1992) Transportation of plant materials out of developing countries. In: Adams, R.P. and Adams, J.E. (eds), *Conservation of Plant Genes: DNA Banking and In Vitro Biotechnology.* Academic Press, San Diego, pp. 257–266.

Alberdi, A., Meza-Basso, L., Fernandez, J., Cristi, R. and Romero, M. (1991) Lipid composition of chloroplasts from cold-acclimated and non-acclimated *Nothofagus dombeyi. Phytochemistry* 30, 763–768.

Al-Khatib, K. and Paulsen, G.M. (1989) Enhancement of thermal injury to photosynthesis in wheat plants and thylakoids by high light intensity. *Plant Physiology* 90, 1041–1048.

Amzallag, G.N., Seligmann, H., and Lerner, H.R. (1993) A developmental window for salt-adaptation in *Sorghum bicolor. Journal of Experimental Botany* 44, 645–652.

Aono, M., Kubo, A., Saji, H., Tanaka, K. and Kondo, N. (1993) Enhanced tolerance to photooxidative stress of transgenic *Nicotiana tabacum* with high chloroplastic glutathione reductase activity. *Plant Cell Physiology* 34, 129–135.

Baertlein, D.A., Lindow, S.E., Panopoulos, N.J., Lee, S.P., Mindrinos, M.N. and Chen, T.H.H. (1992) Expression of a bacterial ice nucleation gene in plants. *Plant Physiology* 100, 1730–1736.

Bassman, J.H. and Zwier, J.C. (1991) Gas exchange characteristics of *Populus trichocarpa, Populus deltoides* and *Populus trichocarpa* × *P. deltoides* clones. *Tree Physiology* 8, 145–159.

Berry, J.A. and Björkman, O. (1980) Photosynthetic response and adaptation to temperature in higher plants. *Annual Review of Plant Physiology* 31, 491–543.

Blits, K.C. and Gallagher, J.L. (1990) Effect of NaCl on lipid content of plasma membranes isolated from roots and cell suspension cultures of the dicot halophyte *Kosteletzkya virginica* (L.) Presl. *Plant Cell Reports* 9, 156–159.

Boardman, N.K. (1977) Comparative photosynthesis of sun and shade plants. *Annual Review of Plant Physiology* 28, 355–377.

Bowler, C., Slooten, L., Vandenbranden, S., De Rycke, R., Botterman, J., Sybesma, C., Van Montagu, M. and Inzé, D. (1991) Manganese superoxide dismutase can reduce cellular damage mediated by oxygen radicals in transgenic plants. *EMBO Journal* 10, 1723–1732.

Brugnoli, E. and Lauteri, M. (1991) Effects of salinity on stomatal conductance, photosynthetic capacity, and carbon isotope discrimination of salt-tolerant (*Gossypium hirsutum* L.) and salt-sensitive (*Phaseolus vulgaris* L.) C_3 non-halophytes. *Plant Physiology* 95, 628–635.

Davies, F.S. and Flore, J.A. (1986) Gas exchange and flooding stress of highbush and rabbiteye blueberries. *Journal of the American Society for Horticultural Science* 111, 565–571.

Downton, W.J.S., Grant, W.J.R. and Robinson, S.P. (1985) Photosynthetic and stomatal responses of spinach leaves to salt stress. *Plant Physiology* 77, 85–88.

Farquhar, G.D. and Sharkey, T.D. (1982) Stomatal conductance and photosynthesis. *Annual Review Plant Physiology* 33, 317–345.

Foyer, C.H., Souriau, N., Perret, S., Lelandais, M., Kunert, K.-J., Pruvost, C. and Jouanin, L. (1995) Overexpression of glutathione reductase but not glutathione synthetase leads to increases in antioxidant capacity and resistance to photoinhibition in poplar trees. *Plant Physiology* 109, 1047–1057.

Genty, B., Brantais, J.-M., Viera Da Silva, J.B. (1987) Effects of drought on primary photosynthetic processes of cotton leaves. *Plant Physiology* 83, 360–364.

Gething, M.-J. and Sambrook, J. (1992) Protein folding in the cell. *Nature* 355, 33–45.

Gupta, A.S., Heinen, J.L., Holaday, A.S., Burke, J.J. and Allen, R.D. (1993) Increased resistance to oxidative stress in transgenic plants that overexpress chloroplastic Cu/Zn superoxide dismutase. *Proceedings of the National Academy Sciences, USA* 90, 1629–1633.

Haque, M.Z. (1988) Effect of nitrogen, phosphorus, and potassium on spikelet sterility induced by low temperature at the reproductive stage of rice. *Plant and Soil* 109, 31–36.

Heitholt, J.J., Johnson, R.C. and Ferris, D.M. (1991) Stomatal limitation to carbon dioxide assimilation in nitrogen- and drought-stressed wheat. *Crop Science* 31, 135–139.

Hetherington, S.E., He, J. and Smillie, R.M. (1989) Photoinhibition at low temperature in chilling-sensitive and -resistant plants. *Plant Physiology* 90, 1609–1615.

Hodgkin, T. and Debouck, D.G. (1992) Some possible applications of molecular genetics in the conservation of wild species for crop improvement. In: Adams, R.P. and Adams, J.E. (eds), *Conservation of Plant Genes: DNA Banking and In Vitro Biotechnology.* Academic Press, San Diego, pp. 153–181.

Hsiao, T.C. (1973) Plant responses to water stress. *Annual Review of Plant Physiology* 24, 519–570.

Iturriaga, G., Schneider, K., Salamini, F. and Bartels, D. (1992) Expression of desiccation-related proteins from the resurrection plant *Craterostigma plantagineum* in transgenic tobacco. *Plant Molecular Biology* 20, 555–558.

Kishor, P.B.K., Hong, Z., Miao, G.-H., Hu, C.-A.A. and Verma, D.P.S. (1995) Overexpression of Δ^1-pyrroline-5-carboxylate synthetase increases proline production and confers osmotolerance in transgenic plants. *Plant Physiology* 108, 1387–1394.

Kobza, J. and Edwards, G.E. (1987) Influences of leaf temperature on photosynthetic carbon metabolism in wheat. *Plant Physiology* 83, 69–74.

Kruse, E., Mock, H.P. and Grimm, B. (1995) Reduction of coproporphyrinogen oxidase level by antisense RNA synthesis leads to deregulated gene expression of plastid proteins and affects the oxidative defense system. *EMBO Journal* 14, 3712–3720.

Lee, J.H., Hubel, A. and Schoffl, F. (1995) Derepression of the activity of genetically engineered heat shock factor causes constitutive synthesis of heat shock proteins and increased thermotolerance in transgenic *Arabidopsis. Plant Journal* 8, 603–612.

Levitt, J. (1972) *Response of Plants to Environmental Stresses*. Academic Press, New York.

Lin, Z.-P., Hu, Z. and Ye, H. (1992) Screening valuable genes from wild species of plants. In: Adams, R.P. and Adams, J.E. (eds), *Conservation of Plant Genes: DNA Banking and In Vitro Biotechnology*. Academic Press, San Diego, pp. 241–246.

Lindstrom, O.M. and Dirr, M.A. (1989) Acclimation and low-temperature tolerance of eight woody taxa. *HortScience* 24, 818–820.

Lynch, D.V. and Steponkus, P.L. (1987) Plasma membrane lipid alterations associated with cold acclimation of winter rye seedlings (*Secale cereale* L. cv Puma). *Plant Physiology* 83, 761–767.

Majeau, N., Aroldo, M.A. and Coleman, J.R. (1994) Modification of carbonic anhydrase activity by antisense and over-expression constructs in transgenic tobacco. *Plant Molecular Biology* 25, 377–385.

Martin, B., Ort, D.R. and Boyer, J.S. (1981) Impairment of photosynthesis by chilling-temperatures in tomato. *Plant Physiology* 68, 329–334.

Masojidek, J., Trivedi, S., Halshaw, L., Alexiou, A. and Hall, D.O. (1991) The synergistic effect of drought and light stresses in sorghum and pearl millet. *Plant Physiology* 96, 198–207.

McKersie, B.D., Chen, Y., de Beus, M., Bowley, S.R., Bowler, C., Inze, D., D'Halluin, K. and Botterman, J. (1993) Superoxide dismutase enhances tolerance of freezing stress in transgenic alfalfa (*Medicago sativa* L.). *Plant Physiology* 103, 1155–1163.

Misra, S., Choudhury, A., Pal, P.K. and Gohosh, A. (1986) Effect on the leaf lipids of three species of mangrove of periodic submergence in tidal water. *Phytochemistry* 25, 1083–1087.

Murata, N., Ishizaki-Nishizawa, O., Higashi, S., Hayashi, H., Tasaka, Y. and Nishida, I. (1992) Genetically engineered alteration in the chilling sensitivity of plants. *Nature* 356, 710–712.

Neven, L.G., Haskell, D.W., Guy, C.L., Denslow, N., Klein, P., Green, L.G. and Silverman, A. (1992) Association of 70-kilodalton heat-shock cognate proteins with acclimation to cold. *Plant Physiology* 99, 1362–1369.

Palta, J.P. and Li, P.H. (1979) Frost-hardiness in relation to leaf anatomy and natural distribution of several *Solanum* species. *Crop Science* 19, 665–671.

Pearcy, R.W. (1978) Effect of growth temperature on the fatty acid composition of the leaf lipids in *Atriplex lentiformis* (Torr.) Wats. *Plant Physiology* 61, 484–486.

Pelham, H.R.B. (1986) Speculations on the functions of the major heat shock and glucose-regulated proteins. *Cell* 46, 959–961.

Pezeshki, S.R. and Chambers, J.L. (1986) Variation in flood-induced stomatal and photosynthetic responses of three bottomland tree species. *Forest Science* 32, 914–923.

Pitcher, L.H., Brennan, E., Hurley, A., Dunsmuir, P., Tepperman, J.M. and Zilinskas, B.A. (1991) Overproduction of petunia chloroplastic copper/zinc superoxide dismutase does not confer ozone tolerance in transgenic tobacco. *Plant Physiology* 97, 452–455.

Powles, S.B. (1984) Photoinhibition of photosynthesis induced by visible light. *Annual Review of Plant Physiology* 35, 15–44.

Powles, S.B. and Björkman, O. (1983) Photoinhibition of photosynthesis: effect on chlorophyll fluorescence at 77K in intact leaves and in chloroplast membranes of *Nerium oleander*. *Planta* 156, 97–107.

Purcell, L.C., Ashley, D.A. and Boerma, H.R. (1987) Effects of chilling on photosynthetic capacity, and leaf carbohydrate and nitrogen status of soybean. *Crop Science* 27, 90–95.

Reynolds, M.R., Ewing, E.E. and Owens, T.G. (1990) Photosynthesis at high temperature in tuber-bearing *Solanum* species. *Plant Physiology* 93, 791–797.

Richer-Leclerc, C., Rioux, J.A., Beaudoin, M.F., Lapointe, D., Guillemette, L. and Drapeau, R. (1993) Potentiel d'adaptation et de croissance d'arbres ornementaux sous les conditions climatiques du Québec et du nord-est ontarien. *Canadian Journal of Plant Science* 73, 557–567.

Slooten, L., Capiau, K., Van Camp, W., Van Montagu, M., Sybesma, C. and Inze D. (1995) Factors affecting the enhancement of oxidative stress tolerance in transgenic tobacco overexpressing manganese superoxide dismutase in the chloroplasts. *Plant Physiology* 107, 737–750.

Stone, J.M., Palta, J.P., Bamberg, J.B., Weiss, L.S. and Harbage, J.F. (1993) Inheritance of freezing resistance in tuber-bearing *Solanum* species: evidence for independent genetic control of nonacclimated freezing tolerance and cold acclimation capacity. *Proceedings of the National Academy of Sciences, USA* 90, 7869–7873.

Sutka, J. (1981) Genetic studies of frost resistance in wheat. *Theoretical and Applied Genetics* 59, 145–152.

Tal, M. (1965) Genetics of salt tolerance in higher plants: theoretical and practical considerations. *Plant and Soil* 89, 199–226.

Tarczynski, M.C., Jensen, R.G. and Bohnert, H.J. (1993) Stress protection of transgenic tobacco by production of the osmolyte mannitol. *Science* 259, 508–510.

Tyystjärvi, E., Ovaska, J., Karunen, P. and Aro, E.-M. (1989) The nature of light-induced inhibition of photosystem II in pumpkin (*Cucurbita pepo* L.) leaves depends on temperature. *Plant Physiology* 91, 1069–1074.

Vierling, E. (1991) The roles of heat shock proteins in plants. *Annual Review of Plant Physiology and Plant Molecular Biology* 42, 579–620.

Vigh, L., Huitema, H., Woltjes, J. and van Hasselt, P.R. (1986) Drought stress-induced changes in the composition and physical state of phospholipids in wheat. *Physiologia Plantarum* 67, 92–96.

Wada, H., Gombos, Z. and Murata, N. (1990) Enhancement of chilling tolerance of a cyanobacterium by genetic manipulation of fatty acid desaturation. *Nature* 347, 200–203.

Wolter, F.P., Schmidt, R. and Heinz, E. (1992) Chilling sensitivity of *Arabidopsis thaliana* with genetically engineered membrane lipids. *EMBO Journal* 11, 4685–4692.

Ziska, L.H., Seemann, J.R. and De Jong, T.M. (1990) Salinity induced limitations on photosynthesis in *Prunus salicina*, a deciduous tree species. *Plant Physiology* 93, 864–870.

Engineering for Bacterial and Fungal Disease Resistance

17

H.J.M. Löffler and D.E.A. Florack
DLO Centre for Plant Breeding and Reproduction Research (CPRO-DLO), PO Box 16, NL 6700 AA Wageningen, The Netherlands

INTRODUCTION

Worldwide, a large and diverse variety of ornamental crops are cultivated. As with other crops, these ornamentals are vulnerable to pests and diseases, many of which are caused by bacterial or fungal pathogens (Larson, 1992). A broad spectrum of diseases is incited by these pathogens. Most notable are the generally occurring fungi *Fusarium oxysporum*, causing wilt (e.g. in carnation and chrysanthemum) or rot (mainly in ornamental bulbs); powdery and downy mildews; rusts; *Botrytis* spp., which cause blight on many ornamentals, and the soil-borne fungi *Pythium* spp. and *Rhizoctonia solani*, responsible for damping-off. There are many more fungi that are specific for one or few crops with devastating effects (e.g. *Stromatinia gladioli* in *Gladiolus*). Generally occurring pathogenic bacteria are *Pseudomonas* spp., *Xanthomonas* spp., *Erwinia* spp. (causing the important bacterial blight in many cotoneasters and the genus *Pyrus*) and *Agrobacterium* spp.

Most diseases can effectively be controlled by chemical pesticides. Although these pesticides are a warranty for an economically sound production of ornamentals, their application has two major disadvantages. First, microorganisms are able to develop resistance to pesticides, urging continually higher doses and the development of new pesticides. Second, the continuous use of such pesticides has a severe impact on the environment. For a sustainable floriculture it is therefore essential that the use of pesticides is reduced drastically.

The main alternative for the use of pesticides is growing cultivars which are (partially) resistant to pathogens. This notion has been generally recognized in many other horticultural or agricultural crops where breeding for resistance has actually resulted in cultivars with resistances to a number of pathogens. In ornamentals, breeding for disease resistance has only recently

 Biotechnology of Ornamental Plants (eds R.L. Geneve, J.E. Preece and S.A. Merkle)

received attention (Sparnaaij, 1991). This can be ascribed to certain features characteristic of ornamentals. These will be described briefly below.

DIVERSIFICATION

Numerous ornamental crops are cultivated and for each crop up to hundreds of different cultivars are grown. Therefore, the ornamental collection is more diverse than that of other horticultural or agricultural crops. Since diversification is essential for ornamentals, introduction of disease resistance is interesting only when it can be achieved in a broad cultivar spectrum. Furthermore, this assortment has a high turnover. New cultivars are being developed constantly, and sometimes cultivars last only a few years. Therefore the budget for the development of a resistant cultivar is only limited compared to that of other crops. Breeding for resistance in ornamentals will be stimulated if the resistance is general and easy to introduce. Biotechnology may offer these opportunities (Dons *et al.*, 1991; Dons, 1992).

LIMITED RESEARCH

Due to the large variety of ornamentals and their relatively limited economic importance, research activities are spread over many different crops. Ornamental plant breeding is mainly done by relatively small private companies which cannot afford high investments in research. Therefore, compared to the basic research effort towards resistance (e.g. in tomato), research towards any ornamental is minimal. Since basic research always provides the foundations for application of techniques, breeding for resistance in ornamentals is hampered by the lack of research on ornamentals.

VEGETATIVE PROPAGATION

In contrast to many other crops, ornamentals are mostly propagated vegetatively. Therefore, breeding to homozygosity is not necessary and hardly occurs. Thus most ornamentals are highly heterozygous which implies that cross-pollination will result in segregation of many of the valuable characteristics of the cultivar. Therefore it is almost impossible to add a resistance trait to a specific cultivar; only new, resistant cultivars can be developed.

USE OF PESTICIDES

For many ornamentals, a large part of the plant or the complete plant is the ultimate product that reaches the consumer. Any damage by pathogens will decrease the ornamental value and thus pesticides are extensively used. Since ornamentals are not consumed, residues of pesticides on ornamental plants do not threaten human health directly, thereby making the use of pesticides more easily accepted.

BREEDING FOR RESISTANCE TO BACTERIAL AND FUNGAL DISEASES

CROSS-BREEDING

A large genetic variation for disease resistance exists for many ornamental species. Desirable characteristics can be achieved by pyramiding resistance genes of different cultivars into a breeding selection (Sparnaaij, 1991).

Sometimes interspecific crosses are utilized to introduce resistance into a crop. The genetic barriers in interspecific crosses can be overcome by using *in vitro* techniques such as *in vitro* pollination, grafted style method and embryo rescue (Van Tuyl *et al.*, 1991, 1993). The application of these techniques in lily resulted in the interspecific hybridization between *Lilium longiflorum* and *L. dauricum* that yielded descendants which were resistant to *Fusarium* (Löffler *et al.*, 1995).

Resistance traits from phylogenetically less related species might also be exploited using somatic hybridization by protoplast fusion. However, post-fertilization barriers based on incongruity of the parental genomes can never be overcome by this approach (Colijn-Hooymans, 1990). In addition, 'somatic incompatibility' resulting in the segregation of chromosomes during mitosis, frequently occurs when combining genomes of less related species (Sink, 1993). Therefore the overall potency of protoplast fusion for developing resistant ornamentals seems to be limited.

MARKER-ASSISTED BREEDING

The greatest successes with cross-breeding have been achieved when monogenic traits were involved, because the transfer of polygenic traits is much more complicated. Selection based on the presence rather than the expression of genes would be advantageous. This selection can be indirect, making use of molecular markers linked to the resistance genes (Lindhout, 1990; Jacobsen, 1993). Such markers are now being developed for *Fusarium* resistance in lily. Preliminary results indicate linkage, but as yet the linked markers account only partially for the *Fusarium* resistance in the population (Straathof *et al.*, 1995).

IN VITRO *SELECTION*

A promising technique aiming at selection for a specific trait at the cellular level is *in vitro* selection. The feasibility of this technique for breeding for disease resistance is now being studied for many crops (reviewed by Van den

Bulk, 1991, and Harms, 1992). Despite large research efforts, only a few well documented reports about successful application of this technique have been published. Wenzel and Foroughi-Wehr (1990) demonstrated that resistances in barley, wheat and potato, obtained by *in vitro* selection, were not transmitted to the progeny and might be of epigenetic origin. This is evidently of little interest for breeding purposes. Still, a number of successes proves that the technique is potentially sound (Van den Bulk, 1991).

Success with *in vitro* selection depends on the availability of (i) a regenerable cell culture of the host plant, (ii) a selective agent and (iii) genetic variation for the desired trait.

Regeneration of plants from cell cultures has been achieved for numerous crops, among which many are ornamentals (Ammirato *et al.*, 1990). The selective agent is usually a microbial toxin or culture filtrate of pathogenic fungi or bacteria. These toxins or components in the filtrate should play an important role in the pathogenesis of the respective pathogens. Such a role has only been proven for a limited number of fungal and bacterial toxins. The utilization of an inadequate or inappropriate selective agent could result in poor *in vitro* selections.

Genetic variation (i.e. somaclonal variation: Bouman and de Klerk, Chapter 10) does occur in tissue cultures of ornamentals (Buiatti and Gimelli, 1993). It is difficult to predict how often this results in a new specific trait. The chance that a new resistance gene will develop in tissue culture is low, but the altered expression of an already present resistance gene is feasible. Genetic modifications in a present resistance gene may shift its specificity, thereby creating a new resistance gene (Phillips *et al.*, 1990). Although some preliminary results about the application of *in vitro* selection for disease resistance in ornamentals have been published (Löffler and Mouris, 1991, 1992; Mercuri *et al.*, 1991; Arai and Takeuchi, 1993; Venturo *et al.*, 1993), this method has not yet led to increased levels of resistance in ornamentals. Promising results with *Gladiolus* cells indicate that fusaric acid could be used as a selective agent for resistance to *Fusarium* (Remotti and Löffler, unpublished results). Cells with decreased sensitivity to fusaric acid were selected from a cell suspension culture of the susceptible cultivar 'Peter Pears' (Fig. 17.1). Callus regenerated from these cells had retained the decreased sensitivity to fusaric acid (Fig. 17.1A,B). In addition, these calli were only slightly invaded by the pathogen upon infection (Fig. 1D) in contrast to the non-selected control tissue which was overgrown by the fungus (Fig. 17.1C). Analyses of resistance of plants regenerated from these calli is still in progress.

ENGINEERING BACTERIAL AND FUNGAL DISEASE RESISTANCE

A novel approach in breeding for resistance to pathogens is the application of genetic engineering technology. This technology is based on the direct transfer of genes from one organism to the other and might be employed for engineer-

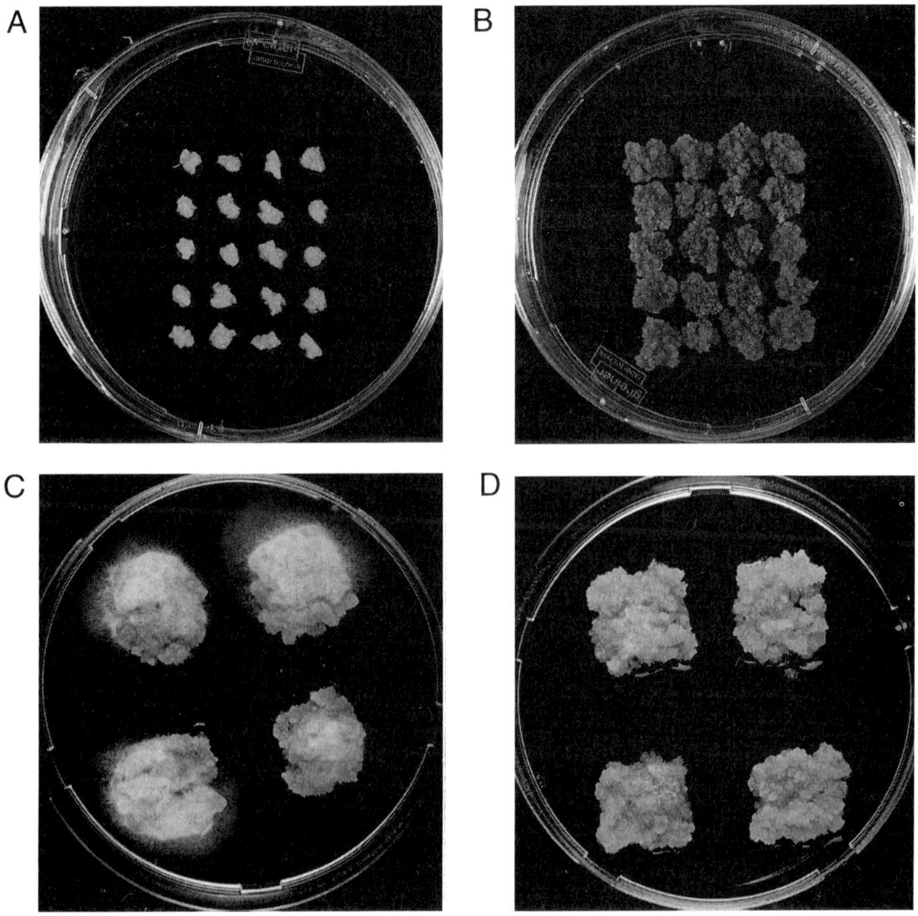

Fig. 17.1. Callus of *Gladiolus* selected for tolerance to fusaric acid (B, D) and control callus (A, C) on MS medium with fusaric acid (A, B) and on plain MS medium but infected with the pathogenic fungus *Fusarium oxysporum* f.sp. *gladioli* (C, D).

ing disease resistance into ornamentals. Prerequisites for a successful application of this technique are: (i) the availability of transformation protocols for ornamental plants, and (ii) the availability of genes that confer resistance. Much effort has been put into the development of transformation techniques, resulting in a wide variety of crops which are accessible for transformation. Recent achievements for ornamentals are discussed in Deroles *et al.*, Chapter 6 of this volume. In a number of crops (but only in one ornamental (Kuehnle *et al.*, 1993), transfer of genes from various sources have led to enhanced resistance to fungi or bacteria. In the following paragraphs some of these genes will be further discussed and examples of how they may be exploited for engineering resistance in ornamental plants will be given.

RESISTANCE GENES

The most obvious genes that may prove effective using genetic engineering technology are those that govern resistance to a specific pathogen in the host. Such genes are present in many plants and many of these are monogenes. The introduction and expression of such genes in ornamentals may confer resistance to the respective pathogen. In tomato, *Verticillium* resistance is known to be a monogenic trait. Isolation of the responsible resistance gene and introduction into a crop like rose may result in a *Verticillium* resistant rose.

In the past few years several disease resistance genes have been isolated from plants. The genes isolated up to now are either involved in a gene-for-gene interaction or in preventing the activity of pathogen-produced pathogenicity factors (Chasan, 1994; Moffat, 1994). The most notable techniques for isolation of such genes are transposon tagging or positional (map-based) cloning.

Johal and Briggs (1992) isolated the *Hm1* gene conferring resistance to *Cochliobolus carbonum (Helminthosporium carbonum)* from maize using the Robertson mutator (Mu) transposon. *The Hm1* gene encodes an NADPH-dependent reductase that inactivates the HC-toxin that is produced by races of the fungus that are pathogenic on *hm1* maize. Other resistance (*R*) genes that have been isolated by transposon tagging are *Cf-9* from tomato and L^6 from flax (Chasan, 1994). The tomato *Cf-9* gene confers resistance to *avr9*-bearing strains of the leaf-mould fungus *Cladosporium fulvum* (*Fulvia fulvum*) and was isolated by the group of Jones (Jones *et al.*, 1994). The flax L^6 gene confers resistance to the flax-rust fungus *Melampsora lini* carrying the corresponding avirulence gene (Lawrence *et al.*, 1994).

Map-based cloning techniques have also been successful. Martin *et al.* (1993) succeeded in cloning the *Pto* gene from tomato, conferring resistance to *avrPto*-bearing *Pseudomonas syringae* pv. *tomato*, the causal agent of bacterial speck. This *R* gene participates in a gene-for-gene interaction and the gene product resembles that of serine/threonine protein kinases. Another *R* gene that has been isolated by positional or map-based cloning is *RPS2* from *Arabidopsis thaliana* (Bent *et al.*, 1994; Mindrinos *et al.*, 1994). This gene confers resistance to *P. syringae* pv. *tomato* strains carrying *avrRpt2*. The derived amino acid sequence of *RPS2* contains leucine-rich repeat, membrane-spanning, leucine zipper and P loop domains (ATP/GTP binding site), and hence most likely is involved in perception of an elicitor produced by the avirulent bacterium.

Interestingly, both the tomato *Cf-9* gene and the flax L^6 gene show strong similarity to the *RPS2* gene with respect to the presence of leucine-rich repeats and P-loops. This could indicate that the respective plant–pathogen combinations perhaps share related mechanisms of recognition and signal transduction (reviewed in Dangl, 1995), which is in favour of a broad application of these genes in other (transgenic) crops. These genes have not been tested in heterologous systems and thus it is not understood whether they can be used to introduce resistance into ornamental species.

PROGRAMMED CELL DEATH

Any pathogen avirulence gene and corresponding plant resistance gene could be used as a so-called two component system. In such a system, simultaneous expression of both genes in a single plant cell will result in a hypersensitive response (HR) and ultimately in death of the cell. One example is the combination *avr9* and *Cf-9* (reviewed in De Wit, 1993; Cornelissen and Melchers, 1993). By providing one or both genes with a pathogen inducible promoter, the system could be used for programming cell death, and hence for engineering disease resistance into crops (De Wit, 1993).

The fundamentals underlying such a two component system were used to isolate the *Cf-9* gene from tomato. To facilitate the isolation of the *Cf-9* gene, Jones *et al.* (1994) introduced the avirulence gene *avr9* of *C. fulvum* into a susceptible *Cf-0*-bearing tomato line. This transgenic line was subsequently crossed to an Ac/Ds-transposon containing, near-isogenic line carrying the resistance gene *Cf-9*. The offspring expressing both *avr9* and *Cf-9* became necrotic and died soon after germination, most likely due to the *avr9–Cf-9* interaction which induces a lethal HR. Only plants that had lost *Cf-9* function, for example by Ac/Ds-transposon insertion, survived, and these mutant plants were subsequently used for the isolation of the respective *Cf-9* gene.

Another approach based on this system was used to prove that a gene cloned from *A. thaliana* was in fact the *RPS2* gene and hence could complement *rps2* mutants (Mindrinos *et al.*, 1994). The cloned *RPS2* gene was introduced by biolistics in *rps2* mutants infiltrated with a strain carrying the *avrRps2* gene. Cells infected with such an *avrRps2*-bearing strain rapidly died upon introduction of the *RPS2* gene as the result of a lethal HR.

Other genes that could be used for programmed cell death are genes encoding cell toxins, or genes involved in so-called accelerated cell death (*ACD*). One such gene, *ACD2*, has recently been identified in *A. thaliana* and most likely is involved in a pathway that negatively regulates a genetically programmed cell death (Greenberg *et al.*, 1994). Mutant *acd2* plants develop spontaneous lesions and show typical HR symptoms, including modification of plant cell walls, resistance to bacterial pathogens, accumulation of defence-related transcripts, the signal molecule salicylic acid and an antimicrobial compound. In the future it may be possible to exploit programmed cell death in ornamentals. The only prerequisite is an activation or suppression of the genes involved upon infection, demanding promoters that are induced upon infection by the target pathogen.

ANTIMICROBIAL PEPTIDES

Upon infection or wounding, most living organisms synthesize small proteins or peptides that exhibit antimicrobial activity. Known antibacterial peptides are the mellitins from wasps (Habermann, 1972), sarcotoxins from the flesh fly *Sarcophaga peregrina* (Okada and Natori, 1985), cecropins and attacins from the giant silkmoth *Hyalophora cecropia* (reviewed in Boman and Hultmark, 1987; Hultmark, 1993), the magainins from the African clawed frog *Xenopus laevis* (Zasloff, 1987), the tachyplesins and polyphemusins from the Southeast Asian horseshoe crab *Tachypleus tridentatus* (Nakamura *et al.*, 1988), the apidaecins and abaecins from the honeybee *Apis mellifera* (Casteels *et al.*, 1989), insect and mammalian defensins (Selsted *et al.*, 1985) and the thionins from monocots (reviewed in Florack and Stiekema, 1994). Known antifungal peptides are the thionins (reviewed by Terras *et al.*, 1993; Florack and Stiekema, 1994), the 2S albumins from *Raphanus sativa* (Terras *et al.*, 1993), zeamatin from *Zea mays* (Roberts and Selitrennikof, 1990), the non-specific lipid transfer proteins from the seeds of barley and maize (Molina *et al.*, 1993) and the antimicrobial peptides that have been isolated from the seeds of *Mirabilis jalapa* (Cammue *et al.*, 1992), *Amaranthus caudatus* (Broekaert *et al.*, 1992) and *Raphanus sativa* (Terras *et al.*, 1992a,b).

Gene constructs encoding such peptides, or precursors thereof, might be potential candidates for engineering bacterial or fungal disease resistance into plants (Destéfano-Beltrán *et al.*, 1993; Jaynes *et al.*, 1993a). Prerequisites for the successful application of such genes and corresponding peptides in engineering disease resistance into ornamental plants are: (i) the peptide should be toxic to the pathogen(s) of interest; (ii) the precursor peptide should be correctly processed in the plant of interest; (iii) the peptide should be expressed in the plant cell up to a level sufficient to kill the pathogen; (iv) the peptide should be correctly targeted to the subcellular compartment where the pathogen resides; (v) the peptide should be resistant to plant proteases and (vi) the peptide should not be toxic to cells or organelles of the plant of interest.

The cecropins A and B from the giant silkmoth (Jaynes *et al.*, 1987, 1993b; Nordeen *et al.*, 1992; Hightower *et al.*, 1994; Florack *et al.*, 1995), and the thionins from wheat and barley (Fernandez de Caleya *et al.*, 1972; Carmona *et al.*, 1993; Florack *et al.*, 1993) have gained a lot of attention. Their applicability in engineering will be discussed in more detail.

Cecropins were originally identified as the major component of the haemocoel of pupae of the giant silkmoth *H. cecropia* upon infection with bacteria, having antibacterial activity (Steiner *et al.*, 1981). Cecropins have also been isolated from many other insects, where they play an important role in the inducible cell-free immunity (Boman and Hultmark, 1987; Boman, 1991). A general feature of cecropins is the processing of a precursor protein in several steps into the biologically active antibacterial peptide. Hence, genes encoding such precursors are only effective in plants when the precursor is correctly processed. To circumvent this, synthetic genes can be constructed that code for precursor cecropins that could be processed by the plant cell machinery. Transgenic tobacco plants expressing such a cecropin analog, 'Shiva-1', were reported to have enhanced resistance to the plant-pathogenic bacterium *Pseudomonas solanacearum* (Jaynes *et al.*, 1993b). In contrast, transgenic tobacco plants harbouring gene constructs coding for unmodified cecropin A or B, were not resistant to bacteria (Florack, 1994; Hightower *et al.*,

1994; Florack *et al.*, 1995). This may be due to the fact that these cecropins are rapidly degraded by plant proteases, as was shown for cecropin B in tobacco cell extracts (Florack *et al.*, 1995), and also in extracts of potato, tomato, willow and chrysanthemum (Florack *et al.*, unpublished results).

Thionins have also been studied extensively. Four different types of thionins are distinguished, based on their origin and specific features (Florack and Stiekema, 1994). Type 1 thionins, derived from the endosperm of monocots, are toxic to a limited number of bacterial species (Fig. 17.2), but also appear to have a broad toxic spectrum against fungi in culture (Cammue *et al.*, 1992; Terras *et al.*, 1993; reviewed in Florack and Stiekema, 1994). Therefore, genes encoding such thionins may be a powerful tool in engineering resistance to bacteria but also to fungi. Unlike cecropins, thionins seem to be very stable in plant cells and accumulate up to high levels in transgenic plants (Carmona *et al.*, 1993; Florack *et al.*, 1994). Expression of hordothionins in tobacco was reported to result in resistance to *P. syringae* pv. *syringae* and pv. *tabaci* (Carmona *et al.*, 1993), although others did not find any resistance as the result of the introduction of hordothionin genes in tobacco, tomato and potato (Florack, 1994; Florack *et al.*, 1994). Hence, the potency of thionins for engineering resistance into plants, especially for engineering fungal disease resistance, still remains to be elucidated.

Since antimicrobial peptides are effective against a range of microbes, they may be a good source for engineering bacterial and/or fungal disease resistance in ornamental species. In a preliminary report, Kuehnle *et al.* (1993) reported that introduction of a gene encoding attacin, another antimicrobial peptide from *H. cecropia*, in *Anthurium* resulted in partial tolerance to *Xanthomonas campestris* pv. *dieffenbachiae* strain D150, the most virulent anthurium blight strain known. This is the only report up to now of a transgenic ornamental with partial resistance to a bacterium. Presently, genes encoding thionins, cecropin B and other antimicrobial peptides are being transferred into rose to confer bacterial resistance to this crop (Florack *et al.*, unpublished results).

LYSOZYMES, LECTINS AND RIBOSOME-INACTIVATING PROTEINS

Lysozyme, an enzyme responsible for the catalytic degradation of microbial cell membranes, also exhibits antibacterial and antifungal activity *in vitro*, but by a different mechanism of action than the peptides described above. Gene constructs coding for lysozymes could also be used for engineering disease resistant plants. In fact, transgenic potato plants expressing a gene coding for the bacteriophage T4 lysozyme fused to the α-amylase signal peptide coding sequence, showed increased resistance to *Erwinia carotovora* species, although the expression levels were very low (Düring *et al.*, 1993). Kuehnle *et al.* (1993) used the T4 lysozyme gene to transform *Anthurium* but have not yet reported about the effects on bacterial resistance of these plants.

A B

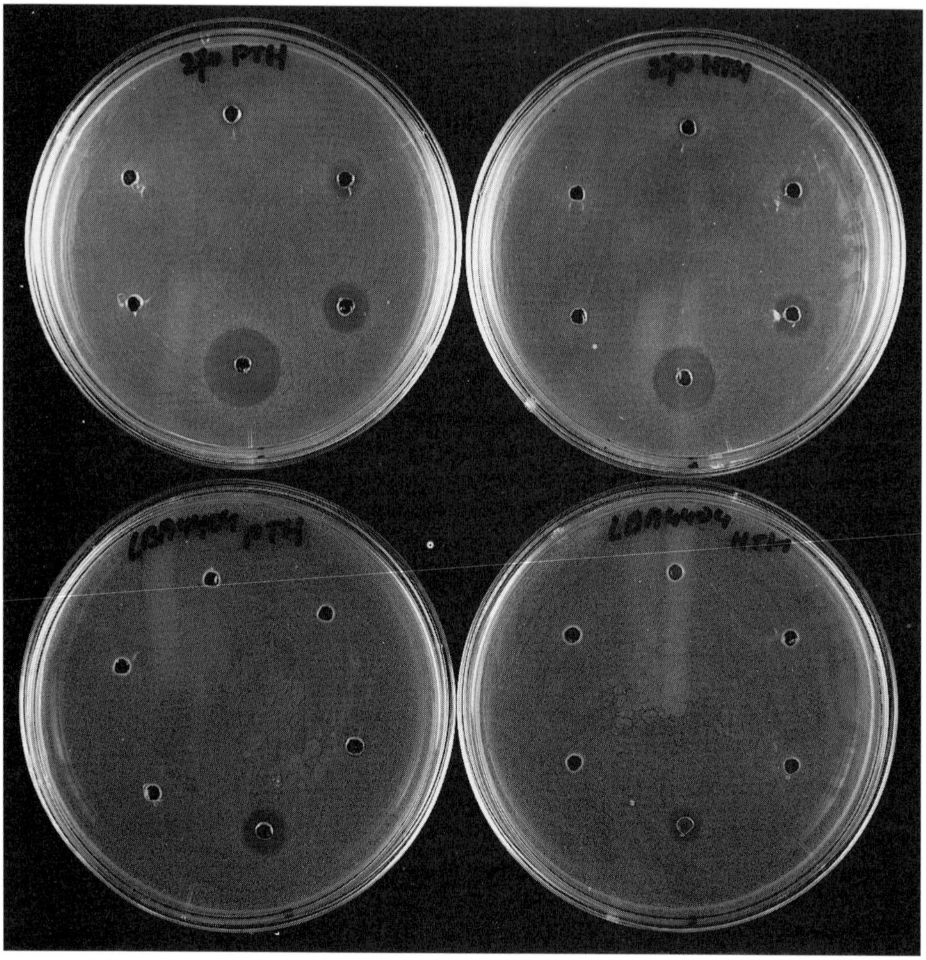

C D

Fig. 17.2. Type 1 thionins from wheat (A, C) and barley (B, D) are toxic to *Clavibacter* spp. (A, B) and *Agrobacterium tumefaciens* (C, D) at relatively low concentrations. Bacteria are mixed through the medium and increasing amounts of proteins (1, 5, 25 and 125 μg amounts clockwise) are applied in wells cut in the medium. After incubation of the plates, growth inhibition zones are visible as the result of diffusion of the proteins into the medium.

Another group of antimicrobial proteins are lectins. Some of these carbohydrate binding proteins have specificity towards substrates containing *N*-acetyl-glucosamine such as chitin, a cell wall compound of many fungi. UDA, isolated from the stinging nettle plant (*Urtica dioica*), and hevein, isolated from the rubber tree (*Hevea brasiliensis*), show antifungal activity (Broekaert *et al.*, 1989; Van Parijs *et al.*, 1991). Presumably the transfer of genes encoding these proteins may enhance the fungal resistance of the transgenic plants.

A completely different group of antimicrobial proteins are the so-called ribosome-inactivating proteins (RIPs). These proteins inactivate to various degrees ribosomes of non-related organisms, among which are those of fungi (Leah *et al.*, 1991). Transgenic tobacco plants expressing a barley RIP had a somewhat higher level of resistance to the fungus *Rhizoctonia solani* than the control plants (Logemann *et al.*, 1992).

PLANTIBODIES

In recent years, immunological techniques have become important tools in many research areas. Monoclonal antibodies can be raised against a wide range of antigens. Introduction and expression of the corresponding genes in plants can result in transgenic plants which produce antibodies. Provided that such an antibody recognizes a factor important in pathogenesis, this might result in resistant plants. Functional antibodies have been produced in leaves (Hiatt *et al.*, 1989) and roots (Van Engelen *et al.*, 1994) of transgenic tobacco plants. Expression of an antibody against a viral coat protein has been reported to result in protection against virus infection (Tavladoraki *et al.*, 1993). When antibodies are directed against antigens of fungal or bacterial pathogens, the interaction between plant and pathogen may be blocked, resulting in a resistant plant. A prerequisite is that the antigen must play an important role in pathogenesis (i.e. a key enzyme or a host-specific toxin). However, knowledge about the importance of various factors in plant--pathogen interactions, especially those involving ornamental crops, is still limited and might delay a successful application of this technique.

PARASITE-DERIVED RESISTANCE

Some microorganisms produce toxins which are of primary importance in pathogenesis (Buiatti and Ingram, 1991) or underly the characteristic symptom development. Elimination of such a toxin would either protect plants completely against the pathogen or would reduce symptom development. Since the toxin may also be lethal for the microorganism itself, the pathogen had to evolve some strategy to overcome the toxic effect of its own toxin. In most cases the pathogen simply detoxifies the toxin or has a toxin-insensitive target enzyme. The genes involved in these mechanisms of self-protection could be used for engineering resistance against these specific pathogens, so-called parasite-derived resistance (Sanford and Johnston, 1985). This strategy was applied by Anzai *et al.* (1989), who isolated the gene encoding the enzyme acetyltransferase (*trr*) from *Pseudomonas syringae* pv. *tabaci*. This enzyme specifically detoxifies the host-specific toxin called tabtoxin which is involved in the interaction between this pathogen and tobacco and underlies the appearance of typical chlorotic haloes. Transgenic tobacco plants expressing this *trr* gene were resistant to infection by the pathogen.

A different strategy was followed by De La Fuente-Martinez *et al.* (1992). They studied the interaction of *Pseudomonas syringae* pv. *phaseolicola* and *Phaseolus vulgaris*. This bacterium produces the toxin phaseolotoxin which inhibits ornithine carbamoyltransferase (OCTase), an enzyme located in the chloroplast. The bacterium itself has an insensitive form of the enzyme for self-protection. The gene encoding this enzyme was isolated from the bacterium and fused to the transit peptide of ribulose biophosphate carboxylase to ensure correct targeting in plants. By introducing this construct in tobacco, the normally sensitive enzyme of the plant was complemented by the insensitive form of the pathogen, thus conferring resistance to the bacterium. Another interesting gene that might be more broadly applicable in this respect is one that codes for an enzyme which can detoxify fusaric acid. This toxin is produced by many *formae speciales* of *Fusarium oxysporum*, the causal agent of one of the most devastating fungal diseases known for many different plant species, including ornamentals. In the search for such a gene, Ouchi *et al.* (1989) isolated a gene from *Cladosporium werneckii* which encoded an enzyme capable of detoxifying fusaric acid. Applicability of this gene in engineering fungal disease resistance into plants remains to be elucidated.

PHYTOALEXINS

Phytoalexins are low molecular weight antimicrobial compounds produced by plants locally upon infection by a pathogen (Paxton, 1981). Since these compounds are also toxic for plant cells, local accumulation in infected tissue results in rapid cell death and hence protects the plant from further spread of the pathogen. In this manner phytoalexins play a role in HR of the plant. Phytoalexins produced by enzymes from the phenylpropanoid pathway are well studied (Lamb *et al.*, 1989). This pathway accounts for the production of isoflavonoid phytoalexins. Key enzymes in this pathway are PAL (phenylalanine ammonia lyase), C4H (cinnamate 4-hydroxylase), 4CL (4-coumarate-CoA ligase), CHS (chalcone synthase) and CHI (chalcone isomerase). Most of these enzymes have been purified and the respective genes have been cloned by now. The ability to study the effect of altered expression of those genes on the plant resistance is now feasible. Overexpression may increase the rate and/or speed of phytoalexin production in plants, thus stimulating the defence response of the plant. The manipulation of the production of isoflavonoid phytoalexins is generally considered to be difficult since an entire pathway is involved. It may well be necessary to alter the level of more than one regulating enzyme to achieve the desired effects. This disadvantage counts less for another group of phytoalexins: the stilbenes. Stilbenes are produced in a limited group of plant families only and exhibit antifungal activity. Resveratrol, one of the stilbenes, was found to be involved in resistance of grape (*Vitis vinifera*) against *Botrytis cinerea* (Stein and Blach, 1985). A key enzyme in the production of resveratrol is stilbene synthase, converting *p*-coumaryl-CoA and malonyl-CoA into 3,4′5-trihydroxystilbene (resveratrol). Although tobacco lacks this enzyme, the substrates are present. The gene encoding the enzyme was isolated from groundnuts (*Arachis hypogaea*) by Hain *et al.* (1990) and transferred to tobacco and tomato. As anticipated, transgenic plants produced resveratrol and exerted an enhanced resistance to *Botrytis cinerea* (Hain, 1993). Since this fungus is a very important pathogen of ornamentals, the stilbene synthase gene would be ideal to transfer to ornamental plants, providing that the necessary substrates are present.

SAR AND PR PROTEINS

The phenomenon of systemic-acquired-resistance (SAR) has been studied extensively over the past few years. SAR coincides with the *de novo* synthesis of proteins in the plant, so-called pathogenesis-related (PR) proteins (Van Loon, 1985), and with a burst of active oxygen species (Chen *et al.*, 1993). Most of the proteins involved have been purified and characterized, and are being evaluated for their antimicrobial activity *in vitro*. The best characterized are chitinases and β-1,3-glucanases, which each on their own, but especially in combination, are toxic to fungi *in vitro* (Mauch *et al.*, 1988). A large number of genes encoding PR proteins have been isolated and are being studied in transgenic plants for their use in engineering disease resistance. Several successes following this strategy have been reported and many more are expected in the coming years. Expression of a chitinase gene from bean (*Phaseolus vulgaris*) under control of the constitutive CaMV 35S promoter resulted in increased levels of chitinase in tobacco and enhanced resistance to the soilborne pathogen *Rhizoctonia solani* (Broglie *et al.*, 1991). As expected, no resistance was found to *Pythium aphanidermatum*, which is an oomycete that lacks chitin in the cell wall. Likewise, Nishizawa *et al.* (1992) found a high level of resistance to the powdery mildew *Erysiphe cichoracearum* after the introduction of a rice-derived chitinase gene into tobacco. Despite these successes, Meins *et al.* (1993) did not find an enhanced resistance of tobacco plants to the fungus *Cercospora nicotianae* after the introduction of a constitutively expressed chitinase or β-1,3-glucanase gene. Similar results were reported for tomato by Van den Elzen *et al.* (1993). They transformed tomato plants with genes encoding chitinase or β-1,3-glucanase. Both enzymes were assessed *in vitro* and found to be toxic to *Fusarium oxysporum* f.sp. *lycopersici*, but did not enhance the resistance of the tomato plant to the fungus when introduced separately. The combination of both genes, however, was effective in protecting tomato against the fungus. Simultaneous expression of chitinase and glucanase genes in tobacco also resulted in increased resistance to *Cercospora nicotianae* (Zhu *et al.*, 1994), whereas also in this case each of the genes on its own was insufficient to produce the desired resistance.

Recently, the expression of various other PR protein genes in transgenic plants has been reported. Ryals *et al.* (1993) reported enhanced resistance of transgenic tobacco plants to *Phytophthora parasitica* upon expression of SAR8.2, and resistance to *Peronospora tabacina* upon expression of PR-I. Van den Elzen *et al.* (1993) found a thaumatin-like protein, toxic for *Phytophthora infestans*, which might be applied for introducing resistance to oomycetes. A related protein, called osmotin and accumulating after virus-infection, was found by Stinzi *et al.* (1991).

PROSPECTS

Ornamentals when cultivated are exposed to many types of diseases caused by bacteria and fungi. Although most of these diseases can be effectively controlled by chemical pesticides, it is obvious that the widespread use of such chemicals creates an enormous environmental threat. Therefore it is essential that resistant genotypes are developed for a sustainable floriculture. In comparison to agricultural crops, only limited efforts in breeding for resistant ornamentals have been undertaken in the past. In the near future it may be possible that cross-breeding could contribute to more resistant ornamental varieties. Many ornamentals, especially bulbous or woody crops, have a long generation time and to introduce new traits by cross pollination is very time consuming. However, by applying indirect selection in a breeding programme using molecular markers linked to the desired traits, the new varieties could be more easily developed. Molecular-marker-aided selection is especially valuable when genomic maps are available which are saturated with markers containing a number of important horticultural traits.

Novel technologies also offer good opportunities for further improvement of ornamentals. Cell biological techniques are now available that may be applied to ornamentals for broadening their genetic base. The application of those techniques makes it possible to introduce traits that are present in related species by circumventing pre- or post-fertilization barriers. *In vitro* selection might become a common tool of plant breeders, provided that the role of the selective agent in disease development is understood.

Although the forementioned techniques will surely contribute to new resistant ornamentals, the largest impact can be expected from genetic engineering. Many different strategies have been developed for engineering disease resistance into plants. In general, two sources of genes are being exploited: (i) 'classical' resistance genes isolated from plants where they actually are involved in resistance and (ii) genes coding for antimicrobial proteins or peptides, plantibodies, RIPs, PR proteins, etc., isolated from many different sources. These genes have proved to generate valuable resistance and are presently being tested for their effectiveness in many heterologous plant species. Many more resistance genes will become available in the near future.

A possible problem may be public acceptance to purchase transgenic ornamentals. Since ornamentals are not grown as a food source, genetically engineered ornamental crops may be more readily accepted by consumers than genetically engineered agricultural crops, especially when the novel trait inserted into the genome of the ornamental plant leads to a reduced use of environmentally hazardous pesticides.

In conclusion, many new techniques for breeding resistant cultivars will become available for breeders of ornamentals in the near future. The development of reliable transformation methods for ornamentals will be the first challenge for the modern breeder. This, and the availability of gene constructs that will confer resistance to a specific pathogen, will determine the success of biotechnology in ornamentals.

ACKNOWLEDGEMENTS

The authors are indebted to Drs J.J.M. Dons, H.M.C. van Holsteijn, W.J. Stiekema and J.M. van Tuyl for critically reading the manuscript and supplying valuable suggestions for improving the text and to K.M. Haymes for valuable suggestions and correcting the English.

REFERENCES

Ammirato, P.V., Evans, D.R., Sharp, W.R. and Bajaj, Y.P.S. (1990) *Handbook of Plant Cell Culture. Vol. V: Ornamental Species.* McGraw-Hill, New York.

Anzai, H., Yoneyama, K. and Yamaguchi, I. (1989) Transgenic tobacco resistant to a bacterial disease by the detoxification of a pathogenic toxin. *Molecular and General Genetics* 219, 492–494.

Arai, M. and Takeuchi, M. (1993) Influence of Fusarium wilt toxin(s) on carnation cells. *Plant Cell, Tissue and Organ Culture* 34, 287–293.

Bent, A.F., Kunkel, B.N., Dahlbeck, D., Brown, K.L., Schmidt, R., Giraudat, J., Leung, J. and Staskawicz, B.J. (1994) *RPS2* of *Arabidopsis thaliana*: a leucine-rich repeat class of plant disease resistance genes. *Science* 265, 1856–1860.

Boman, H.G. (1991) Antibacterial peptides: key components needed in immunity. *Cell* 65, 205–207.

Boman, H.G. and Hultmark, D. (1987) Cell-free immunity in insects. *Annual Review of Microbiology* 41, 103–126.

Broekaert, W.F., Van Parijs, J. Leyns, F., Joos, H. and Peumans, W.J. (1989) A chitin binding lectin from stinging nettle rhizomes with antifungal properties. *Science* 245, 1100–1102.

Broekaert, W.F., Mariën, W., Terras, F.R.G., De Bolle, M.F.C., Proost, P., Van Damme, J., Dillen, L., Claeys, M., Rees, S.B., Vanderleyden, J. and Cammue, B.P.A. (1992) Antimicrobial peptides from *Amaranthus caudatus* seeds with sequence homology to the cysteine/glycine-rich domain of chitin-binding proteins. *Biochemistry* 31, 4308–4314.

Broglie, K., Chet, I., Holliday, M., Gressman, R., Diddle, P., Knowlton, S., Mauvais, C.J. and Broglie, R. (1991) Transgenic plants with enhanced resistance to the fungal pathogen *Rhizoctonia solani*. *Science* 254, 1194–1196.

Buiatti, M. and Gimelli, F. (1993) Somaclonal variation in ornamentals. In: Schiva, T. and Mercuri A. (eds), *Creating Genetic Variation in Ornamentals*. EUCARPIA, Italy, pp. 5–24.

Buiatti, M. and Ingram, D.S. (1991) Phytotoxins as tools in breeding and selection of disease-resistant plants. *Experientia* 47, 811–819.

Cammue, B.P.A., De Bolle, M.F.C., Terras, F.R.G., Proost, P., Van Damme, J., Rees, S.B., Vanderleyden, J. and Broekaert, W.F. (1992) Isolation and characterization of a novel class of plant antimicrobial peptides from *Mirabilis japala* L. seeds. *Journal of Biological Chemistry* 267, 2228–2233.

Carmona, M.J., Molina, A., Fernandez, J.A., Lopez-Fando, J.J. and Garcia-Olmedo, F. (1993) Expression of the alfa-thionin gene from barley in tobacco confers enhanced resistance to bacterial pathogens. *The Plant Journal* 3, 457–462.

Casteels, P., Ampe, C., Jacobs, F., Vaeck, M. and Tempst, P. (1989) Apidaecins: antibacterial peptides from honeybees. *EMBO Journal* 8, 2387–2391.

Chasan, R. (1994) Plant–pathogen encounters in Edinburgh. *Plant Cell* 6, 1332–1342.

Chen, Z.X., Silva, H. and Klessig, D.F. (1993) Active oxygen species in the induction of plant systemic acquired resistance by salicylic acid. *Science* 262, 1883–1886.

Colijn-Hooymans, C.M. (1990) The use of protoplasts for the improvement of ornamentals. In: De Jong, J. (ed.), *Integration of In Vitro techniques in Ornamental Plant Breeding*. EUCARPIA, Wageningen, pp. 116–123.

Cornelissen, B.J.C. and Melchers, L.S. (1993) Strategies for control of fungal diseases with transgenic plants. *Plant Physiology* 101, 709–712.

Dangl, J.L. (1995) Pièce de résistance: novel classes of plant disease resistance genes. *Cell* 80, 363–366.

De La Fuente-Martinez, J.M., Mosqueda-Cano, G., Alvarez-Morales, A. and Herrera-Estrella, L. (1992) Expression of a bacterial phaseolotoxin-resistant ornithyl transcarbamylase in transgenic tobacco plants confers resistance to *Pseudomonas syringae* pv *phaseolicola*. *Bio/Technology* 10, 905–909.

Destéfano-Beltrán, L., Nagpala, P.G., Cetiner, S.M., Denny, T. and Jaynes, J.M. (1993) Using genes encoding novel peptides and proteins to enhance disease resistance in plants. In: Chet, I. (ed.), *Biotechnology in Plant Disease Control*. Wiley-Liss, New York, pp. 175–189.

De Wit, P.J.G.M. (1993) Key genes in plant–pathogen interactions and their application in control of plant diseases. In: Zadoks, J. (ed.), *Modern Crop Protection: Development and Perspectives*. Wageningen Press, Netherlands, pp. 91–98.

Dons, J.J.M. (1992) Genetic modification of ornamental crops: procedures and applications. *Mededelingen van de Faculteit Landbouwwetenschappen Gent*, 57/4a, 1471–1476.

Dons, J.J.M., Mollema, C., Stiekema, W.J. and Visser, B. (1991) Routes to the development of disease resistant ornamentals. In: Harding, J., Singh, F. and Mol, J.N.M. (eds), *Genetics and Breeding of Ornamental Species*. Kluwer Academic Publishers, Dordrecht, pp. 387–417.

Düring, K., Porsch, P., Fladung, M. and Lörz, H. (1993) Transgenic potato plants resistant to the phytopathogenic bacterium *Erwinia carotovora*. *The Plant Journal* 3, 587–598.

Fernandez de Caleya, R., Gonzalez-Pascual, B., Garcia-Olmedo, F. and Carbonero, P. (1972) Susceptibility of phytopathogenic bacteria to wheat purothionins *in vitro*. *Applied Microbiology* 23, 998–1000.

Florack, D.E.A. (1994) Application of hordothionins and cecropin B for engineering bacterial disease resistance into plants. PhD thesis, Wageningen Agricultural University, The Netherlands.

Florack, D.E.A. and Stiekema, W.J. (1994) Thionins: properties, possible biological roles and mechanisms of action. *Plant Molecular Biology* 26, 25–37.

Florack, D.E.A., Visser, B., De Vries, Ph.M., Van Vuurde, J.W.L. and Stiekema, W.J. (1993) Analysis of the toxicity of purothionins and hordothionins for plant pathogenic bacteria. *Netherlands Journal of Plant Pathology* 99, 259–268.

Florack, D.E.A., Dirkse, W.G., Visser, B., Heidekamp, F. and Stiekema, W.J. (1994) Expression of biologically active hordothionins in tobacco. Effects of pre- and pro-sequences at the amino and carboxyl termini of the hordothionin precursor on mature protein expression and sorting. *Plant Molecular Biology* 24, 83–96.

Florack, D.E.A., Allefs, S., Bollen, R., Bosch, D., Visser, B. and Stiekema, W.J. (1995) Expression of giant silkmoth cecropin B genes in tobacco. *Transgenic Research* 4, 132–141.

Greenberg, J.T., Guo, A., Klessig, D.F. and Ausubel, F.M. (1994) Programmed cell death in plants: a pathogen-triggered response activated coordinately with multiple defense functions. *Cell* 77, 551–563.

Haberman, E. (1972) Bee and wasp venoms. *Science* 177, 314–322.

Hain, R. (1993) Foreign phytoalexin expression in plants results in disease resistance. *Agro Food Industry Hi Tech* 4, 33.

Hain, R., Bieseler, B., Kindl, H., Schröder, G. and Stöcker, R. (1990) Expression of a stilbene synthase gene in *Nicotiana tabacum* results in synthesis of the phytoalexin resveratrol. *Plant Molecular Biology* 15, 325–335.

Harms, C.T. (1992) Engineering genetic disease resistance into crops: biotechnological approaches to crop protection. *Crop Protection* 11, 291–306.

Hiatt, A., Cafferkey, R. and Bowdish, K. (1989) Production of antibodies in transgenic plants. *Nature* 342, 76–78.

Hightower, R., Baden, C., Penzes, E. and Dunsmuir, P. (1994) The expression of cecropin peptide in transgenic tobacco does not confer resistance to *Pseudomonas syringae* pv. *tabaci*. *Plant Cell Reports* 13, 295–299.

Hultmark, D. (1993) Immune reactions in *Drosophila* and other insects: a model for immunity. *Trends in Genetics* 9, 178–183.

Jacobsen, E. (1993) Resistance breeding using molecular biological techniques. In: Zadoks, J. (ed.), *Modern Crop Protection: Development and Perspectives*. Wageningen Press, The Netherlands, pp. 83–90.

Jaynes, J.M., Xanthopoulos, K.G., Destéfano-Beltrán, L. and Dodds, J.H. (1987) Increasing bacterial resistance utilizing antibacterial genes from insects. *BioEssays* 6, 263–270.

Jaynes, J.M., Destéfano-Beltran, L., Nagpala, P. and Cetiner, S. (1993a) Lytic peptides and their potential for enhancing plant disease resistance. In: Bruening, G., Garcia-Olmedo, F. and Ponz, F. (eds), *Proceedings of the Workshop on Engineering Plants against Pests and Pathogens*. Instituto Juan March de Estudios e Investigaciones, Madrid, pp. 37–41.

Jaynes, J.M., Nagpala, P., Destéfano-Beltran, L., Huang, J.H., Kim, J.H., Denny, T. and Cetiner, S. (1993b) Expression of a cecropin B lytic peptide in transgenic tobacco confers enhanced resistance to bacterial wilt caused by *Pseudomonas solanacearum*. *Plant Science* 89, 43–53.

Johal, G.S. and Briggs, S.P. (1992) Reductase activity encoded by the *Hm1* disease resistance gene in maize. *Science* 258, 985–987.

Jones, D.A., Thomas, C.M., Hammond-Kosack, K.E., Balint-Kurti, P.J. and Jones, J.D.G. (1994) Isolation of the tomato *Cf-9* gene for resistance to *Cladosporium fulvum* by transposon tagging. *Science* 266, 789–793.

Kuehnle, A.R., Chen, F.C. and Jaynes, J.M. (1993) Engineering bacterial blight resistance into *Anthurium*. In: Schiva, T. and Mercuri A. (eds), *Creating Genetic Variation in Ornamentals*. EUCARPIA, Italy, pp. 127–129.

Lamb, C.J., Lawton, M.A., Dron, M. and Dixon, R.A. (1989) Signals and transduction mechanisms for activation of plant defenses against microbial attack. *Cell* 56, 215–224.

Larson, R.A. (1992) *Introduction to Floriculture*, 2nd edition. Academic Press, San Diego.

Lawrence, G.J., Ellis, J.G. and Finnegan, E.J. (1994) Cloning a rust-resistance gene in flax. In: Daniels, M.J., Downie, J.A. and Osbourn, A.E. (eds), *Advances in Molecular Genetics of Plant–Microbe Interactions Vol. 3*. Kluwer Academic Publishers, Dordrecht, pp. 303–306.

Leah, R., Tommerup, H., Svendsen, I. and Mundy, J. (1991) Biochemical and molecular characterization of three anti-fungal proteins from barley seed. *Journal of Biological Chemistry* 266, 1564–1573.

Lindhout, P. (1990) The application of molecular markers for indirect selection. In: De Jong, J. (ed.), *Integration of In Vitro Techniques in Ornamental Plant Breeding*. EUCARPIA, Wageningen, pp. 173–178.

Löffler, H.J.M. and Mouris J.R. (1991) Prospects of *in vitro* selection for resistance against *Fusarium oxysporum* in lily. In: De Jong, J. (ed.), *Integration of In Vitro Techniques in Ornamental Plant Breeding*. EUCARPIA, Wageningen, pp. 80–85.

Löffler, H.J.M. and Mouris J.R. (1992) Fusaric acid: phytotoxicity and *in vitro* production by *Fusarium oxysporum* f.sp. *lilii*, the causal agent of basal rot in lilies. *Netherlands Journal of Plant Pathology* 98, 107–115.

Löffler, H.J.M., Meijer, H,. Straathof, Th.P. and Van Tuyl, J.M. (1995) Segregation of *Fusarium* resistance in an interspecific cross between *Lilium longiflorum* and *Lilium dauricum*. *Acta Horticulturae* 414, 203–208.

Logemann, J., Jach, G., Tommerup, H., Mundy, J. and Schell, J. (1992) Expression of a barley ribosome-inactivating protein leads to increased fungal protection in transgenic tobacco plants. *Bio/Technology* 10, 305–308.

Martin, G.B., Broomonschenkel, S.H., Chunwongse, J., Frary, A., Ganal, M.W., Spivey, R., Wu, T., Earle, E.D. and Tanksley, S.D. (1993) Map-based cloning of a protein kinase gene conferring disease resistance in tomato. *Science* 262, 1432–1436.

Mauch, F., Mauch-Mani, B. and Boller, T. (1988) Antifungal hydrolases in pea tissue. II. Inhibition of fungal growth by combinations of chitinas and β-1,3-glucanase. *Plant Physiology* 88, 936–942.

Meins, F., Belfa, R. and Kunz, C. (1993) The biology of plants transformed with chimeric class I chitinase and β-1,3-glucanase genes. In: Bruening, G., Garcia-Olmedo, F. and Ponz, F. (eds), *Proceedings of the Workshop on Engineering Plants against Pests and Pathogens*. Instituto Juan March de Estudios e Investigaciones, Madrid, p. 27.

Mercuri, A., Schiva, T., Baratta, G., Feoglio, G. and Burchi, G. (1991) Pollen selection in carnation for resistance to *Fusarium oxysporum* f.sp. *dianthi*. *Acta Horticulturae* 307, 225–232.

Mindrinos, M., Katagiri, F., Yu, G.-L. and Ausubel, F.M. (1994) The *A. thaliana* disease resistance gene *RPS2* encodes a protein containing a nucleotide-binding site and leucine-rich repeats. *Cell* 78, 1089–1099.

Miyata, T., Tokunaga, F., Yoneya, T., Yoshikawa, K., Iwanaga, S., Niwa, M., Takao, T.

and Shimonishi, Y. (1989) Antimicrobial peptides, isolated from horseshoe crab hemocytes, tachyplesin II, and polyphemusins I and II: chemical structures and biological activity. *Journal of Biochemistry* 106, 663–668.

Moffat, A.S. (1994) Mapping the sequence of disease resistance. *Science* 265, 1804–1805.

Molina, A., Segura, A. and Garcia-Olmedo, F. (1993) Lipid-transfer proteins (nsLTPs) from barley and maize leaves are potent inhibitors of bacterial and fungal plant pathogens. *FEBS Letters* 316, 119–122.

Nakamura, T., Furunaka, H., Miyata, T., Tokunaga, F., Muta, T., Iwanaga, S., Takao, T. and Shimonishi, Y. (1988) Tachyplesin, a class of antimicrobial peptide from the hemocytes of the horseshoe crab (*Tachypleus tridentatus*). Isolation and chemical structure. *Journal of Biological Chemistry* 263, 16709–16713.

Nishizawa, Y., Kondo, K., Akutsu, K. and Hibi, T. (1992) Disease-resistant transgenic tobacco expressing a rice chitinase gene. *Abstract Papers of the Annual Meeting of the Phytopathological Society, Japan* 1992, 45.

Nordeen, R.O., Sinden, S.L., Jaynes, J.M. and Owens, L.D. (1992) Activity of cecropin SB37 against protoplasts from several plant species and their bacterial pathogens. *Plant Science* 82, 101–107.

Okada, M. and Natori, S. (1985) Primary structure of sarcotoxin I, an antibacterial protein induced in hemolymph of *Sarcophaga peregrina* (flesh fly) larvae. *Journal of Biological Chemistry* 260, 7174–7177.

Ouchi, S., Toyoda, H., Utsumi, R., Hashimoto, H. and Hamda, T. (1989) A promising strategy for the control of fungal diseases by the use of toxin-degrading microbes. In: Graniti, A., Durbin, R.D. and Ballis, A. (eds), *Phytotoxins and Plant Pathogenesis*. Springer-Verlag, Berlin, pp. 301–317.

Paxton, J.D. (1981) Phytoalexins – a working redefinition. *Phytopathologische Zeitschrift* 101, 106–109.

Phillips, R.L., Kaeppler, S.M. and Peschke, V.M. (1990) Do we understand somaclonal variation? In: Nijkamp, H.J.J., van der Plas, L.H.W. and van Aartrijk, J. (eds). *Progress in Plant Cellular and Molecular Biology*. Kluwer Academic Publishers, Dordrecht, pp. 131–141.

Roberts, W.K. and Selitrennikoff, C.P. (1990) Zeamatin, an antifungal protein from maize with membrane-permeabilizing activity. *Journal of General Microbiology* 136, 1771–1778.

Ryals, J., Ward, E., Uknes, S,. Gaffney, T., Friedrich, L., Alexander, D. and Gut Rella, M. (1993) Towards the genetic engineering of fungal resistance in crops. In: Bruening, G., Garcia-Olmedo, F. and Ponz, F. (eds), *Proceedings of the Workshop on Engineering Plants against Pests and Pathogens*. Instituto Juan March de Estudios e Investigaciones, Madrid, pp. 37–41.

Sanford, J.C. and Johnston, S.A. (1985) The concept of parasite-derived resistance–deriving resistance genes from the parasite's own genome. *Journal of Theoretical Biology* 113, 395–405.

Selsted, M., Brown, D.M., Delange, R.J., Harwig, S.L and Lehrer, R.I. (1985) Primary structure of six antimicrobial peptides of rabbit peritoneal neutrophils. *Journal of Biological Chemistry* 260, 4579–4584.

Sink, K.C. (1993) Cell fusion: opportunities to exploit distant germplasm for improvement of ornamentals. In: Schiva, T. and Mercuri, A. (eds), *Creating Genetic Variation in Ornamentals*. EUCARPIA, Italy, pp. 47–61.

Sparnaaij, L.D. (1991) Breeding for disease and insect resistance in flower crops. In:

Harding, J., Singh, F. and Mol, J.N.M. (eds), *Genetics and Breeding of Ornamental Species*. Kluwer Academic Publishers, Dordrecht, pp. 179–211.

Stein, U. and Blach, R. (1985) Untersuchungen über Stilbenproduktion und Botrytisanfälligkeit bei Vitis-Arten. *Vitis* 24, 75–87.

Steiner, H., Hultmark, D., Engström, Ä., Bennich H. and Boman, H.G. (1981) Sequence and specificity of two antibacterial proteins involved in insect immunity. *Nature* 292, 246–248.

Stinzi, A., Heitz, T., Kauffmann, S., Legrand, M. and Fritig, B. (1991) Identification of a basic pathogenesis-related, thaumatin-like protein of virus-infected tobacco as osmotin. *Physiological and Molecular Plant Patholology* 38, 137–146.

Straathof, Th.P., Van Tuyl, J.M., Dekker, B., Van Winden, M.J.M. and Sandbrink, J.M. (1995) Genetic analysis of inheritance of partial resistance to *Fusarium oxysporum* in Asiatic hybrid lily using RAPD markers. *Acta Horticulturae* 414, 209–218.

Tavladoraki, P., Benvenuto, E., Trinca, S., De Marlini, D., Cattanea, A. and Galeffi, P. (1993) Transgenic plants expressing a functional single-chain Fv antibody are specifically protected from virus attack. *Nature* 366, 469–472.

Terras, F.R.G., Goderis, I.J., Van Leuven, F., Vanderleyden, J., Cammue, B.P.A. and Broekaert, W.F. (1992a) *In vitro* antifungal activity of a protein from radish (*Raphanus sativa* L.) seeds homologous to nonspecific lipid transfer proteins. *Plant Physiology* 100, 1055–1058.

Terras, F.R.G., Schoofs, H.M.E., De Bolle, M.F.C., Van Leuven, F., Rees, S.B., Vanderleyden, J., Cammue, B.P.A. and Broekaert, W.F. (1992b) Analysis of two novel classes of antifungal proteins from *Brassicaceae* species. *FEBS Letters* 316, 233–240.

Terras, F.R.G., Schoofs, H.M.E., Thevissen, K., Osborn, R.W., Vanderleyden, J., Cammue, B.P.A. and Broekaert, W.F. (1993) Synergistic enhancement of the antifungal activity of wheat and barley thionins by radish and oilseed rape 2S albumins and by barley trypsin inhibitors. *Plant Physiology* 103, 1311–1319.

Van den Bulk, R.W. (1991) Application of cell and tissue culture and *in vitro* selection for disease resistance breeding – a review. *Euphytica* 56, 269–285.

Van den Elzen, P.J.N., Jongedijk, E., Melchers, L.S. and Cornelissen, B.J.C. (1993) Virus and fungal resistance: from laboratory to field, In: Bevan, M.V., Harrison, B.D. and Leaver, C.J. (eds), *The Production and Uses of Genetically Transformed Plants*. Chapman & Hall, London, pp. 83–90.

Van Engelen, F., Schouter, A., Molthoff, J.W., Salinas, J., Dinkse, W.G., Schots, A., Bakker, J., Gommas, F.J., Gangsma, M.A., Bosch, D. and Stichema, W. (1994) Coordinate expression of antibody subunit genes yields high levels of functional antibodies in roots of transgenic tobacco. *Plant Molecular Biology* 26, 1701–1710.

Van Loon, L.C. (1985) Pathogenesis-related proteins. *Plant Molecular Biology* 4, 111–116.

Van Parijs, J., Broekaert, W.F., Goldstein, I.J. and Peumans, W.J. (1991) Hevein: an antifungal protein from rubber-tree (*Hevea brasiliensis*) latex. *Planta* 183, 258–264.

Van Tuyl, J.M., Van Diën, M.P., Van Creij, M.G.M., Van Kleinwee, T.C.M., Franken, J. and Bino, R.J. (1991) Application of *in vitro* pollination, ovary culture, ovule culture and embryo rescue for overcoming incongruity barriers in interspecific *Lilium* crosses. *Plant Science* 74, 115–126.

Van Tuyl, J.M., Van Creij, M.G.M., Van Diën, M.P., Eikelboom, W., Kerckhoffs, D.M.F.J. and Meijer, B. (1993) New genetic variation in *Lilium* and *Tulipa* by wide hybridization. In: Schiva, T. and Mercuri A. (eds), *Creating Genetic Variation in Ornamentals*. EUCARPIA, Italy, pp. 141–149.

Venturo, R., Beruto, M., Martini, P. and Gimelli, F. (1993) Organogenesis and *in vitro* selection for Fusarium resistance in carnation. In: Schiva, T. and Mercuri A. (eds), *Creating Genetic Variation in Ornamentals*. EUCARPIA, Italy, pp. 297–300.

Wenzel, G. and Foroughi-Wehr, B. (1990) Progeny tests of barley, wheat, and potato regenerated from cell cultures after *in vitro* selection for disease resistance. *Theoretical and Applied Genetics* 80, 359–365.

Zasloff, M. (1987) Magainins, a class of antimicrobial peptides for *Xenopus laevis* skin: isolation, characterization of two active forms, and partial cDNA sequence of a precursor. *Proceedings of the National Academy of Sciences, USA* 84, 5449–5453.

Zhu, Q., Mahler, E.A., Masoud, S., Dixon, R.A. and Lamb, C.J. (1994) Enhanced protection against fungal attack by consitutive co-expression of chitinase and glucanase genes in transgenic tobacco. *Bio/Technology* 12, 807–812.

Biotechnological Approaches for Virus Resistance in Floral Crops

18

Margaret E. Daub, Ronald K. Jones and James W. Moyer
Department of Plant Pathology, North Carolina State University, Raleigh, North Carolina 27695, USA

Production of floral crops has frequently been severely affected by virus and viroid diseases. Floral crop production is conducive to the proliferation of virus diseases because many of the crops are asexually propagated, and a few major propagators supply most of the plants to finish growers over very large geographic areas. Virus diseases have not only caused losses to the floral crop industry, but floral crops have also served as a source of introducing viruses into other crops and other countries. Due to the importance of viruses to the floral industry, practices to control virus diseases and their spread account for a large portion of the costs of floral crop production. Traditional virus control measures have focused on efforts to produce and disseminate 'virus-free' propagation materials, i.e. stock generated through meristem tip culture and indexed for the presence of known viruses. The increasing importance of insect-vectored viruses, however, has pointed out the limitations in traditional control strategies and has focused interest on the development of virus-resistant plants through molecular techniques.

The use of disease-resistant cultivars is one of the most commonly used strategies for controlling plant viruses in many crop species. However, development of virus-resistant plants through conventional plant breeding strategies has had limited use in floral crops due primarily to the diversity of species involved, the overriding importance of horticultural and quality traits, and the difficulty in identifying virus resistance genes. Since 1986 there has been considerable interest in a form of resistance which is based on introducing genes from a virus into plants (Powell-Abel *et al.*, 1986; Beachy *et al.*, 1990). We are using this strategy to develop resistance to tomato spotted wilt virus (TSWV) and the related impatiens necrotic spot virus (INSV) in floral crops. Using chrysanthemum as a model system, our goal is to test the usefulness of genetic engineering strategies for protection of floral crops against damaging virus diseases.

HISTORICAL IMPORTANCE OF VIRUSES IN FLORAL CROPS

Throughout history, viruses have impacted the floral industry in diverse ways, from the generation of horticulturally desirable plants to forcing alterations in industry practices. For example, while most viruses of floral crops have severe negative impacts on crop production, tulip breaking virus greatly increased the market value of tulips during the sixteenth and seventeenth centuries. Tulip breaking virus is known to have occurred in Europe since tulips were first imported from Asia Minor and Persia in the mid 1500s. Even though it was recognized that virus-infected bulbs slowly degenerated, bulbs producing broken flowers (yellow, pink or red flowers with white stripes) were highly prized and sometimes commanded high prices. Financial speculation in tulip bulbs was common in the seventeenth century, and 1637 was the year of 'tulipmania' when some broken tulips were sold for 3000–4000 guilders each in the Netherlands. Between 1820 and 1850 in Britain, £100–150 was the price of a well-broken tulip (Dickens, 1979). Shortly after this period it was learned that the striking flower colour could be graft-transmitted to ordinary bulbs. This discovery destroyed this bulb market, and as such is probably the first example of the adverse impact of viruses on the floral crop industry!

Virus disease problems have driven the industry to alter its production practices. The history of commercial lily production, for example, is one of repeated adaptation to virus problems. Easter lilies became an important commercial crop around 1870. At that time, the bulbs were propagated in the Azores and Bermuda. By 1896, Bermuda was exporting some 14,000 cases of bulbs yearly. Around 1893, viruses began attacking the Easter lily bulbs propagated in Bermuda, almost ending bulb production on the island. This disease, known as the Bermuda lily disease, is now known to be caused by two or more viruses. Bulb propagation was moved to Japan. By the early 1930s, more than 30 million bulbs were exported to the United States yearly. However, virus diseases once again became a common occurrence in Japanese-grown lily bulbs, and bulb production was moved to the United States in the late 1930s due to the virus problem and impending war. Since that time, bulb propagation in the United States has been moved periodically to avoid virus diseases (Kiplinger and Langhans, 1967).

In vitro propagation technology adopted by the floral industry was also initially an effort to control virus and viroid disease problems in several important crops. During the early to mid 1940s, chrysanthemum production in the United States was severely affected by chrysanthemum stunt viroid. Infected plants showed mottling of leaves, general stunting, pale green stems and leaves, and flowers which formed prematurely and were smaller than normal. In addition, flower stems would not absorb water after harvest and therefore had no shelf life. In the mid-1950s, aspermy virus threatened chrysanthemum production in the USA. The threat of these diseases to the commercial production of chrysanthemums led to the development of methods for obtaining and maintaining indexed chrysanthemum propagation stocks free of stunt, aspermy, and several other viruses (Raju and Olson, 1985). This technology is still the primary approach used to control viruses in chrysanthemum and many other floral crops.

The need to control viruses in floral crops has importance beyond the floral industry itself, as floral crops have been responsible for introduction and spread of viruses to other crops and into new geographical locations. Dahlias, for example, are planted worldwide for their spectacularly large flowers and because of their large fleshy roots which allow for easy propagation and transport over long distances. Dahlias are commonly infected with TSWV, but infection causes little damage other than chlorotic ring patterns in the leaves. TSWV-infected dahlias have been implicated or proven to be the source of damaging outbreaks of TSWV on other crops. TSWV was introduced into tobacco-growing areas in Rhodesia (Zimbabwe) in dahlia tubers commonly grown in town gardens (Hopkins, 1943). Quarantine restrictions were placed on nurseries, and in response to appeals and house-to-house inspections, most gardeners destroyed all their dahlias. Despite these efforts, the wide host range of the virus permitted it to became established in several other garden plants and weeds, and it continues to infect tobacco crops in Zimbabwe. In California, TSWV infection of lettuce has been shown to be greatest in fields near urban areas where perennial floral crops such as calla lily, nasturtium, dahlia and chrysanthemum abound (Grogan *et al.*, 1955). Problems with tomato production have also been tied to the presence of TSWV-infected floral crops and perennial weeds in New Jersey (Holmes, 1948).

VIRUS CONTROL PROCEDURES

In 1965, Hollings noted, 'The intensive monoculture of crop plants today invites the epidemic spread of many diseases and pests, among the most important of which are viruses. Rapid vegetative propagation and a flourishing international trade have made matters worse' (Hollings, 1965). Thirty years later, production of floral crops in the USA has greatly increased. Floral crop production is very intensive, and much of the production is based on international movement of bulbs, offshore production of propagation material, and international shipment of cut flowers and other materials. International free trade agreements, like the recently enacted North American Free Trade Agreement (NAFTA), will only continue to increase international floral crop movement and increase movement and outbreak of virus diseases. Thus economical and highly effective virus control procedures are increasingly needed.

The major means of control of virus diseases of floral crops has traditionally been through the production and distribution of propagation materials which have been indexed for the presence of known viruses (so called 'virus-free' stock). Initial efforts to generate virus-free materials by meristem tip culture and root tip cuttings were made with dahlia and date back to the early 1950s (Morel and Martin, 1952; Holmes, 1954). A 1965 review (Hollings, 1965) listed four main techniques to rid plants of viruses: (i) exploiting erratic virus distribution within the plant, (ii) chemotherapy, (iii) heat treatment, and (iv) meristem-tip culture. These techniques increase the probability of obtaining a plant free of virus, although often a significant percentage of the treated plants remain infected. Therefore, a critical component is the implementation of testing (indexing) procedures to select the healthy plants. Heat treatment and meristem-tip culture along with virus indexing systems are currently in use for the major vegetatively propagated commercial floral crops such as chrysanthemum, geranium, orchid and carnation. Combining these two techniques with tissue culture to rapidly multiply the virus-indexed clones to commercial quantities is standard practice in the floral industry today. This methodology, together with strategies to maintain healthy source plants, has greatly reduced the threat of virus diseases in the floral industry over the past several decades in the United States.

During the mid- to late 1980s, however, the pesticide-resistant western flower thrips, a vector of TSWV, became widely distributed in the floral industry. During the late 1980s and early 1990s, TSWV and the closely related INSV caused severe damage to floral crops such as begonia, chrysanthemum, exacum, gloxinia and impatiens. The rapid spread and severe economic impact of these insect-vectored viruses vividly point out the weakness in virus-free propagation schemes, i.e. the need for virus resistance. This need for resistance has spawned programmes to develop genetically engineered virus-resistant floral crops in recent years.

TOMATO SPOTTED WILT-LIKE VIRUSES

TSWV and the related INSV (Law and Moyer, 1990) are members of a group of unusual plant viruses that are collectively known as tospoviruses. These viruses are characterized by having a particle consisting of three linear single-stranded RNA molecules, each encapsidated by the nucleocapsid or 'N' protein. These nucleocapsids are then surrounded by a membrane of host origin which also contains two virus-encoded proteins (for review see German *et al.*, 1992). Tospoviruses are further distinguished by their ambisense and negative-sense coding strategy, requiring subgenomic messenger RNA species for translation and protein expression.

TSWV and INSV are distributed worldwide and have extremely broad host ranges, being capable of infecting over 500 different species of both monocots and dicots in over 50 plant families (German *et al.*, 1992). These include important field, fruit, vegetable and floral crops, in addition to many weed species. Among floral crops, chrysanthemum, gloxinia, standard and New Guinea impatiens, exacum, and begonia are particularly affected. Losses due to these viruses by the floral industry have been severe.

TSWV and INSV are vectored by thrips, the predominant one in floral crops being the western flower thrips, *Frankliniella occidentalis* (Best, 1968). The relationship between these viruses and thrips is unusual in that while only the larvae can acquire the virus, it is transmitted only by the adult stages. In general, incidence of these viruses increases with increasing populations of thrips. Thus large thrips populations not only have a direct effect on crop production by feeding injury, but, much more importantly, they increase the probability of TSWV spread within the crop.

CONVENTIONAL CONTROL STRATEGIES FOR TSWV-LIKE VIRUSES

Conventional control strategies for TSWV-like viruses in floral crops have emphasized early detection and removal of infected planting material from production areas as well as preventative measures to control the thrips vector. The lack of accurate serological diagnosis of plants seemingly infected with TSWV based on symptom expression and presence of the thrips vector led to the identification of INSV, a closely related but serologically distinct virus which is rare in field crops but widespread in the floral industry (Law and Moyer, 1990; German *et al.*, 1992). The development of antisera for both TSWV and INSV has allowed for accurate indexing programmes and diagnosis of tospovirus-infected stock, and has significantly improved the industry's ability to accurately and rapidly diagnose these viruses in floral crops.

Control of the western flower thrips vector is difficult as these insects have developed resistance to many insecticides (Robb *et al.*, 1988) and breed successfully in greenhouses all year long. In contrast to the lack of control of virus spread by insect vectors in the field through insecticide applications, population suppression by frequent insecticide applications is effective and essential to prevent spread of TSWV-like viruses by the western flower thrips in greenhouses. Screening of greenhouses, isolation of propagation stock from production areas, eliminating weeds and non-production reservoir plants from greenhouses, monitoring of thrips populations, and judicious use of pesticides coupled with the monitoring and removal of infected material, are all required for control. However, these measures are often not adequate for complete control. Even low but persistent populations of thrips and a few infected reservoir plants can maintain the virus in greenhouses for years. Many growers have been unsuccessful in eradicating the viruses from their greenhouses. A three to four month period of total elimination of TSWV and INSV-susceptible plants (e.g. production of just poinsettia in the autumn) greatly helps reduce or eliminate virus reservoirs.

BIOTECHNOLOGY APPROACHES FOR THE DEVELOPMENT OF DISEASE RESISTANCE

The difficulty in controlling TSWV-like viruses in floral crops has underscored the need for the development of virus-resistant plants. Conventional breeding strategies require the identification of sources of disease-resistance genes, a difficult task given the diversity of floral crop species that are susceptible to these viruses. Also, the overriding importance of appearance and general horticultural traits, the large number of cultivars which are produced per crop, and the rapid turnover in cultivars have made breeding for disease resistance of secondary importance (Sparnaaij, 1991). Breeding for virus resistance in particular, has not been widely used with floral crops. However, the recent development of technologies for gene identification and gene transfer into plants has provided the opportunity for engineering disease resistance into horticulturally desirable cultivars without altering critical quality traits. To date, the greatest success with genetic engineering for plant disease resistance has been with virus resistance.

A number of approaches have been investigated for the development of virus resistance in plants. These include: (i) cloning of natural virus resistance genes from plants, (ii) altered expression of genes which are induced during resistance reactions, (iii) expression of antibodies which would bind critical viral proteins, and (iv) expression of virus genes in plants.

CLONING OF NATURAL VIRUS-RESISTANCE GENES

Virus-resistant plants are routinely used in breeding programmes to develop high-quality cultivars of important crop species. Given the difficulty in controlling plant viruses using chemical or cultural methods, the use of disease-resistant cultivars is the primary means of control of viruses in many field and vegetable crops. All virus-resistant crops grown today are based on genes for resistance which occur naturally in plants. However, for reasons described previously, there are relatively few examples of virus resistance in floral crops.

One approach to engineering virus-resistant plants may be to transfer naturally occurring virus resistance genes into unrelated crop species. This approach requires the identification and isolation of the resistance gene from the plant genome, followed by transfer and expression in alternative plants. Genes which specify resistance to fungal pathogens of maize (Meeley *et al.*, 1992), tomato (Jones *et al.*, 1994) and flax (Lawrence *et al.*, 1995), and bacterial pathogens of tomato (Martin *et al.*, 1993) and *Arabidopsis* (Bent *et al.*, 1994) have already been isolated, as has the tobacco *N* gene, which encodes resistance to tobacco mosaic virus (Whitham *et al.*, 1994). Further studies are needed to define the effectiveness of such genes in conditioning disease resistance in heterologous plant species.

ALTERED EXPRESSION OF PATHOGENESIS-RELATED PROTEINS

Pathogenesis-related or 'PR' proteins are proteins produced in plants in response to pathogen infection and abiotic stress, and induction of these proteins is generally correlated with disease resistance (Linthorst, 1991). PR proteins fall into ten major classes, some of which have been identified as chitinases and β-1,3 glucanases. One strategy for engineering disease resistance has been to alter the expression of these genes so that they are constitutively expressed rather than having to be induced following infection by pathogens. Constitutive, high-level expression of some PR proteins has been shown to provide low levels of protection against some fungal pathogens (Alexander *et al.*, 1993; Broglie and Broglie, 1993). PR proteins are also accumulated during resistance responses to viruses, suggesting that some of them may have antiviral properties. However, efforts to produce virus resistance by constitutive expression of PR proteins (particularly the PR1 group) were not successful (Cutt *et al.*, 1989; Linthorst *et al.*, 1989).

ANTIBODY-MEDIATED RESISTANCE

In 1989, Hiatt and co-workers (Hiatt *et al.*, 1989) transformed plants with cDNAs made to mouse hybridoma messenger RNA and successfully expressed antibodies in plants. This accomplishment opened the possibility of transforming plants with genes for antibodies which can bind to and thus interfere with critical pathogenesis-required proteins. This strategy has been used to engineer resistance to artichoke mottled crinkle virus (AMCV) (Tavladoraki *et al.*, 1993). A gene for a monoclonal antibody which had affinity for the AMCV coat protein was transformed into *Nicotiana benthamiana*. Symptom expression was delayed in transgenic plants expressing the antibody genes.

PATHOGEN-DERIVED RESISTANCE

Pathogen-derived resistance is a concept with origins in both plant and animal pathogens. In plants, the concept dates back to the 1920s, when it was discovered that plants could be protected from infection by severe strains of a virus by prior inoculation with a mild or latent strain of the same virus. This process was named 'cross protection' (McKinney, 1929), and the viral components responsible have been a source of considerable speculation and study. The concept of pathogen-derived resistance was described in a molecular context in a theoretical paper (Sanford and Johnston, 1985). The basic concept behind pathogen-derived resistance is that all organisms have genes that regulate critical stages in their life cycle. If genes can be identified which negatively regulate critical pathogenesis processes, then it should be possible to use those genes to impart resistance. The technologies derived from advances in molecular biology have made it possible to begin to dissect plant viruses to determine which parts of the viral genome are responsible for the virus-derived resistance (cross protection) phenomenon. In the 1980s it was reported that the capsid protein of tobacco mosaic virus (TMV) was responsible for cross protection (Sherwood and Fulton, 1982). This finding was confirmed when it was discovered that the TMV capsid protein gene, when cloned and integrated into the plant's own genome, resulted in delayed symptom expression upon TMV infection (Powell-Abel *et al.*, 1986). Since then it has been discovered that other parts of the viral genome are also capable of imparting resistance. At the present time, there are several strategies which involve isolating some portion of the viral genetic material and subcloning into an appropriate vector that can be used to transform host plants so that the plants then express a resistant phenotype (for reviews see Hanley-Bowdoin and Hemenway, 1992; Wilson 1993; Grumet, 1995). Approaches tested to date include the use of capsid protein genes, replicase genes, defective interfering (DI) and satellite RNAs, ribozymes and antisense strategies.

Capsid protein genes

Capsid-protein-mediated resistance has received far and away the greatest attention. The capsid protein is the protein which protects the viral genome. It is also known to be important in many other viral functions such as the elicitation of host responses and in vector transmission. This form of pathogen-derived resistance is based upon isolating the gene which codes for the capsid protein and using it to transform plants. Capsid-protein-mediated resistance may be somewhat misleading in that, although the protein itself may be involved in some resistance responses, it has also been demonstrated that the protein is not necessary in all instances where resistance has been observed (Lindbo and Dougherty, 1992). Even though the mechanism responsible for imparting the resistance phenotype to the host is not yet understood, the capsid protein gene is the most frequent source of pathogen-derived resistance. It has been tested in most of the major crops and is now being extended to minor crops such as various floral crops and exotics such as papaya. The technical limitation for floral crops is not the viral resistance strategy but rather the gene delivery (transformation) system that must be developed for each crop.

Replicase genes

Replicase genes code for the protein or one of the components of the protein that is responsible for replicating the viral genome. A 54 kDa portion of the replicase gene for tobacco mosaic virus was used to impart resistance in tobacco to tobacco mosaic virus (Golemboski *et al.*, 1990). Interestingly, in this initial example of replication-mediated resistance, no protein has been associated *in vivo* with the particular genome segment used to impart the resistance. The replicase-mediated strategy of resistance has also been reported for protection against other viruses such as potato virus X (Braun and Hemenway, 1992).

Defective-interfering (DI) and satellite RNAs

Defective-interfering (DI) RNAs are genomic segments from the virus that contain internal deletions. These may be of variable length, but all contain significant portions of the 3′ and 5′ ends of the particular segment (Hillman *et al.*, 1987). Naturally occurring DIs have the capacity to modulate disease development. Most frequently disease development is attenuated. However, there are examples of some DIs which may enhance disease development. Attempts have been made to characterize those DIs which down-regulate disease expression and insert these into the host genome, similar to other viral genes (Jones *et al.*, 1990).

Satellite RNAs are viral RNAs which are distinct from the viral genome in that they do not share any sequence homologies (Waterworth *et al.*, 1979). However, satellites are dependent upon the viral genome for replication and are encapsidated along with the viral genome. Once introduced into plants in nature, the satellite can be spread along with the virus and become established in the natural environment. In spite of the difference in origin, satellites may have similar phenotypic effects to that of DI RNAs. Thus whole satellite RNAs have been used in attempts to control disease (for review see Wilson, 1993). This particular strategy has currently fallen from favour. As relatively minor base substitutions are responsible for major shifts in symptom-modulating capabilities, there is a relatively high risk that mild strains may mutate to more severe strains when put in nature in an uncontrolled manner.

Ribozymes

A strategy which has been tried, but is as yet without success, is the use of ribozymes (Gerlach *et al.*, 1987). Ribozymes are small RNA molecules which have enzymatic-like nuclease activity that cleaves RNA molecules at specific nucleotide sequences. Critical sequences used in ribozymes are derived from some satellites. Theoretically, ribozymes have considerable potential in that they are capable of cleaving RNA molecules, and this cleavage is site specific. Therefore, ribozymes can be altered so as to target specific viral sequences resulting in cleavage of RNAs required for specific viral functions. While *in vitro* experiments have held some promise, *in vivo* experiments have not achieved an acceptable level of success (for review see Wilson, 1993).

Antisense

The concept of antisense is to transform plants with the strand of the viral nucleic acid that is complementary to the strand of nucleic acid that codes for a particular gene. There are reports of the successful use of antisense constructs. In general, though, the frequency of identifying transformants with the desired level of resistance is much less than with many of the other strategies. It has been generally accepted that antisense is most effective where the replicative process of the pathogen (in this case the virus) occurs in the nucleus, as opposed to the situation with many plant RNA viruses, where replication occurs in the cytoplasm (Hull and Davies, 1992; Wilson, 1993).

DISEASE RESISTANCE IN FLORAL CROPS THROUGH BIOTECHNOLOGY

For efficient use of genetic engineering, floral crops pose a challenge both because of the diversity of species and cultivars available as well as the lack of extensive studies of tissue culture protocols. None the less, progress is being made in the development of transformation protocols. *Agrobacterium*-mediated transformation of the three major cut flower crops (rose, chrysanthemum and carnation) has been reported, and progress is being made with a number of other floral crops such as *Gerbera* and orchid (see Robinson and Firoozabady, 1993, for review; Firoozabady *et al.*, 1994).

The first genetically engineered floral crops reported were petunia (Jorgensen and Mol, 1991; Meyer, 1991) and chrysanthemum (Courtney-Gutterson *et al.*, 1993, 1994), both with altered flower colour. A number of reports have documented efforts to develop virus and other disease resistance in floral crops through genetic engineering (Dons *et al.*, 1991; Persoon, 1992; Urban *et al.*, 1994; Yepes *et al.*, 1995). However, there are no reports in the literature to date of virus-resistant floral crops developed through genetic engineering. The development of insect resistance through genetic engineering is also being investigated extensively (Boulter, 1993), and there have been efforts with chrysanthemum to express genes for either the *Bacillus thuringiensis* endotoxin or for proteinase inhibitors (Dons *et al.*, 1991; van Wordragen *et al.*, 1993). Given the increasing importance of insect-vectored viruses, engineering plants for insect resistance may be a useful approach for virus control.

TRANSFORMATION OF CHRYSANTHEMUM FOR RESISTANCE TO TSWV

The focus of our laboratories for the last few years has been control of TSWV and INSV in floral crops. Efforts have been directed towards characterization of virus isolates (Law and Moyer, 1990; German *et al.*, 1992) and on accurate diagnosis and control of this disease in greenhouses. As conventional control measures have proved inadequate for effective control of these viruses, we have recently begun focusing on the development of resistance through genetic engineering. We chose chrysanthemum as a model system for this work since it is economically important, is highly susceptible to TSWV, and extensive studies have been done on *in vitro* culture and transformation technology. As a source of virus resistance we chose to use the nucleocapsid ('*N*') gene of TSWV, as efforts in several laboratories with expression of the TSWV N gene in tobacco and tomato have documented its utility in the engineering of disease resistance (Gielen *et al.*, 1991; de Haan *et al.*, 1992; MacKenzie and Ellis, 1992; Pang *et al.*, 1992, 1994; Kim *et al.*, 1994; Prins *et al.*, 1995; Vaira *et al.*, 1995).

For our studies, the *N* gene was cloned from a highly virulent dahlia isolate of TSWV (Urban *et al.*, 1994). The gene was cloned from the small RNA by reverse-transcriptase polymerase chain reaction (Ausubel *et al.*, 1989). The vector plasmid used was pBI121 (Clontech Laboratories), a commercially available vector which contains the CaMV 35S promoter for expression of the transgene in plants, a gene for β-glucuronidase (GUS) production, and a gene encoding kanamycin resistance for selection of transformants. Five different constructions of the *N* gene were made. These included the full-length gene with viral flanking sequences, the full-length gene without flanking sequences, a mutated version of the gene which produces a truncated protein in plants, an antisense version of the gene, and the gene driven by a double 35S promoter. To test the effectiveness of the constructs, they were transformed into tobacco cultivar 'Burley 21', and plants were screened for TSWV resistance following mechanical inoculation. All constructs with the exception of the one containing viral flanking sequences successfully protected tobacco against infection (Perez *et al.*, 1993). Resistance expressed in tobacco was by an apparent lack of systemic movement of the virus in the plant; mechanically inoculated leaves developed lesions. However, the virus did not spread systemically into the rest of the plant. Resistance was also shown to be effective against heterologous strains of the virus, but not to all isolates in our collection.

Efforts with chrysanthemum initially focused on the development of efficient regeneration and transformation protocols. For regeneration, we were most successful using a modification of a protocol developed by May and Trigiano (1991) for chrysanthemum. Our regeneration procedure involved the use of leaf explant tissue plated on Murashige and Skoog medium (Murashige and Skoog, 1962) amended with 11.5 μM indoleacetic acid (IAA) and 0.5–1.0 μM benzyladenine (BA) plus 20 μM $CoCl_2$ to suppress ethylene synthesis (Urban *et al.*, 1994). Cultures were also incubated under yellow plexiglas filters which absorb wavelengths below 450 nm, as blue wavelengths of light have been shown to catalyse the photooxidation of EDTA resulting in the production of toxic levels of formaldehyde and iron-deficient medium (Hangarter and Stasinopoulos, 1991). Using this procedure, we were successful in regenerating both pot and cut flower chrysanthemum cultivars.

Agrobacterium tumefaciens strain specificities were also tested, as extensive experiments using the *A. tumefaciens* vector strain LBA4404 failed to yield transformed plants. We tested three wild type strains, Chry5, a strain initially isolated from chrysanthemum (Bush and Pueppke, 1991), Ach5, the parent strain of the commonly used disarmed vector strain LBA4404, and A281, a wide-host-range strain which is highly virulent on many plant species (van Wordragen *et al.*, 1991). Strain virulence was measured by the ability to induce hormone-autonomous callus formation on leaf explants *in vitro* (Urban *et al.*, 1994). Using this assay, Ach5 was shown to be unable to transform three different chrysanthemum cultivars. However, Chry5 and A281 were highly virulent. A disarmed version of A281 (EHA105) was then used for transformation experiments.

Our initial efforts to transform chrysanthemum utilized the pot mum cultivar 'Iridon', as this cultivar showed the highest frequency of regeneration.

Transformants were isolated using the above regeneration protocol coupled with a three to five day co-culture period with EHA105 containing pBI121 or pBI121 containing the TSWV *N* gene (Urban *et al*., 1994). Transformation was confirmed by PCR and Southern analysis, and the genes were shown to be stable through vegetative propagation.

We have now transformed two cultivars of chrysanthemum with the *N* gene of TSWV, and have recovered TSWV-resistant plants (Sherman *et al*., 1993; Sherman *et al*., 1996). For these studies we used both 'Iridon' and the highly TSWV-susceptible cut flower mum cultivar 'Polaris'. Plants were transformed using a modification of the above transformation protocol, and were selected on medium containing paromomycin. Transformation was confirmed by PCR analysis, and then transformed plants were screened for resistance using a highly virulent isolate of TSWV (TSWV-GB) isolated from an infected plant of chrysanthemum cultivar 'Golden Bounty' in a commercial greenhouse. A colony of TSWV-infected thrips was established in a greenhouse chamber by allowing thrips to feed on TSWV-infected chrysanthemum and *Datura stramonium* plants. Cuttings of transformed plants were then challenged with TSWV by incubating them in the chamber for ten weeks. As with tobacco, all plants developed localized lesions at the point of infection; all challenged chrysanthemum plants showed small chlorotic lesions at the initial thrips feeding sites on older leaves. Susceptible plants went on to develop systemic symptoms and stem cankers typical of TSWV infection, whereas resistant plants developed no systemic symptoms or stem lesions. Putatively resistant plants were also assayed by ELISA for the presence of the virus in upper leaves of the plants. A small percentage of the plants which lacked systemic symptoms possessed positive ELISA values; these plants were scored as susceptible. Overall, 12.7% of the transgenic plants were resistant, defined as lacking systemic symptoms and having negative ELISA values. Resistant plants are currently being screened to determine the stability of resistance through multiple cycles of vegetative propagation and through meristem tip culture.

FUTURE PROSPECTS

In many ways the outlook for development and use of genetically engineered floral crops is positive, given the progressiveness and innovativeness of the industry as well as the industry's needs. Due to problems such as pesticide resistance, Occupational Health and Safety Administration (OSHA)-imposed re-entry restrictions, and loss of pesticide registrations, the floral industry is interested in pest control techniques that reduce or eliminate the need to use pesticides, and there is enough profit in major floral crops to pay for expensive but necessary pest control. The floral industry has always been a leader in agriculture in adopting new and novel pest control practices. Examples include the development of *in vitro* techniques for virus elimination, virus-indexing procedures, low-volume spray technology, and integrated pest management techniques to reduce pesticide use. Indexing for viral, bacterial and fungal pathogens started with the floral industry as far back as the 1950s. The industry's increasing use of *in vitro* culture for propagation of planting material puts it in a unique position to utilize genetic engineering, given access to horticulturally useful genes. Also, since floral crops are not food crops, it should be easier to release and gain public acceptance of genetically engineered plants.

Many steps are still needed, however, for genetic engineering to be an efficient and readily utilized technique for the development of virus-resistant floral crops. First among the needs are truly novel transformation technologies. Current technologies are limited by the need to develop regeneration protocols for each crop species and cultivar. Quantum leaps in transformation technology that bypass conventional *in vitro* culture limitations are needed. More information is needed on the commercial acceptability of virus-resistant transgenics. It is not yet clear how durable or broad-spectrum this type of resistance will be or what is required to create a commercially acceptable phenotype. More fundamental research is needed on the cause of a resistance phenotype. We and others have made a number of observations regarding the development of truly resistant plants using genetic engineering, most of them generated from experience with capsid-protein-mediated resistance. The level of resistance of plants generated in this way varies from individual plant to individual plant, with only a very small percentage being sufficiently resistant to warrant further studies. At the present time it is not possible to adequately predict the level of resistance to be expressed by a particular transformant based upon a molecular characteristic such as the number of insertion sites or level of protein or RNA synthesis. In a practical sense, this means that it is necessary to evaluate a large number of plants to obtain ones suitable for commercialization. Finally, due to the current emphasis by most granting agencies on fundamental research, there is a need for a partnership between industry, government and academia if the promises of this technology are to be realized.

REFERENCES

Alexander, D., Goodman, R.M., Gut-Rella, M., Glascock, C., Weymann, K., Friedrich, L., Maddox, D., Ahl-Goy, P., Luntz, T., Ward, E. and Ryals, J. (1993) Increased tolerance to two oomycete pathogens in transgenic tobacco expressing pathogenesis-related protein 1a. *Proceedings of the National Academy of Sciences, USA* 90, 7327–7331.

Ausubel, F.M., Brent, R., Kingston, R.E., Moore, D.D., Seidman, J.G., Smith, J.A. and Struhl, K. (1989) *Current Protocols in Molecular Biology*. John Wiley, New York, Section 15.4.

Beachy, R.N., Loesch-Fries, S. and Tumer, N.E. (1990) Coat-protein mediated resistance against virus infection. *Annual Review of Phytopathology* 28, 451–474.

Bent, A.F., Kunkel, B.N., Dahlbeck, D., Brown, K.L., Schmidt, R., Giraudat, J., Leung, J. and Staskawicz, B.J. (1994) *RPS2* of *Arabidopsis thaliana*: a leucine-rich repeat class of plant disease resistance genes. *Science* 265, 1856–1860.

Best, R.J. (1968) Tomato spotted wilt virus. *Advances in Virus Research* 13, 65–146.

Boulter, D. (1993) Insect pest control by copying nature using genetically engineered crops. *Phytochemistry* 34, 1453–1466.

Braun, C.J. and Hemenway, C. (1992) Expression of amino-terminal portions or full-length viral replicase genes in transgenic plants confers resistance to potato virus X infection. *Plant Cell* 4, 735–744.

Broglie, R. and Broglie, K. (1993) Chitinase gene expression in transgenic plants: a molecular approach to understanding plant defense responses. *Philosophical Transactions of the Royal Society London, Series B* 342, 265–270.

Bush, A.L. and Pueppke, S.G. (1991) Characterization of an unusual new *Agrobacterium tumefaciens* strain from *Chrysanthemum morifolium* Ram. *Applied Environmental Microbiology* 57, 2468–2472.

Courtney-Gutterson, N., Firoozabady, E., Lemieux, C., Nicholas, J., Morgan, A., Robinson, K., Otten, A. and Akerboom, M. (1993) Production of genetically engineered color-modified chrysanthemum plants carrying a homologous chalcone synthase gene and their field performance. *Acta Horticulture* 336, 57–62.

Courtney-Gutterson, N., Napoli, C., Lemieux, C., Morgan, A., Firoozabady, E. and Robinson, K.E.P. (1994) Modification of flower color in florist's chrysanthemum: production of a white-flowering variety through molecular genetics. *Bio/Technology* 12, 268–271.

Cutt, J.R., Harpster, M.H., Dixon, D.C., Carr, J.P., Dunsmuir, P. and Klessig, D.F. (1989) Disease response to tobacco mosaic virus in transgenic tobacco plants that constitutively express the pathogenesis-related PR1b gene. *Virology* 173, 89–97.

de Haan, P., Gielen, J.J.L., Prins, M., Wijkamp, I.G., van Schepen, A., Peters, D., van Grinsven, M.J.Q.M. and Goldbach, R. (1992) Characterization of RNA-mediated resistance to tomato spotted wilt virus in transgenic tobacco plants. *Bio/Technology* 10, 1133–1137.

Dickens, J.S.W. (ed.) (1979) *Diseases of Bulbs*. HMSO, London, 205 pp.

Dons, J.J.M., Mollema, C., Stiekema, W.J. and Visser, B. (1991) Routes to the development of disease resistant ornamentals. In: Harding, J., Singh, F. and Mol, J.N.M. (eds), *Genetics and Breeding of Ornamental Species*. Kluwer Academic Publishers, Dordrecht, pp. 387–417.

Firoozabady, E., Moy, Y., Courtney-Gutterson, N. and Robinson, K. (1994) Regeneration of transgenic rose (*Rosa hybrida*) plants from embryonic tissue. *Bio/Technology* 12, 609–613.

Gerlach, W.L., Llewellyn, D. and Haseloff, J. (1987) Construction of a plant disease resistance gene from the satellite RNA of tobacco ringspot virus. *Nature* 328, 802–805.

German, T., Ullman, D. and Moyer, J.W. (1992) Tospoviruses: diagnosis, molecular biology, phylogeny, and vector relationships. *Annual Review of Phytopathology* 30, 315–348.

Gielen, J.J.L., de Haan, P., Kool, A.J., Peters, D., van Ginsven, M.Q.J.M. and Goldbach, R.W. (1991) Engineered resistance to tomato spotted wilt virus, a negative-strand RNA virus. *Bio/Technology* 9, 1363–1367.

Golemboski, D.B., Lomonossoff, G.P. and Zaitlin, M. (1990) Plants transformed with tobacco mosaic virus nonstructural gene sequence are resistant to the virus. *Proceedings of the National Academy of Sciences, USA* 87, 6311–6315.

Grogan, R.G., Snyder, W.C. and Bardin, R. (1955) Diseases of lettuce. *Circular of the California Agricultural Experiment Station Extension Service* p. 448.

Grumet, R. (1995) Genetic engineering for crop virus resistance. *HortScience* 30, 449–456.

Hangarter, R.P. and Stasinopoulos, T.C. (1991) Repression of plant tissue culture growth by light is caused by photochemical change in the culture medium. *Plant Science* 79, 253–257.

Hanley-Bowdoin, L. and Hemenway, C. (1992) Expression of plant viral genes in transgenic plants. In: Wilson, T.M.A. and Davies, J.W. (eds), *Genetic Engineering with Plant Viruses*. CRC Press, Boca Raton, pp. 251–295.

Hiatt, A., Cafferkey, R. and Bowdish, K. (1989) Production of antibodies in transgenic plants. *Nature* 342, 76–78.

Hillman, B.I., Carrington, J.C. and Morris, T.J. (1987) A defective interfering RNA that contains a mosaic of a plant virus genome. *Cell* 51, 427–433.

Hollings, M. (1965) Disease control through virus-free stock. *Annual Review of Phytopathology* 3, 367–396.

Holmes, F.O. (1948) Resistance to tomato spotted wilt in tomato. *Phytopathology* 38, 467–473.

Holmes, F.O. (1954) Elimination of spotted wilts from dahlias by propagation of tip cuttings. *Phytopathology* 45, 224–226.

Hopkins, J.C.F. (1943) Mycological notes 16. The campaign against the kromneck virus. *Rhodesia Agriculture Journal* 40, 47–49.

Hull, R. and Davies, J.W. (1992) Approaches to nonconventional control of plant virus diseases. *Critical Reviews in Plant Science* 11, 17–33.

Jones, D.A., Thomas, C.M., Hammond-Kosack, K.E., Balint-Kurti, P.J. and Jones, J.D.G. (1994) Isolation of tomato *Cf-9* gene for resistance to *Cladosporium fulvum* by transposon tagging. *Science* 266, 789–793.

Jones, R.W., Jackson, A.O. and Morris, T.J. (1990) Defective interfering RNAs and elevated temperatures inhibit replication of tomato bushy stunt virus in inoculated protoplasts. *Virology* 176, 539–545.

Jorgensen, R. and Mol, J. (1991) Modulation of flower color and its intensity via directed gene manipulation. In: Harding, J., Singh, F. and Mol, J.N.M. (eds), *Genetics and Breeding of Ornamental Species*. Kluwer Academic Publishers, Dordrecht, pp. 309–316.

Kim, J.W., Sun, S.S.M. and German, T.L. (1994) Disease resistance in tobacco and tomato plants transformed with the tomato spotted wilt virus nucleocapsid gene. *Plant Disease* 78, 615–621.

Kiplinger, D.C. and Langhans, R.W. (eds) (1967) *Easter Lilies. Bulletin of the New York Lily School*, 158 pp.

Law, M.D. and Moyer, J.W. (1990) A tomato spotted wilt-like virus with a serologically distinct N protein. *Journal of General Virology* 71, 2597–2601.

Lawrence, G.J., Finnegan, E.J., Ayliffe, M.A. and Ellis, J.G. (1995) The *L6* gene for flax rust resistance is related to the *Arabidopsis* bacterial resistance gene *RPS2* and the tobacco viral resistance gene *N*. *Plant Cell* 7, 1195–1206.

Lindbo, J.A. and Dougherty, W.G. (1992) Pathogen derived resistance to a potyvirus: immune and resistant phenotypes in transgenic tobacco expressing altered forms of a potyvirus coat protein nucleotide sequence. *Molecular Plant–Microbe Interactions* 5, 144–153.

Linthorst, H.J.M. (1991) Pathogenesis-related proteins of plants. *Critical Reviews in Plant Science* 10, 123–150.

Linthorst, H.J.M., Meuwissen, R.L.J., Kauffmann, S. and Bol, J.F. (1989) Constitutive expression of pathogenesis-related proteins PR-1, GRP, and PR-S in tobacco has no effect on virus infection. *Plant Cell* 1, 285–291.

MacKenzie, D.J. and Ellis, P.J. (1992) Resistance to tomato spotted wilt virus infection in transgenic tobacco expressing the viral nucleocapsid gene. *Molecular Plant–Microbe Interactions* 5, 34–40.

Martin, G.B., Brommonschenkel, S.H., Chunwongse, J., Frary, A., Ganal, M.W., Spivey, R., Wu, T., Earle, E.D. and Tanksley, S.D. (1993) Map-based cloning of a protein kinase gene conferring disease resistance in tomato. *Science* 262, 1432–1436.

May, R.A. and Trigiano, R.N. (1991) Somatic embryogenesis and plant regeneration from leaves of *Dendranthema grandiflora*. *Journal of the Americam Society for Horticultural Science* 116, 366–371.

McKinney, H.H. (1929) Mosaic disease in the Canary Islands, West Africa, and Gibraltar. *Journal of Agricultural Research* 39, 557–578.

Meeley, R.B., Johal, G.S. and Briggs, S.P. (1992) A biochemical phenotype for a disease resistance gene of maize. *Plant Cell* 4, 71–77.

Meyer, P. (1991) Engineering of novel flower colours. In: Harding, J., Singh, F. and Mol, J.N.M. (eds), *Genetics and Breeding of Ornamental Species*. Kluwer Academic Publishers, Dordrecht, pp. 285–307.

Morel, G. and Martin, C. (1952) Guerison de dahlias atteints d'une maladie a virus. *Comptes Rendus* 235, 1324–1325.

Murashige, T. and Skoog, F. (1962) A revised medium for rapid growth and bioassays with tobacco tissue cultures. *Physiologia Plantarum* 15, 473–497.

Pang, S.Z., Nagpala, P., Wang, M., Slightom, J. and Gonsalves, D. (1992) Resistance to heterologous isolates of tomato spotted wilt virus in transgenic tobacco expressing its nucleocapsid protein gene. *Phytopathology* 82, 1223–1229.

Pang, S.Z., Bock, J.H., Gonsalves, C., Slightom, J.L. and Gonsalves, D. (1994) Resistance of transgenic *Nicotiana benthamiana* plants to tomato spotted wilt and impatiens necrotic spot tospoviruses: evidence of the involvement of the N protein and N gene RNA in resistance. *Phytopathology* 84, 243–249.

Perez, R., Geske, S.M., Speck, J., Reece, P., Moyer, J.W. and Daub, M.E. (1993) Transformation of tobacco with four different nucleocapsid constructs of tomato spotted wilt virus. *Phytopathology* 83, 1355, Abstract.

Persoon, J. (1992) Florigene innovates old varieties. *Prophyta* 46, 42–43.

Powell-Abel, P., Nelson, R.S., De, B., Noffman, N., Rogers, S.G., Fraley, R.T. and Beachy, R.N. (1986) Delay of disease development in transgenic plants that express the tobacco mosaic virus coat protein gene. *Science* 232, 738–743.

Prins, M., de Haan, P., Luyten, R., van Veller, M., van Grinsven, M.Q.J.M. and Goldback, R. (1995) Broad resistance to tospoviruses in transgenic tobacco plants expressing three nucleoprotein gene sequences. *Molecular Plant–Microbe Interactions* 8, 85–91.

Raju, B.C. and Olson, C.J. (1985) Indexing systems for producing clean stock for disease control in commercial floriculture. *Plant Disease* 69, 189–192.

Robb, K., Parrella, M.P. and Newman, J.P. (1988) The biology and control of the western flower thrips, Part 1. *Ohio Florists' Association Bulletin* 699, 2–5.

Robinson, K.E.P. and Firoozabady, E. (1993) Transformation of floriculture crops. *Scientia Horticulturae* 55, 83–99.

Sanford, J.C. and Johnston, S.A. (1985) The concept of parasite-derived resistance deriving resistance genes from the parasites' own genome. *Journal of Theoretical Biology* 113, 395–405.

Sherman, J.M., Moyer, J.W. and Daub, M.E. (1996) Genetically engineered resistance to tomato spotted wilt virus in chrysanthemum, *Dendranthema grandiflora*: a model system for virus protection in ornamental crops. *Acta Horticulturae* (in press).

Sherman, J.M., Urban, L.A., Geske, S.A., Speck, J., Moyer, J.W. and Daub, M.E. (1993) Transformation of chrysanthemum with the nucleocapsid gene from tomato spotted wilt virus. *Phytopathology* 83, 1355, Abstract.

Sherwood, J.L. and Fulton, R.W. (1982) The specific involvement of coat protein in tobacco mosaic virus cross protection. *Virology* 119, 150–158.

Sparnaaij, L.D. (1991) Breeding for disease and insect resistance in flower crops. In: Harding, J., Singh, F. and Mol, J.N.M. (eds), *Genetics and Breeding of Ornamental Species*. Kluwer Academic Publishers, Dordrecht, pp. 179–211.

Tavladoraki, P., Benvenuto, E., Trinca, S., Martinis, D.D., Cattaneo, A. and Galeffi, P. (1993) Transgenic plants expressing a functional single-chain Fv antibody are specifically protected from virus attack. *Nature* 366, 469–472.

Urban, L.A., Sherman, J.M., Moyer, J.W. and Daub, M.E. (1994) High frequency shoot regeneration and *Agrobacterium*-mediated transformation of chrysanthemum (*Dendranthema grandiflora*). *Plant Science* 98, 69–79.

Vaira, A.M., Semeria, L., Crespi, S., Lisa, V., Allavena, A. and Accotto, G.P. (1995) Resistance to tospoviruses in *Nicotiana benthamiana* transformed with the N gene of tomato spotted wilt virus: correlation between transgene expression and protection in primary transformants. *Molecular Plant–Microbe Interactions* 8, 66–73.

van Wordragen, M.F., de Jong, J., Huitema, H.B.M. and Dons, H.J.M. (1991) Genetic transformation of chrysanthemum using wild type *Agrobacterium* strains: strain and cultivar specificity. *Plant Cell Reports* 8, 692–695.

van Wordragen, M.F., Honee, G. and Dons, H.J.M. (1993) Insect-resistant chrysanthemum calluses by introduction of a *Bacillus thuringiensis* crystal protein gene. *Transgenic Research* 2, 170–180.

Waterworth, H.E., Kaper, J.M. and Tousignant, M.E. (1979) CARNA 5, the small cucumber mosaic virus-dependent replicating RNA, regulates disease expression. *Science* 204, 845–847.

Whitham, S., Dinesh-Kumar, S.P., Choi, D., Hehl, R., Corr, C. and Baker, B. (1994) The product of the tobacco mosaic virus resistance gene N: similarity to toll and the interleukin-1 receptor. *Cell* 78, 1101–1115.

Wilson, T.M.A. (1993) Strategies to protect crop plants against viruses: pathogen-derived resistance blossoms. *Proceedings of the National Academy of Sciences, USA* 90, 3134–3141.

Yepes, L.M., Mittak, V., Pang, S.Z., Gonsalves, C., Slightom, J.L. and Gonsalves, D. (1995) Biolistic transformation of chrysanthemum with the nucleocapsid gene of tomato spotted wilt virus. *Plant Cell Reptorts* 14, 694–698.

Biotechnological Approaches to Study and Improve Insect Resistance of Woody Plants

19

Ned B. Klopfenstein[1] and Elwood R. Hart[2]

[1]*USDA Forest Service, Rocky Mountain Forest and Range Experiment Station, National Agroforestry Center, East Campus, University of Nebraska, Lincoln, Nebraska 68583, USA;* [2]*Departments of Entomology and Forestry, 401 Science II, Iowa State University, Ames, Iowa 50011, USA*

Insects attack nearly all tissues of woody plants selected for ornamental purposes except for gingko and a few other notable exceptions. Because ornamental plantings usually are close to human activities, the means and techniques for pest management in these plantings are more apparent than are those used for agricultural and forest pests. Consequently, societal awareness and related regulatory pressures are greater in the urban environment. Local ordinances on permissible pest management practices are often added to state and federal regulations, and pressure is increasing to develop management tools other than insecticides. A promising alternative approach to pest management of woody ornamentals is using pest-resistant selections for plantings.

Classical breeding and selection programmes are an important traditional means of producing pest-resistant selections of woody ornamentals with diverse genetic backgrounds. Because such programmes are very time consuming, faster means of obtaining resistant varieties are being explored through biotechnological applications. Genetic engineering, one means of applying more recent scientific advances to improving host plant resistance to insects, is the major focus of biotechnology in this area. Most of the genetic engineering efforts have concentrated on producing and using transgenic materials for agricultural crops (e.g. James *et al.*, 1993; Koziel *et al.*, 1993) or forestry (e.g. Klopfenstein *et al.*, 1993; Robison *et al.*, 1994) applications. Comparatively fewer efforts are under way to increase insect resistance in ornamental plants (Dons *et al.*, 1991). Funds for biotechnical research on ornamental plants are limited because it is perceived that higher economic and ecological returns are associated with investments in agricultural research applications.

Concerns that have driven biotechnological research for agriculture and forestry also are applicable to ornamental selections. In both agricultural and ornamental crops, dependency on insecticidal chemicals is still relatively high.

Although increased awareness and regulations have improved pesticide practices, potentially undesirable effects remain. Two major continuing concerns are environmental pollution and increasing insecticide resistance in pest populations. Both concerns are warranted if one considers that 25% of all pesticides used in the USA in 1992–1993 were for non-agricultural applications (Aspelin, 1994).

The economics of forestry systems have resulted in less dependency on pesticides in forestry than in agricultural or ornamental uses. More recently, concerns regarding biodiversity and multiple-use applications of wooded areas have further distanced pesticides from consideration as a viable insect control option. Most of these concerns are equally applicable to ornamental crops and plantings, especially woody perennials. In working with trees or shrubs in either system, several similarities are evident between tree selections developed for ornamental plantings and those developed for plantation forestry. In both systems, breeding and selection programmes traditionally concentrated on characteristics other than insect resistance. Trees were bred or selected to an elite ideotype based upon form, growth characteristics, phenology, and tolerance to a range of environmental factors. In ornamental selections, aesthetic characteristics, such as colour, often were also considered desirable. As in agronomic crops, insect resistance is often inadvertently lost from subsequent generations when it is not amplified. Both plantation trees and woody ornamentals are committed to a site for a perennial duration, increasing the probability that pest insects will discover the planting and subsequently become adapted to utilize the resource. Also, both types of plantings are likely to be clonally propagated to ensure adherence to the ideotype.

There are, of course, differences between commercial plantations and ornamental plantings. Outside a nursery or greenhouse environment, ornamental trees are less likely to be planted in a comparative monoculture than are plantation trees. Also, damage tolerance of ornamental trees is considerably less than would be acceptable for plantation trees.

Similar approaches to incorporate biotechnological advances can be followed for both plantation forestry and ornamental plantings of woody perennials. If one assumes that a favourable selection has been developed in all aspects except resistance to insect pests, then the process is a matter of overlaying pest resistance on an existing genome. To accomplish this, several biotechnological approaches may be taken. In parallel with classical approaches to developing pest-resistant selections, the objective may be antibiosis, antixenosis, or tolerance, all of which would leave intact those desirable characteristics already incorporated into the genome.

Rapid progress in tissue culture techniques, transformation procedures, and defence gene characterization has provided great potential for understanding and improving woody plant resistance to insects. The *Populus* genus serves as a a model system for molecular biological studies of woody plants, and thereby is the focus of many initial biotechnological applications. Features of *Populus* spp. that have contributed to its emergence as a model system include: (i) *Populus* spp. are grown worldwide and are valued for diverse purposes; (ii) physiological and genetic traits of the genus are well studied; (iii) tissue culture systems for *in vitro* propagation and regeneration are well defined for many *Populus* spp. (Chun *et al.*, 1988); (iv) genetic transformation of *Populus* spp. can be achieved by various methods (McCown *et al.*, 1991; Devantier *et al.*, 1993; Confalonieri *et al.*, 1994); (v) screening at the molecular level is facilitated by the small genome of *Populus* spp. (Dhillon *et al.*, 1984; Parsons *et al.*, 1989); and (vi) a genomic map for a hybrid poplar pedigree is nearly completed (Bradshaw, 1994). Biotechnological approaches to study and improve insect resistance in *Populus* and other woody plants are summarized below.

TECHNOLOGIES

TISSUE CULTURE

Various aspects of plant tissue culture such as somatic embryogenesis, axillary shoot proliferation, adventitious shoot organogenesis, and *in vitro* rooting are reviewed elsewhere in this volume. Tissue culture methods are a valuable asset to programmes aimed at increasing insect resistance of woody ornamentals. Such methods can provide a means to clonally propagate large numbers of desirable plants with specific insect-resistance characteristics. Test plantings using clonal material at several locations help define the genetic contribution to insect resistance and the interacting role of the environment.

Somaclonal variation can be exploited via *in vitro* culture and regeneration to identify new sources of insect resistance. Differential susceptibility to insect pests was described in somaclonal variants for herbaceous plant species (Isenhour *et al.*, 1991). This approach has received little attention in the development of insect resistance in woody ornamentals, and it may represent an inefficient means to define resistant selections (Gatehouse, 1991). However, *in vitro* regeneration methods that are developed to exploit somaclonal variation are also potentially useful for genetic engineering studies.

In plant improvement programmes, abortion of the developing embryo is one of the barriers to introducing desirable traits through interspecific breeding. Techniques for embryo rescue were developed to circumvent this problem for several genera that include ornamental selections, such as *Taxus* (Flores and Sgrignoli, 1991) and *Populus* (Kang, 1994).

A significant barrier to developing resistant selections of woody plants is the great length of time required for the plant to develop to a stage where resistance may be tested. This testing time period may be shortened by using tissue-cultured plantlets in a preliminary screen for insect resistance prior to field testing. An early screening method to evaluate the resistance of a transgenic hybrid poplar (*Populus deltoides* × *P. nigra*) to cottonwood leaf beetle (*Chrysomela scripta* F.) was developed using *in vitro* feeding (Kang, 1994).

GENETIC ENGINEERING

One of the first applications of genetic engineering to woody plants was to study and improve insect resistance. Rapid improvement in plant transformation technologies have expanded the range of plant species that can be genetically engineered. The repertoire of available genes for use in genetic engineering is also rapidly expanding as new insect resistance genes are identified. Basic biological information on insect/host interactions is needed, however, to assure the efficacy of genetic engineering to improve insect resistance.

The most effective strategy for managing insects is the use of resistant planting stock in combination with other biological control measures. Potential advantages of host plant resistance include: (i) host tissue can be continually protected across variable weather conditions; (ii) plant defence mechanisms can be directed towards the sensitive stage of insect development; (iii) only plant-invading or -consuming insects are affected; and (iv) plant defence compounds can be restricted to plant tissues, are biodegradable, and can be selected for non-toxicity to other organisms (Boulter *et al.*, 1990). Efforts to use genetic engineering techniques for insect resistance are focused on genes encoding *Bacillus thuringiensis* endotoxin, proteinase inhibitors, or chitinases.

Bacillus thuringiensis

Bacillus thuringiensis (*B.t.*) is a Gram-positive bacterium that produces proteinaceous crystalline inclusions during sporulation. After ingestion by insects, the crystalline inclusions are solubilized and proteolytically cleaved in the alkaline conditions of the midgut, releasing active f-endotoxins (Brain *et al.*, 1991; Gill *et al.*, 1992). Activated toxins bind to the larval midgut epithelium and disrupt the cellular osmotic balance, eventually causing cellular lysis, feeding disruption and death. Although specific *B.t.* toxins display activity against a narrow range of target insects, an array of *B.t.* toxins were identified that are active against various insect species within the Lepidoptera, Diptera, and Coleoptera (Strittmatter and Wegener, 1993).

Insect toxicity tests of transgenic *Populus* spp. containing *B.t.* toxin genes were initiated (Strauss *et al.*, 1991; Tian *et al.*, 1993; Howe *et al.*, 1994). Transgenic poplar expressing a *B.t.* δ-endotoxin gene significantly reduced feeding and weight gain by gypsy moth (*Lymantria dispar* L.) and forest tent caterpillar (*Malacosoma disstria* Hubner). In addition, exposure to transgenic δ-endotoxin significantly increased mortality of forest tent caterpillar (McCown *et al.*, 1991; Robison *et al.*, 1994). Other deciduous trees, walnut and apple, have been transformed with a truncated *cryIA(c)* gene that encodes *B.t.* intracrystalline protein targeted at lepidopteran insects such as the codling moth (*Cydia pomonella* L.) (James *et al.*, 1993; Dandekar *et al.*, 1994). Among numerous transgenic walnut lines tested, codling moth toxicity ranged from 98% mortality to no significant effect. Field trials are under way to assess foreign gene effects on morphological, structural, and physiological properties, and to monitor the stability of expression of the introduced genes.

Conifers genetically engineered with *B.t.* toxin genes were also reported. Transgenic white spruce (*Picea glauca*) containing the *B.t. cryIA* endotoxin gene under the control of a CaMV 35S promoter were regenerated and evaluated for toxicity to spruce budworm (*Choristoneura fumiferana* Clemens) (Ellis *et al.*, 1993). In addition, an *Agrobacterium rhizogenes*-mediated transformation system was used to transform larch (*Larix decidua*) with a *B.t.* toxin gene that conferred resistance to gypsy moth (Karnosky *et al.*, 1994).

Plants transformed with *B.t.* toxin genes are generally considered to be environmentally safe because of the high specificity of the toxin and its short persistence in nature. A major issue for consideration is the potential for the emergence of toxin-resistant insects. This issue can be addressed by regulating expression of the *B.t.* toxin gene, so that the gene is only expressed when the plant has sustained a specific threshold of damage (Klopfenstein *et al.*, 1991), and by interplanting with host plants that do not contain the *B.t.* toxin gene (Raffa, 1989).

Proteinase inhibitors

Proteinase inhibitors are plant defence proteins that can contribute resistance to insect herbivory (Wolfson, 1991). Proteinase inhibitors may act by binding digestive proteinases of insects, thereby depressing the level of available amino acids or effective nutrient levels of consumed plant tissue. Alternatively, proteinase inhibitors could deplete nutrient reserves of insects by causing an overproduction of insect proteinases (Ryan, 1990). If proteinase inhibitors can delay pest growth, development and reproduction without placing a strong selection pressure on the pest, these defensive compounds may contribute a pest resistance that is reasonably stable. Specific proteinase inhibitors also may potentiate the insecticidal activity of *B.t.* toxins (MacIntosh *et al.*, 1990). Proteinase inhibitor-based resistance also is being considered for its potential compatibility with other pest management strategies such as other host resistance mechanisms and biological control.

Transgenic hybrid poplars (*Populus alba* L. × *P. grandidentata* Michx.) were produced that express potato proteinase inhibitor II (*pin2*) under the control of a CaMV 35S or a bacterial nopaline synthase (*nos*) promoter (Klopfenstein *et al.*, 1993). In leaf feeding bioassays with imported willow leaf beetle (*Plagiodera versicolora* Laicharting), *pin2*-expressing poplars showed significant differences in insect development and leaf consumption relative to the untransformed control poplar (Klopfenstein *et al.*, 1994). Additional studies on *pin2* expression in other transgenic poplar hybrids (*Populus* × *euramericana*) are under way with the intent of evaluating resistance to cottonwood leaf beetle (Heuchelin and McNabb, 1994; Kang, 1994). Other hybrid poplar (*Populus tremula* × *P. tremuloides*) trees have been transformed with a cysteine proteinase inhibitor (OCI) gene from rice, and subsequently displayed toxicity to two coleopterous leaf feeders, *Chrysomela populi* and *C. tremulae* (Pilate *et al.*, 1994). Ingested foliage from selected transgenic lines expressing OCI reduced larval growth and development while increasing mortality.

Work to transform other woody species with proteinase inhibitor genes is also under way. Transgenic apple trees containing the CpTi gene encoding cowpea trypsin inhibitor were produced (James *et al.*, 1993), and birch (*Betula pendula*) has been transformed with *pin2* to study its effect on insect resistance (Keinonen-Mettala *et al.*, 1994). Transformation of a coniferous species with proteinase inhibitor genes has not yet been reported.

Chitinases

Chitinases are believed to have a prominent role in plant defence against fungal attack (Collinge *et al.*, 1993). Chitinases catalyse the hydrolysis of chitin, a polymer of β-1,4-linked *N*-acetylglucosamine residues that is common in fungal cell walls and arthropod exoskeletons. Plant chitinase genes were identified that are induced by pathogen challenge or various stresses; however, the role of chitinases in insect resistance is yet to be defined. It is postulated that chitinases could aid plant defence against insects by slowing insect development or by acting to regulate other defence genes via the breakdown products of chitinase (Strauss *et al.*, 1991). Previous work has demonstrated a potential influence of chitinases on *B.t.* toxin activity against spruce budworm (Gunner *et al.*, 1985). To further characterize the defence role of chitinases in woody plants, birch (*Betula pendula*) was transformed with sugar beet chitinase (*Ch4*) genes to study its effect on insect resistance (Keinonen-Mettala *et al.*, 1994).

Other insect resistance genes

Options to genetically engineer insect resistance into woody plants will continue to increase as more useful genes are characterized. The potential sources and activities of such genes are practically limitless. Potentially useful genes exist in a wide variety of organisms, and genes can also be synthesized or modified in the laboratory to encode a product with unique activity. Gene products could act in the production of antifeedants, inhibitors of digestive enzymes, repellents, metabolic inhibitors, toxins, pheromones, hormones, secondary products, lytic enzymes, signals for other defence genes, ribosome-inactivating proteins, or products with countless other activities (Gatehouse *et al.*, 1991; Strauss *et al.*, 1991). Transgenic approaches to insect resistance have focused on systems in which a single gene product could affect the target insect. It is much more difficult to genetically manipulate resistance mechanisms that involve complex interactions of multigenic traits. Such multigenic interactions can be approached by sequential genetic manipulation of a few genes at a time; however, these resistance mechanisms could also be approached by processes of marker-aided selection.

GENE REGULATION

The regulation of plant defence genes is a critical factor in host plant resistance to insects. To be effective, plant defence genes that target a specific insect must be adequately expressed in the tissue under insect attack, at the appropriate developmental stage, and under the proper environmental conditions. For some plant defence genes, expression is effectively regulated by an inducible promoter that responds to specific elicitors or mechanical wounding such as that caused by insect attack.

Optimal regulation of a defence gene is dependent on the specific insect problem that is to be managed. For example, defence genes must be expressed in newly formed leaves during warm temperatures in mid-spring to manage insects that feed on those leaves at that time and temperature. Managing insects that are a problem under conditions of tree stress will require defence genes that are active during plant stress. To avoid placing a strong selection pressure on insects, some defence genes should be expressed only after a critical threshold of damage occurs. Effects of defence gene products on beneficial organisms must also be considered when evaluating optimal spatial and temporal expression.

When trees are genetically engineered to improve insect resistance, transgene promoters should be carefully selected to provide optimal expression of the resistance gene. Information on promoter function in transgenic trees is becoming more available, but continued studies are necessary to understand long-term promoter function in older trees grown under various environmental conditions. Initial studies on promoter function in transgenic trees are usually conducted in the greenhouse to evaluate tissue-specific expression; however, studies on transgene expression in field-grown trees are under way (Ellis *et al.*, 1994). Such tests under natural conditions are essential to understand fully the potential effects of transgenes on insect resistance over the life-span of the tree.

Regulation of native defence genes is critical to insect resistance of woody plants. Induced defence genes can be isolated by comparing gene expression before and after an inducing stimulus such as wounding or pest challenge is applied. Putative defence genes with homology to chitinase, proteinase inhibitor, and vegetative storage protein genes, were isolated from wound-induced cDNA libraries of *Populus* sp. (Bradshaw *et al.*, 1989, 1991; Parsons *et al.*, 1989; Davis *et al.*, 1991, 1993; Hollick and Gordon, 1993). As the expression of native defence genes becomes better understood, promoters can be isolated that efficiently regulate the expression of introduced defence genes. In addition, studies on native defence gene expression can provide a basis to select trees with robust native defence systems that respond appropriately to pest challenge.

CHEMICAL ECOLOGY

Host plant resistance to insects also must include studies of the chemical interactions between the host and the pest insects with which it is associated. These studies define, at the molecular level, those host factors associated with acceptance or rejection of the plant by the insect, or those chemicals ultimately responsible for resistance. The genus *Populus* also is considered a model system for these studies. Phenolic glycosides (Lindroth and Hemming, 1990; Bingaman and Hart, 1993) and long-chain alcohols and aldehydes (S. Lin, Iowa State University, personal communication) were associated with insect feeding behaviour on *Populus*. Elucidation of the chemicals affecting biological activity in phytophagous insects is a powerful first step towards determining the role that these chemicals play in the physiology of the plant and of the insect. This can lead in turn to a determination of the genetic mechanisms controlling these chemicals.

GENETIC SELECTION

Genomic mapping and marker-aided selection are being developed to provide a system to characterize and select woody plants based on characteristics of their DNA (Neale *et al.*, 1992; reviewed by Whetten, Chapter 8, this volume). As correlations are developed between DNA characteristics and physical traits of woody plants, it will become possible to use DNA from seedling tissue to predict traits of the future mature plants. Genetic markers associated with insect resistance should become available after genetically mapped, pedigreed plant materials are scored for characteristics of insect resistance.

In addition, specific defence genes that may encode resistance to insects are being identified in woody plants. Examples of such putative defence genes include chitinase and trypsin inhibitor-like genes from poplar (Bradshaw *et al.*, 1989; Davis *et al.*, 1991) and a trypsin inhibitor gene from *Acacia* sp. (Hung *et al.*, 1992). Work is currently under way to clone a native proteinase inhibitor gene from birch (Keinonen-Mettala *et al.*, 1994).

GENERAL CONSIDERATIONS

INSECT BIOLOGY

No application of biotechnology is expected to produce resistant plant selections without sufficient understanding of the insect pest biology. For example, relationships between the phenology of the insect and that of the plant determine the state of the target tissue at the time of attack. Also, the quantity and quality of chemicals in a plant tissue may change with physiological development (Bingaman and Hart, 1993). Therefore, any resistance mechanism must be expressed in the susceptible plant tissues at an effective level when the insect is likely to feed on those tissues. For these reasons, an insect's physiological susceptibility to the proposed resistance mechanism must be understood before modifications are made to the plant genome. For example, a pest insect's digestive proteinase system should be understood before proposing the transformation of a host plant species with a specific proteinase inhibitor gene.

One of the more notable features of insect populations is the ability to adapt to environmental changes. Sibling species or biotypes that are more highly adapted to host plants or are resistant to specific management measures have occurred in many instances. Often, these populations are extremely difficult to differentiate from related forms. However, determining and following such population or species differentiation is necessary for predicting outbreak potential among some pest complexes and for selecting management options. Molecular methods based on DNA or protein profiles can help differentiate biologically distinct pest populations that are morphologically similar (e.g. Soto-Adames, 1994; Wells, 1994). Methods are being developed to characterize populations of woody plant pests, such as Nantucket pine tip moth, pine coneworm species, sawfly species, and the balsam woolly adelgid (Richmond *et al.*, 1992). As pest biotype populations become better characterized, the role of specific defence genes can be better understood.

RESISTANCE STABILITY

Although stability of any resistance factor is considered as an ideal, its achievement is not a simple feat. Resistance to insecticides is not uncommon when there has been prolonged exposure of a pest species to a chemical. Similarly, pest insects are known to evolve adapted biotypes to resistant plant selections. Judicious use of selected material could reduce the probability of such adaptation to newly developed resistant selections. Practices that avoid strong selection pressures are necessary to avoid or reduce adaptation to the resistant selections. Mechanisms that reduce the overall success of a pest population without acting as an intense selection factor are a primary consideration for increasing the useful life-span of resistance.

ECOSYSTEM INTERACTIONS

Factors other than efficacy must be considered before utilizing specific mechanisms of host resistance. If biotechnology is used as a tool in pest management systems, then its impact on the ecosystem must be considered, according to the best tenets of pest management theory and practice. For example, if the genome is modified to include a resistance mechanism, then traits that previously were selected to approach those of an ideotype must be maintained in an acceptable condition. This would include aspects such as growth, form, colour, climatic tolerance, and other resistance mechanisms. Also, pest management involves the use of tools other than resistance to reduce phytophagous insect impact on host plants. With an increasing interest in the use of predator and parasite species to reduce insect populations, studies must be performed at the tritrophic level to be certain that the resistance mechanisms are compatible with the beneficial complex. As indicated earlier, plants rarely exist with only a single pest species; thus, the impact of any resistance mechanisms must be measured against their effect on other members of the phytophagous complex associated with the host plant. Another pest management consideration must be utilization of material in a fashion that reduces the probability of inducing insect resistance to the transgenic selections. Because there is less tendency toward large-scale, monocultural ornamental plantings, there should be less selection pressure in these plantings than in agricultural or forestry systems.

Protection of conspecific germplasm, both native and that used in breeding and selection programmes, also is an ethical concern. If necessary, genetically modified material could be limited to reproductively sterile selections, or regulated in such a fashion to assure that plants are removed from an environment prior to becoming sexually mature. Although neither approach would be easily achieved, the first may be the most reasonable from a regulatory standpoint.

Nutrient cycling is another important consideration. All plant tissues will, at some time, die and be subjected to organisms that are responsible for the decay process. No host plant resistance factor that interferes drastically with this process should be considered acceptable.

ETHICS

Legal and ethical considerations that must be addressed in working with transgenic material and other newer technologies are discussed elsewhere in this volume. An early response by concerned individuals and groups was hesitation, and in some cases fear, about the potential impact of genetically transformed selections on the environment or human health. This has led, in many instances, to increased local, state, or national restrictions governing research and development of transgenic plants, including trees. While the value of many specific restrictions is arguable, one benefit from the controversies was a focus on ethical issues regarding the judicious use of transgenic materials. A major concern is the potential escape of transgenic plants or germplasm into native populations. As a result, most research personnel now recognize both the legal and ethical obligations to take all reasonable precautions to prevent such unplanned escape.

REFERENCES

Aspelin, A.L. (1994) *Pesticide Industry Sales and Usage, 1992 and 1993 Market Estimates*. US Environmental Protection Agency, Washington, DC.

Bingaman, B.R. and Hart, E.R. (1993) Clonal and leaf age variation in *Populus* phenolic glycosides: implications for host selection by *Chrysomela scripta* (Coleoptera: Chrysomelidae). *Environmental Entomology* 22, 397–403.

Boulter, D., Gatehouse, J.A., Gatehouse, A.M.R. and Hilder, V.A. (1990) Genetic engineering of plants for insect resistance. *Endeavour* (Oxford) 14, 185–190.

Bradshaw, H.D., Jr (1994) Genetic mapping and trait improvement in forest trees. In: Michler, C.H., Becwar, M.R., Cullen, D., Nance, W.L., Sederoff, R.R. and Slavicek, J.M. (eds), *Applications of Biotechnology to Tree Culture, Protection, and Utilization*. USDA Forest Service, North Central Forest Experiment Station, General Technical Report NC-175, p. 34.

Bradshaw, H.D., Jr, Hollick, J.B., Parsons, T.J., Clarke, H.R.G. and Gordon, M.P. (1989) Systemically wound-responsive genes in poplar trees encode proteins similar to sweet potato sporamins and legume Kunitz trypsin inhibitors. *Plant Molecular Biology* 14, 51–59.

Bradshaw, H.D., Jr, Parsons, T.J. and Gordon, M.P. (1991) Wound-responsive gene expression in poplars. *Forest Ecology and Management* 43, 211–224.

Braun, C.J., Jilka, J.M., Hemenway, C.L. and Tumer, N.E. (1991) Interactions between plants, pathogens and insects: possibilities for engineering resistance. *Current Opinion in Biotechnology* 2, 193–198.

Chun, Y.W., Klopfenstein, N.B., McNabb, H.S., Jr and Hall, R.B. (1988) Biotechnical applications in *Populus* species. *Journal of the Korean Forestry Society* 77, 467–483.

Collinge, D.B., Kragh, K.M., Mikkelsen, J.D., Nielsen, K.K., Rasmussen, U. and Vad, K. (1993) Plant chitinases. *The Plant Journal* 3, 21–40.

Confalonieri, M., Balestrazzi, A. and Bisoffi, S. (1994) Genetic transformation of *Populus nigra* by *Agrobacterium tumefaciens*. *Plant Cell Reports* 13, 256–261.

Dandekar, A.M., McGranahan, G.H., Vail, P.V., Driver, J., Uratsu, S.L., Leslie, C., Tebbets, J.S. and Viss, P. (1994) Engineering resistance to the insect pest codling moth in apple and walnut. In: Michler, C.H., Becwar, M.R., Cullen, D., Nance, W.L., Sederoff, R.R. and Slavicek, J.M. (eds), *Applications of Biotechnology to Tree Culture, Protection, and Utilization*. USDA Forest Service, North Central Forest Experiment Station, General Technical Report NC-175, p. 25, Abstract.

Davis, J.M., Clarke, H.R.G., Bradshaw, H.D., Jr and Gordon, M.P. (1991) *Populus* chitinase genes: structure, organization, and similarity of translated sequences to herbaceous plant chitinases. *Plant Molecular Biology* 17, 631–639.

Davis, J.M., Egelkrout, E.E., Coleman, G.D., Chen, T.H.H., Haissig, B.E., Riemenschneider, D.E. and Gordon, M.P. (1993) A family of wound-induced genes in *Populus* shares common features with genes encoding vegetative storage proteins. *Plant Molecular Biology* 23, 135–143.

Devantier, Y.A., Moffatt, B., Jones, C. and Charest, P.J. (1993) Microprojectile-mediated DNA delivery to the Salicaceae family. *Canadian Journal of Botany* 71, 1458–1466.

Dhillon, S.S., Miksche, J.P. and Cecich, R.A. (1984) DNA changes in senescing leaves of *Populus deltoides*. *Plant Physiology* (suppl.) 75, 120.

Dons, J.J.M., Mollema, C., Stiekema, W.J. and Visser, B. (1991) Routes to the development of disease resistant ornamentals. In: Harding, J., Singh, F. and Mol, J.N.M.

(eds), *Genetics and Breeding of Ornamental Species*. Kluwer Academic Publishers, Dordrecht, pp. 387–417.

Ellis, D.D., McCabe, D.E., McInnis, S., Ramachandran, R., Russell, D.R., Wallace, K.M., Martinell, B.J., Roberts, D.R., Raffa, K.F. and McCown, B.H. (1993) Stable transformation of *Picea glauca* by particle acceleration. *Bio/Technology* 11, 84–89.

Ellis, D.D., Rintamaki, J.A., Wraith, S.J., McCown, B.H., Kleiner, K.W. and Raffa, K.F. (1994) Expression of introduced genes in trees: practical and environmental considerations. In: *Proceedings of the TAPPI 1994 Biological Sciences Symposium*, 3–6 October, Minneapolis. TAPPI Press, Atlanta, pp. 139–146.

Flores, H.E. and Sgrignoli, P.J. (1991) *In vitro* culture and precocious germination of *Taxus* embryos. *In Vitro Cellular and Developmental Biology* 27, 139–142.

Gatehouse, J.A. (1991) Breeding resistance to insects. In: Murray, D.R. (ed.), *Advanced Methods in Plant Breeding and Biotechnology*. CAB International, Wallingford, UK, pp. 250–276.

Gatehouse, J.A., Hilder, V.A. and Gatehouse, A.M.R. (1991) Genetic engineering of plants for insect resistance. *Plant Biotechnology* 1, 105–135.

Gill, S.S., Cowles, E.A. and Pietrantonio, P.V. (1992) The mode of action of *Bacillus thuringiensis* endotoxins. *Annual Review of Entomology* 37, 615–636.

Gunner, H.B., Zimet, M. and Berger, S. (1985) Chitinase producing *Bt* strains. In: *Proceedings of the Symposium on Microbial Control of Spruce Budworm and Gypsy Moths*. USDA Forest Service General Technical Report NE-100, pp. 103–108.

Heuchelin, S.A. and McNabb, H.S., Jr (1994) Expression studies of proteinase inhibitor II in transgenic *Populus*. *Proceedings of the International Union of Forestry Research Organizations (IUFRO) Molecular Genetics (S2.04.06) Workshop*, 20–23 May 1994, Prouts Neck, Scarborough, Maine, Abstract.

Hollick, J.B. and Gordon, M.P. (1993) A poplar tree proteinase inhibitor-like gene promoter is responsive to wounding in transgenic tobacco. *Plant Molecular Biology* 22, 561–572.

Howe, G.T., Goldfarb, B. and Strauss, S.H. (1994) *Agrobacterium*-mediated transformation of hybrid poplar suspension cultures and regeneration of transformed plants. *Plant Cell, Tissue and Organ Culture* 36, 59–71.

Hung, C.-H., Lee, M.-C. and Lin, J.-Y. (1992) Nucleotide sequence of cDNA for *Acacia confusa* trypsin inhibitor and implication of post-translation processing. *Biochemical and Biophysical Research Communications* 184, 1524–1528.

Isenhour, D.J., Duncan, R.R., Miller, D.R., Waskom, R.M., Hanning, G.E., Wiseman, B.R. and Nabors, M.W. (1991) Resistance to leaf-feeding by the fall armyworm (Lepidoptera: Noctuidae) in tissue culture derived sorghums. *Journal of Economic Entomology* 84, 680–684.

James, D.J., Passey, A.J., Webster, A.D., Barbara, D.J., Dandekar, A.M., Uratsu, S.L. and Viss, P. (1993) Transgenic apples and strawberries: advances in transformation, introduction of genes for insect resistance and field studies of tissue cultured plants. *Acta Horticulturae* 336, 179–184.

Kang, H.-D. (1994) Advanced biotechnical approaches and gene manipulation in *Populus* species. Unpublished PhD thesis, Iowa State University, Ames, 107 pp.

Karnosky, D.F., Podila, G.K., Shin, D.-I., Tsai, T.J. and Chiang, V.L. (1994) Progress in production of transgenic trees with value-added genes: results with larch and aspen. In: *Proceedings of the TAPPI 1994 Biological Sciences Symposium*, 3–6 October, Minneapolis. TAPPI Press, Atlanta, pp. 157–160.

Keinonen-Mettala, K., Pappinen, A., Susi, A., and von Weissenberg, K. (1994) Molecular biology of resistance to insects and diseases in birch and pine. In: Michler,

C.H., Becwar, M.R., Cullen, D., Nance, W.L., Sederoff, R.R. and Slavicek, J.M. (eds), *Applications of Biotechnology to Tree Culture, Protection, and Utilization.* USDA Forest Service, North Central Forest Experiment Station, General Technical Report NC-175, p. 31, Abstract.

Klopfenstein, N.B., Shi, N.Q., Kernan, A., McNabb, H.S., Jr, Hall, R.B., Hart, E.R. and Thornburg, R.W. (1991) Transgenic *Populus* hybrid expresses a wound-inducible potato proteinase inhibitor II – CAT gene fusion. *Canadian Journal of Forest Research* 21, 1321–1328.

Klopfenstein, N.B., McNabb, H.S., Jr, Hart, E.R., Hall, R.B., Hanna, R.D., Heuchelin, S.A., Allen, K.K., Shi, N.-Q. and Thornburg, R.W. (1993) Transformation of *Populus* hybrids to study and improve pest resistance. *Silvae Genetica* 42, 86–90.

Klopfenstein, N.B., Martinez, J., Avila, F.J., Chun, Y.W., Allen, K.K., Hart, E.R., Hall, R.B. and McNabb, H.S., Jr (1994) Proteinase inhibitor II gene expression and pest resistance of transgenic poplar. In: Michler, C.H., Becwar, M.R., Cullen, D., Nance, W.L., Sederoff, R.R. and Slavicek, J.M. (eds), *Applications of Biotechnology to Tree Culture, Protection, and Utilization.* USDA Forest Service, North Central Forest Experiment Station, General Technical Report NC-175, p. 30, Abstract.

Koziel, M.G., Beland, G.L., Bowman, C., Carozzi, N.B., Crenshaw, R., Crossland, L., Dawson, J., Desai, N., Hill, M., Kadwell, S., Launis, K., Lewis, K., Maddox, D., McPherson, K., Meghji, M.R., Merlin, E., Rhodes, R., Warren, G.W., Wright, M. and Evola, S.V. (1993) Field performance of elite transgenic maize plants expressing an insecticidal protein derived from *Bacillus thuringiensis*. *Bio/Technology* 11, 194–200.

Lindroth, R.L. and Hemming, D.C. (1990) Responses of the gypsy moth (Lepidoptera: Lymantriidae) to tremulacin, an aspen phenolic glycoside. *Environmental Entomology* 19, 842–847.

MacIntosh, S.C., Kishore, G.M., Perlak, F.J., Marrone, P.G., Stone, T.B., Sims, S.R. and Fuchs, R.L. (1990) Potentiation of *Bacillus thuringiensis* insecticidal activity by serine proteinase inhibitors. *Journal of Agriculture and Food Chemistry* 38, 1145–1152.

McCown, B.H., McCabe, D.E., Russell, D.R., Robison, D.J., Barton, K.A. and Raffa, K.F. (1991) Stable transformation of *Populus* and incorporation of pest resistance by electric discharge particle acceleration. *Plant Cell Reports* 9, 590–594.

Neale, D.B., Devey, M.E., Jermstad, K.D., Ahuja, M.R., Alosi, M.C. and Marshall, K.A. (1992) Use of DNA markers in forest tree improvement research. *New Forests* 6, 391–407.

Parsons, T.J., Bradshaw, H.D., Jr and Gordon, M.P. (1989) Systemic accumulation of specific mRNAs in response to wounding in poplar trees. *Proceedings of the National Academy of Sciences, USA* 86, 7895–7899.

Pilate, G., Leple, J.-C., Bonade-Bottino, M., Augustin, S., Delplanque, A., Dumanois Le Tan, V. and Jouanin, L. (1994) Transgenic poplars expressing a cysteine proteinase inhibitor are toxic to coleopterous insects. In: Michler, C.H., Becwar, M.R., Cullen, D., Nance, W.L., Sederoff, R.R. and Slavicek, J.M. (eds), *Applications of Biotechnology to Tree Culture, Protection, and Utilization.* USDA Forest Service, North Central Forest Experiment Station, General Technical Report NC-175, p. 82, Abstract.

Raffa, K.F. (1989) Genetic engineering of trees to enhance resistance to insects: evaluating the risks of biotype evolution and secondary pest outbreak. *BioScience* 39, 524–534.

Richmond, J.A., Page, M., Drooz, A.T. and Davis, E.J. (1992) Looking at forest insects by contemporary methods. In: *Proceedings of the Society of American Foresters, 1992 National Convention*, 25–28 October 1992. Richmond, Virginia, pp. 519–521.

Robison, D.J., McCown, B.H. and Raffa, K.F. (1994) Responses of gypsy moth (Lepidoptera: Lymantriidae) and forest tent caterpillar (Lepidoptera: Lasiocampidae) to transgenic poplar, *Populus* spp. , containing a *Bacillus thuringiensis d*-endotoxin gene. *Environmental Entomology* 23, 1030–1041.

Ryan, C.A. (1990) Protease inhibitors in plants: genes for improving defences against insects and pathogens. *Annual Review of Phytopathology* 28, 425–449.

Soto-Adames, F.N., Robertson, H.M. and Berlocher, H.S. (1994) Phylogenetic utility of partial DNA sequences of G6pdh at different taxonomic levels in hexapods with emphasis on Diptera. *Annals of the Entomological Society of America* 87, 723–726.

Strauss, S.H., Howe, G.T. and Goldfarb, B. (1991) Prospects for genetic engineering of insect resistance in forest trees. *Forest Ecology and Management* 43, 181–209.

Strittmatter, G. and Wegener, D. (1993) Genetic engineering of disease and pest resistance in plants: present state of the art. *Zeitschrift für Naturforschung* 48, 673–688.

Tian, Y.C., Li, T.Y., Mang, K.Q., Han, Y.F., Li, L., Wang, X.P., Lu, M.Z., Dai, L.Y., Han, Y.N. and Yan, J.J. (1993) Insect tolerance of transgenic *Populus nigra* plants transformed with *Bacillus thuringiensis* toxin gene. *Agricoltura Ricerca* 15, 22, Abstract.

Wells, M.M. (1994) Small genetic distances among populations of green lacewings of the genus *Chrysoperla* (Neuroptera: Chrysopidae). *Annals of the Entomological Society of America* 87, 737–744.

Wolfson, J.L. (1991) The effects of induced plant proteinase inhibitors on herbivorous insects. In: Tallamy, D.W. and Raupp, M.J. (eds), *Phytochemical Induction by Herbivores*. John Wiley, New York, pp. 223–243.

Genetic Engineering of Horticultural and Forestry Crops for Herbicide Tolerance

20

Don E. Riemenschneider
Forestry Sciences Laboratory, 5985 Highway K, Rhinelander, Wisconsin 54501, USA

INTRODUCTION

Many agronomic, horticultural and forestry species have been genetically selected to increase their utility to humans. However, traditional selective breeding methods may be limited for several reasons, one being that genetic improvement is limited to those traits for which allelic variants exist in the population under selection or can be introduced from a sexually compatible population. For example, limited genetic variation significantly affected the development and use of herbicides (Comstock, 1989a) because variation in physiological response to herbicides does not always exist in crop populations. Thus, herbicide development strategies remained essentially unchanged for years. Crop cultivar development was the starting point; chemicals that killed associated weeds while leaving the crop undamaged (i.e. selectivity) were developed second (Comstock, 1989a).

Tolerance to herbicides was one of the first useful traits to be genetically engineered into crops, mostly because the prerequisites to progress (Jorgensen, 1993) were readily met. Herbicide tolerance can be achieved by altering the target enzyme of the herbicide, detoxifying the herbicide or overexpressing the target enzyme (Stalker, 1989), and all of the above strategies have been successfully pursued. Genes encoding tolerant, detoxifying or overproduced enzymes have been isolated from both microbes and higher plants. An acceptable level of gene expression in most higher plants has been achieved with promoters derived from *Agrobacterium* opine genes or from the cauliflower mosaic virus 35S promoter. Selection for the introduced gene has been accomplished either by the use of linked selectable markers or by using the herbicide itself as the selective agent. Each of the above factors will be discussed in considerable detail in the following sections.

An additional reason for the early and continued successes of genetically engineered herbicide-tolerant crops is the potential economic value of the product. Modern crops may be poor competitors with weeds and some method of weed control is obligatory. Herbicide treatments have become an integral part of agriculture because they are the most cost-effective way to control weeds, thereby increasing plant growth and quality (Mazur and Falco, 1989). Weed competition can be especially damaging in intensively grown plantings of poplar (*Populus sp.*) (Hansen and Netzer, 1986) and willows (*Salix discolor* Muhl., S. *viminalis L.*) (Labrecque *et al.*, 1994).

In the following sections I will discuss the identification and characterization of genes that confer tolerance to several herbicides. The potential risks of genetically engineered herbicide-tolerant crops will also be discussed because, in my opinion, a strong argument can be presented that social acceptance of such crops may be the factor that most limits the current deployment of herbicide-tolerant crops. The reader is directed to several current books and review articles that offer more detail on herbicide tolerance through genetic engineering (Mazur and Falco, 1989; Holt *et al.*, 1993), on natural selection for herbicide tolerance (Warwick, 1991), and on the risks and social concerns associated with genetic engineering (Levin and Strauss, 1990; Fincham and Ravetz, 1991; Ginzburg, 1991; Kung and Wu, 1993; Wohrmann and Tomiuk, 1993). I use the term 'tolerance' in this chapter, except in quotation or citation, because its definitions 'physiological resistance to poison' or 'capacity to absorb a drug continuously or in large doses without adverse effect' is more precise than 'resistance' in reference to a plant's response to chemical treatment. The literature is about evenly split in this regard.

Methods to genetically transform plant tissues and to regenerate whole plants are prerequisite to achieving herbicide tolerance through genetic engineering. Overall, the ability to transform and regenerate important horticultural crops seems adequate for the engineering of of herbicide tolerance. For example, many vegetable crops have been transformed, in many cases with herbicide-tolerance genes (Table 20.1), and whole plants have been regenerated (Wijbrandi and deBoth, 1993). Methods to accomplish the transformation of floricultural species, including rose, chrysanthemum and carnation, which together comprise 60% of cut flower production, have recently been developed (Robinson and Firoozabady, 1993). Several woody plants, including horticultural and forestry species (Gardner, 1993; Schuerman and Dandekar, 1993) have also been transformed and regenerated, although effective transformation and regeneration methods still do not exist for our most important forest trees. Most successful transformation protocols have involved the use of either *Agrobacterium tumefaciens* or *A. rhizogenes*, but the use of particle acceleration methods has proved effective in some cases with species that lie outside the host range of the aforementioned biological vectors. Several additional non-biological transformation technologies appear promising but await certain biological and technological developments. The development of herbicide-tolerant horticultural crops lags behind the overall pace of developing transformation methodologies, but that lag does not seem attributable, even in part, to an inability to achieve transformation and regeneration. Genetic transformation

Table 20.1. Genetic transformation and regeneration of whole plants has been accomplished for many of the most important vegetable crops. Examples of transformation with genes that confer herbicide tolerance are few in number and species distribution compared to overall transformation research. Data on production, and transformation and regeneration examples summarized from Fisk and Dandekar (1993) and Wijbrandi and deBoth (1993).

Crop	Production (1000 t)	Number of examples of transformation and regeneration	Herbicide tolerance
Tomato	36,200	13	De Block *et al.* (1987). Phosphinothricin, degration by bacterial[a] phosphinothricin acetyltransferase Fillatti *et al.* (1987a). Glyphosate, mutant bacterial *aroA* gene encoding a modified, herbicide insensitive 5-enolpyruvoyl-shikimate-3-phosphate synthase
Cabbage	20,921	6	De Block *et al.* (1990). Phosphinothricin, degration by bacterial[a] phosphinothricin acetyltransferase
Onion	9,630	None	None
Watermelon	9,105	None	None
Carrot	7,720	6	Donn *et al.* (1990). Phosphinothricin, degration by bacterial[b] phosphinothricin acetyltransferase
Cucumber	4,777	2	None
Pea	3,497	3	None
Muskmelon	2,853	2	Donn *et al.* (1990). Phosphinothricin, degration by bacterial[b] phosphinothricin acetyltransferase
Pepper	2,480	2	None
Snap bean	1,354	No plants	None
Squash	1,263	1	None
Artichoke	1,114	None	None
Garlic	655	None	None
Eggplant	518	3	None

[a] *Streptomyces hygroscopicus*.
[b] *Streptomyces viridochromogenes*.

and regeneration methods are discussed in detail elsewhere in this volume and will not be further addressed in this chapter.

HERBICIDE TOLERANCE GENES

GLUTAMINE SYNTHETASE INHIBITORS

Glutamine synthetase (GS, E.C. 6.3.1.2) catalyses the formation of glutamine from glutamate and ammonia (Dixon and Webb, 1979). GS activity can be inhibited by glutamine analogues (Leason *et al.*, 1982), including the herbicide phosphinothricin (PPT, Basta™).

Another herbicide, bialaphos (Herbiace™), is a tripeptide antibiotic produced by *Streptomyces hygroscopicus*. Bialaphos consists of PPT plus two L-alanine residues. The alanine residues are removed from PTT by plant intracelluar peptidases leaving active herbicide. Inhibition of GS activity results in the accumulation of ammonia leading to death of the affected cell and plant. PTT-tolerant plants were first obtained by selection among alfalfa cell lines (Donn *et al.*, 1984; Mazur and Falco, 1989). Molecular analysis demonstrated that the coding region for GS had been amplified approximately ten-fold (Tischer *et al.*, 1986). Transformation of tobacco plants with the amplified sequence under control of a cauliflower mosaic virus 35S promoter resulted in increased GS tRNA and protein (Eckes, 1987) and conferred tolerance to PTT, but only at low levels (Eckes and Wengenmayer, 1987). Overall, the development of herbicide-tolerant transgenic plants using amplified or mutant GS genes has been difficult, possibly because GS can be encoded by several genes, each of which may have organ or organelle-specific activity (Tingey *et al.*, 1987).

Tolerance to PTT and bialaphos has also been achieved by detoxification (Mazur and Falco, 1989). The *bar* gene encodes phosphinothricin acetyltransferase (PAT) in *S. hygroscopicus* as part of bialaphos biosynthesis to prevent autotoxicity (Murakami *et al.*, 1986). The *bar* gene was cloned from S. *hygroscopicus* and inserted into an *Agrobacterium* plant transformation vector with a 35S promoter. Transgenic *Nicotiana tabacum* plants subsequently produced PAT enzyme activity in leaf extracts at levels that varied 30-fold among transgenic plants. Transgenic plants did not accumulate ammonia and even the lowest detectable amount of PAT conferred tolerance to both PTT and bialaphos (De Block *et al.*, 1987). PTT tolerance through detoxification was further demonstrated in transgenic *Lycopersicum esculentum* and *Solanum tuberosum* (DeBlock *et al.*, 1987). Versatility of the detoxification strategy was also demonstrated in *Atropa belladona* transformed with 35S-bar using a binary *Agrobacterium rhozogenes* vector (Saito *et al.*, 1992), in *Beta vulgaris* using a binary *A. rhizogenes* vector (D'Halluin *et al.*, 1992), and in *Oryza indica* using particle bombardment (Christou *et al.*, 1991).

AMINO ACID BIOSYNTHESIS INHIBITORS

Glyphosate

The sole member of this class of herbicides is glyphosate (*N*-[phosphonomethyl]-glycine). Glyphosate (Roundup^{TM}) inhibits the enzyme 5-enolpyruvoyl-shikimate-3-phosphate synthase (EPSPS, E.C. 2.5.1.19), an enzyme of the shikimic acid pathway that catalyses the conversion of 5-phosphoshikimate to 3-enolpyruvoyl-shikimate-5-phosphate (Dixon and Webb, 1979). Genes that confer tolerance to glyphosate have been achieved from at least two sources, a mutant gene from *Salmonella typhimurium (Salmonella aro*A) (Comai *et al.*, 1983, 1985) and an amplified plant *aro*A from *Petunia hybrida* (Amrhein *et al.*, 1983; Steinrucken *et al.*, 1986). Mutant *Salmonella* tolerant to glyphosate were obtained by mutagenizing cultures with ethylmethanesulphonate and then screening for bacterial colonies able to grow on medium containing the herbicide (Comai *et al.*, 1985). Subsequent sequence analysis demonstrated that the decreased affinity of the mutant enzyme to glyphosate was caused by a single base-pair substitution. The substitution resulted in·a single amino acid change in the mutant bacterial EPSPS (Stalker *et al.*, 1985). Early transformation research was conducted with the mutant structural *aro*A gene linked to an octopine or mannopine synthase promoter (mas-*aro*A). Insertion of the mas-*aro*A construct into tobacco using *Agrobacterium tumefaciens* mediated transformation resulted in tobacco plants with increased but incomplete tolerance to glyphosate (Comai *et al.*, 1985). Insertion of the mas-*aro*A construct into hybrid poplar *Populus alba* × *P. grandidentata* also resulted in plants with increased tolerance to glyphosate compared to non-transformed control plants (Fillatti *et al.*, 1987b). However, transformed poplars were still damaged by subcommercial dosages of the herbicide (Riemenschneider *et al.*, 1988).

One reason for incomplete tolerance conferred by mas-*aro*A is that native EPSPS is active in the chloroplast. Thus, plant EPSPS genes include a transit-peptide coding sequence that mediates translocation from the cytoplasm (site of synthesis) to the chloroplast (site of action) (Mousdale and Coggins, 1987). Subsequent experiments utilized constructs that included a 35S cauliflower mosaic virus promoter and transit-peptide sequences from pea and soybean (Fig. 20.1) (Riemenschneider and Haissig, 1990; Donahue *et al.*, 1994). Tolerance to glyphosate in hybrid poplar was increased in comparison to non-transformed controls and in comparison to mas-*aro*A (Donahue *et al.*, 1994). Interestingly, the transit-peptide coding sequence from pea and soybean may successfully mediate chloroplastic translocation in *Larix* (D. Karnosky, personal communication). Such conservation of function across plant taxa is encouraging for the further improvement of horticultural and forestry crops.

Glyphosate tolerance has also been achieved by challenging cell lines of *Petunia hybrida* (Steinrucken and Amrhein, 1980; Steinrucken *et al.*, 1986) and *Daucus carota* (Nafziger *et al.*, 1984). Isolated lines produced EPSPS protein levels as much as 20-fold greater than controls (Shah *et al.*, 1986). The cloned *Petunia* EPSPS gene, under control of a 35S promoter, increased the tolerance

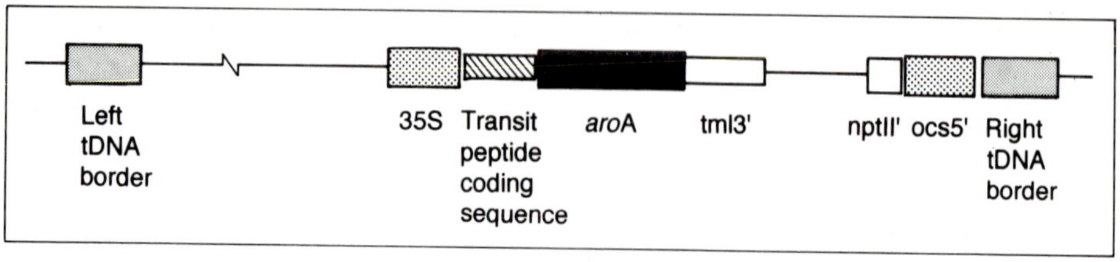

Fig. 20.1. Restriction endonuclease map of the T-DNA of plasmid pCGN1107 showing *Eco*RI and *Sal*I cleavage sites. The T-DNA contained two chimeric genes ocs-NPT II′ and 35S-*aroA*. The *aroA* chimeric gene also contained transit and mature small subunit RuBP carboxylase sequences.

of transgenic petunia plants four-fold compared to non-transformed controls (Shah *et al.*, 1986). Transgenic tomato and oil seed rape plants containing glyphosate-tolerance constructs have been tested in the United States and Canada (Fischhoff, 1989). Selection of glyphosate-tolerant cell lines has also been accomplished for a hybrid poplar *(Populus nigra* × *P. maximowiczii).* Selected cells survived in the presence of 0.35 mM glyphosate but regeneration could not be achieved without subculture onto glyphosate-free media (Klimaszewska and Cheliak, 1987). Stability of glyphosate tolerance among regenerated plants was unknown at the time of publication (Klimaszewska and Cheliak, 1987).

Sulphonylureas and imidazolinones

Acetolactate synthase (ALS, E.C. 4.1.3.18) catalyses the carboxylation of 2-acetolactate to form 2-pyruvate (Dixon and Webb, 1979). Inhibition of ALS activity is the primary mode of action of two major groups of herbicides, the sulphonylureas and imidazolinones. Plants with herbicide-inhibited ALS activity are unable to synthesize the branched amino acids leucine, valine and isoleucine (Shaner *et al.*, 1984), which leads to death. The first tolerance to sulphonylurea herbicides was obtained by selection among *Nicotiana tabacum* cells (Chaleff and Ray, 1984). *Nicotiana* cell cultures were exposed to 5.6×10^{-9} M chlorsulphuron or sulphometuron methyl (sulphonylureas) and isolates that tolerated both herbicides were obtained. Notably, isolates selected for one herbicide demonstrated tolerance to the other, which was expected given the similar structures of the two sulfonylureas (Chaleff and Ray, 1984). The mutations were stable over sexual generations and inherited as semidominant, single-gene Mendelian traits (Chaleff and Ray, 1984). Similar experiments with *Salmonella typhimurium* (LaRossa and Schloss, 1984), *Saccharomyces* (Falco and Dumas, 1985), *Nicotiana* (Chaleff and Mauvais, 1984), and *Arabidopsis* (Haughn and Somerville, 1986) followed by biochemical analysis, led to the identification of ALS as the target enzyme. Plant ALS genes contain several sites within conserved regions where mutations causing single amino acid substitutions can confer herbicide tolerance (Mazur and Falco, 1989). Single mutations can result in selective tolerances to sulphonylureas, imidazolinones, or other ALS inhibitors; or in cross-tolerances (Hattori *et al.*, 1992). Multiple mutations in the same gene can result in cross-tolerances to several herbicides (Hattori *et al.*, 1992).

The use of microbial genes to produce plants tolerant to ALS inhibitors has not been much pursued (Mazur and Falco, 1989), the reasons for which are many. First, ALS, like EPSPS, (Mousdale and Coggins, 1987) is active in plant chloroplasts (Miflin, 1974; Jones *et al.*, 1985), which requires a transit-peptide coding sequence. Second, ALS, unlike EPSPS, is active as a tretrameric enzyme, which requires post-transport assembly of two different subunits (Eoyang and Silverman, 1984). Overall, the complex biochemistry involved in the expression of ALS led to the use of plant ALS genes as the basis for genetic engineering of plant sulphonylurea and imidazolinone tolerance (Mazur and Falco, 1989). Plant ALS genes that conferred tolerance to sulphonylurea have been isolated from *Nicotiana* (Chaleff and Ray, 1984; Chaleff *et al.*, 1987) and *Arabidopsis* (Haughn *et al.*, 1988). Tolerance genes from *Nicotiana* have been inserted into several horticultural and agronomic crop species (Falco *et al.*, 1987; Mazur *et al.*, 1987; Smith *et al.*, 1988) with results that varied from low to potentially useful. More recent studies have demonstrated that simple over-expression of native genes may not confer useful levels of sulphonylurea tolerance because increased native ALS mRNA may not direct proportional increases in ALS enzyme activity (Odell *et al.*, 1990). A lack of knowledge regarding post-translational protein processing, intracellular translocation of pre-ALS, and the role of a second ALS subunit in plant ALS activity, make interpretation of expression studies difficult (Odell *et al.*, 1990). Yet, a lack of detailed knowledge of post-transcriptional processing of a mutant *Arabidopsis* ALS gene in tobacco plants did not prevent the production of transgenic plants with a 1000-fold increase in chlorsulphuron tolerance compared to control plants (Odell *et al.*, 1990).

In contrast to genetic engineering approaches, imidazolinone-tolerant cell lines of maize have been isolated (Anderson and Georgeson, 1986) and the tolerance genes transferred to commercial lines of maize through breeding. This strategy, although necessitating laborious backcross breeding, avoids current regulatory restrictions on commercial release because genetic engineering is not used.

PHOTOSYNTHESIS INHIBITORS

Triazines

The *S*-triazines (atrazine, simizine) and ureas (diuron) inhibit photosystem II (Mazur and Falco, 1989). Atrazine has been shown to displace plastoquinone from its binding site on a membrane protein known as the Q_b protein (Mathis, 1987). The Q_b protein is encoded in plants by a chloroplast gene termed *psbA*. Selectivity in crop plants is mostly related to the presence in tolerant crops of the enzyme glutathione-*S*-transferase that detoxifies atrazine by conjugation (Mazur and Falco, 1989). Most work on triazine tolerance has focused on two strategies: (i) breeding with tolerant species and (ii) genetic engineering with mutant chloroplast *psbA* genes. First, triazine-tolerant weeds have been used as a source of tolerance by backcrossing into related crop species. For example, atrazine tolerance has been introduced into *Brassica* crops by crossing with tolerant *Brassica campestris* (Beversdorf *et al.*, 1980). An atrazine-tolerant cultivar of canola resulted from the backcross breeding and has been widely deployed in Canada despite reduced yields compared to the parent cultivar (Mazur and Falco, 1989).

A second strategy has been to genetically engineer susceptible plants with a chloroplast *psbA* gene that has been cloned from a triazine-tolerant plant (Cheung *et al.*, 1988). Genes that encode a triazine-tolerant Q_b protein have been found in algae, bacteria and plants (Hirschberg *et al.*, 1987). The structural *psbA* gene from triazine-tolerant *Amaranthus hybridus* was spliced to the regulatory and transit-peptide coding sequences of the small subunit of ribulose bisphosphate carboxylase/oxygenase of pea. Transformation of *Nicotiana tabacum* via an *Agrobacterium* vector led to transgenic plants that had increased tolerance to atrazine compared to non-transformed controls (Cheung *et al.*, 1988). Chloroplast membranes from transgenic tobacco plants contained the modified Q_b protein indicating that the translocation strategy had been successful (Cheung *et al.*, 1988). But, no transformed plants were completely immune to atrazine when tested *in vitro*.

Bromoxynil

Bromoxynil (3,5-dibromo-4-hydroxybenzonitril) inhibits photosynthetic electron transport in photosystem II (Sanders and Pallett, 1985). Crop plant selectivity is based on the ability of the tolerant crop to hydrolyse the herbicide using a nitrilase (Buckland *et al.*, 1973). Bromoxynil has a short half-life in the soil because many microbes also hydrolyse the herbicide (Stalker and McBride, 1987; Stalker *et al.*, 1988). One microbial bromoxynil-specific nitrilase gene (*bxn*) was isolated from *Klebsiella pneumoniae* ssp. *ozaenae* (McBride *et al.*, 1986; Stalker and McBride, 1987). The structural gene was cloned into an *Agrobacterium* transformation vector with a mannopine synthase promoter and inserted into *Lycopersicon*. Second-generation progeny were tested for bromoxynil tolerance (Stalker *et al.*, 1988). Transgenic tomatoes were tolerant to applications of a commercial bromoxynil formulation (Buctril™) 8- to 16-fold in excess of normal field application rates. Further, use of the relatively inefficient mannopine synthase promoter suggested that tolerance could be achieved at low levels of expression and that directed translocation to the chloroplast membrance (site of photosystem II activity) was not needed (Stalker *et al.*, 1988). Confirming studies were conducted with the same *bxn* structural gene in combination with the cauliflower mosaic virus 35S promoter and also with promoter and transit-peptide sequences from the small subunit of ribulose bisphosphate carboxylase/oxygenase of sunflower and maize (Leroux *et al.*,

1990). Transgenic *Nicotiana tabacum* were tolerant to bromoxynil when the *bxn* was controlled by either 35S or sunflower small subunit scenario (SSU) promoters, confirming that translocation to the chloroplast was unneeded (Leroux *et al.*, 1990). Further, the sunflower SSU promoter was tissue specific in *Nicotiana,* causing highest expression in green tissues. This demonstrated that bromoxynil detoxification within photosynthesizing tissues was sufficient to confer whole-plant tolerance (Leroux *et al.*, 1990). The maize SSU promoter was ineffective (Leroux *et al.*, 1990).

ISSUES SURROUNDING THE DEPLOYMENT OF ENGINEERED CROPS

Genetically engineered herbicide-tolerant crops are a new technology. Like any new technology, the mere existence of genetically engineered crops will cause debate as to how, or even whether, they should be deployed. The possible secondary effects of herbicide-tolerant crops may be manifold and involve biological, socioeconomic and ethical issues (Comstock, 1989a). For example, it has been argued that genetically engineered herbicide-tolerant crops may increase the use of herbicides (Sagoff, 1991). Increased use of herbicides may, in turn, lead to environmental degradation or the accelerated appearance of herbicide-tolerant weed populations, which are already numerous (Holt *et al.*, 1993). Many of the arguments advanced in opposition to genetic engineering for herbicide tolerance contain similar interwoven components. Gressel (1992, p. 517) noted that:

> The area of genetic engineering of herbicide-resistant crop varieties has gathered a most contentious confluence of anti-chemical, anti-genetic engineering forces that have used this as part of their political agenda ... By bulking all crops as if they are one and all herbicides as if they were one, it is simple to come up with conclusions that industry is trying to develop a carcinogen-resistant crop so they can exorbitantly and monopolistically sell seed and chemical that the farmer does not need. They then expect the tolerance genes will go from crop to weed. The resulting resistant super weeds will either ruin agriculture (in one scenario) or are being 'developed' so that there will be a market for the next generation of herbicides (in another scenario).

The literature on risks associated with the deployment of genetically engineered crops is formidable and growing. For example, the computerized literature search conducted in the course of preparing this manuscript yielded over 500 references on the subject. The literature on the risks of herbicide-tolerant crops far outweighs the literature on the biology of the crops themselves, especially in the period from 1990 to 1994. The scientific challenge for those engaged in research on genetically engineered herbicide-tolerant crops is critically to examine the various claims and counter-claims surrounding future deployment, determine where gaps in necessary knowledge exist, and conduct research designed to fill those gaps. Some knowledge gaps might be filled by logical analysis of existing biological, economic or ethical principles and information. Others may require novel experimentation. I suggest here that at least three areas merit future critical study or experimentation:

1. Would the deployment of genetically engineered herbicide-tolerant crops increase herbicide use?
2. Would potentially increased herbicide use related to deployment of genetically engineered herbicide-tolerant crops *necessarily* accelerate the development of herbicide-tolerant weeds?
3. Would the deployment of herbicide-tolerant crops result in flow of the transgene into neighbouring populations?

INCREASED HERBICIDE USE AND CHEMICAL DAMAGE

Opponents of herbicide-tolerant crops often raise the objection that the deployment of those crops will result in increased herbicide use leading, in turn, to a chemically contaminated and degraded environment. While increased use of herbicides *may* follow the deployment of tolerant crops, the proposition that increased herbicide use *must* follow is far from proven. Comstock (1989a,b) in a fascinating treatise on the ethics of genetically engineered herbicide tolerance, notes that the war against weeds has followed a time-honoured pattern. You (meaning the farmer) '. . . start with seeds that have traits you desire, and then you find chemicals or cultural practices to get rid of everything else' (Comstock, 1989a, p. 264). It is further argued that 'the time honored technique is about to be reversed', and that 'Recent scientific developments have opened the possibility of permitting us to start with chemicals and then move on to seeds, (Comstock, 1989a, p. 264).

The new strategy may be superior to the old for several reasons. First, the ability to engineer herbicide tolerance means that more opportunity exists to select chemicals for the absence of adverse effects on the health of humans and other animals. New herbicides, such as glyphosate, do seem far safer for humans and the environment than older herbicides such as the phenoxy acids and triazines. Second, the ability to engineer tolerance means that chemical 'hammers', herbicides that are effective at very low dose rates, may be developed followed by engineered crop tolerance. It seems reasonable to infer that any technology that facilitates the development of low-dose herbicides would lead to a decline in the industry-wide application of active ingredient. Third, the ability to engineer tolerance means that the spectrum of tolerances in crops can be broadened, which can lead to the use of many *different* herbicides. This will decrease potential problems from any given herbicide (Gressel, 1992). The aforementioned arguments seem sufficient to render debatable the propositions that herbicide use *must* increase because of genetically engineered tolerances and that increased use *must* lead to environmental damage. Whether herbicide use *will* increase seems to rest more on the rate of world population growth than on the availability of genetic engineering technology.

ACCELERATED DEVELOPMENT OF HERBICIDE TOLERANT WEEDS

Repeated application of a herbicide can result in the development of herbicide-tolerant weeds (Warwick, 1991; Holt *et al.*, 1993) because of selection among individuals in a population that vary genetically in tolerance. The first herbicide-tolerant weed was a triazine-tolerant variety of *Senecio vulgaris* reported in the United States in 1968 (Ryan, 1970). Since then, thousands of herbicide-tolerant weed populations have developed in response to the application of 15 different herbicides or classes of herbicides (Table 20.2). In addition, weed biotypes have developed that are tolerant to two or more herbicides (Solymosi and Lehoczki, 1989; Powles and Howat, 1990). The proposition has been put forward that deployment of genetically engineered herbicide-tolerant crops may accelerate the development of herbicide-tolerant weeds in response to increased herbicide use, *apart* from any flow of the engineered gene to a weed population. One argument has it that herbicide-tolerant crops will place intense selective pressure on weed populations because they will promote the 'persnickety aesthetic that accompanies monocultural farm practices' (Comstock, 1989a, p. 292). Alternative methods of weed control have been suggested. They include rotation of intensively managed fields to reduce the severity of weed infestations typical of monocultures, weed control by

Table 20.2. Populations of herbicide-resistant plants have evolved in response to herbicide application and now include at least 57 species (Holt *et al.*, 1993). Data extracted from Table 1 in Holt *et al.* (1993).

Herbicide class	Number of species	Number of locations	Engineered tolerance
ACCase inhibitors	8	>1000	–
ALS inhibitors	8	>1000	Yes
Amides	2	2	–
Aminotriazoles	2	2	–
Arsenicals	1	2	–
Benzonitriles	1	1	–
Bipyridiliums	18	>50	–
Carbamates	2	70	–
Dinitroanilines	5	>20	–
Glyphosate	0	0	Yes
tGS inhibitors	0	0	Yes
Phenoxyacetic acids	6	5	Yes
Picloram	1	1	–
Pyridazinones	1	3	–
Substituted ureas	7	>50	–
Triazines	57	>1000	Yes
Uracils	2	1	–

mechanical cultivation, and 'as a last resort, present generation chemicals may be handsprayed on specific spots when rotation and cultivation fail to offer sufficient control' (Comstock, 1989b, p. 129).

What may the effect of herbicide-tolerant crops be on the evolution of tolerant weeds and the efficacy of alternative weed control methods? First, some herbicide-tolerant crops might contribute to selection for herbicide-tolerant weeds. For example, herbicides that inhibit ALS activity have given rise to eight species and over 1000 populations of tolerant weeds (Table 20.2). Also, application of triazine herbicides have produced over 1000 populations of tolerant weeds representing 57 species (Table 20.2). Deployment of crops with engineered tolerance to the aforementioned herbicides may aggravate an already disturbing problem. But, the application of such herbicides could be reduced if crops tolerated an alternative contact herbicide that might be applied as a 'reactive' response to a weed problem. The reactive strategy would be especially attractive if the alternative herbicide were one that has not been associated with the development of tolerant weeds. Herbicides amenable to reactive weed control strategies exist. Herbicide-tolerant weed populations developed in response to the herbicide glyphosate have not been reported as of 1993 (Table 20.2, Holt *et al.*, 1993). The same is true for the GS inhibitors, phosphinothricin and biaphalos (Table 20.2). Thus, it seems that genetically engineered herbicide-tolerant crops might well reduce the development of herbicide-tolerant weeds by offering the farmer an alternative method of weed control. Such a strategy has been proposed for the control of parasitic witchweeds and broomrapes, the Third World's worst weeds (Gressel, 1992). The effect of newer herbicides and mixtures of herbicides on the development of herbicide-tolerant weed populations could be determined by appropriate experimentation.

FLOW OF TRANSGENES TO WILD RELATIVES

Genes from cultivated crops, including genes inserted by genetic engineering, can theoretically move from the cultivar to a sexually compatible wild relative. The potential spread of transgenes and the likely ecological outcome of that spread depends on several factors that vary with crop and environment. For example, deployment of genetically engineered maize in California would result in low risk of gene transfer because maize has no sexually compatible wild relatives in California (Ellstrand and Hoffman, 1990). Other crops in the same enviromnent may have several compatible relatives including onion (37 relatives), potato (21 relatives), broccoli and cauliflower (9 relatives), and lettuce (6 relatives) (Ellstrand and Hoffman, 1990). The presence of compatible relatives may be most problematic with native crops such as ornamentals and forest trees because they often have very short histories of genetic selection and are little diverged from their wild progenitors. The probable ecological outcome of gene transfer would also depend of the nature of the gene. For example, a gene that confers no adaptive advantage on the receptor wild population would be unlikely to persist in the environment (Ellstrand and Hoffman, 1990). Clearly, genetically engineered herbicide-tolerant crops should be carefully evaluated for potential gene flow to wild, weedy relatives.

Methods exist to quantify the potential spread of engineered genes into compatible populations (Manasse and Kareiva, 1991). Rate of gene flow can be expressed as a statistic known as ARS (assymptotic rate of spread), the rate at which an organism's or gene's range increases in radius (Manasse and Kareiva, 1991). ARS can be modelled using a variety of mathematical systems that consider the effects of pollen and seed dispersal distances and patterns, the relative fitness of the engineered gene, and the spatial distribution of a compatible population (Manasse and Kareiva, 1991). Pollen and seed dispersal can be determined experimentally for any species. The change in plant fitness because of an engineered gene can also be experimentally determined in herbicide-treated and herbicide-free environments. The spatial distribution of compatible populations is also knowable. For agronomic and vegetable crops such data are available from USDA records. The distribution of species that are compatible with engineered forestry crops, such as hybrid poplars, would be more complex. However, modern geographic information systems, in addition to demonstrating the inevitable complexity of spatial distributions, can provide the coordinate locations of compatible species that might be used as input to gene flow models.

CONCLUSIONS

In this chapter I have reviewed the development of herbicide-tolerance genes and some issues surrounding the deployment of genetically engineered herbicide-tolerant crops. Genes that confer tolerance to several herbicides are available including tolerances to the sulphonylureas, imidazolinones, glyphosate, triazines, bromoxynil and phosphinothricin. Levels of tolerance range from commercially insignificant to complete immunity depending on the specific instance. Clearly, the ability to insert genes into important crops has outstripped the number of instances in which that ability has been used to confer tolerance to herbicides. It is arguable that the lag between ability and accomplishment can be attributed, at least in part, to public opposition to the deployment of herbicide-tolerant crops, which has become an important issue in publicly funded research. Arguments against the deployment of such crops include possible damage to the environment because of increased herbicide use and the accelerated development of herbicide-tolerant weeds because of increased selection pressure or unwanted movement of transgenes from the crop to wild relatives. Some arguments against deployment may have merit. Other arguments, based neither on reason nor empirical evidence, thrive because we lack a detailed understanding of how chemicals, crops, wild-crop relatives, and the physical environment respond interactively in the face of a new technology. Much of that understanding is attainable by empirical experimentation and theoretical studies designed to estimate the probability of occurrence of hypothetically unfavourable outcomes. The lack of such research may be the current limiting factor to the deployment of genetically engineered herbicide-tolerant crops.

REFERENCES

Amrhein, N.D., Schab, J.J. and Schulz, A. (1983) Biochemical basis for glyphosate tolerance in a bacterium and a plant tissue culture. *FEBS Letters* 157, 191–196.

Anderson, P.C. and Georgeson, M. (1986) Selection and characterization of imidazolinone tolerant mutants of maize. In: *Biochemical Basis of Herbicide Action*. Wye College, Ashford, UK, Abstract.

Beversdorf, W.D., Weiss-Lerman, J., Erickson, L.R. and Souze Machado, V. (1980) Transfer of cytoplasmically-inherited triazine resistance from bird's rape to cultivated *Brassica campestris* and *Brassica napus*. *Canadian Journal of Genetics and Cytology* 22, 167–172.

Buckland, J.L, Collins, R.E. and Pullin, E.M. (1973) Metabolism of bromoxynil octanoate in growing wheat. *Pesticide Science* 4, 149–162.

Chaleff, R.S. and Mauvais, C.J. (1984) Acetolactate synthase is the site of action of two sulphonylurea herbicides in higher plants. *Science* 224, 1143–1145.

Chaleff, R.S. and Ray, T.B. (1984) Herbicide-resistant mutants from tobacco cell cultures. *Science* 223, 1148–1151.

Chaleff, R.S., Sebastian, S.A., Creason, G.L., Mazur, B.J. and Falco, S.C. (1987) Developing plant varieties resistant to sulphonylurea herbicides. In: Arntzen, C.J. and Ryan, C. (eds), *Molecular Strategies for Crop Protection*. A.R. Liss, New York.

Cheung, A.Y., Bogorad, L., Van Montagu, M. and Schell, J. (1988) Relocating a gene for herbicide-tolerance: a chloroplast gene is converted into a nuclear gene. *Proceedings of the National Academy of Sciences, USA* 85, 391–395.

Christou, P., Ford, T.L. and Kofron, M. (1991) Production of transgenic rice *(Oryza sativa* L.) plants from agronomically important *indica* and *japonica* via electric discharge particle acceleration of exogenous DNA into immature zygotic embryos. *Bio/Technology* 9, 957–962.

Comai, L., Sen, L. and Stalker, D. (1983) An altered *aro*A gene product confers resistance to the herbicide glyphosate. *Science* 221, 370–371.

Comai, L., Facciotti, D., Hiatt, W.R., Thompson, G., Rose, R. and Stalker, D. (1985) Expression in plants of a mutant *aro*A gene from *Salmonella typhimurium* confers tolerance to glyphosate. *Nature* 317, 741–744.

Comstock, G. (1989a) Genetically engineered herbicide resistance, part one. *Journal of Agricultural Ethics* 2, 263–306.

Comstock, G. (1989b) Genetically engineered herbicide resistance, part two. *Journal of Agricultural Ethics* 3, 114–146.

DeBlock, M. Botterman, J., Vandewiele, M., Dockx, J., Thoen, C., Gossele, V., Movva, N.R., Thompson, C., VanMontagu, M. and Leemans, J. (1987) Engineering herbicide resistance in plants by expression of a detoxifying enzyme. *EMBO Journal* 6, 2513–2518.

DeBlock, M., De Brouwer, D. and Tenning, P. (1990) Transformation of *Brassica napus* and *Brassica oleraceae* using *Agrobacterium tumefaciens* and the expression of the bar and neo genes in the transgenic plants. *Plant Physiology* 91, 694–701.

D'Halluin, K., Bossut, M., Bonne, E., Mazur, B., Leemans, J. and Botterman, J. (1992)

Transformation of sugarbeet (*Beta vulgarisi L.*) and evaluation of herbicide resistance in transgenic plants. *Bio/Technology* 10, 309–314.
Dixon, M. and Webb, E.C. (1979) *Enzymes*. Academic Press, New York.
Donahue, R.A., Davis, T.D., Marquardt, P.M., Michler, C.H., Carter, D.R., Riemenschneider, D.E., Haissig, B.E. and Isebrands, J.G. (1994) Photosynthetic and growth analysis of genetically modified hybrid poplar. *Canadian Journal of Forest Research* 24, 2377–2383.
Donn, G., Tischer, E., Smith, J.A. and Goodman, H.M. (1984) Herbicide-resistant alfalfa cells: an example of gene amplification in plants. *Journal of Molecular and Applied Genetics* 2, 621–635.
Donn, G., Dirks, R., Eckes, P. and Uijtewaal, B. (1990) Transfer and expression of modified PTT-acetyltransferase gene from *Streptomyces viridochromogenes* in tomato, melon, carrot and strawberry. In: *Abstracts VIIth International Congress on Plant Tissue and Cell Culture*, Amsterdam, p. 176.
Eckes, P. (1987) Isolation of L-phosphinothricin resistant plants. In: Wettstein, D. and Chua, N.H. (eds), *Proceedings of a NATO Advanced Study Institute,* Hoechst, Frankfurt. p. 635.
Eckes, P. and Wengenmayer, F. (1987) Overproduction of glutamine synthetase in transgenic plants. In: (*Regulation of Plant Gene Expression, 29th Harden Conference Program Abstracts,* Wye College, Ashford, UK.
Ellstrand, N.C. and Hoffman, C.A. (1990) Hybridization as an avenue of escape for engineered genes. *BioScience* 40, 438–442.
Eoyang, L. and Silverman, P.M. (1984) Purification and subunit composition of acetohydroxyacid synthase I from *Escerichia coli* K-12. *Journal of Bacteriology* 157, 184–189.
Falco, S.C. and Dumas, K.S. (1985) Genetic analysis of mutants of *Saccharomyces cerevisiae* resistant to the herbicide sulphometuron methyl. *Genetics* 109, 21–35.
Falco, S.C., Knowlton, S., LaRossa, R.A., Smith, J.K. and Mazur, B.J. (1987) Herbicides that inhibit amino acid biosynthesis: the sulphonylureas – a case study. In: *1987 British Crop Protection Conference*, PCPC Publications, Surrey, UK, pp. 149–158.
Fillatti, J.J., Kiser, J., Rose, R. and Comai, L. (1987a) Efficient transfer of a glyphosate tolerance gene into tomato using a binary *Agrobacterium tumefaciens* vector. *Bio/Technology* 5, 726–730.
Fillatti, J.J., Sellmer, J., McCown, B., Haissig, B. and Comai, L. (1987b) *Agrobacterium* mediated transformation and regeneration of *Populus. Molecular and General Genetics* 206, 192–199.
Fincham, J.R.S. and Ravetz, J.R. (1991) *Genetically Engineered Organisms: Benefits and Risks.* Open University Press, Buckinghamshire, UK.
Fischhoff, D.A. (1989) Applications of plant genetic engineering to crop protection. *Phytopathology* 79, 38–40.
Fisk, H.J. and Dandekar, A.M. (1993) The introduction and expression of transgenes in plants. *Scientia Horticulturae* 55, 5–36.
Gardner, R.C. (1993) Gene transfer into tropical and subtropical crops. *Scientia Horticulturae* 55, 65–82.
Ginzburg, L.R. (1991) *Assessing Ecological Risks of Biotechnology.* Butterworth-Heinemann, Boston.
Gressel, J. (1992) Addressing real weed science needs with innovations. *Weed Technology* 6, 509–525.
Hansen, E.A. and Netzer, D.A. (1986) *Weed Control Using Herbicides in Short-Rotation Intensively Cultured Poplar Plantations.* Research Paper NC-260, USDA Forest Service, North Central Forest Experiment Station, St Paul, Minnesota.

Hattori, J., Rutledge, R., Labbe, H., Brown, D., Sunohara, G. and Miki, B. (1992) Multiple resistance to sulphonylureas and imidazolinones conferred by an acetohydroxyacid synthase gene with separate mutations for selective resistance. *Molecular and General Genetics* 232, 167–173.

Haughn, G.W. and Somerville, C.R. (1986) Sulphonylurea-resistant mutants of *Arabidopsis thaliana. Molecular and General Genetics* 204, 430–434.

Haughn, G.W., Smith, J., Mazur, B. and Somerville, C. (1988) Transformation with mutant *Arabidopsis* acetolactate synthase gene renders tobacco resistant to sulphonylurea herbicides. *Molecular and General Genetics* 211, 266–271.

Hirschberg, J. Yehuda, A.B., Pecker, I. and Ohad, N. (1987) Mutations resistant to photosystem II herbicides. In: Wettstein, D. and Chua, N.H. (eds), *Plant Molecular Biology, Proceedings of a NATO Advanced Study Institute.* Plenum Press, New York, pp. 357–366.

Holt, J.S., Powles, S.B. and Holtum, J.A.M. (1993) Mechanisms and agronomic aspects of herbicide resistance. *Annual Review of Plant Physiology and Plant Molecular Biology* 44, 203–229.

Jones, A.V., Young, R.M. and Leto, K.J. (1985) Subcellular localization and properties of acetolactate synthase, target site of the sulphonylurea herbicides. *Plant Physiology* (Supplement) 77, p. 55.

Jorgensen, R. (1993) The modification of horticultural plant phenotypes by direct gene transfer. *Scientia Horticulturae* 55, 1–4.

Klimaszewska, K. and Cheliak, W.M. (1987) Selection for glyphosate-tolerant cell cultures in poplar. In: *Proceedings of the IEA Task II Meeting and Workshops on Cell Culture and Coppicing.* The Finnish Forest Research Institute, Kannus Research Station, Box 44, SF-69101 Kannus, Finland, pp. 17–23.

Kung, S.D. and Wu, R. (1993) *Transgenic Plants. Vol. 2. Present Status and Social and Economic Impacts*. Academic Press, San Diego.

Labrecque, M., Teodorescu, T.I., Babeux, P., Cogliastro, A. and Daigle, S. (1994) Impact of herbaceous competition and drainage conditions on the early productivity of willows under short-rotation intensive culture. *Canadian Journal of Forest Research* 24, 493–501.

LaRossa, R.A. and Schloss, J.V. (1984) The sulphonylurea herbicide sulphometuron methyl is an extremely potent and selective inhibitor of acetolactate synthase in *Salmonella typhimurium. Journal of Biological Chemistry* 259, 8753–8757.

Leason, M., Dunliffe, D., Parkin, D., Lea, P.J. and Miflin, B.J. (1982) Inhibition of pea leaf glutamine synthetase by mithionine sulphoximine, phosphinothricin and other glutamate analogues. *Phytochemistry* 21, 855–857.

Leroux, B., Lebrun, M., Garnier, P., Sailland, A., Pelissier, B. and Freyssinet, G. (1990) Engineering herbicide resistance in tobacco plants by expression of a bromoxynil specific nitrilase. *Bulletin de la Société Botanique de France, Actualites Botaniques* 137, 65–78.

Levin, M.A. and Strauss, H.S. (1990) *Risk Assessment in Genetic Engineering.* McGraw-Hill, New York.

Manasse, R. and Kareiva, P. (1991) Quantifying the spread of recombinant genes and organisms. In: Ginzburg, L.R. (ed.), *Assessing Ecological Risks of Biotechnology*. Butterworth-Heinemann, Stoneham, Massachusetts, pp. 215–231.

Mathis, P. (1987) Primary reactions of photosynthesis: discussion of current issues. In: Biggens, J. (ed.), *Progress in Photosynthesis Research, Proceedings of the International Congress of Photosynthesis.* Martinus Nijhoff, Dordrecht, pp. 151–160.

Mazur, B.J. and Falco, S.C. (1989) The development of herbicide resistant crops. *Annual Review of Plant Physiology and Plant Molecular Biology* 40, 441–470.

Mazur, B.J., Falco, S.C., Knowlton, S. and Smith, J.K. (1987) Aceolactate synthase, the target enzyme of the sulphonylurea herbicides. *Plant Molecular Biology* 140, 339–349.

McBride, K.E., Kenny, J.W. and Stalker, D.M. (1986) Metabolism of the herbicide bromoxynil by *Klebsiella pneumoniae* subsp. *ozaenae*. *Applied and Environmental Microbiology* 52, 325–330.

Miflin, B.J. (1974) The location of nitrite reductase and other enzymes related to amino acid biosynthesis in the plastids of root and leaves. *Plant Physiology* 54, 550–555.

Mousdale, D.N. and Coggins, J.R. (1987) Subcellular localization of the common shikimate-pathway enzyme in *Pisum sativum*. *Planta* 163, 241–249.

Murakami, T. Anzai, H., Imai, S., Satoh, A., Nagoaka, K. and Thompson, C.J. (1986) The bialaphos biosynthetic genes of *Streptomyces hygroscopicus:* molecular cloning and characterization of the gene cluster. *Molecular and General Genetics* 205, 42–50.

Nafziger, E.D., Widholm, J.M., Steinrucken, H.C., and Killmer, J.L. (1984) Selection and characterization of a carrot cell-line tolerant to glyphosate. *Plant Physiology* 76, 571–574.

Odell, J.T., Caimi, P.G., Yadav, N.S. and Mauvais, C.J. (1990) Comparison of increased expression of wild-type and herbicide-resistant acetolactate synthase genes in transgenic plant, and indication of postranscriptional limitation on enzyme activity. *Plant Physiology* 94, 1647–1654.

Powles, S.B. and Howat, P.D. (1990) Herbicide-resistant weeds in Australia. *Weed Technology* 4, 178–185.

Riemenschneider, D.E. and Haissig, B.E. (1990) Producing herbicide-tolerant poplars using genetic transformation mediated by *Agrobacterium tumefaciens* C58: a summary of recent research. In: Ahuja, M.R. (ed.), *Biotechnology of Woody Plants*. Plenum Press, New York, pp. 247–264.

Riemenschneider, D.E., Haissig, B.E., Sellmer, J. and Fillatti, J.J. (1988) Expression of an herbicide-tolerance gene in young plants of a transgenic hybrid poplar clone. In: Ahuja, M.R. (ed.), *Somatic Cell Genetics of Woody Plants*. Kluwer Academic Publishers, Dordrecht, pp. 73–80.

Robinson, K.E.P. and Firoozabady, E. (1993) Transformation of floriculture crops. *Scientia Horticulturae* 55, 83–99.

Ryan, G.F. (1970) Resistance of common groundsel to simizine and atrazine. *Weed Science* 18, 614–616.

Sagoff, M. (1991) On making nature safe for biotechnology. In: Ginzburg, L.R. (ed.), *Assessing Ecological Risks of Biotechnology*. Butterworth-Heinemann, Boston, pp. 341–365.

Saito, K., Yamazake, M., Anzai, H., Yoneyama, K. and Murakoshi, I. (1992) Transgenic herbicide-resistant *Atropa belladonna* using an Ri binary vector and inheritance of the transgenic trait. *Plant Cell Reports* 11, 219–224.

Sanders, G.E. and Pallett, K.E. (1985) *In vitro* activity and binding characteristics of the hydroxybenzonitriles in chloroplasts isolated from *Matricaria inodora* and *Viola arvensis*. *Pesticide Biochemistry and Physiology* 24, 317–325.

Schuerman, P.L. and Dandekar, A.M. (1993) Transformation of temperate woody crops: progress and potentials. *Scientia Horticulturae* 55, 101–124.

Shah, D.M., Horsch, R.B., Klee, H. J., Kishore, G.M., Winter, J.A. and Tumer, N.E. (1986) Engineering herbicide tolerance in transgenic plants. *Science* 233, 478–481.

Shaner, D.L., Anderson, P.C. and Stidham, M.A. (1984) Imidazolinones (potent inhibitors of acetohydroxyacid synthase). *Plant Physiology* 76, 545–546.

Smith, J.K., Mauvais, C.J., Knowlton, S. and Mazur, B.J. (1988) Molecular biology of resistance to sulphonylurea herbicides. In: *ACS Symposium on Biotechnology of Crop Protection.* ACS Books, Washington, DC, pp. 25–36.

Solymosi, P. and Lehoczki, E. (1989) Characterization of a triple (atrizine–pyrazon–pyridate) resistant biotype of common lambsquarters (*Chenopodium album* L.). *Plant Physiology* 134, 685–690.

Stalker, D.M. (1989) Producing herbicide-resistant plants by gene transfer technology. In: Boger, P. and Sandmann, G. (eds), *Target Sites of Herbicide Action*. CRC Press, Boca Raton, pp. 147–163.

Stalker, D.M. and McBride, K.E. (1987) Cloning and expression in *Eschericia coli* of a *Klebsiella ozaenae* plasmid-borne gene encoding a nitrilase specific for the herbicide bromoxynil. *Journal of Bacteriology* 169, 955–960.

Stalker, D.M., Hyatt, W.R. and Comai, L. (1985) A single amino acid substitution in the enzyme 5-enolpyruvylshikimate-3-phosphate synthase confers resistance to the herbicide glyphosate. *Journal of Biological Chemistry* 260, 4724–4728.

Stalker, D.M., McBride, K.E. and Malyi, L.D. (1988) Expression in plants of a bromoxynil-specific bacterial nitrilase that confers herbicide resistance. In: Fraley, R.T., Frey, N.M. and Schell, J. (eds), *Genetic Improvements of Agriculturally Important Crops: Progress and Issues.* Cold Spring Harbor Laboratory, Cold Spring Harbor, New York, pp. 37–39.

Steinrucken, H.C. and Amrhein, N. (1980) The herbicide glyphosate is a potent inhibitor of 5-enolpyrovylshikimic acid-3-phosphate synthase. *Biochemical and Biophysical Research Communications* 94, 1207–1212.

Steinrucken, H.C., Schulz, A., Amrhein, N., Porter, C. and Fraley, R.T. (1986) Overproduction of a 5-enol-pyruvylshikimate-3-phosphate synthase in a glyphosate-tolerant *Petunia hybrida* cell line. *Archives of Biochemistry and Biophysics* 244, 169–178.

Tingey S.V., Walker, E. and Crouzzi, G.M. (1987) Glutamine synthetase genes of pea encode distinct polypeptides which are differentially expressed in leaves, roots and nodules. *EMBO Journal* 6, 1–9.

Tischer, E., DasSarma, S. and Goodman, H.M. (1986) Nucleotide sequence of an alfalfa glutamine synthetase gene. *Molecular and General Genetics* 203, 221–229.

Warwick, S.I. (1991) Herbicide resistance in weedy plants: physiology and population biology. *Annual Review of Ecoloy and Systematics* 22, 95–114.

Wijbrandi, J. and deBoth, M.T.J. (1993) Temperate vegetable crops. *Scientia Horticulturae* 55, 37–63.

Wohrmann, K. and Tomiuk, J. (1993) *Transgenic Organisms: Risk Assessment of Deliberate Release.* Birkhauser Verlag, Boston.

Index